Peptide Therapeutics
Strategy and Tactics for Chemistry, Manufacturing and Controls

Drug Discovery Series

Titles in the Series:

1: Metabolism, Pharmacokinetics and Toxicity of Functional Groups
2: Emerging Drugs and Targets for Alzheimer's Disease; Volume 1
3: Emerging Drugs and Targets for Alzheimer's Disease; Volume 2
4: Accounts in Drug Discovery
5: New Frontiers in Chemical Biology
6: Animal Models for Neurodegenerative Disease
7: Neurodegeneration
8: G Protein-Coupled Receptors
9: Pharmaceutical Process Development
10: Extracellular and Intracellular Signaling
11: New Synthetic Technologies in Medicinal Chemistry
12: New Horizons in Predictive Toxicology
13: Drug Design Strategies: Quantitative Approaches
14: Neglected Diseases and Drug Discovery
15: Biomedical Imaging
16: Pharmaceutical Salts and Cocrystals
17: Polyamine Drug Discovery
18: Proteinases as Drug Targets
19: Kinase Drug Discovery
20: Drug Design Strategies: Computational Techniques and Applications
21: Designing Multi-Target Drugs
22: Nanostructured Biomaterials for Overcoming Biological Barriers
23: Physico-Chemical and Computational Approaches to Drug Discovery
24: Biomarkers for Traumatic Brain Injury
25: Drug Discovery from Natural Products
26: Anti-Inflammatory Drug Discovery
27: New Therapeutic Strategies for Type 2 Diabetes: Small Molecules
28: Drug Discovery for Psychiatric Disorders
29: Organic Chemistry of Drug Degradation

30: Computational Approaches to Nuclear Receptors
31: Traditional Chinese Medicine
32: Successful Strategies for the Discovery of Antiviral Drugs
33: Comprehensive Biomarker Discovery and Validation for Clinical Application
34: Emerging Drugs and Targets for Parkinson's Disease
35: Pain Therapeutics; Current and Future Treatment Paradigms
36: Biotherapeutics: Recent Developments using Chemical and Molecular Biology
37: Inhibitors of Molecular Chaperones as Therapeutic Agents
38: Orphan Drugs and Rare Diseases
39: Ion Channel Drug Discovery
40: Macrocycles in Drug Discovery
41: Human-Based Systems for Translational Research
42: Venoms to Drugs: Venom as a Source for the Development of Human Therapeutics
43: Carbohydrates in Drug Design and Discovery
44: Drug Discovery for Schizophrenia
45: Cardiovascular and Metabolic Disease: Scientific Discoveries and New Therapies
46: Green Chemistry Strategies for Drug Discovery
47: Fragment-Based Drug Discovery
48: Epigenetics for Drug Discovery
49: New Horizons in Predictive Drug Metabolism and Pharmacokinetics
50: Privileged Scaffolds in Medicinal Chemistry: Design, Synthesis, Evaluation
51: Nanomedicines: Design, Delivery and Detection
52: Synthetic Methods in Drug Discovery: Volume 1
53: Synthetic Methods in Drug Discovery: Volume 2
54: Drug Transporters: Role and Importance in ADME and Drug Development
55: Drug Transporters: Recent Advances and Emerging Technologies
56: Allosterism in Drug Discovery
57: Anti-Aging Drugs: From Basic Research to Clinical Practice
58: Antibiotic Drug Discovery: New Targets and Molecular Entities
59: Peptide-Based Drug Discovery: Challenges and New Therapeutics
60: Drug Discovery for Leishmaniasis
61: Biophysical Techniques in Drug Discovery
62: Acute Brain Impairment Through Stroke: Drug Discovery and Translational Research
63: Theranostics and Image Guided Drug Delivery
64: Pharmaceutical Formulation: The Science and Technology of Dosage Forms
65: Small-Molecule Transcription Factor Inhibitors in Oncology
66: Therapies for Retinal Degeneration: Targeting Common Processes

67: Kinase Drug Discovery: Modern Approaches
68: Advances in Nucleic Acid Therapeutics
69: MicroRNAs in Diseases and Disorders: Emerging Therapeutic Targets
70: Emerging Drugs and Targets for Multiple Sclerosis
71: Cytotoxic Payloads for Antibody–Drug Conjugates
72: Peptide Therapeutics: Strategy and Tactics for Chemistry, Manufacturing, and Controls

How to obtain future titles on publication:
A standing order plan is available for this series. A standing order will bring delivery of each new volume immediately on publication.

For further information please contact:
Book Sales Department, Royal Society of Chemistry, Thomas Graham House, Science Park, Milton Road, Cambridge CB4 0WF, UK
Telephone: +44 (0)1223 420066, Fax: +44 (0)1223 420247
Email: booksales@rsc.org
Visit our website at books.rsc.org

Peptide Therapeutics
Strategy and Tactics for Chemistry, Manufacturing and Controls

Edited by

Ved Srivastava
Intarcia Therapeutics, USA
Email: vedsrivastava1@yahoo.com

ROYAL SOCIETY OF CHEMISTRY

Drug Discovery Series No. 72

Print ISBN: 978-1-78801-433-5
PDF ISBN: 978-1-78801-644-5
EPUB ISBN: 978-1-78801-869-2
Print ISSN: 2041-3203
Electronic ISSN: 2041-3211

A catalogue record for this book is available from the British Library

The Royal Society of Chemistry is a charity, registered in England and Wales, Number 207890, and a company incorporated in England by Royal Charter (Registered No. RC000524), registered office: Burlington House, Piccadilly, London W1J 0BA, UK, Telephone: +44 (0) 20 7437 8656.

For further information see our web site at www.rsc.org

Preface

Peptide therapy has become a key strategy in innovative drug development. However, one of the potential barriers for the development of novel peptide drugs is that they may lack a clearly defined Chemistry, Manufacturing and Controls (CMC) strategy encompassing pre-clinical development through to commercialization. CMC can often become a rate-limiting step for peptide-based drug development owing to a deficiency in knowledge and/or a lack of formal policies or CMC guidelines.

More than 70 peptide drugs have been approved for marketing and more than 150 peptide molecules have entered into clinical trials for a wide variety of therapeutic indications, including metabolic, cardiovascular, oncological and central nervous system diseases. Most of the commercial successes have been in metabolic diseases and oncology and include the insulin analogs NovoLog® (insulin aspart), Humalog® (insulin lispro), Victoza® (liraglutide), Byetta® (exenatide), Lupron® (leuprolide), Sandostatin® (octreotide) and Forteo® (teriparatide). In recent years, peptidomimetic inhibitors of intracellular proteases such as Kyprolis® (bortezomib) and Ninlaro® (ixazomib) have also been approved for commercialization. Despite these successes, specific regulatory challenges remain that are often related to managing quality standards. These challenges are due in part to a lack of official regulatory guidelines for peptide drugs, as there are no current FDA or ICH guidelines that address the quality of pharmaceutical peptide products. Regulators frequently use a risk-based approach applied on a case-by-case basis when reviewing NDAs (New Drug Applications) or ANDAs (Abbreviated New Drug Applications).

The system for regulating medicines in Europe is unique. It is based on a closely coordinated regulatory network of competent national authorities in the Member States of the European Economic Area (MS-EEA) working together with the European Medicines Agency (EMA) and the

Drug Discovery Series No. 72
Peptide Therapeutics: Strategy and Tactics for Chemistry, Manufacturing and Controls
Edited by Ved Srivastava

Published by the Royal Society of Chemistry, www.rsc.org

European Commission. A network of more than 4500 experts appointed by the EU Member States guarantees the high-standard of scientific assessment of medicines.

This book on peptide therapeutics provides a holistic approach to strategy and tactics for CMC. The book is focused on covering four key areas in 14 chapters: (1) the regulatory process in both the USA and Europe and guidelines including immunogenicity evaluation strategy (Chapters 1–3), (2) challenges in the manufacturing process, CMC overview and economic and environmental impact (Chapters 4–6), (3) holistic quality control strategy, novel techniques for characterization of peptide impurities and stability testing (Chapters 7–11) and (4) peptide design, the impact of functional excipients on peptide formulation and nanoparticles for peptide delivery (Chapters 12–14). The extensive references cited in each chapter offer additional detail on the subject matter.

This book is an essential reference for students and researchers, in both academia and industry, to facilitate development and manufacturing of peptide-based drugs using regulatory requirements. The goal of this book is to provide a valuable resource and reference, not only for the peptide researcher engaged in peptide-based drug development and peptide manufacturing, but also for young scientists learning the CMC and regulatory process as it relates to the development of peptide-based medicines.

The chapters are written by well-known key opinion leaders from the US Food and Drug Administration (FDA), the Federal Institute for Drugs and Medical Devices in Europe, Lilly, Merck, Amgen, Contract Manufacturing Organizations (CMOs) – Polypeptide and Bachem, and analytical chemistry Contract Research Organization (CROs) – Teva and Almac, *etc.* Short synopses of each chapter are presented below.

1. The chapter on *Regulatory Considerations for Peptide Therapeutics* in the USA presents an up-to-date overview of the approval process for New Drug Applications (NDAs) and Abbreviated New Drug Applications (ANDAs) as it pertains to peptide drugs at the FDA. The chapter also provides a review of practice, format and content of NDAs and ANDAs, including challenges with Drug Master Files (DMFs). It emphasizes the special approval pathways such as orphan drug designation programs and pediatric exclusivity, post-approval activities and life-cycle management of peptide drugs.
2. The chapter on *Regulatory Perspective on Synthetic Peptides in Europe* addresses European Quality guidelines and International Quality guidelines, including European Pharmacopoeia (Ph. Eur.) requirements. The chapter also covers control strategy and the requirements for clearly differentiated relevant data and documentation for synthetic peptides for a clinical trial application or a marketing authorization dossier for Investigational Medicinal Products (IMPs).

3. The *Biological and Immunogenicity Evaluation Strategy for Therapeutic Peptides: CMC Perspective* chapter covers immunogenicity evaluation as an essential part of peptide-based therapeutic development and assessment of the immune response against therapeutic biologics using various methodologies such as enzyme-linked immunosorbent assay (ELISA), surface plasmon resonance (SPR), chemiluminescence and flow cytometry assays for binding antibodies and cell-based assays for neutralizing antibodies.
4. The *Chemistry, Manufacturing and Controls: Active Pharmaceutical Ingredient (API) and Drug Product (DP)* chapter summarizes the CMC topics for active pharmaceutical ingredients and drug products in general and discusses the distinct activities required, including the risk management at each clinical phase (Phase I, Phase II and Phase III) from development to commercialization.
5. The chapter *Peptide Manufacturing Methods and Challenges* illustrates different manufacturing technologies on a commercial scale, addressing the challenges of meeting regulatory standards while satisfying commercial viability and challenges in the upstream synthesis and downstream purification process of manufacturing.
6. The *Economic and Environmental Factors Affecting the Sustainability of Peptide Manufacturing* chapter provides an analysis of the causes of industry consolidation through an examination of current economic and environmental issues that affect peptide therapeutic manufacturing. The chapter also discusses the generic and NCE markets trend for peptide therapeutics, advances in commercial manufacturing methods, the economics of peptide manufacturing, factors for sustaining a profit and green chemistry, and waste disposal. Finally, it emphasizes the need for a more efficient and rational use of resources.
7. The chapter *A Holistic Quality Control Strategy for Peptide APIs* outlines the principles of process design and development as a comprehensive control strategy and emphasizes how to achieve critical quality attributes (CQAs) consistently given the complexity and variability of the manufacturing process. It addresses the idea that "consistent quality cannot be tested-in, but rather must be built-in by design," according to FDA guidance. The approach utilizes process design through the practical application of a quality target product profile (QTPP), quality by design (QbD) and identification and mitigation of risks.
8. The chapter on *Higher-Order Structure Characterization and Comparability Assessments for Peptides* addresses the advantages and limitations of different analytical methodologies for the study of higher-order peptide structures. The application of orthogonal techniques such as intrinsic fluorescence spectroscopy and differential scanning calorimetry for secondary and tertiary structure and the application of techniques such as self-association, micro-flow imaging

(MFI), asymmetric flow field-flow fractionation (AF4) and compositional gradient multi-angle light scattering (CG-MALS) for elucidating quaternary structure are reviewed.

9. The chapter on *Impurity Characterization and Quantification by LC-HRMS* covers liquid chromatography coupled with high-resolution mass spectrometry (LC-HRMS) as an approach that can be used as part of the analytical framework to ensure proper quality control of peptide drug products, including the assessment of peptide-related impurities. The chapter illustrates the suitability of LC-HRMS with three case studies: salmon calcitonin, bivalirudin and exenatide drugs.
10. The chapter on *Mass Spectrometric Analysis for the Quality Control of Peptide Conjugate Therapeutics* summarizes advances in HRMS and provides an overview of special techniques (*e.g.* ion stripping and isotopic labeling) that are employed for the analytical characterization and quantitation of peptides, including conjugates such as PEGylated peptides and antibody–drug conjugates.
11. The chapter on *Identity, Content and Purity of Therapeutic Peptides by NMR Spectroscopy* describes the application of 1D and 2D NMR spectroscopy for the determination of the identity, content, structure and purity of peptides and how NMR is used for the determination of unrelated impurities such as process-related impurities and extractable/leachable compounds.
12. The chapter on *Transport, Stability and Delivery Considerations for the Design of Peptide Drugs* addresses key stability and pharmacokinetic considerations for the development of peptide drugs and identifies approaches and methodologies for their assessment in preclinical research and development. The chapter also covers current and emerging chemistry strategies, assumptions for druggability with a peptide modality and alternative routes of administration, such as oral, pulmonary, nasal and transdermal, using cost-effective dosage forms.
13. The chapter *Assessing the Impact of Functional Excipients on Peptide Drug Product Attributes During Pharmaceutical Development* discusses drug product formulation with appropriate choices of excipient(s) and composition(s) impacting the Quality Target Product Profile (QTPP) and mitigating peptide-specific CMC liabilities. This chapter highlights guidance and regulations related to the selection and evaluation of novel excipients, quality risk assessment and management and risk evaluation and mitigation in formulation development from discovery through development phases.
14. The final chapter, on *Formulations of Microspheres or Nanoparticles for Peptide Delivery*, describes the significance of nanoparticles for an alternative delivery system for peptide and protein drugs. Different types of microspheres, their preparation, characterization, factors affecting drug delivery and mechanism of drug delivery are described.

Case studies using peptide drugs such as leuprolide, triptorelin, octreotide, lanreotide, human growth hormone, buserelin, abarelix and exenatide are provided.

I would like to thank and express my sincere gratitude to all the authors who have contributed to this edition of *Peptide Therapeutics: Strategy and Tactics for Chemistry, Manufacturing and Controls* for their hard work in writing each chapter and most importantly for sharing their expertise with a broad spectrum of readers. Thanks are also due to Drew Gwilliams and Katie Morrey at the Royal Society of Chemistry for their guidance and support.

My special thanks go to Professor Bradley Pentelute, Massachusetts Institute of Technology, for providing editorial recommendations for the book. I dedicate this book to my parents (Shravan and Kusum) for their encouragement and my wife (Nisha) and children (Aaron and Nikita) for their continued support.

Ved Srivastava
Intarcia Therapeutics
Research Triangle Park, Durham, NC, USA

Editor Biography

Dr Ved Srivastava is Vice President of Chemistry at Intarcia Therapeutics and President of the American Peptide Society. Prior to that, he was the Head of Peptide Chemistry at GlaxoSmithKline (USA) and in the senior leadership role at Amylin Pharmaceuticals (now AstraZeneca US). He co-founded Phoundry Pharmaceuticals, a biotech company focused on the discovery of peptide hormone therapeutics. Phoundry was acquired by Intarcia Therapeutics in 2015.

Ved spent several years in a leadership role with Amylin Pharmaceuticals where he focused on the discovery and development of novel peptide hormones for diabetes, obesity and neuropsychiatric therapies. He has participated in the development and commercialization of SYMLIN®, Byetta® and BYDUREON® first-in-class medicines for the treatment of diabetes.

Ved has over 25 years of experience with expertise in drug discovery and development in the area of metabolic diseases, CNS and inflammation with major emphasis in peptide medicinal chemistry, chemistry manufacturing and control (CMC) and peptide drug delivery. He has numerous scientific disclosures including patents, scientific articles and invited lectures.

Ved is the Editor of three other books: (1) *Peptide-based Drug Discovery: Challenges and New Therapeutics* (Royal Society of Chemistry), (2) *Comprehensive Medicinal Chemistry III*, Volume 6, Biologics Medicine (Elsevier) and (3) *Peptide 2015* (American Peptide Society). He is an Editorial Board member of *The FASEB Journal* of the Federation of American Societies for

Drug Discovery Series No. 72
Peptide Therapeutics: Strategy and Tactics for Chemistry, Manufacturing and Controls
Edited by Ved Srivastava

Published by the Royal Society of Chemistry, www.rsc.org

Experimental Biology and the Editor-in-chief of the *Drug Development Series* books of the Royal Society of Chemistry.

Ved serves in the governance and leadership team of the American Peptide Society (APS) and the American Chemical Society (ACS) and other peptide societies. Ved is also an appointed member of the BIO1 – Peptides and Insulins Expert Committee and the Therapeutic Peptides Expert Panel of the US Pharmacopeia (USP) in partnership with the FDA. He earned a PhD in organic chemistry from the University of Lucknow, India, and had subsequent postdoctoral appointments at the University of Georgia and the University of Colorado Medical School.

Contents

Drug Discovery Series No. 72
Peptide Therapeutics: Strategy and Tactics for Chemistry, Manufacturing and Controls
Edited by Ved Srivastava

Published by the Royal Society of Chemistry, www.rsc.org

CHAPTER 1

Regulatory Considerations for Peptide Therapeutics

LARISA WU

Office of Pharmaceutical Quality, Center for Drug Evaluation and Research, US Food and Drug Administration, Silver Spring, MD 20993, USA
Email: larisa.wu@fda.hhs.gov

1.1 Introduction

Being at the borderline between typical small molecules and large proteins, peptides have raised a series of regulatory challenges. Although the use of the term "peptide" varies in the scientific literature outside the regulatory framework, the currently used regulatory definition delineates that peptides are α-amino acid polymers with specific defined sequences, 40 amino acids or fewer in size, and regardless of their production method (synthetic or of recombinant DNA origin). Despite the pharmaceutical industry's interest in peptides as drug candidates, specific regulatory challenges persist, especially related to establishing their quality standards. Moreover, current US Food and Drug Administration (FDA) and International Council for Harmonisation of Technical Requirements for Pharmaceuticals for Human Use (ICH) guidelines regarding the quality of pharmaceutical products do not include peptides. This chapter presents an overview of the approval process for New Drug Applications (NDAs) and Abbreviated New Drug Applications (ANDAs) as pertaining to the quality of peptide drugs.[1,2]

Drug Discovery Series No. 72
Peptide Therapeutics: Strategy and Tactics for Chemistry, Manufacturing and Controls
Edited by Ved Srivastava

Published by the Royal Society of Chemistry, www.rsc.org

1.2 Peptide Quality Assessment at the FDA

The FDA Center for Drug Evaluation and Research (CDER) is in charge of timely assessment of all data submitted by applicants relevant to the safety, efficacy and quality of peptide drug products. Within the CDER, the quality information in marketing applications [also known as the Chemistry, Manufacturing and Controls (CMC) section] is under the purview of the Office of Pharmaceutical Quality (OPQ). The OPQ assesses the identity, strength, quality and purity of peptide drug substances and products by integrating their review, inspection, surveillance, policy and research and creating the parity of quality between new drug products, generic drug products.[3]

1.3 Overview of the Current Drug Approval Process Employed for Peptide Applications

When a peptide drug product application is submitted to the FDA, it is the responsibility of the applicant to provide evidence that the peptide product under evaluation is safe, effective, and of high quality. Once an application has been submitted for review, OPQ regulators have the responsibility to assess the quality of the peptide drug with respect to its impact on the safety and efficacy by taking a stepwise risk analysis approach. This approach includes (1) understanding the complexity of the peptide and its clinical use, (2) evaluating the process- and product-related factors that may impact the safety and efficacy of the proposed peptide product and (3) determining whether additional studies (*in vitro* or *in vivo*) are needed to address any residual uncertainty related to the safety of the peptide product. This is not a "one size fits all" approach, but rather a case-by-case analysis depending on factors characteristic of each different peptide.[2]

As with any other drug product, the overall peptide drug development and approval process generally follows pre-defined steps: (1) preclinical investigation, (2) clinical investigation, (3) post-approval marketing surveillance and (4) life-cycle management.[1] An overview of the stages of drug development, relevant application submissions and corresponding regulatory paths as applicable to peptides is provided below.[4]

Before any clinical investigations regarding a new peptide may begin, the applicant must submit an Investigational New Drug (IND) application. Clinical investigations, as planned and documented through the IND process, generally consist of three phases: Phase I, in which safety studies are conducted in a small number of individuals (20–200 people); Phase II, in which efficacy studies begin in volunteers (up to several hundred people) of the target population; and Phase III, in which human testing continues in a significant number of patients (several hundred to several thousand people).[4] Once Phase III is complete, the applicant will submit an NDA under Section 505(b)(1) of the Federal Food, Drug, and Cosmetic Act (FDCA) including all collected clinical and nonclinical data. If an applicant intends to rely, to any extent, on safety or effectiveness investigations that were not conducted directly by or for the applicant, the NDA may be submitted in accordance with FDCA Section 505(b)(2).[1]

Upon approval, the drug product can be marketed. Any subsequent modifications to an approved application should be submitted to the FDA for assessment as post-approval changes or supplements. After NDA approval, the next stage in the life cycle of a drug product happens when all forms of exclusivity (*e.g.* patent protection) have expired for the "innovator" or "reference listed" drug. At that point, under FDCA Section 505(j), the Reference Listed Drug (RLD) may be cited in an ANDA submitted for marketing a generic version of the drug product. An ANDA contains information to show that the proposed generic product is therapeutically equivalent and thus interchangeable with the RLD, specifically in terms of active ingredient, dosage form, strength, route of administration, labeling, quality, performance characteristics and intended use. "Bioequivalence," provides the basis for establishing product interchangeability. Through reliance on bioequivalence, generally, ANDAs do not require significant amounts of time and money that would be normally needed for the submission of NDAs.[1]

1.3.1 The New Drug Application Assessment Process

NDAs submitted to FDA for assessment should comprehensively provide all clinical and nonclinical data collected during drug development efforts that were undertaken by or on behalf of the applicant. CDER assessors will evaluate these data to decide whether adequate evidence has been established with regard to safety, efficacy, risk–benefit profile, proposed labeling and quality.[6]

1.3.1.1 INDs and Presubmission of NDAs

Before starting any of the clinical trials essential for providing data to be included in the NDA, applicants are required to have submitted an IND; the IND should summarize evidence of safety and efficacy from preclinical studies and should demonstrate the preparedness of investigators for clinical trials.[7] The most relevant INDs to broad discussions of drug development and regulations are "investigator" INDs. An investigator IND is initiated by a physician who leads the investigation and decides how the investigational drug is administered. An investigator IND can be submitted for an unapproved drug or to investigate an approved drug for a new indication or new patient population. Other IND types, such as emergency use INDs (for patients who do not meet enrollment criteria for an existing study protocol) and treatment INDs (for patients with serious and life-threatening conditions), are also essential in terms of public health as they expand access to experimental drugs.[1]

Good communication between an applicant and the FDA can greatly facilitate the process of "filing" an IND. A "fileable" IND must provide information about the investigational drug manufacture, data from animal pharmacology and toxicology studies, and the proposed clinical investigation protocols, including investigator information. The IND should be submitted to the FDA at least 30 days before the in-human trials are initiated. If an answer is not received from the FDA within 30 days, the applicant may proceed with clinical trials according to the IND. If the FDA has any concerns about the way clinical trials are conducted or about their resultsin the IND, a "clinical hold" may be imposed. Successfully elaborated clinical development, per IND regulations, will typically originate from complete Phase I, II, and III clinical investigations.[5] The FDA assessment of Phase I focuses on product safety, whereas assessment of Phase II and III submissions includes evaluation of efficacy, clinical investigations, and the prospect of meeting the regulatory standards for marketing approval.[1]

In a peptide IND, FDA regulators evaluate the quality information by assessing whether chemistry and manufacturing data related to the peptide drug substance and peptide drug product may indicate health risks to subjects enrolled in IND trials.[8] Quality information related to peptide drug substance and peptide drug product is provided in a summary report, containing physical, chemical and biological characteristics relevant to the peptide, composition of the peptide drug product, a brief description of the manufacturing process, and acceptable controls and analytical methods used to assess peptide identity, strength, quality, purity and stability.[1] The applicant should also demonstrate that it can adequately produce and supply batches of the peptide drug product, that the proposed peptide drug product is reasonably safe for initial testing in humans, and that the clinical investigators are qualified to perform their responsibilities as assigned during the trial. The amount of the quality information usually varies with the phase and scope of the clinical investigation, but it is expected that this information will be thoroughly documented upon submission of a peptide NDA.[1]

1.3.1.2 Format and Content of the NDA

The intent of the NDA is to provide FDA with documentation of data gathered during animal and human clinical investigations. The format and content of NDAs submitted to the FDA are specified in regulations (21 CFR 314.50) and reflected in the form FDA 356(h), used for all drug products including peptides.[9] The contents and the specifics of a peptide NDA are as follows.[1]

1. *Index* – The index guides the FDA regulators through the peptide NDA and describes the contents and location of each section.
2. *Labeling* – Peptide drug product labeling is a critical information and it must be submitted electronically using structured product labeling (SPL) formatting in compliance with the Physician's Labeling Rule (PLR) as described in 21 CFR 201.56 and 201.57.[10]
3. *Summary* – The peptide NDA summary delivers a general understanding of the peptide drug product. It provides an overview of the safety and efficacy of the peptide drug product for its proposed useand an indication of the overall quality and accuracy of the NDA.

4. *Chemistry, manufacturing and controls/pharmaceutical quality* – The pharmaceutical quality section provides detailed information on the composition, manufacture, test methods, specifications and stability of the peptide drug substance and the final peptide drug product.

   The peptide drug substance subsection should detail all relevant physical and chemical properties and tests performed to demonstrate the identity, purity, potency and stability of the peptide, as well as methods of its manufacture.[1] Information to be submitted in an application for a peptide drug substance is addressed in Section 1.3.2.3.2 (Drug Master Files).

   Similarly, the peptide drug product subsection must present a rigorous account of peptide drug product characterization, manufacture and packaging, peptide drug product specifications, microbiology, container closure system and stability. Recommendations on the quality information that should be included in a peptide drug application[11] are listed below.

   - *Description and composition of the peptide drug product* – Lists the active and inactive ingredients in the peptide drug product together with their functions and amounts and provides a description of the dosage form and container closure system proposed for marketing of the peptide drug product.
   - *Pharmaceutical development report* – Contains details of the development studies conducted by the applicant to establish that the proposed peptide dosage form, formulation, manufacturing process, container closure system, microbiological attributes and instructions for use of the product are appropriate for the purpose specified in the application.
   - *Manufacture* – Includes a list of the manufacturers with their addresses and responsibilities, and describes the manufacturing and packaging processes, by providing flowcharts and details on the equipment, materials, and in-process controls (*e.g.* operating parameters, environmental controls, process tests and in-process material tests) for the finished peptide drug dosage form.
   - *Control of excipients* – Describes the excipients and provides the proposed specifications and tests completed to confirm their quality. Also, it identifies novel excipients and excipients of human or animal origin.
   - *Control of the peptide drug product* – Provides the specifications and tests selected to assure the peptide drug product quality and lot-to-lot consistency of the finished product. It also presents sampling procedures and validation results, and it lists all peptide drug product impurities including expected peptide impurities, degradation products, and residual solvents.
   - *Reference standards* – Provides information on the reference standard(s) used to test the peptide drug product and its impurities.

- *Container closure systems* – Describes the proposed container closure system(s) in which the peptide drug product will be marketed, including the identity of materials of construction and their compatibility with the peptide.
- *Stability* – Describes the stability protocols and stability results supporting the proposed expiration date and storage conditions of the peptide, and provides the post-approval stability protocol and stability commitment.

In addition to the general CMC information as outlined above, an NDA for a peptide product should also include specific data capturing quality factors that affect the safety (including immunogenicity) of the peptide.[2]

The immunogenicity of peptide products can have clinical consequences, including production of antibodies working against the therapeutic peptide, loss of peptide efficacy, neutralization of the human peptide counterpart, and general effects impacting the immune system (including allergy and anaphylaxis).[12]

In general, the peptide primary structure (amino acid sequence) can be a determining factor for the peptide immunogenicity. Although immunogenicity inherent to the peptide sequence is unavoidable, peptide candidates with minimal potential to elicit an immune response can be selected during drug development through sequence analysis, used to predict the peptide immunogenicity.[2] A vast array of analytical methods can be used to capture the complexity of the peptide structure. Analysis of the primary structure of a peptide is, in fact, elucidation of its amino acid sequence using amino acid chromatographic analysis, or in conjunction with sequencing methods such as by Edman degradation (N-terminal residue identification) or peptide mapping coupled with mass spectrometry.[13,14] Furthermore, Edman sequencing and tandem mass spectrometry can be used together to elucidate the primary structure of complex peptides having disulfide bonds, which may not be identified from simple amino acid analysis and sequencing.[15] Peptides with less than 40 amino acids generally exhibit high conformational flexibility, triggered by a secondary structured consisting predominantly of random coils and limited degree of secondary structures (*e.g.* α-helices and β-sheets). For some peptides, their structural ordering may impact the biological activity; therefore, elucidation of secondary structure is needed when performing peptide structure characterization.[2] Commonly used methods to investigate peptide secondary structure include circular dichroism (CD), Fourier transform infrared (FTIR), Raman, fluorescence and nuclear magnetic resonance (NMR) spectroscopy and X-ray crystallography. Far-UV CD (170–250 nm) can be applied to analyze α-helices, whereas near-UV CD (250–320 nm) can detect environmental changes around aromatic amino acids.[16,17] FTIR spectroscopy can be used to estimate the β-sheet formation and further differentiate between parallel and antiparallel forms, and also aggregates.[18,19] Raman spectroscopy is less affected by water absorption and can be used to study local interactions (involving Cys or Tyr groups) in addition to the peptide conformation. Furthermore, NMR spectroscopy and X-ray crystallography are currently the only methods that allow the characterization of the entire molecular structure of the peptide by elucidating the relative positions of all atoms (and thus all amino acids) within the peptide chain.[2]

Peptide structure can be characterized by various methods, including but not limited to those illustrated above; however, taken individually, each method can only capture limited features of higher-order structure, therefore orthogonal characterization methods are critical to correctly elucidate the peptide secondary structure. Moreover, an *in vitro* bioassay may be included as part of the characterization of higher-order structure and biological activity of complex peptides as it provides essential information on the peptide structure–activity relationship.[2]

In addition to the structure, peptide purity also impacts its safety.[20] Table 1.1 provides a compilation of peptide-related impurities resulting from synthesis (process related) or arising from degradation (product related) that may occur in peptide drug products regardless of the manufacturing process.

Peptide-related impurities and degradants are tightly controlled during the development and manufacturing of the peptide products. In general, the recommendations for reporting, identifying, and qualifying impurities for peptides differ from those cited in the guidelines for small molecules because peptide active pharmaceutical ingredients (APIs) can have many peptide-related impurities. Proposed thresholds for peptide-related impurities resulting from the peptide synthesis or arising from its degradation are usually established on a case-by-case basis considering the risk of immunogenicity and depending on the clinical use of the peptide(*e.g.*, therapeutic use, vaccine, *in vivo* or *in vitro* diagnostics). As such, ICH guidelines for impurities may be suitable for peptides with fewer than 10 amino acids in length. Such peptides usually can be treated as small molecules as they are highly pure, and less likely to cause severe adverse immune events.[2] For all others, levels of impurities should be justified [by clinical and toxicology data for innovator peptides, or by reference to available United States Pharmacopeia (USP), European Pharmacopoeia (EP), British Pharmacopoeia (BP) and Japanese Pharmacopoeia (JP) monographs, the scientific literature, development and stability data, toxicological data, significant metabolites of the drug substance and/or RLD for generic drugs].[2]

A special case of peptide degradants is aggregates. Aggregates of therapeutic peptides/proteins that are made of minimum 10–20 epitopes at a repetitive spacing of ∼100 Å and having a molecular weight greater than 100 kDa can induce immune responses by efficiently activating T-cell help for antibody production.[21,22] Aggregates lead to serious problems in the development of peptide drugs, and they may form as a result of various stress factors (*e.g.* temperature fluctuations, light, shaking, surfaces, pH adjustments) during manufacturing, processing, and shipping. Aggregates may also appear upon storage as a result of denaturation (through thermal, pH, dielectric constant

Table 1.1 Process-related impurities and degradants that may occur in peptide drug products. Reproduced from ref. 2 with permission from Elsevier, Copyright 2017.

Impurity	Mechanism
Denaturation	Modification of the peptide structure that changes its physical, chemical and biological properties; occurs in the presence of denaturing agents
Proteolysis	Upon exposure to harsh conditions, extreme pH, high temperature or enzymes
Aggregation and precipitation	Association of hydrophobic amino acid residues to form aggregates; precipitation occurs if on a macroscopic scale
Deamination	Hydrolysis of the side-chain amide linkage of an amino acid residue with formation of a free carboxylic acid
Oxidation and reduction	Induced by temperature, pH, metal ions and buffers during synthesis and/or storage
Disulfide exchange	Change in conformation due to reaction between disulfides within a peptide chain
Diastereoisomerization (racemization)	Alteration of L-amino acids to D,L mixtures, with formation of peptide bonds sensitive to proteolytic enzymes
β-Elimination	Proceeds through a carbanion intermediate; susceptible residues under alkaline conditions are Cys, Lys, Phe, Ser and Thr
Deletion (incomplete coupling)	Incomplete coupling due to incomplete removal of the protecting group of the last coupled amino acid or insufficient activation of the incoming amino acid
Truncation	Missing one or more amino acid residues at either the N- or C-terminal end due to precipitation of resin beads during synthesis
Amino acid insertion	Insertion of an additional amino acid into the peptide sequence due to excess of 9-fluorenylmethoxycarbonyl (Fmoc)-protected amino acids and improper washing
Dimers	Self-association of hydrophobic amino acids residues in peptides to form aggregates
By-products generated by incomplete deprotection	Results in protecting groups covalently attached to peptide sequences

and/or ionic strength changes), or peptide modification (through oxidation, deamidation and primary sequence changes). Some of these factors (*e.g.* temperature, pH, light) can be easily controlled during manufacture and storage, but control is more challenging for others (*e.g.* mechanical stress, surfaces[23–26]). Knowing that a high content of aggregates can be correlated with increased safety risks, including immunogenicity, it becomes necessary to evaluate the aggregate profiles by using a combination of orthogonal analytical methods to cover a wide range of aggregates sizes, such as size-exclusion chromatography (SEC), analytical ultracentrifugation (AUC), dynamic light scattering (DLS) and field-flow fractionation (FFF).[2]

Excipients can also play an important role in the safety of peptide drug products. It has been reported that interactions of some peptides and small proteins with preservatives (such as phenylethyl alcohol, chlorobutanol, benzalkonium chloride and benzyl alcohol) may be responsible for conformational changes and aggregation,[27] which may lead to increased immunogenicity, allergic reactions, and even anaphylactic shocks (*e.g.*, salmon calcitonin peptide).[28] Also, peptide interaction with Tween 80 (polysorbate) micelles and exposure of peptide epitope to the micellar surface may induce antibodies.[29] Based on these observations, it can be concluded that the formulation of a peptide drug product can change the propensity of the peptide to aggregate, deaminate, oxidize, *etc.*; therefore, it can enhance or reduce its immunogenic potential. An appropriate design of the peptide product formulation may affect the product safety and immunogenicity by stabilizing the peptide against physical and chemical degradation, and minimizing its interactions with the excipients.[2,30]

Another factor that affects the safety of the peptide drug product is the interaction of the drug product formulation with the container closure system (especially rubber components), which may produce extractables and leachables. To this regard, organic compounds leached from the uncoated chlorobutyl rubber plunger of a syringe in contact with erythropoietin formulations (Eprex®) containing polysorbate 80 may have been a factor enhancing immunogenicity.[31] Leachables from the rubber components of the container closure system may impact peptide immunogenicity by acting as adjuvants for peptide crosslinking and antibody induction. Leachables must be studied as part of stability studies according to ICH and FDA stability guidance for drug products.[32]

The remaining information in the pharmaceutical quality section of a peptide NDA refers to samples of the peptide drug substance and/or finished peptide product, validation of analytical methods, and environmental impact information provided with adequate detail so that FDA quality assessors can thoroughly evaluate any impact on the peptide drug product performance.[1]

The rest of the peptide NDA contents are outlined below.

5. *Nonclinical pharmacology and toxicology* – This section includes all relevant animal and laboratory studies to be evaluated by the FDA with respect to inconsistent or inadequate toxic effects, acute, subacute and chronic toxicity, pharmacological activities and potential carcinogenicity and teratogenicity.[1,33]
6. *Human pharmacokinetics and bioavailability* – This section of the peptide NDA includes clinical pharmacokinetic and bioavailability data and analyses, and relevant analytical and statistical methods.[1]
7. *Clinical microbiology* – This section is provided only when effects on the physiology of a targeted microorganism are relevant to the NDA.[1] It reports the drug action on microbial physiology, the antimicrobial spectrum affected, resistance mechanisms, and laboratory methods used.[34]

8. *Clinical data section* – The clinical dataset is the basis of the efficacy and safety of the proposed peptide drug product. A wide range of FDA and ICH guidelines address different aspects of the clinical studies in NDA submissions that cannot be covered in detail here.
9. *Safety update* – The first safety update report is usually filed as an amendment to the NDA at 4 months after the submission of the original NDA and contains safety information that arises from ongoing studies, animal studies, and other sources.[1]
10. *Statistical* – This section of the peptide NDA is closely linked to the clinical data section and contains statistical analyses with a format and content agreed upon by the applicant and the FDA.[1]

11–12. *Case reports* – This section includes tabulations of cases of patient data and data elements with proper case report forms (CRFs). The CRFs for patients who died during a clinical study and for patients who discontinued because of an adverse event should also be included.[1,35]

13–14. *Patent information and certification* – Patent certification is mandatory for any relevant patents that claim the listed peptide drug on which the investigations relied, or that claim a use for the listed peptide drug.[1]

15. *Establishment description* – This section provides general information about the organization, physical plant and major equipment, in addition to quality assurance functions.[1]
16. *Debarment certification* – The FDA is authorized to debar individuals convicted of crimes related to the development, approval, or regulation of drugs; these individuals should not provide any type of services to the applicants of a peptide drug application.[1,36]
17. *Field copy certification* – Applicants based in the USA must submit a "field" copy of the pharmaceutical quality section, application form, and summary of the peptide NDA to FDA. This information will be used during the preapproval inspection (PAI) at the manufacturing site.[1]
18. *User fee cover sheet* – Form FDA 3397 is used to determine the applicability of a user fee and indicate whether a payment has been made to an FDA account concurrent with the submission of the peptide NDA.[1,37]
19. *Financial information* – This section includes clinical investigator financial disclosures (form FDA 3455) and certification (form FDA 3454), describing financial interests of the investigator and applicant–investigator financial arrangements.[1,38]
20. *Other information* – Any information that was submitted to FDA before the peptide NDA submission should be referenced in this section. An applicant may propose other uses of this section at the pre-NDA meeting.

1.3.1.3 NDA Assessment

1.3.1.3.1 Filing. Prior to submitting a peptide NDA to the FDA, applicants are strongly encouraged to request a meeting with the FDA to discuss the planned content of the application and ensure that the submitted application is complete and fileable. The date of receipt by the FDA "central document room" of the peptide NDA starts the review time clock for that application. When the NDA is ready for assessment, a regulatory project manager (RPM) manages the assessment process and coordinates all communications with the applicant. After payment of all required fees and submission of complete administrative information, the peptide NDA is assigned to a multidisciplinary team of FDA assessors, who by day 45 of the review (day 30 for priority reviews) will discuss their respective decisions regarding the fileability of the application. All filing review issues identified by the assessment team are communicated in writing to the applicant. Applications failing to provide necessary information may result in a refuse-to-file (RTF) letter by day 60. Otherwise, the application is accepted for filing and the applicant receives a 74 day letter, confirming the application action date and the review choice (standard or accelerated).[1,39]

1.3.1.3.2 Assessment. The overall NDA assessment and approval process (or "cycle") usually takes place in six major steps: (1) presubmission activities, (2) process submission, (3) review plan, (4) conduct review, (5) take official action and (6) post-action feedback.[40,41]

Once the application is filed, a planning meeting is held with the assessment team to discuss timelines, high-level labeling revisions, need for advisory committees, and/or inspections and review activities. The review timeline for NDAs, including New Molecular Entities (NMEs), is 10 months from receipt for standard reviews, and 6 months for priority reviews. During the main review phase, each assigned FDA assessor, according to discipline (*e.g.* medical officer, pharmacologist, chemist, statistician, microbiologist, clinical pharmacologist), evaluates the pertinent portion of the peptide application, proposes labeling revisions, and writes an assessment document.[1]

All approvability and high-level labeling issues are identified by the time the mid-cycle meeting is held (*i.e.* month 5 for standard reviews, and month 3 for priority reviews). Deficiencies identified by respective discipline assessors are usually communicated to the applicant after the mid-cycle meeting through a mid-cycle Information Request (IR). The conclusions of all assessment and inspection activities are integrated during the wrap-up meeting; the medical officer takes the lead to reconcile observations and written summaries, after which the action package is finalized.[1]

In the event that the peptide NDA is approved, the applicant receives a letter that authorizes the manufacturer to distribute the peptide product in accordance with any pre-established agreements and post-marketing commitments. If the agency decides on the basis of its assessment not to issue an approval letter, a complete response (CR) letter will be issued, citing deficiencies and offering recommendations to the applicant. Applicants who receive a CR letter may request an end-of-review meeting with the FDA to discuss deficiencies and further steps to be taken, including resubmission, withdrawal, or request an opportunity for a hearing.[1,42]

1.3.1.3.3 Labeling. Final discussions about the package insert will usually take place near the end of the NDA assessment, possibly 2–5 weeks before the PDUFA (Prescription Drug User Fee Act) action date. At this time, the FDA will communicate recommendations concerning the peptide product label to the applicant; the agency and applicant will work to reach agreement on the final wording of the package insert prior to the action date.[1]

1.3.1.3.4 FDA–Applicant Communications During NDA Review. While evaluating the quality, safety and efficacy of the peptide drug product, based on a thorough scientific examination of the submitted application, the FDA may choose to seek clarification from the applicant through an information request (IR) email or teleconference, discipline review letter (DRL), or face-to-face meeting. Any information needed in support of the peptide application review must be officially submitted as an amendment to the application.[1]

1.3.1.3.5 Advisory Committee. If significant issues arise during the peptide NDA assessment, the FDA can seek advice from an advisory committee. An advisory committee comprises a specialized group of external experts, identified by the FDA, that convenes to provide advice to the agency concerning any number of matters,[43] including NME, novel clinical trial designs or the use of surrogate endpoints, particular safety issues or drug effectiveness.[1]

1.3.1.4 Special Approval Pathways

1.3.1.4.1 Expedited Approval Pathways. In order to increase the availability of drugs intended to treat serious and/or life-threatening conditions, four expedited drug approval pathways have been created: Priority Review Designation, Accelerated Approval, Fast Track Designation, and Breakthrough Therapy Designation.[44] Drug products evaluated and approved by these expedited approval pathways have efficacy, safety, and quality held at the same rigorous regulatory standards as drugs evaluated and approved in the FDA regular review programs.

1.3.1.4.2 Orphan Drug Designation Program. An orphan designation signifies that a drug indication applies to a rare disease, which by definition affects fewer than 200 000 American patients per year. The Orphan Drug Act (1983) and amendments (1992 and 2013) established several incentives to encourage the development of orphan drugs, including market exclusivity for 7 years after approval, fee waivers, tax credit on clinical research, technical assistance during the review process, and other federal research grants.[45]

1.3.1.4.3 Pediatric Exclusivity. There have been a few FDA initiatives to improve medical product research and availability for children, including the Food and Drug Administration Modernization Act (FDAMA) of 1997,

Best Pharmaceuticals for Children Act of 2002, Food and Drug Administration Amendments Act (FDAAA) of 2007 and Food and Drug Administration Safety and Innovation Act (FDASIA) of 2012. Pediatric exclusivity is a 6-month extension of exclusivity, added to existing periods of marketing exclusivity or patent protection.[46]

1.3.2 Abbreviated New Drug Application Assessment Process

Abbreviated New Drug Applications (ANDAs) contain information submitted to CDER for the assessment of generic drug products. Generic drug applications are "abbreviated" because they generally allow the applicant to rely on preclinical (animal) and clinical (human) data used to establish the safety and effectiveness of an approved drug product (RLD). Generic products must be therapeutically equivalent to the RLD, as established through the combined criteria of pharmaceutical equivalence and bioequivalence.[1] For a generic peptide drug product, pharmaceutical equivalence signifies that the generic product and the RLD contain the identical peptide API and its strength, dosage form, route of administration, quality, performance characteristics and intended use are the same for the generic product and the RLD.[47] Bioequivalence signifies that the generic product exposes the patient to the same amount of API in the same time course as does the RLD; in general, generic peptide drugs formulated as parenterals with the same active and inactive ingredients in the same concentration [qualitatively and quantitatively (Q1/Q2) the same] as the RLD receive a biowaiver.[48] Generic peptide applications are filed under FDCA Section 505(j) and authorized for marketing only after all forms of exclusivity for the innovator peptide drug product have expired.[1]

1.3.2.1 Format and Content of the ANDA

The format and content of ANDA submissions to the FDA are specified in regulations (21 CFR 314.94) and reflected in the "harmonized" application form FDA 356(h). The list of contents of a peptide ANDA and the difference between the required contents of peptide NDA and ANDA submissions are provided below.[1]

1. *Index/table of contents.*
2. *Basis for ANDA submission* – This section shows that the peptide ANDA refers to the RLD selected by the agency and documents a comparison between the generic peptide drug and the RLD, including a statement that the peptide API, the route of administration, dosage form, and strength of the proposed peptide drug product are the same as those of the RLD.
3. *Labeling* – According to 21 CFR 314.94(8), it includes a side-by-side comparison of the applicant's proposed labeling (container and carton) and the labeling approved for the RLD for each strength and package size; any differences should be explained in annotation.[1]

4. *Chemistry, manufacturing and controls/pharmaceutical quality* – The quality section contains information as required under 21 CFR 314.50(d) and detailed below as applicable for generic peptides.

The quality section of a peptide ANDA, like that of a peptide NDA, covers raw materials, manufacturing facility, manufacturing of the peptide API and in-process controls, a master production record and a description of the equipment to be used in the manufacture of the commercial batch of the peptide drug product, materials used in peptide API packaging, controls for the finished dosage form, analytical methods, stability of the finished dosage form, sterility assurance data, and samples and environmental considerations.[1] Stability requirements for generic and new drug peptides are identical, according to ICH and FDA guidelines.[49,50]

Differences between peptide ANDA and NDA quality sections are usually related to the peptide API information; because the peptides that are used in the manufacture of generic peptide drug products are usually obtained from third-party external manufacturers, the ANDA applicant must typically reference a Drug Master File (DMF), previously submitted to the FDA, containing proprietary information (see Section 1.3.2.3.2). Additional requirements for peptide ANDAs as compared with peptide NDAs include but are not limited to the following:[1]

- Any inactive ingredient in the proposed peptide generic product that differs from the RLD must be identified and demonstrated to have no effect on safety and efficacy.
- Specifications should be justified with reference to compendia (*e.g.* USP, JP) and/or analysis of the RLD.
- Release characteristics of the proposed generic peptide must parallel those of the RLD.

Similarly to peptide drugs submitted as NDAs, generic peptides undergo extensive characterization with respect to primary and secondary structure, oligomer and aggregation states, and biological activity. For generic peptides, the primary structure and the position of disulfide bonds are critical determinants of the peptide API sameness as compared with the RLD. When comparing peptides, the two-dimensional NMR methods can be used to identify the fingerprint peptide structure and detect any alterations in the peptide structure caused by environmental changes (such as new excipients), and demonstrate sameness of higher-order structure between the generic peptide and its RLD.[2] Most generic peptide drug products have formulations that are Q1/Q2 the same as the RLD, thus presenting a low risk. However, there are situations when differences in formulations introduced by new excipients may affect peptide conformation or interact with the peptide. These new excipients could lead to the formation of dimers and higher-order aggregates, which may alter the stability, safety (including immunogenicity), and efficacy of the peptide drug product, rendering a peptide drug product essentially different from the RLD. Nonetheless, for generic peptides that have the same amino acid sequence and are formulated Q1/Q2 the same as the RLD, it can be generally inferred that they have the same higher-order structure as the RLD. Assessment of the biological activity is particularly important in cases where a synthetic peptide is used as the API, if an additional confirmatory test is needed for demonstrating sameness between the proposed generic and RLD products.[2]

Different factors including modification and degradation resulting from the manufacture or storage conditions, residual solvents, formulations or leachables from the container closure system can potentially change the quality of the peptide API and affect the safety (including immunogenicity) of the peptide products.[51] Clearly, the potential for immunogenicity should be taken into consideration but, especially for generic peptides, the safety concerns can be mitigated by controlling the peptide drug quality through optimizing the manufacturing, drug formulation, and assuring proper storage conditions.

For the peptide API in a proposed generic drug product, a one-time study making a direct comparison of peptide-related impurity profiles between the proposed generic drug product and multiple batches of its RLD is beneficial to show comparably low levels of peptide-related impurities under long-term storage and accelerated/stress conditions. In particular, when evaluating the quality of a generic peptide, it is expected that impurities will differ between a generic peptide and its RLD, especially if the peptide manufacture pathways use different synthetic methods (chemical synthesis *versus* recombinant DNA technology).[2] To diminish the risks to patients, these specified impurities, different from those of the RLD, should be strictly controlled at low levels established based on prior knowledge and supported by risk assessment, and/or, depending on the peptide, as indicated by guidelines.[52] Recently, the FDA developed a draft guidance according to which, for synthetic peptide drug products (including glucagon, liraglutide, nesiritide, teriparatide and teduglutide) referencing a recombinant-derived peptide drug product, the risk of immunogenicity due to impurities in the generic product should not differ from that of the RLD.[52] Mitigation of the immunogenicity risk can be achieved by providing data to demonstrate that the impurities in the generic peptide product, different from those in the RLD, do not produce a stimulation of innate immune activity greater than that caused by the RLD. Such data can be obtained from *in silico* and *in vitro* studies that determine the binding affinities of the peptide to the major histocompatibility complex (MHC), which can be used to identify the peptide impurities responsible for prompting an immune response and, hence, to inform the relative immunogenicity risk of these impurities different from those of the RLD.[53–55]

Comparison of the aggregation profiles between a proposed generic peptide drug product and its RLD should be done throughout the product shelf-life, and also in accelerated/stress conditions. This comparison uses orthogonal analytical methods (such as SEC, AUC, DLS, and FFF) to cover a wide range of aggregates sizes.[2] Demonstrating the comparability between the generic product and its RLD in peptide aggregate profiles in terms of aggregates sizes and levels can sufficiently address the safety/immunogenicity concerns due to aggregation.

Moreover, the peptide–excipient interactions and the impact of excipients on the peptide stability should be assessed in real-time and forced degradation studies. In this regard, in generic peptide products that have formulations that are Q1/Q2 the same as the RLD, the peptide–excipient interactions are expected to be the same as for the RLD. However, when Q1/Q2 the same formulations of a proposed generic peptide and its RLD differ with respect to preservative, buffer, and/or antioxidant (as an allowed difference in excipients of injectable generic products),[56] the effect of the new excipient on the peptide stability is to be evaluated by real-time and/or forced degradation stability studies performed in comparison with the RLD. The possible interactions between the peptide and the new excipient can be determined using a range of techniques including surface plasmon resonance (SPR), NMR spectroscopy, differential scanning calorimetry (DSC), and/or bioassay.[2]

5. *Human pharmacokinetics and bioavailability/bioequivalence* – This section contains information that shows that the drug product is bioequivalent to the RLD, but waivers of *in vivo* bioequivalence studies are available in certain cases.[57] Most of the injectable peptide drugs are subject to receiving a biowaiver if their formulation is Q1/Q2 the same as the RLD.
6. *Case report forms* – See case reports for NDAs (Section 1.3.1.2).
7. *Patent information and certification* – This section contains all relevant patents for the peptide drug product, method of use, licensing agreements, disputed patent information, amended certifications, *etc.* Brand peptide NDA applicants must confirm patent certification under one of the following criteria (*i.e.* Paragraphs):
   I. patent information on the peptide drug product that is the subject of the ANDA has not been submitted to the FDA;
   II. the patent has expired;
   III. the date the patent expires; or
   IV. the patent is invalid or not infringed by the peptide drug product proposed in the submitted ANDA.[1]
8. *Financial certification or disclosure statement* – See certification and financial information for NDAs (Section 1.3.1.2).

An applicant may amend a peptide ANDA under review to revise submitted information, or respond to information requests, which generally tend to involve the quality section of the application.

1.3.2.2 ANDA Assessment

1.3.2.2.1 Pre-ANDA (pANDA). The demonstration of equivalence for complex generic products including peptides is challenging enough that many complex products currently do not have generic competition. The FDA established the Regulatory Science Program in the Generic Drug

User Fee Amendments of 2012 (GDUFA) to address this challenge through research and funding of studies specifically addressing difficult problems in demonstration of equivalence. The GDUFA II program established a pre-ANDA (pANDA) meeting pathway route as a means through which industry could meet with the FDA to clarify regulatory expectations, assist applicants to develop more complete submissions and discuss new or alternative methods for establishing bioequivalence for complex generic drug products. The ultimate goal of this effort is to increase efficiency and effectiveness of the ANDA assessment process and to reduce the number of review cycles required to obtain ANDA approval for complex drug products. Another modality of discussing with the FDA prior to ANDA submission is in the form of a controlled correspondence requesting information on product development, especially for first generics.[58]

1.3.2.2.2 Filing. Once a peptide ANDA is submitted, the FDA determines whether that ANDA is sufficiently complete to allow a substantive assessment. If the submission is incomplete, the FDA can issue a refuse-to-receive (RTR) letter, which prevents a lengthy assessment period, multiple cycles, and/or waste of resources. Complete peptide ANDA submissions may be filed under Paragraph I, II, III and IV certifications depending on the related patent status as listed in the "Orange Book" for the brand peptide RLD. An ANDA applicant who files a Paragraph IV certification must, within 20 days of filing, notify the innovator, who will have 45 days to take action upon receiving the notification. The FDA may hold the peptide ANDA up to 30 months, depending on the outcome of the litigation between the innovator and the ANDA applicant, if any.[1,59]

1.3.2.2.3 Assessment. As more and more peptide RLDs have come off patent and as more companies have begun generic peptide drug manufacture, the volume of peptide ANDAs has been growing over the past decade. According to the GDUFA II, the FDA is committed to review and act, within 10 months from submission, on 90% of those original ANDAs from the date of submission, and within the applicable review goals on 90% of certain priority ANDAs and their amendments.[60] To assist ANDA applicants in improving the quality of submissions, the FDA has issued a number of GDUFA-related guidance documents[61] and is working on issuing product-specific guidelines.[62]

The quality section of a peptide ANDA is the major component of the peptide ANDA assessment. Quality reviewers organized within a number of disciplines work as a team to assess the quality of the peptide drug substance and drug product, manufacturing and controls, batch formulation and records, facilities, product specifications, packaging and stability. A number of assessment enhancement initiatives are under way to ensure the timely review of ANDAs; these focus on efficient communication with the applicant or on efficient assessment (*e.g.* through knowledge management). If all disciplines find the ANDA acceptable and if all facilities are in satisfactory standing as evaluated and inspected, the ANDA receives approval. If the ANDA assessment is completed prior to the expiration of innovator peptide drug product exclusivity, the ANDA applicant will receive tentative approval (TA) but is not allowed to distribute the peptide product. An approval letter is contingent upon the resolution of litigation and issues related to peptide product exclusivity.[1]

1.3.2.3 Special Considerations for ANDAs

1.3.2.3.1 ANDA Filing and Market Exclusivity. New drug product exclusivity is provided by FDCA Sections 505(c)(3)(E) and 505(j)(5)(F). Peptide NDAs seeking to market a new peptide receive a 5 year period of exclusivity; but after 4 years, information from such an NDA approval may be used as part of the submission of Section 505(b)(2) NDA or an ANDA, on condition that the new submission supplies evidence of no infringement.[1] Follow-on peptide drug products containing a previously approved active moiety and approved on the basis of new clinical investigations are granted a 3 year period of exclusivity, whereas the first generic version of a peptide drug product may be eligible for a 180-day period of exclusivity. Orphan drug exclusivity extends 7 years, and pediatric exclusivity provides a 180 day extension period to existing patents or exclusivity.[1]

1.3.2.3.2 Drug Master Files (DMFs). A Drug Master File (DMF) is a document prepared and submitted by a manufacturer to support an IND, NDA, ANDA, another DMF, or their amendents and supplements. A type II DMF is commonly used to support the peptide API information in ANDAs; although it provides detailed information about the facilities, manufacturing process, packaging, and storage of a peptide API, the peptide DMF is neither a regulatory requirement nor a substitute for a marketing application.[1] Before the FDA can assess the peptide DMF submitted in support of an application, the DMF holder must provide a letter of authorization (LOA) that allows the applicant to reference that DMF. Because the information submitted in the peptide DMF is proprietary, deficiencies as identified during assessment of that DMF are communicated by the FDA to the DMF holder only, and cannot be disclosed to the applicant. Instead, the applicant referencing that peptide DMF is notified by the FDA if the DMF is deficient; it is then the DMF holder's responsibility to inform the applicant if the deficiencies in the DMF have been resolved. A DMF submission is never "approved"; it is assessed to determine whether it is adequate to support a particular application that references it.[1]

A type II peptide DMF is subject to GDUFA fees, an initial "Completeness Assessment" (CA), and communications with DMF holders.[62] The CA determines whether the peptide DMF is substantially complete and ready to undergo a comprehensive scientific evaluation, but does not guarantee that the DMF will be found adequate.[1] A type II peptide DMF must cover characterization of the peptide API, manufacturing and controls of the intermediates and final peptide API, including a list of the organic impurities, reference standards or materials, container closure system used for packaging the peptide API, and stability data and storage conditions. A type II peptide DMF must pass the initial CA to be referenced in an ANDA.[1] A list of DMFs that have passed the CA and are available for reference by ANDAs under GDUFA is available on the FDA website.[63]

As pertaining to the peptide DMF, an increased regulatory focus has been placed lately on the control of starting materials used in the manufacture of synthetic peptides, and also control of the synthetic steps.[64] Some details regarding information submitted in a type II DMF for a chemically synthesized peptide are presented below.

Control of Starting Materials. Starting materials for solid-phase peptide synthesis (SPPS) include resins used as solid supports, amino acids and protected amino acid derivatives, peptide fragments, reagents used in coupling, deprotection and cleavage steps of the synthesis and solvents.[2]

Resins

The selection of the resin itself affects the outcome of the SPPS. The resin should have an appropriate particle size that allows ease of handling and removal of solvents used during the SPPS, and it should be inert. The extent of the amino acid initial loading onto the resin is determined by the resin swelling (mL g^{-1}) properties, and also by its substitution ratio (mmol g^{-1}); an inefficient loading may lead to C-terminal deletion/truncated peptide impurities, and may directly impact the purity and the yield of the final peptide drug substance.[65] The substitution of the resin can be determined after loading of the first amino acid, for example, by common color or UV tests.[2]

Amino Acids and Amino Acid Derivatives

With respect to the quality of the amino acids and amino acid derivatives, particular attention is given to their chiral properties and impurities. Impurities in raw materials, including (i) free amino acids or amino acid derivatives, (ii) amino acid contaminants, (iii) incorrect enantiomers, *i.e.* D-isomer,[66] (iv) dipeptides or oligopeptides, and (v) β-alanine containing contaminants,[67] are likely carried over into the final peptide drug substance and, therefore, should be tightly controlled.[2]

Reagents and Solvents

Generally, there is limited concern regarding the quality attributes of the reagents and solvents used in the initial stages of the peptide manufacturing process as they are unlikely to be incorporated in the final peptide drug substance. These materials are tested in-house for vendor qualification and at the minimum should be pure to avoid side reactions. Nevertheless, solvents and reagents (*e.g.* acetonitrile, methanol, 2-propanol, acetic acid, trifluoroacetic acid) typically used in the purification or precipitation of peptides can be carried over and should be quantified in the final peptide drug substance.[2]

In-process Control Strategies. Special attention is given to each step of the SPPS, including initial coupling, final deprotection and cleavage, and in-process controls. Appropriate in-process controls are important for SPPS in order to avoid augmentation of peptide impurities from one synthesis step to the next. The control strategy developed during manufacturing process design refers to the identification of potential side reactions and impurities, and steps undertaken to minimize their levels.[2] Following the selection and control of the raw materials and attachment of the first amino acid, the peptide chain is built through a repeated cycle, involving first the deprotection of the N^{α}-amino protecting group (Fmoc), activation, and coupling of the next amino acid.[2]

Deprotection Control Strategies

The removal of the Fmoc protecting group is normally achieved through short exposure to a solution of 20–30% piperidine in dimethylformamide. Although in most cases deprotection strategies are effective, for long peptides incomplete deprotection leading to deletion peptide impurities can occur even at high concentrations of piperidine and after repeated trials. In this case, increased time allocated to the deprotection step or a stronger base, *e.g.* DBU (1,8-diazabicyclo[5.4.0]undec-7-ene) are used.[2] Moreover, a washing step is introduced between deprotection and the next coupling step to remove the piperidine reagent and thus avoid premature deprotection of the next amino acid and generation of insertion impurities. In-process tests to ensure complete removal of piperidine are usually employed, including measurement of the pH of the washing solvent and a chloranil test.[68]

Coupling Control Strategies

In the coupling step, the coupling reagents are transforming the free carboxyl terminus of the incoming amino acid into an active electrophilic form. The selection of the activation reagent is a critical variable; an optimal reagent allows efficient acylation of the amino function, thus increasing the coupling efficiency and limiting the side reactions, such as racemization. In principle, racemization can occur in any step of the SPPS; however, the risk is higher during activation and coupling.[2] Extra caution is still needed for certain amino acids that are prone to racemization (*e.g.* Hys, Cys, Asp[69–71]) and also under reaction conditions with elevated temperature[72] or with excess of base.[73] Racemization impurities may not be readily detected and separated by high-performance liquid chromatography (HPLC); therefore, it is important to minimize their levels as they may cause aggregation of the peptide product, and also potential biological consequences.[74–76]

Another side reaction that is common during coupling is insertion. An insertion peptide impurity can be generated when an incoming amino acid first acylates an amide bond of the resin bond peptide, and then inserts itself into the peptide sequence during the next round of coupling.[77] The insertion can be caused by an excess amount of Fmoc-protected amino acid and is more likely to occur at locations on the peptide sequence where the amide bond is further away from bulky side chains that introduce steric hindrance.[78] For certain difficult-to-react amino acids, recoupling may be necessary to reach completion of the reaction; however, if needed, a capping step may also be introduced to prevent the generation of deletion sequences. Capping itself will generate a truncated sequence; but truncated sequences are much easier to separate than the deletion sequences.[2] The completion of coupling can be determined and controlled using suitable in-process color tests such as Kaiser, trinitrobenzenesulfonate (TNBS), chloranil or Bromophenol Blue tests, which are able to detect the presence of residual amine. Each of these tests has its limitations; therefore, two independent tests are recommended to monitor the completion of the coupling reaction against negative controls (with protected amino function) and positive controls (with deprotected amino function).[2]

The deprotection–coupling cycle is repeated until the last amino acid is attached to the resin-bound peptide.

Cleavage Control Strategies

During the cleavage step, the entire peptide chain, together with the side-chain protecting groups of the amino acid residues, are cleaved simultaneously. Concentrated trifluoroacetic acid (TFA) (95%) cocktail containing water and triisopropylsilane (TIS) scavengers is the most commonly used cleavage reagent in Fmoc-protected SPPS. Scavengers are added to protect sensitive nucleophilic side groups (such as Trp, Met, Tyr and Cys) by converting the cations generated during the reaction into corresponding hydrocarbons.[79] Scavengers containing sulfur, such as ethanedithiol (EDT), dithioerythritol (DTE), and thioanisole, are helpful for trapping carbocations and preventing oxidation and disulfide bond formation for Met and Cys residues,[80,81] whereas TIS can replace EDT to protect Arg and Trp residues.[82] Failure to properly shield these sensitive amino acids may cause unwanted impurities, which can further produce peptide aggregation.[83] The cleavage time and temperature need to be optimized during the development stage and be strictly monitored in peptide production.[2]

Post-cleavage Work-up Control Strategies

Certain peptides may need to undergo cyclization through the formation of oxidative intramolecular disulfide bridges. In this regard, a regioselective strategy with orthogonal Cys protecting groups can be employed for peptides with multiple disulfide bonds[84] and the oxidation conditions (temperature, duration) can be controlled to minimize the formation of intermolecular disulfide bonds responsible for the formation of dimers and aggregates.[85,86] In addition, controls (*e.g.* Ellman test) can be established to assess the extent of formation of disulfide linkages in these peptides.[2]

Finally, selection of the peptide purification route takes into account the size, polarity, and ionic character of the desired peptide and process-related impurities, including diastereomeric, deletion, truncated, deaminated, isomerized and oxidized peptides, disulfide exchange products, oligomers/aggregates, by-products generated by incomplete deprotection of amino acid side-chain protecting groups, and reagents and solvents used in the peptide synthesis. Commonly used separation methods are reversed-phase chromatography (RPC) and ion-exchange chromatography (IEC).[2]

Owing to concerns regarding stability in aqueous solutions, many peptides are supplied as lyophilized (freeze-dried) products. Lyophilization conditions (temperature, pressure, rate and duration) in all stages and the residual moisture in the lyophilized peptide are strictly controlled throughout the product shelf-life because high concentrations (5–20%) of residual moisture reduce the glass transition temperature (T_g) of the product, resulting in instability issues.[2]

As an alternative to SPPS, small peptides composed of several amino acids can be efficiently prepared by solution-phase peptide synthesis (SPS) through the coupling of single orthogonally protected amino acids in solution, thus avoiding unwanted side reactions and overcoming synthetic difficulties.[87,88] The core chemistry of SPS, including deprotection, activation, and coupling, is similar to that of SPPS. The prime advantages of SPS are related to ease of scale-up, fewer side reactions, high purity of the final peptide, and convenient, flexible control methods applicable throughout the entire synthetic process.[2,89,90] However, the long reaction time in SPS,[91] combined with the need for purification steps after each coupling and deprotection reaction are major disadvantages. Because no solid support is used in SPS, extensive extraction, precipitation, and washing are needed to yield a relatively pure peptide intermediate. Consequently, the number of solution-phase stages and isolations needed increases with the length of the peptide, which considerably complicates its synthesis process.[92] To this end, combination of SPPS and SPS methods is the most effective approach to synthesize long peptides; specifically, short fragments of the required peptide are first synthesized by SPPS, then coupled together in solution to form the final long peptide. In this case, quality control of the intermediate peptide segments becomes critical and quantitative HPLC methods should be established to ensure segment purity.[2]

1.3.3 Post-approval Activities and Life-cycle Management

Whereas "amendments" are submitted to update or modify unapproved peptide applications, "supplements" are submitted to modify approved peptide applications. Supplements must address post-approval changes to peptide drug product components and composition, manufacturing sites and/or processes, specifications, container closure systems, stability protocols and expiration dates, or labeling. Supplement submissions list all changes that may affect the peptide drug product, with sufficient detail for the FDA to determine the scope of change readily.[1] Post-approval changes to an approved peptide NDA or ANDA are classified as major, moderate, or minor and should be reported using one of four reporting categories:[93] prior-approval supplements, Changes Being Effected (CBE-30 and CBE) supplements, or annual reports.[1]

1.3.3.1 *Prior-approval Supplements (PASs)*

PASs describe major changes, having the substantial potential to affect adversely peptide drug product quality, safety, and effectiveness; such changes require approval by the FDA before they are implemented in any distributed peptide drug product. Prior-approval changes described in a PAS include changes in the formulation (*e.g.* dosage or ingredient-release form), components and composition of the peptide drug product, changes in the manufacturing site that could trigger an FDA inspection, or placement of aseptic sterilization or various manufacturing processes (including packaging and closure systems) within new locations.[1] Other changes described in a PAS include changes affecting the impurity profile or the physical, chemical, or biological properties of the peptide drug product, changes affecting specifications or analytical processes, or changes to peptide drug product labeling (*e.g.* based on post-marketing study results).[1]

1.3.3.2 *Changes Being Effected (CBE-30 and CBE) Supplements*

CBE-30 supplements to an approved peptide application describe changes having moderate potential to impact the safety and effectiveness of the peptide drug product adversely; if, within 30 days of receiving a CBE-30 supplement, the FDA informs the applicant that a PAS is required, no peptide drug product affected by the change may be distributed. CBE supplements without the 30-day specification may also be used to describe moderate changes. Moderate changes reported in the CBE-30/CBE supplements may pertain to manufacturing facilities or packaging sites for the peptide API or peptide drug product, manufacturing processes, equipment, starting materials, specifications, analytical procedures, container closure systems or labeling. Examples can be found in FDA guidelines.[1,93]

1.3.3.3 *Annual Reports*

Annual reports submitted to an approved peptide application report minor changes, with little potential to affect the peptide drug product adversely. Changes reported in annual reports may be related to new manufacturing sites for peptide intermediates, new packaging sites for secondary packaging, new labeling sites, new equipment of the same design and operating principles, new specifications made to comply with an official peptide compendium, tightening of specifications, and minor labeling edits. Examples of each type of change that is appropriate for reporting in an annual report can be found in FDA guidelines.[1,93]

Disclaimer

The views and opinions expressed in this chapter are only those of the author and do not necessarily reflect the views and policies of the FDA.

Acknowledgements

The author would like to thank Dr Lawrence Yu for useful comments and invaluable support during the drafting of this chapter.

References

1. L. Wu, H. Smith, H. Zhang and L. X. Yu, Drug Product Approval in the United States and International Harmonization, in *Developing Solid Oral Dosage Forms: Pharmaceutical Theory and Practice*, ed. Y. Qiu, Y. Chen, G. Zhang, L. X. Yu and R. V. Mantri, Elsevier, 2017, pp. 1049–1077.
2. L. Wu, F. Chen, S. L. Lee, A. Raw and L. X. Yu, Building Parity between Brand and Generic Peptides: Regulatory and Scientific Considerations for Quality of Synthetic Peptides, *Int. J. Pharm.*, 2017, **518**, 320–334.
3. FDA Pharmaceutical Quality Oversight – One Quality Voice (2015). http://www.fda.gov/downloads/AboutFDA/CentersOffices/OfficeofMedicalProductsandTobacco/CDER/UCM442666.pdf as of July 17, 2018.
4. The FDA's Drug Review Process: Ensuring Drugs are Safe and Effective (2014). http://www.fda.gov/Drugs/ResourcesForYou/Consumers/ucm143534.htm as of July 17, 2018.
5. W. M. Troetel, The Investigational New Drug Application and the Investigator's Brochure, in *New Drug Approval Process*, ed. A. Guarino, Dekker, New York, 2004.
6. New Drug Application (NDA) (2014). http://www.fda.gov/drugs/developmentapprovalprocess/howdrugsaredevelopedandapproved/approvalapplications/newdrugapplicationnda/default.htm as of July 17, 2018.
7. Investigational New Drug (IND Application) (2015). http://www.fda.gov/drugs/developmentapprovalprocess/howdrugsaredevelopedandapproved/approvalapplications/investigationalnewdrugindapplication/default.htm as of July 17, 2018.
8. FDA, Guidance for Industry Content and Format of Investigational New Drug Applications (INDs) for Phase 1 Studies of Drugs, Including Well Characterized, Therapeutic, Biotechnology-derived Products, Nov 1995.
9. 21 CFR 314.50(c) Content and Format of an Application, Form FDA-356h Application to Market a New Drug, Biologic or an Antibiotic Drug for Human Use. http://www.fda.gov/downloads/AboutFDA/ReportsManualsForms/Forms/UCM082348.pdf as of July 17, 2018.
10. 21 CFR 201.56 (General) and 201.57 (Specific) Final Rule: Requirements on the Content and Format of Labeling for Human Prescription Drug and Biological Products, April 1, 2017, https://www.accessdata.fda.gov/scripts/cdrh/cfdocs/cfCFR/CFRSearch.cfm?fr=201.57 as of July 30, 2018.
11. FDA, Guidance for Industry Drug Product Chemistry, Manufacturing and Controls Information, Jan 2003.

12. D. Verthelyi and V. V. Wang, Trace Levels of Innate Immune Response Modulating Impurities (IIRMIs) Synergize to Break Tolerance to Therapeutic Proteins, *PLoS One*, 2010, **5**(12), e15252.
13. R. L. Heinrikson, Application of automated sequence analysis to the understanding of protein structure and function, *Ann. Clin. Lab. Sci.*, 1978, **8**(4), 295–301.
14. N. Lundell and T. Schreitmuller, Sample preparation for peptide mapping–A pharmaceutical quality-control perspective, *Anal. Biochem.*, 1999, **266**(1), 31–47.
15. H. Xu, L. W. Zhang and M. A. Freitas, Identification and characterization of disulfide bonds in proteins and peptides from tandem MS data by use of the MassMatrix MS/MS search engine, *J. Proteome Res.*, 2008, **7**(1), 138–144.
16. S. M. Kelly and N. C. Price, The Use of Circular Dichroism in the Investigation of Protein Structure and Function, *Curr. Protein Pept. Sci.*, 2000, **1**(4), 349–384.
17. C. H. Li, *et al.*, Applications of Circular Dichroism (CD) for Structural Analysis of Proteins: Qualification of Near- and Far-UV CD for Protein Higher Order Structural Analysis, *J. Pharm. Sci.*, 2011, **100**(11), 4642–4654.
18. Y. J. Jiang, *et al.*, Qualification of FTIR Spectroscopic Method for Protein Secondary Structural Analysis, *J. Pharm. Sci.*, 2011, **100**(11), 4631–4641.
19. A. Natalello and S. M. Doglia, Insoluble protein assemblies characterized by fourier transform infrared spectroscopy, *Methods Mol. Biol.*, 2015, **1258**, 347–369.
20. M. P. C. Ratnaparkhi and S. P. Pandya, Peptides and Proteins in Pharmaceuticals, *Int. J. Curr. Pharm. Res.*, 2011, **3**(2), 1–9.
21. H. M. Dintzis, R. Z. Dintzis and B. Vogelstein, Molecular Determinants of Immunogenicity - Immunon Model of Immune-Response, *Proc. Natl. Acad. Sci. U. S. A.*, 1976, **73**(10), 3671–3675.
22. A. S. Rosenberg, Effects of protein aggregates: An immunologic perspective, *AAPS J.*, 2006, **8**(3), E501–E507.
23. M. Rabe, D. Verdes and S. Seeger, Understanding protein adsorption phenomena at solid surfaces, *Adv. Colloid Interface Sci.*, 2011, **162**(1–2), 87–106.
24. A. W. P. Vermeer and W. Norde, CD spectroscopy of proteins adsorbed at flat hydrophilic quartz and hydrophobic Teflon surfaces, *J. Colloid Interface Sci.*, 2000, **225**(2), 394–397.
25. L. Ashton, *et al.*, Shear-Induced Unfolding of Lysozyme Monitored In Situ, *Biophys. J.*, 2009, **96**(10), 4231–4236.
26. J. S. Philo and T. Arakawa, Mechanisms of Protein Aggregation, *Curr. Pharm. Biotechnol.*, 2009, **10**(4), 348–351.
27. X. M. Lam, T. W. Patapoff and T. H. Nguyen, The effect of benzyl alcohol on recombinant human interferon-gamma, *Pharm. Res.*, 1997, **14**(6), 725–729.

28. H. R. Costantino, *et al.*, Development of Calcitonin Salmon Nasal Spray: Similarity of Peptide Formulated in Chlorobutanol Compared to Benzalkonium Chloride as Preservative, *J. Pharm. Sci.*, 2009, **98**(10), 3691–3706.
29. S. Hermeling, *et al.*, Micelle-associated protein in epoetin formulations: A risk factor for immunogenicity?, *Pharm. Res.*, 2003, **20**(12), 1903–1907.
30. F. A. Green, Interactions of a Nonionic Detergent .3. Further Observations on Hydrophobic Interactions, *J. Colloid Interface Sci.*, 1972, **41**(1), 124.
31. K. Boven, *et al.*, The increased incidence of pure red cell aplasia with an Eprex formulation in uncoated rubber stopper syringes, *Kidney Int.*, 2005, **67**(6), 2346–2353.
32. FDA Guidance for Industry Q1A(R2) Stability Testing of New Drug Substances and Products, N., http://www.fda.gov/downloads/drugs/guidancecomplianceregulatoryinformation/guidances/ucm073369.pdf, as of June 30, 2018.
33. ICH M3(R2) Guidance on Nonclinical safety studies for the conduct of human clinical trials and marketing authorization for pharmaceuticals (2009) http://www.ema.europa.eu/docs/en_GB/document_library/Scientific_guideline/2009/09/WC500002720.pdf as of June 30, 2018.
34. FDA, Guidance for Industry Guideline for the Format and Content of the Microbiology Section of an Application, prior to Feb. 1997 http://www.fda.gov/downloads/Drugs/GuidanceComplianceRegulatoryInformation/Guidances/ucm075101.pdf as of June 30, 2018.
35. FDA, Guidance for Industry Guideline for the Format and Content of the Clinical and Statistical Sections in an Application, Jul. 1988, http://www.fda.gov/downloads/Drugs/Guidances/UCM071665.pdf as of June 30, 2018.
36. FDA, Guidance for Industry Submitting Debarment Certification Statement (Draft, 1998). http://www.fda.gov/downloads/drugs/guidancecomplianceregulatoryinformation/guidances/ucm080584.pdf as of June 30, 2018.
37. PDUFA User Fee Cover Sheet (2014). http://www.fda.gov/ForIndustry/UserFees/PrescriptionDrugUserFee/ucm119184.htm as of July 17, 2018.
38. FDA, Guidance for Clinical Investigators, Industry and FDA Staff Financial Disclosure by Clinical Investigators (2013). http://www.fda.gov/downloads/RegulatoryInformation/Guidances/UCM341008.pdf as of July 17, 2018.
39. FDA MAPP 6010.5 Policy and Procedures Office of New Drugs NDAs and BLAs: Filing Review Issues http://www.fda.gov/downloads/AboutFDA/CentersOffices/OfficeofMedicalProductsandTobacco/CDER/ManualofPoliciesProcedures/UCM081990.pdf as of July 17, 2018.
40. For Industry: What information is provided to applicants during the human drug product application review process? (2015). http://www.fda.gov/ForIndustry/FDABasicsforIndustry/ucm238030.htm as of July 17, 2018.

41. FDA, CDER 21st Century Review Process Desk Reference Guide, NDA/BLA ReviewProcess (2015). http://www.fda.gov/downloads/AboutFDA/CentersOffices/OfficeofMedicalProductsandTobacco/CDER/ManualofPoliciesProcedures/UCM218757.pdf as of June30, 2018.
42. 21 CFR 314.110 Complete response letter to the applicant.
43. Advisory Committees: Human Drugs Advisory Committees (2017). http://www.fda.gov/AdvisoryCommittees/CommitteesMeetingMaterials/Drugs/default.htm as of July 17, 2018.
44. FDA Guidance for Industry Expedited Programs for Serious Conditions – Drugs and Biologics (2014). http://www.fda.gov/downloads/drugs/guidancecomplianceregulatoryinformation/guidances/ucm358301.pdf as of July 17, 2018.
45. For Industry: Designating an Orphan Product: Drugs and Biological Products (2013). http://www.fda.gov/ForIndustry/DevelopingProductsforRareDiseasesConditions/HowtoapplyforOrphanProductDesignation/default.htm as of July 17, 2018.
46. Qualifying for Pediatric Exclusivity Under Section 505(A) of the Federal Food, Drug and Cosmetic Act: Frequently Asked Questions on Pediatric Exclusivity (505A) (2016). http://www.fda.gov/Drugs/DevelopmentApprovalProcess/DevelopmentResources/ucm077915.htm as of July 17, 2018.
47. Drugs: Abbreviated New Drug Application (ANDA) (2018). http://www.fda.gov/Drugs/DevelopmentApprovalProcess/HowDrugsareDevelopedandApproved/ApprovalApplications/AbbreviatedNewDrugApplicationANDAGenerics/default.htm as of July 17, 2018.
48. 21 CFR 320.22 Criteria for waiver of evidence of *in vivo* bioavailability or bioequivalence.
49. FDA Guidance for Industry: ANDAs Stability Testing of Drug Substances and Products (2013); https://www.fda.gov/downloads/drugs/guidancecomplianceregulatoryinformation/guidances/ucm320590.pdf as of July 17, 2018.
50. ICH Q1A(R2) Stability Testing of New Drug Substances and Products (2003); http://www.ich.org/fileadmin/Public_Web_Site/ICH_Products/Guidelines/Quality/Q1A_R2/Step4/Q1A_R2__Guideline.pdf, as of July 17, 2018.
51. H. Schellekens, Bioequivalence and the immunogenicity of biopharmaceuticals, *Nat. Rev. Drug Discovery*, 2002, **1**(6), 457–462.
52. FDA Guidance for Industry: ANDAs for certain Highly Purified Synthetic Peptide Drug Products That Refer to Listed Drugs of rDNA Origin (draft 2017) https://www.fda.gov/downloads/Drugs/GuidanceComplianceRegulatoryInformation/Guidances/UCM578365.pdf, as of July 17, 2018.
53. V. A. Walshe, *et al.*, Integrating In Silico and In Vitro Analysis of Peptide Binding Affinity to HLA-Cw*0102: A Bioinformatic Approach to the Prediction of New Epitopes, *PLoS One*, 2009, **4**(11), e8095.

54. L. A. Haile, *et al.*, Detection of Innate immune Response Modulating Impurities in Therapeutic Proteins, *PLoS One*, 2015, **10**(4), DOI: 10.1371/journal.pone.0125078.
55. V. Jawa, *et al.*, T-cell dependent immunogenicity of protein therapeutics: Preclinical assessment and mitigation, *Clin. Immunol.*, 2013, **149**, 534–555.
56. 21 C.F.R. § 314.94(a)(9)(iii)128.
57. 21 CFR 320.22 Criteria for waiver of evidence of *in vivo* bioavailability or bioequivalence.
58. Drugs: Pre-ANDA Program (2018), https://www.fda.gov/Drugs/ResourcesForYou/Consumers/BuyingUsingMedicineSafely/GenericDrugs/ucm578012.htm, as of July 17, 2018.
59. MAPP 5200.14 Filing Review of Abbreviated New Drug Applications (2017) https://www.fda.gov/downloads/AboutFDA/CentersOffices/OfficeofMedicalProductsandTobacco/CDER/ManualofPoliciesProcedures/UCM574493.pdf as of July 17, 2018.
60. GDUFA II Commitment Letter https://www.fda.gov/downloads/ForIndustry/UserFees/GenericDrugUserFees/UCM525234.pdf, as of July 17, 2018.
61. For Industry: GDUFA Guidances and MAPPs https://www.fda.gov/forindustry/userfees/genericdruguserfees/ucm316678.htm, as of July 17, 2018.
62. FDA/CDER SBIA Chronicles:DMF 101 (2014). http://www.fda.gov/downloads/drugs/developmentapprovalprocess/smallbusinessassistance/ucm438678.pdf as of July 17, 2018.
63. For Industry: Generic Drug User Fee Amendments (2018). http://www.fda.gov/ForIndustry/UserFees/GenericDrugUserFees/default.htm as of July 17, 2018.
64. I. Eggen, B. Gregg, H. Rode, A. Swietlow, M. Verlander and A. Szajek, Control Strategies for Synthetic Therapeutic Peptide APIs Part II – Raw Material Considerations, *Pharm. Technol.*, 2014, **38**(4), 48–52.
65. V. Sanz-Nebot, *et al.*, Separation and characterization of complex crude mixtures produced in the synthesis of therapeutic peptide hormones by liquid chromatography coupled to electrospray mass spectrometry (LC-ES-MS), *Anal. Chim. Acta*, 2004, **521**(1), 25–36.
66. M. A. Elsawy, C. Hewage and B. Walker, Racemisation of N-Fmoc phenylglycine under mild microwave-SPPS and conventional stepwise SPPS conditions: attempts to develop strategies for overcoming this, *J. Pept. Sci.*, 2012, **18**(5), 302–311.
67. E. Hlebowicz, *et al.*, Identification of Fmoc-beta-Ala-OH and Fmoc-beta-Ala-amino acid-OH as new impurities in Fmoc-protected amino acid derivatives, *J. Pept. Res.*, 2005, **65**(1), 90–97.
68. T. Vojkovsky, Detection of Secondary-Amines on Solid-Phase, *Pept. Res.*, 1995, **8**(4), 236–237.
69. S. A. Palasek, Z. J. Cox and J. M. Collins, Limiting racemization and aspartimide formation in microwave-enhanced Fmoc solid phase peptide synthesis, *J. Pept. Sci.*, 2007, **13**(3), 143–148.

70. M. D'Hondt, *et al.*, Related impurities in peptide medicines, *J. Pharm. Biomed. Anal.*, 2014, **101**, 2–30.
71. T. Geiger and S. Clarke, Deamidation, Isomerization and Racemization at Asparaginyl and Aspartyl Residues in Peptides - Succinimide-Linked Reactions That Contribute to Protein-Degradation, *J. Biol. Chem.*, 1987, **262**(2), 785–794.
72. M. P. Souza, M. F. M. Tavares and M. T. M. Miranda, Racemization in stepwise solid-phase peptide synthesis at elevated temperatures, *Tetrahedron*, 2004, **60**(21), 4671–4681.
73. L. A. Carpino, D. Ionescu and A. ElFaham, Peptide coupling in the presence of highly hindered tertiary amines, *J. Org. Chem.*, 1996, **61**(7), 2460–2465.
74. V. H. Finder, *et al.*, The Recombinant Amyloid-beta Peptide A beta 1-42 Aggregates Faster and Is More Neurotoxic than Synthetic A beta 1-42, *J. Mol. Biol.*, 2010, **396**(1), 9–18.
75. T. S. Haque, J. C. Little and S. H. Gellman, Stereochemical requirements for beta-hairpin formation: Model studies with four-residue peptides and depsipeptides, *J. Am. Chem. Soc.*, 1996, **118**(29), 6975–6985.
76. N. Benkirane, *et al.*, Antigenicity and immunogenicity of modified synthetic peptides containing D-amino acid residues. Antibodies to a D-enantiomer do recognize the parent L-hexapeptide and reciprocally, *J. Biol. Chem.*, 1993, **268**(35), 26279–26285.
77. A. Isidro-Llobet, *et al.*, Fmoc-2-mercaptobenzothiazole, for the introduction of the Fmoc moiety free of side-reactions, *Biopolymers*, 2007, **88**(5), 733–737.
78. N. L. Benoiton, *Chemistry of Peptide Synthesis*, Taylor & Francis, Boca Raton, FL, 2006.
79. R. Subiros-Funosas, *et al.*, Oxyma: An Efficient Additive for Peptide Synthesis to Replace the Benzotriazole-Based HOBt and HOAt with a Lower Risk of Explosion([1]), *Chem. – Eur. J.*, 2009, **15**(37), 9394–9403.
80. G. B. Fields and R. L. Noble, Solid-Phase Peptide-Synthesis Utilizing 9-Fluorenylmethoxycarbonyl Amino-Acids, *Int. J. Pept. Protein Res.*, 1990, **35**(3), 161–214.
81. E. Atherton, R. C. Sheppard and P. Ward, Peptide-Synthesis .7. Solid-Phase Synthesis of Conotoxin-G1, *J. Chem. Soc., Perkin Trans. 1*, 1985, (10), 2065–2073.
82. C. G. Fields and G. B. Fields, Minimization of Tryptophan Alkylation Following 9-Fluorenylmethoxycarbonyl Solid-Phase Peptide-Synthesis, *Tetrahedron Lett.*, 1993, **34**(42), 6661–6664.
83. J. W. de Beukelaar, *et al.*, The impact of impurities in synthetic peptides on the outcome of T-cell stimulation assays, *Rapid Commun. Mass Spectrom.*, 2007, **21**(7), 1282–1288.
84. M. Gongora-Benitez, *et al.*, Optimized Fmoc Solid-Phase Synthesis of the Cysteine-Rich Peptide Linaclotide, *Biopolymers*, 2011, **96**(1), 69–80.
85. B. De Spiegeleer, *et al.*, Impurity profiling quality control testing of synthetic peptides using liquid chromatography-photodiode array-fluorescence and

liquid chromatography-electrospray ionization-mass spectrometry: the obestatin case, *Anal. Biochem.*, 2008, **376**(2), 229–234.
86. S. Ridge and K. Hettiarachchi, Peptide purity and counter ion determination of bradykinin by high performance liquid chromatography and capillary electrophoresis, *J. Chromatogr. A*, 1998, **817**(1–2), 215–222.
87. G. W. Anderson and A. C. Mcgregor, Tert-Butyloxycarbonylamino Acids and Their Use in Peptide Synthesis, *J. Am. Chem. Soc.*, 1957, **79**(23), 6180–6183.
88. G. Barany and R. B. Merrifield, New Amino Protecting Group Removable by Reduction - Chemistry of Dithiasuccinoyl (Dts) Function, *J. Am. Chem. Soc.*, 1977, **99**(22), 7363–7365.
89. L. A. Carpino, *et al.*, Rapid, continuous solution-phase peptide synthesis: Application to peptides of pharmaceutical interest, *Org. Process Res. Dev.*, 2003, **7**(1), 28–37.
90. Y. Nishiuchi, *et al.*, Chemical synthesis of the precursor molecule of the Aequorea green fluorescent protein, subsequent folding and development of fluorescence, *Proc. Natl. Acad. Sci. U. S. A.*, 1998, **95**(23), 13549–13554.
91. S. Chandrudu, P. Simerska and I. Toth, Chemical Methods for Peptide and Protein Production, *Molecules*, 2013, **18**(4), 4373–4388.
92. B. L. Bray, Large-scale manufacture of peptide therapeutics by chemical synthesis, *Nat. Rev. Drug Discovery*, 2003, **2**(7), 587–593.
93. Guidance for Industry: Changes to an Approved NDA or ANDA (2004). http://www.fda.gov/downloads/drugs/guidancecomplianceregulatoryinformation/guidances/ucm077097.pdf as of July 17, 2018.

CHAPTER 2

Regulatory Perspective on Synthetic Peptides in Europe

RENÉ THÜRMER

BfArM, Federal Institute for Drugs and Medical Devices,
Kurt-Georg-Kiesinger-Allee 3, 53175 Bonn, Germany
Email: rene.thuermer@bfarm.de

2.1 Introduction

Peptide drugs have become an attractive substance class. They offer certain advantages as drugs, including their biological activity, high specificity and low toxicity. Unlike synthetic substances, peptides are degraded into their component proteinogenic amino acids without leading to toxic metabolites.

Synthetic peptides are an exciting class of therapeutic products that are under development for a variety of indications. Many of them have been developed to address urgent medical needs that are so far unmet. Peptide therapeutics that are already on the market target a broad range of indications and are administered either intravenously, subcutaneously and even orally (desmopressin, linaclotide).

Many depot formulations employing biodegradable polymers have been commercialized for decades. Innovative development programmes include, among others, inhalative and transdermal dosage forms. However, a main focus with regard to the nonparenteral route of administration is the evolution of novel oral dosage forms.

The first chemical synthesis of a therapeutic peptide was that of oxytocin in 1953. Recombinant human insulin, the first approved recombinant

Drug Discovery Series No. 72
Peptide Therapeutics: Strategy and Tactics for Chemistry, Manufacturing and Controls
Edited by Ved Srivastava

Published by the Royal Society of Chemistry, www.rsc.org

peptide therapeutic to be manufactured by fermentation, was introduced in 1982 in the USA. To date, over 60 peptide drugs have been approved in the USA, Europe and Japan; over 150 are in active clinical development and an additional 260 have been tested in human clinical trials.[1–3]

Synthetic peptides are a unique class of molecules with diverse mechanisms of action different from those of small molecules but also different from those of biopharmaceuticals. Therefore, synthetic peptides are at the crossroads of small molecules and proteins and, from a pharmaceutical quality regulatory point of view, specific considerations apply to this class of therapeutics.

2.2 European Regulatory Network

The system for regulating medicines in Europe is unique in the world. It is based on a closely coordinated regulatory network of national competent authorities in the Member States of the European Economic Area (EEA) working together with the European Medicines Agency (EMA) and the European Commission. By working closely together, this network ensures that safe, effective and high-quality medicines are authorized throughout the European Union (EU) and that patients, healthcare professionals and citizens are provided with adequate and consistent information about medicines. A network of more than 4500 experts appointed by EU Member States guarantees the high standard scientific assessment of medicines.[4]

New active substance synthetic peptides approved in Europe *via* the centralized procedure (since 2000) are listed in Table 2.1.

Table 2.1 New active substance synthetic peptides approved in Europe *via* the centralized procedure (since 2000).

Year of approval	Active substance	Indication	Number of amino acids
2003	Enfurvitide	HIV	36
2004	Bivalirudin	Anticoagulant	20
2005	Ziconotide	Severe, chronic pain	25
2008	Degarelix	Advanced hormone-dependent prostate cancer	10
2008	Icatibant	Hereditary angioedema	10
2009	Mifamurtide (lipophilic peptide analogue)	Osteosarcoma	N/A
2011	Exenatide	Type 2 diabetes	39
2012	Linaclotide	Irritable bowel syndrome	14
2012	Pasireotide	Cushing disease	6
2013	Lixisenatide	Type 2 diabetes	44
2015	Afamelanotide	Erythropoietic protoporphyria	13
2016	Etelcalcetide	Hyperparathyroidism	8
2018	Macimorelin (peptidomimetic)	Diagnosis of growth hormone deficiency	3

2.3 Guidelines on the Quality of Human Medicines

The EMA's scientific guidelines on the quality of human medicines help applicants prepare marketing authorization applications. Guidelines reflect a harmonized approach of the EU Member States and the EMA on how to interpret and apply the requirements for the demonstration of quality, safety and efficacy set out in Community directives. Applicants and marketing authorization holders are strongly encouraged to follow these guidelines. Applicants need to justify deviations from the guidelines fully in their applications at the time of submission.[5]

Guidelines on the quality of medicines that are harmonized between Europe, Japan and the USA by the International Council on Harmonisation of Technical Requirements for Registration of Pharmaceuticals for Human Use (ICH) are a basis for regulatory decisions. The purpose of the ICH is to reduce or eliminate the need to duplicate the testing carried out during the research and development of new medicines by recommending ways to achieve greater harmonization in the interpretation and application of technical guidelines and requirements for product registration. Now in its third decade of activity, ICH's attention is directed towards extending the benefits of harmonization beyond the founding ICH regions and the number of members and observers has increased tremendously in recent years.[6]

No official guidelines for synthetic peptides are currently available. Synthetic peptides are excluded from the ICH "Guidelines on Impurities in New Drug Substances and New Drug Products" (ICH Q3A and Q3B).[7,8] In the scope of the ICH Q6A "Guideline on Specifications: Test Procedures and Acceptance Criteria for New Drug Substances and Drug Products: Chemical Substances" it is stated that this guideline may be applicable to synthetic peptides of low molecular weight, but it is not sufficient to describe adequately specifications of higher molecular weight peptides and polypeptides.[9] The ICH Q6B "Guideline on Specifications: Test Procedures and Acceptance Criteria for Biotechnological/Biological Products" clearly states that this document does not cover synthetic peptides and polypeptides.[10] Consequently, synthetic peptides are formally excluded from the scope of ICH Q11 "Guideline on Development and Manufacture of Drug Substances (Chemical Entities and Biotechnological/Biological Entities)."[11] Nevertheless, in ICH Q11 it is clearly stated that this guideline might also be appropriate for other types of products, such as synthetic peptides and oligonucleotides, following consultation with the appropriate regulatory authorities.

In contrast to other regions, the European Pharmacopoeia (Ph. Eur.) defines legally binding acceptance criteria for one of the most important critical quality attributes of drug substances, namely purity. The Ph. Eur. monograph "Substances for Pharmaceutical Use" defines requirements for the reporting, identification and qualification threshold for peptides obtained by chemical synthesis.

The applicability of other European and international guidance documents to synthetic peptides is discussed in the following sections.

2.4 Overall Control Strategy

Identifying critical quality attributes (CQAs) is vital during pharmaceutical development. A critical quality attribute is a physical, chemical, biological or microbiological property or characteristic that should be within an appropriate limit, range or distribution to ensure the desired product quality. This is described in detail in the ICH quality guidelines on "Pharmaceutical Development" (ICH Q8) and "Development and Manufacture of Drug Substances" (ICH Q11).[12]

The requirements for synthetic peptides used as active substances are the focus of this chapter. Nevertheless, the resulting medicinal product is key with respect to efficacy and safety for the patient. Its clinical use, including indication, route of administration, treatment duration and intended dose, predetermines the Quality Target Product Profile (QTPP), which is a prospective summary of the quality characteristics of a medicinal product that ideally will be achieved to ensure the desired quality, taking into account the safety and efficacy of the medicinal product.

Critical quality attributes are part of the overall target product profile that is based on the desired clinical performance. The extent of characterization is linked to the level of risk associated with each phase of drug development. It is expected that the synthetic peptide molecule will have been well characterized before a marketing authorization application is submitted to the regulatory agencies.

What to control is one of the key questions in connection with regulatory submissions during clinical development and marketing authorization applications. Therefore, an overall control strategy should be developed to ensure consistent quality. An integrated control strategy incorporates procedural control and also material, in-process, specification and stability testing. The integrated control strategy reflects knowledge of the product attributes, in addition to an understanding of the means by which these attributes are controlled through integrated control elements during drug substance manufacture. Quality risk management tools are a valuable instrument in developing an integrated control strategy to determine CQAs and to identify manufacturing process risks that may have an impact on the CQAs. Control elements to reduce these risks should be established.

Characterization and quality control testing are required throughout the clinical development of synthetic peptides intended for therapeutic use. Detailed information on structural characterization studies, including the impurity profile, should be provided in submissions to regulatory agencies worldwide. Impurities are an important class of potential drug substance CQAs because of their potential impact on medicinal product safety and efficacy. Consequently, a major focus is knowledge and control of impurities.

Clinical qualification is considered the most important aspect when setting acceptance criteria in specifications. For critical attributes, the acceptance criteria should not be wider than what has been qualified to yield a safe and efficacious product. Therefore, identifying differences in impurity profiles using orthogonal analytical methods for the detection and quantification of impurities in early stages of drug development and the application of good science will be extremely helpful in the course of development.

New and improved state-of-the-art analytical technologies and techniques are becoming available continuously and should be applied during development for both characterization and routine control. Different aspects of control strategies for synthetic peptides have been published elsewhere.[13–15]

2.5 Documentation in Dossier of Marketing Authorization

In the European regulatory framework, active substances can be classified into

- new active substances, used for the first time in a medicinal product;
- existing active substances not described in the Ph. Eur. or the pharmacopoeia of an EU Member State;
- existing active substances described in the Ph. Eur. or in the pharmacopoeia of an EU Member State.

Different options for the information to be provided in a marketing authorization application (MAA) are described in the "Guideline on Summary of Requirements for Active Substances in the Quality Part of the Dossier."[16] For new active substances and existing active substances not described in the Ph. Eur. or the pharmacopoeia of an EU Member State, the applicant may submit full details of manufacture as outlined in the "Guideline on the Chemistry of Active Substances" (EMA/454576/2016).[17] Full details of the chemistry, manufacturing process and quality controls during manufacture for the active substance may also be submitted in an Active Substance Master File (ASMF) as outlined in the "Guideline on Active Substance Master File Procedure" (EMEA/CVMP/134/02 or CHMP/QWP/227/02).[18] In such cases, the applicant's part of the ASMF needs to be included in the marketing authorization application.

In addition, a third option is possible for existing active substances described in the Ph. Eur. or in the pharmacopoeia of an EU Member State. Certificates of Suitability (CEPs) certify that the quality of a given substance can be suitably controlled by the Ph. Eur. monograph – with additional tests, if necessary (stated on the CEP). These CEPs are issued by the European Directorate for the Quality of Medicines and HealthCare (EDQM) and the applicant should include a copy of the most current CEP in the dossier, together with a written assurance that no significant changes in the manufacturing method have taken place following the granting of the certificate or its last revision.

Numerous CEPs have been issued for synthetic peptides and the information on the status of the certificate is available in the certification database on the website of the EDQM. Additional information on CEPs is available on the "Knowledge Database." This database enables information to be obtained on a given substance or general method of analysis. The database contains information such as the state of work, the last issue of the Pharmeuropa in which the draft of the monograph was published, the most recent publication of the text in the Ph. Eur., whether the monograph is currently undergoing revision, the chromatogram in pdf format and the trade names of some reagents, such as for chromatographic columns.[19]

2.6 European Pharmacopoeia (Ph. Eur.)

The Council of Europe is Europe's leading human rights organization. It comprises 47 Member States, 28 of which are members of the European Union. The EDQM is a Directorate of the Council of Europe. Its mission is to contribute to the basic human right of access to good-quality medicines and healthcare and to promote and protect public health.[20]

The EDQM publishes the Ph. Eur., which is Europe's legal and scientific benchmark for pharmacopoeial standards that contribute to delivering high-quality medicines in Europe and beyond. The ninth edition (including Supplement 9.8) contains 2406 monographs (including dosage forms), 365 general texts (including general monographs and methods of analysis) and around 2730 descriptions of reagents. Table 2.2 lists the existing monographs for peptides in the Ph. Eur. and the current work programme relevant for synthetic peptides.

The Ph. Eur. also contains several general chapters that are specifically relevant for synthetic peptides:

- 2.2.55. Peptide Mapping (relevant only for longer peptides).
- 2.2.56. Amino Acid Analysis.
- 2.2.64. Peptide Identification by Nuclear Magnetic Resonance Spectrometry.
- 2.5.34. Acetic Acid in Synthetic Peptides.

2.7 Technical Guide for the Elaboration of Monographs on Synthetic Peptides and Recombinant DNA Proteins

The "Technical Guide for the Elaboration of Monographs on Synthetic Peptides and Recombinant DNA Proteins" published by the EDQM is intended to provide guidance to authors, contributors and users of the Ph. Eur. on the elaboration of active substance monographs for synthetic peptides and products of recombinant DNA (rDNA) technology.[21] This applies in particular to EDQM expert group members, authorities responsible for granting marketing authorization for synthetic peptides and recombinant DNA proteins, Official Medicines Control Laboratories (OMCLs),

Table 2.2 Monographs for peptides in the Ph. Eur. (as of December 2018).

Existing monographs	Current work programme
Buserelin	Octreotide (adopted March 2018 – publication postponed)
Desmopressin	Lanreotide
Felypressin	Triptorelin
Glucagon (recombinant)	Atosiban
Gonadorelin	Glatiramer acetate/glatiramer injection
Goserelin	*Peptide-based radiopharmaceutical preparations*
Leuprorelin	Edotreotide (radiopharmaceutical preparations; ^{68}Ga, ^{90}Y)
Oxytocin	Indium (^{111}In) pentetreotide
Protirelin	Technetium (^{99m}Tc) mertiatide
Salmon calcitonin (recombinant and synthetic)	
Somatostatin	
Teriparatide (recombinant)	
Terlipressin	
Tetracosactide	
Vasopressin	

manufacturers of synthetic peptides and rDNA proteins and public and private analytical laboratories working for one of the above.

In Chapter 6 of the Technical Guide, the requirements for synthetic peptides are described. The document states that synthetic peptides may be considered to differ from products of recombinant DNA technology in two structural aspects:

- they are usually small, typically below 5000 Da;
- they may have chemical structures that do not occur naturally in proteins or peptides.

As a consequence, they can generally be sufficiently characterized by a battery of physicochemical tests, which is reflected in the structure of the monograph for synthetic peptides.

Title. The title of the monograph reflects the International Non-proprietary Name (INN) of the substance established by the World Health Organization (WHO). As a general rule, the salt of the peptide is not specified in the title of the monograph. If required, appropriate clarifications are given in the Definition section.

Definition. The Definition section states

- molecular formula and the corresponding CAS number;
- molecular mass;
- physical form;
- structural formula;
- identity and biological activity of the substance and, where appropriate, its naturally occurring analogue;

- assay specifications;
- route of production;
- salt form;
- any chemical modifications, such as esterification or amidation.

Characters. The appearance of the solid synthetic peptide should be described. The solubility, if applicable, should be provided.

Identification. Two methods are generally considered sufficient for identification purposes. Typically, the Identification section of the monograph comprises

- amino acid analysis or nuclear magnetic resonance (NMR) spectrometry;
- reversed-phase liquid chromatography (LC).

Although both amino acid analysis and NMR spectrometry may be described in the monograph in addition to the other identification test, flexibility is given to users to perform only one of those tests. Identification by NMR spectrometry applies to peptides comprising up to approximately 15 amino acids.

Where appropriate, for instance where the peptide comprises unnatural amino acids or when prescribed in individual monographs, the use of NMR spectrometry may also be required in addition to conventional methods based on analysis of the amino acid composition or sequence. In many cases it may be appropriate to include more than one LC step. The inclusion of multiple chromatographic identification tests is particularly important if a spectroscopic method is not included.

Where NMR spectrometry is prescribed for identification, a separate reference standard for NMR identification is established from a bulk of the substance; no assigned content is necessary.

Tests – Related Peptides. Typically, monographs for synthetic peptides contain a reversed-phase LC test for related peptides. Such tests are validated for specified impurities known to be potential contaminants and are transparent with respect to these impurities. The provisions for synthetic peptides in the general monograph "Substances for Pharmaceutical Use" (2034) apply to all synthetic peptides unless stated otherwise.

Monographs should include acceptance criteria for

- each specified impurity;
- unspecified impurities, normally set at the identification threshold;
- total impurities.

Reference standards for specified impurities, mixtures of impurities or, where appropriate, the reference peptide spiked with the specified impurities are provided. Where necessary, specified impurities may have to be quantified separately using independent methods. Where a monograph depends on a single purity test, the capacity of the method to measure all relevant impurities should be demonstrated.

Tests – Optical Rotation and Absorbance. These useful tests should be retained where appropriate. Chiral chromatography may offer some potential to replace optical rotation.

Tests – Counter Ion. Counter ion determination is a general requirement for peptides. In the case of peptides with acetate as the counter ion, the method used for the determination of acetic acid is that described in general chapter 2.5.34, "Acetic Acid in Synthetic Peptides," unless it has been demonstrated that this method is not appropriate for the peptide to be tested.

Tests – Loss on Drying, Water Content. Loss on drying is used less frequently because it requires large quantities of material. It is often replaced by a water content determination. The method for determining water content is that described in general chapter 2.5.12, "Water: Semi-micro Determination (Karl Fischer)," or in general chapter 2.5.32, "Water: Micro Determination (Coulometric Titration)."

Tests – Bacterial Endotoxins. When a substance for pharmaceutical use is intended for parenteral use, it has to comply with the test for bacterial endotoxins (2.6.14, "Bacterial Endotoxins"). As compliance with a test is required *via* the general monograph "Substances for Pharmaceutical Use" (2034), this requirement may not be repeated in the individual monographs.

Assay. Assays for synthetic peptides are generally comparative chromatographic procedures, performed using a defined chemical reference substance as the standard. Results are normally expressed in terms of the counter ion-free, anhydrous substance. The permitted limits for such tests are typically asymmetric, the upper limit being 100% + the permitted assay reproducibility (usually ±2.0%) and the lower limit being 100% – (the permitted assay reproducibility + the maximum permitted level of impurity).

Storage. Although the statements given under this heading in a Ph. Eur. monograph do not constitute pharmacopoeial requirements, the appropriate information to safeguard the quality of a pharmacopoeial material during storage is to be given here where appropriate.

Labelling. The status of the labelling section is defined in the General Notices: "In general, labelling is subject to supranational and national regulation and to international agreements. The statements under the heading Labelling therefore are not comprehensive and, moreover, for the purposes of the pharmacopoeia only those statements that are necessary to demonstrate compliance or non-compliance with the monograph are mandatory. Any other labelling statements are included as recommendations. When the term 'label' is used in the pharmacopoeia, the labelling statements may appear on the container, the package, a leaflet accompanying the package or a certificate of analysis accompanying the article, as decided by the competent authority."

The parameters included in this section are determined on a case-by-case basis.

2.8 Information to be Provided in Module 3 for Synthetic Peptides (New Active Substances)

The following content of this chapter is organized according to the structure of Module 3 of the Common Technical Document for the drug substance. The agreement to assemble all the quality, safety and efficacy information in a common format (called CTD – Common Technical Document) has revolutionized the regulatory review processes, led to harmonized electronic submissions that, in turn, enabled the implementation of good review practices. For industries, it has eliminated the need to reformat the information for submission to the different ICH regulatory authorities.[22]

3.2.S.1. General Information. The nomenclature, structure and general properties of the peptide are described in this section of Module 3. This includes the INN, CAS name and other names such as company code names used in development. The structural formula and the sequence formula including the three-letter amino acid code should be presented in addition to the molecular formula, the molecular weight and the relative molecular mass. In the case that D-amino acids or non-natural amino acids are included in the sequence, this should also be stated.

3.2.S.2. Manufacture.
3.2.S.2.1. Manufacturer(s). All drug substance contract testing laboratories should also be included in this section.

3.2.S.2.2. Description of Manufacturing Process and Process Controls. The description of the manufacturing process for the active substance represents the applicant's commitment for the manufacture of the active substance. Information should be provided to describe the manufacturing process adequately, including special unit operations and process controls. Optional processes, alternative processes and reprocessing with associated controls that may be completed by the intermediate or active substance manufacturer should also be described. Particular emphasis should be placed on steps of the process that have an impact on the quality of the active substance or intermediates and that are classified as "critical."

Schematic representation of the manufacturing process. Graphical representations of the synthetic process(es) should be provided. These should comprise reaction schemes that include the chemical structures and molecular formulae of starting materials, intermediates and the active substance, in addition to the reagents, catalysts and solvents used as applicable. It should be clear whether intermediates are isolated or nonisolated. A block flow diagram that identifies operating conditions, unit operations, weights, yield ranges, *etc.*, can be provided optionally.

Sequential procedural narrative. A sequential procedural narrative of the manufacturing process should be submitted. This narrative should include the quantities (or ranges) of materials, (starting materials, intermediates, solvents, catalysts, reagents and process aids), used in a current representative production-scale batch. The narrative should describe each step in the manufacturing process and identify critical steps, critical process parameters, process controls employed and ranges for process parameters.

The description should include but not be limited to reaction times (*e.g.* for coupling or other stirring times), amounts of excess of amino acids coupled to the resin or other reagents, temperatures, chromatographic parameters during purification, gradients and lyophilization process parameters.

Synthetic peptide manufacturing processes often use splitting and pooling to create a drug substance. A clear batch definition and information on the maximum batch size should be provided.

The cases where routine reprocessing is carried out during the purification process should be identified and justified. Any data to support this justification should be provided. The reprocessing method should be clearly described and the criteria for deciding when reprocessing can be performed should be given. Other approaches and classifications for steps that are repeated in the process may be applicable. Reprocessing steps should be validated.

Reworking procedures are not routine options and should not be included in the dossier. Reworking should be carried out according to ICH Q7.[23]

3.2.S.2.3. Control of Materials. The quality and the classification of starting materials currently play a major role in regulatory discussions and the number of scientific advice procedures on this issue has increased tremendously over the last few years. The EMA "Reflection Paper on the Requirements for Selection and Justification of Starting Materials for the Manufacture of Chemical Active Substances" was first published in 2014 and an updated version appeared in July 2017. However, these documents were superseded by the ICH Q11 Q&As "Questions & Answers: Selection and Justification of Starting Materials for the Manufacture of Drug Substances," which was published in August 2017 and implemented in

Europe in March 2018 and in the USA in February 2018.[24] The Q&As are intended to provide additional clarification and to promote convergence on the considerations for the selection and justification of starting materials and on the information that should be provided in marketing authorization applications and/or Master Files.

Currently protected amino acid derivatives are acceptable as starting materials (terminal and side-chain protection if relevant) in the manufacturing process of synthetic peptides. Nevertheless, in all applications a justification on the classification of starting materials needs to be provided. For D-amino acids and unnatural amino acid derivatives, more detailed information is requested. Numerous development programmes utilize the advantage of conjugation and other derivatizations of peptides. In these cases, more complex structures may be incorporated in the peptide sequence. The classification of these materials will be handled on a case-by-case basis and early interaction with the regulatory agencies is recommended.

Proposals by applicants have been made that define linear polypeptides as starting materials prior to cyclization, which is not acceptable to regulatory agencies. In care cases, protected dipeptides may be accepted as starting materials when a sound justification (*e.g.* reduction of impurities for a difficult coupling step resulting in by-products) is provided.

Typical specification attributes for amino acid derivatives are appearance, appearance of solution, identification, overall purity, related impurities, isomer content, assay, water content and counter-ion content. A short description of the analytical methods used for the control of the starting materials should be available.

During manufacturing process development, the impurity profile of the starting materials is often investigated followed by a criticality assessment of which impurities may have an impact on the structure of the peptide during the solid-phase synthesis process.

It is a current trend that amino acids and amino acid derivatives are of non-animal/non-human origin. Nevertheless, monograph 5.2.8 of the Ph. Eur., "Minimizing the Risk of Transmitting Animal Spongiform Encephalopathy Agents *via* Medicinal Products," and the "Note for Guidance on Minimizing the Risk of Transmitting Animal Spongiform Encephalopathy Agents *via* Human and Veterinary Medicinal Products" (EMA/410/01) should be considered and TSE safety should be addressed by the submission of certificates.

Resins preloaded with linkers coupled to the first amino acid of a peptide sequence are also considered starting materials. Identification, substitution/loading and swelling properties are important specification parameters for such synthesis resins.

The names and addresses of the suppliers of the starting materials should be included in the dossier. The addition of alternative vendors for the starting materials needs to be approved by a variation according to European legislation. Information, in the form of a flowchart, indicating the synthetic process of the starting materials should be provided.

A list of all other reagents, solvents and chromatographic materials used in the manufacturing process of a synthetic peptide should be provided. Specifications for all materials should be laid down. Benzene could be present as a contaminant of solvents used in the manufacturing process and therefore the depletion of benzene during the manufacturing process should be demonstrated.

3.2.S.2.4. Controls of Critical Steps and Intermediates. The criticality of the manufacturing steps for peptides made by solid-phase synthesis should be evaluated during development according to the principles described in ICH Q9–Q11.[25,26] In-process controls should be defined.

The control of critical steps can be achieved by a combination of analytical tests and process control. During a solid-phase synthesis of a peptide this could include, *e.g.*, 9-fluorenylmethoxycarbonyl (Fmoc) deprotection, control of washing steps, coupling or acetylation reaction monitoring, drying steps and control of cleavage steps. The most common test for the monitoring of coupling reactions is the Kaiser test, which is based on the reaction of ninhydrin with primary amines.

During the purification process, individual fractions are combined into a pool of fractions. The pooling strategy should be defined and acceptance criteria for the purity of individual fractions and the main pool should be stated. These criteria for purity usually include overall purity and criteria for individual impurities in addition to a limit for unspecified impurities. In the case that secondary purification steps are used in the manufacturing process, the criteria should also be defined. Filtration and lyophilization steps are also adequately controlled.

Intermediates as defined in the manufacturing process are tested by in-process controls before use in the next stage of the manufacturing process. The methods used for in-process control and/or intermediate testing should be described in the dossier.

For solution-phase synthesis or fragment condensation processes, other requirements may apply.

3.2.S.2.5. Process Validation and/or Evaluation. Even if no process validation data are provided in the application, the active substance manufacturing process must be validated before commercial distribution. Process validation data and/or evaluation studies for aseptic processing and sterilization should be provided.

3.2.S.2.6. Manufacturing Process Development. At least a description and discussion of any significant changes made to the manufacturing process and/or manufacturing sites of the active substance used in producing nonclinical, clinical, scale-up, pilot and, if available, production-scale batches should be provided in this section of Module 3.[17]

However, ideally this section could be used as an extremely important section of Module 3. The significance of a drug substance manufacturing change during development should be assessed by evaluating its potential to impact the quality of the drug substance (and/or intermediate, if appropriate). As described in ICH Q11, in this case the information provided on the development of the drug substance manufacturing process should identify significant changes during process development, link relevant drug substance batches with the developmental stage of the manufacturing process used to prepare them and explain how prior knowledge, risk assessments and experimental studies were used to establish important aspects of the manufacturing process and control strategy.

It is recommended that the manufacturing process development section begins with a narrative summary that describes important milestones in the development of the process and explains how they are linked to assuring that the intended quality of the drug substance is achieved.

The CQAs of the drug substance should be listed and the rationale for designating these properties or characteristics as CQAs should be provided. In some cases, it might be appropriate to explain why other properties or characteristics that might be considered potential CQAs are not included in the list of CQAs.

Links or references should be provided to information submitted elsewhere in the submission that supports the designation of these properties or characteristics as CQAs.

A description and discussion should be provided of significant changes made to the manufacturing process or site of manufacture of drug substance batches used in support of the marketing application (*e.g.* those used in nonclinical or clinical studies or stability studies in support of a marketing authorization) and, if available, production-scale batches. The description should follow a chronological sequence ending with the proposed commercial process.

The reason for each significant change should be explained, together with an assessment of its potential to impact the quality of the drug substance (and/or intermediate, if appropriate). Batch information [batch size or scale, site and date of manufacture, route and process used and intended purpose (*e.g.* in a specified toxicological or clinical study)] and supporting data from comparative analytical testing on relevant drug substance batches should be provided.

3.2.S.3. Characterization.

3.2.S.3.1. Elucidation of Structure and Other Characteristics. The primary structure (sequence) of the peptide should be confirmed by analytical data. The technological achievements over recent decades are tremendous and mass spectrometry (MS) is a powerful analytical tool for the structure elucidation of peptides. Variants of the MS technique can be used to determine the molecular mass of a peptide and to confirm its amino acid sequence. Typical representative spectra and interpretation of the fragmentation data, including assignments and tables with theoretical and observed mass values, should be provided.

Amino acid analysis as described in the Ph. Eur. general chapter 2.2.56 usually complements the characterization of synthetic peptides. Elemental analysis may be used to determine the level of nitrogen in a specific sequence.

NMR spectrometry is described in Ph. Eur. general chapter 2.2.64, "Peptide Identification by Nuclear Magnetic Resonance Spectrometry." However, the scope described in that general chapter is qualitative and consists of comparing the NMR spectrum of a test sample with that of a reference sample acquired under identical conditions. Furthermore, the scope is restricted to one-dimensional NMR spectrometry.

NMR experiments are recommended to be part of the characterization studies. One- and two-dimensional techniques can be used to assign the structure by means of ^{1}H, ^{13}C and ^{15}N NMR data.

NMR can be used for

- determination of the number and types of proton nuclei;
- determination of the peptide sequence;
- identification of amino acids;
- assignment of carbon atoms;
- assignment of nitrogen atoms;
- secondary and tertiary structure elucidation.

Chiral gas chromatography (GC) is often used to identify and quantify the enantiomers of the different amino acids after acid hydrolysis. As the hydrolysis is known to induce some level of racemization, it is carried out in deuterated hydrochloric acid, yielding deuterated amino acids if the racemization takes place at this stage; deuterated and nondeuterated amino acid residues are detected separately by a mass spectrometric detector placed in tandem with the chiral GC system. Enantiomeric purity can be controlled by several means during the manufacture of a synthetic peptide. However, it should be justified that it is sufficient to perform the test on enantiomeric purity as a characterization test and that no routine release control is required.

Ultraviolet (UV) and infrared (IR) spectroscopy are part of the standard characterization programme. Additional information on the secondary structure can be gathered from these techniques.

Circular dichroism (CD) spectroscopy can be used to determine the absorption, *e.g.* in the presence of chromophores such as tryptophan, tyrosine, phenylalanine, disulfide bonds and peptide bonds. It measures differences in absorbance between left and right circularly polarized light and hence asymmetric properties of the chromophores. Changes in the structure and hence aromatic environments result in different CD spectra. Near-UV CD spectroscopy determines the tertiary structure due to asymmetric environments of tryptophan, tyrosine, phenylalanine and disulfide. Far-UV CD spectroscopy determines the secondary structure due to asymmetric environments of the peptide.

Usually no bioassay is required for the routine release of synthetic peptides. Nevertheless, bioassays can serve as additional tools for the characterization of synthetic peptides.

Peptide mapping may be applicable for longer peptides. Ph. Eur. general chapter 2.2.55, "Peptide Mapping," should be considered.

Peptides can also be characterized by Thioflavin T (ThT) dye assays and intrinsic tryptophan fluorescence to investigate whether a peptide can form fibrillary aggregates.

Physicochemical characterization of the drug substance could include solubility and hygroscopicity studies, determination of the isoelectric point (pI) and thermogravimetric studies. The morphology may be examined by powder X-ray diffraction (PXRD), Raman spectroscopy, differential scanning calorimetry (DSC) and light microscopy.

3.2.S.3.2. Impurities. Purity is one of the most important CQAs for synthetic peptides. Impurities are often categorized as either peptide-related impurities or nonpeptide impurities. Peptide-related impurities contain structural elements of the synthetic peptide sequence. Nonpeptide impurities include process reagents and their potential by-products, residual solvents, elemental impurities and potential genotoxic impurities.

Peptide-related impurities and their presence in peptide therapeutics are well described in the literature and only a short summary of different impurities classes is provided here.[27,28] Peptide-related impurities may originate from different sources:

- starting materials;
- formation during the manufacturing process;
- resulting from degradation during the manufacturing process or during storage.

The levels of certain related substances may be the result of combined contributions from these sources.

Related substances resulting from starting materials. Impurities present in the starting materials may cause the formation of peptide-related impurities. Examples of such impurities include incorrect enantiomers, β-Ala residues, dipeptides, single amino acid derivatives in dipeptides and amino acids with incorrect or without protecting groups. These impurities may be incorporated in the sequence during the assembly of the peptide by solid-phase peptide synthesis (SPPS). Narrow acceptance criteria for those impurities should be set for each starting material used in the manufacture of the peptide.

Related substances formed during the manufacturing process. Related substances designated as process impurities may be a result of undesired or incomplete reactions during SPPS or cleavage.

Stereoisomers. With the exception of glycine, all amino acids constituting synthetic peptides are optically active with at least one stereogenic centre each. The manufacturing process can likewise lead to the formation of stereoisomers by epimerization. For each of these amino acids, racemization during the manufacturing process is possible. Hence, in theory, a multitude of different diastereoisomers of the target peptide can be formed.

However, optimized SPPS conditions ensure minimized racemization. Amino acid derivatives with urethane-type N^{α} protecting groups (*e.g.* Fmoc) are rather resistant to oxazolone formation which leads to racemization at the stereocentre. Additionally, additives and bases used in the manufacturing process also suppress the potential racemization of utilized amino acids.

Deletion sequences and truncated sequences. Deletion sequences are peptides with one or several amino acids missing. Their formation is often prevented by repetition of the coupling procedure to complete the coupling reaction and systematic acetylation to cap unreacted coupling sites. The acetylation procedure results in acetylated peptide fragments (truncated sequences).

Insertion sequences. If a particular amino acid is coupled more than once during one coupling step, an insertion sequence is generated. Insertion sequences may occur as the result of premature N^{α}-deprotection during prolonged coupling reactions or incomplete removal of excess amino acid derivative prior to the next N^{α}-deprotection step.

Related substances formed during cleavage. Certain side-chain protecting groups are released as reactive cationic species during cleavage. These can react with the nucleophilic side chains of sensitive amino acids (*e.g.* Trp, Tyr, Met). To prevent the formation of related substances through such side reactions, suitable scavengers (*e.g.* water) are added to the cleavage mixture. Potential impurities could also result from incomplete cleavage of the amino acid side-chain protecting groups.

Degradation products. Degradation products of synthetic peptides may also occur as process impurities. Their content may increase during storage. Generally, the following pathways can contribute to the formation of degradation products of synthetic peptides:

- oxidation;
- hydrolysis;
- deamidation;
- diketopiperazine and pyroglutamic acid formation;
- β-elimination;
- condensation and formation of dehydropeptides;
- disulfide cleavage/exchange.

For regulatory filings, it is important to provide detailed information on the impurity profile of the specific synthetic peptide. Potential and observed impurities should be discussed. As part of the control strategy, actions and measures to prevent the formation of the different types of impurities should be described. The suitability of analytical methods and techniques to detect and quantitate the levels of impurities should be addressed. In some cases, peptide-related impurities coelute with the main peak or other impurities. To improve the control in these cases, alternative methods that are capable of separating the impurity in question from other impurities should be developed and validated.

Aggregation may occur for synthetic peptides, mainly due to changes in secondary structures. Since aggregation could potentially be related to safety issues, including immunogenicity, it is extremely important to study the potential to form aggregates. A combination of analytical characterization methods should be used. The potential to aggregate may be an intrinsic physical property of the related sequence; nevertheless, control and understanding are most important at the level of the resulting medicinal product formulation and stability properties. Successful understanding of aggregation can minimize challenges in both drug substance and medicinal product manufacture. Strategies to avoid aggregation should be developed.

Nonpeptide impurities/process-related impurities. Nonpeptide impurities include process reagents, by-products, residual solvents and elemental impurities. The SPPS process requires extensive washing of the resin with solvents. Reagents and solvents used for the coupling steps are washed with incremental quantities of a suitable solvent. Nevertheless, for all reagents and solvents used in the manufacturing process, the depletion should be addressed in the dossier by either data and/or risk analysis.

Trifluoroacetic acid (TFA) is often used in the manufacturing process. TFA is removed as a counter ion during the chromatographic desalting procedure. Usually residual TFA is controlled in the specification of a synthetic peptide.

In most cases, no metals or metal catalysts are used in the manufacturing process. The raw materials, utilities and equipment are controlled to limit the introduction of elemental impurities to the drug substance. Data on exemplary batches showing levels below the limit of ICH Q3D can be used to justify the omission of this attribute in the specification for routine release. In the rare cases where palladium is used for the cleavage of protecting groups, the absence of palladium should be demonstrated by data and the requirements of ICH Q3D should be considered.[29]

Leachables from the synthesis resin and column materials are further potential sources of nonpeptide process-related impurities.

The principles of assessment described in ICH Guideline M7(R1) apply for process-related impurities. This also includes benzene, which may be present as a contaminant in solvents used in the synthesis of the peptide.

Genotoxicity and control of mutagenic impurities. Genotoxicity testing is an important part of drug development and safety evaluation of new pharmaceuticals. Guidance is provided in ICH Guideline M7(R1), "Assessment and Control of DNA Reactive (Mutagenic) Impurities in Pharmaceuticals to Limit Potential Carcinogenic Risk."[30] However, synthetic peptides are formally excluded from the scope of this guideline and it is stated that assessment of the mutagenic potential of impurities as described in the guideline is not intended for the following types of drug substances and medicinal products: biological/biotechnological, peptide, oligonucleotide, radiopharmaceutical, fermentation products, herbal products and crude products of animal or plant origin.

ICH Guideline S6(R1), "Preclinical Safety Evaluation of Biotechnology-Derived Pharmaceuticals," is intended primarily to recommend a basic framework for the preclinical safety evaluation of biotechnology-derived pharmaceuticals.[31] It applies to products derived from characterized cells through the use of a variety of expression systems including bacteria, yeast, insect, plant and mammalian cells. The active substances include proteins and peptides, their derivatives and products of which they are components. The principles outlined in this guidline may also be applicable to recombinant DNA protein vaccines, chemically synthesized peptides, plasma-derived products, endogenous proteins extracted from human tissue and oligonucleotide drugs.

ICH Guideline S6 states: "The range and type of genotoxicity studies routinely conducted for pharmaceuticals are not applicable to biotechnology-derived pharmaceuticals and therefore are not needed. Moreover, the administration of large quantities of peptides/proteins may yield uninterpretable results. It is not expected that these substances would interact directly with DNA or other chromosomal material. Studies in available and relevant systems, including newly developed systems, should be performed in those cases where there is cause for concern about the product (*e.g.* because of the presence of an organic linker molecule in a conjugated protein product). The use of standard genotoxicity studies for assessing the genotoxic potential of process contaminants is not considered appropriate. If performed for this purpose, however, the rationale should be provided."

A group of authors mainly in industry recently addressed some issues related to genotoxicity testing of peptide/protein-related biotherapeutics and reviewed the current practice.[32] Their paper provides valuable insights but is not a substitute to regulatory decisions. The authors state: "For biotechnology-derived peptides, the application of ICH S6(R1) rather than ICH S2(R1) for genotoxicity assessment is obvious and omission of genotoxicity testing for such peptides can be easily justified by referring to the ICH S6(R1) recommendations.[33] This may be less clear for chemically synthesized peptides. In fact, chemically synthesized peptides are often evaluated for genotoxicity like small drug substances and ICH S2(R1) is applied."[34]

The five main conclusions in this paper are as follows:

- From a scientific point of view, there is no reason for testing peptides containing exclusively natural amino acids irrespective of the manufacturing process (biological or synthetic).
- If non-natural amino acids, organic linkers and other chemical entities have already been tested for genotoxicity, there is no need to re-evaluate them when used in different peptide/protein-related biotherapeutics.
- Unless the peptides have been modified to be able to enter the cells, it is generally more appropriate to evaluate the peptides containing the non-natural amino acids *in vivo* where the cleavage products can be formed.
- For linkers, it is important to determine if exposure to reactive forms is likely to occur and from what origin. When the linkers are anticipated to be potential mutagenic impurities, they should be evaluated according to ICH M7(R1).
- If linkers are expected to be catabolic products, it is recommended to test the entire conjugate *in vivo*, as this would ensure that the relevant "free" linker forms stemming from *in vivo* catabolism are tested.

As stated above, this paper does not have any regulatory impact and in each submission the chosen strategy on genotoxicity testing has to be justified on a case-by-case basis.

3.2.S.4. Control of Drug Substance.

3.2.S.4.1. Specification. The specification should cover the critical quality attributes that assure adequate quality at release and throughout retesting. It is one part of the comprehensive control strategy for the drug substance.[35]

Typical specification tests included as an attribute in the specification are as follows:

- appearance;
- identification by mass;
- identification by relative retention time (RRT);
- amino acid analysis;
- purity/total impurities;
- individual impurities;
- unspecified/unidentified impurities;
- assay/content;
- acetic acid content/counter-ion content;
- TFA content;
- pH;
- water;
- mass balance;
- residual solvents;
- elemental impurities;

- specific optical rotation;
- bacterial endotoxins;
- microbiology.

Ph. Eur. methods should be used and a reference to Ph. Eur. should be included where relevant. Additional tests may be required depending on the subsequent use of the drug substance. Justification for the drug substance specification should be provided in 3.2.S.4.5, "Justification of Specification." The acceptance criteria laid down in the drug substance specification are identical with the limits that apply for stability studies.

Synthetic peptides are excluded from the scope of ICH Guideline Q3A, "Impurities in New Drug Substances," and consequently the limits laid down in this guideline are not applicable. In contrast to other regions in Europe, specific limits for peptide-related impurities are defined in the Ph. Eur. According to the Ph. Eur. monograph "Substances for Pharmaceutical Use," peptide-related impurities should be reported above 0.1%, identified above 0.5% and qualified above 1.0% (see Table 2.3).

3.2.S.4.2. Analytical Procedures. Details of the analytical procedures used for testing the active substance should be provided. They should be described in such a way that they can be repeated by an Official Medicines Control Laboratory (OMCL).

A broad range of analytical methods are used for the routine release of a synthetic peptide.[13] Several methods or their combinations are usually utilized for routine identity testing, such as reversed-phase high-performance liquid chromatography (RP-HPLC), MS and amino acid analysis (AAA).

Assay is usually performed using HPLC or nitrogen content determination. The purity of a synthetic peptide is usually determined using HPLC. Ultra-performance liquid chromatography (UPLC) is becoming more and more a standard method in purity determination. The purity method should be capable of separating impurities introduced through starting materials, process-related impurities and impurities resulting from degradation of the peptide. One or more analytical methods may be required to accomplish this objective. A single HPLC method may be appropriate based on the sequence

Table 2.3 Peptide impurity limits provided in the Ph. Eur. monograph "Substances for Pharmaceutical Use".

Limits	Reporting threshold/%	Identification threshold/%	Qualification threshold/%
Impurity limits for synthetic peptides (Ph. Eur. Monograph "Substances for Pharmaceutical Use")	0.1	0.5	1.0
Impurity limits for small molecules (ICH Q3A)[a]	0.05	0.1	0.15

[a]Maximum daily dose less than 2 g.

of the peptide and knowledge of its potential impurities. However, sufficient selectivity/specificity has to be demonstrated. Otherwise, orthogonal/ additional purity methods should be used. In addition, use of LC-MS compatible methods may be advantageous for the identification of impurities.

For counter ion, water, residual solvents, elemental impurities, specific optical rotation, bacterial endotoxins and microbiology, Ph. Eur. methods or suitable in-house methods such as GC for residual solvent are employed.

3.2.S.4.3. Validation of Analytical Procedures. Analytical validation data, including experimental results for the analytical procedures used for the control of the active substance, should be provided unless methods in the respective drug substance monograph in the Ph. Eur. are referred to and the tests in the monograph have been demonstrated to be suitable to control the substance. Validation of analytical tests concerning the active substance should be performed according to the requirements of ICH Q2.[36]

3.2.S.4.4. Batch Analyses. A description of batches and results of batch analyses should be provided as follows:

- Batches of material used in the preclinical tests and clinical studies reported in support of the application.
- Data illustrating the actual results obtained from routine quality control of the active substance. Results from at least three recent consecutive batches from each manufacturing site, manufactured according to the proposed process at not less than 10% of the maximum production scale at the time of submission, should be provided. These results should demonstrate that routine production material falls within the specification limits cited for the purpose covered by the marketing authorization.

The results should include the following:

- date of manufacture;
- batch size and number;
- place of manufacture (data from all manufacturing sites must be provided);
- results of analytical determinations;
- use of batches.

Presentation of this information in tabular form is recommended for improved clarity. Test results should be expressed numerically, *e.g.* impurity levels. The batch analyses should include all the tests in the specification. There may, however, be cases where previous batches were tested using a slightly different specification. In these cases, a brief explanatory note should be included. Any apparently inconsistent or anomalous results in the batch analyses should be explained.

3.2.S.4.5. Justification of Specification. Justification of the control strategy and active substance specification should be provided. The specification should be based on results from preclinical, clinical and, where applicable, production-scale batches and taking into account the qualification of impurities and the overall control strategy.

The requirements of the general monograph of the Ph. Eur. "Substances for Pharmaceutical Use" (2034) should be met, where applicable. For existing active substances, the respective monograph of the Ph. Eur. or, in default of this, the respective monograph of the pharmacopoeia of an EU Member State should be the basis of the active substance specification. Supplementation by additional tests (*e.g.* impurity tests) might be necessary.

3.2.S.5. Reference Standards or Materials. Information on the reference standards or reference materials used for testing of the active substance should be provided: specifications, full analytical and physicochemical characterizations, impurities profile, *etc.* Chemical reference substances (Ph. Eur. CRS) are qualified as primary reference standards and do not need to be further qualified, provided they are used for their intended purpose. The criteria for establishing the primary reference substances should be given with full analytical profiles. The procedure for establishing secondary reference standards or materials normally used for routine analysis should be stated. In the case that impurity standards have been established, information on these standards should be provided.

3.2.S.6. Container Closure System. There are no specific requirements for synthetic peptides with respect to the documentation to be provided in the section on the container closure system. Compliance with Commission Regulation (EU) No. 10/2011 on plastic materials and articles intended to come into contact with food and its latest amendment (EU) 2018/79 and the EMA "Guideline on Plastic Primary Packaging Materials" should be demonstrated.[37,38]

3.2.S.7. Stability. The types of studies conducted, protocols used and the results of the studies should be summarized (3.2.S.7.1: "Stability Summary and Conclusions"). The summary should include results, for example, from forced degradation studies and stress conditions (light stress, higher temperature, *etc.*), and also conclusions with respect to storage conditions and retest date or expiry date as appropriate. Information on the analytical procedures used to generate the data and validation of these procedures if different from the release methods should be included. Changes in analytical methods during stability studies should be justified and indicated in an appropriate format. The principles of the ICH Q1A Guideline "Stability Testing of New Drug Substances and Drug Products" should be considered.[39]

2.9 Considerations for Peptides Conjugates

Conjugation has emerged as a popular mechanism to alter or enhance the properties of peptide and protein drug candidates. The proportion of conjugated peptides and the variety of conjugated moieties have increased over time and many peptide conjugates are used in current development programmes.

Conjugation to poly(ethylene glycol) (PEG), lipids and proteins such as Fc fragments has been used as a half-life extension strategy. Conjugation can also be used to deliver a cytotoxic payload or imaging agent to specific cell types targeted by the peptide.[1] However, there is added complexity with respect to the characterization and control of these conjugates. The control of the unconjugated peptide as an intermediate is essential. Adequate specifications and control methods should be established for these intermediates.

The conjugation chemistry should be described in the pharmaceutical development section and good process understanding and knowledge are expected. Conjugatable *versus* nonconjugatable impurities should be identified by means of a risk analysis and the incorporation into the target molecule should be investigated.

An additional attribute for conjugated peptides is the amount of the free unconjugated peptide and the free form of the conjugate moiety (*e.g.* free PEG/linker). Di-PEGylation or multi-PEGylation may also occur and should be adequately controlled.

The choice of the starting material of the conjugation component needs to be justified according to ICH Q11, "Questions & Answers: Selection and Justification of Starting Materials for the Manufacture of Drug Substances."[24]

The designation of the starting materials should be discussed with the regulatory agencies. An activated PEG form (*e.g.* in the form of an aldehyde) may not be suitable as starting material and is considered to be an intermediate itself. It has to be assured that all steps of the intermediate synthesis starting from the defined starting material are performed under good manufacturing practice (GMP). Consequently, the activation of the suitable PEG starting material is considered a part of the manufacturing process. Full information should be provided in Section 3.2.S.3.2 of Module 3, including flowchart, process description with all process steps, raw materials and critical and noncritical manufacturing process controls. In most cases, the conjugation moiety is manufactured at a different contract manufacturer as the synthetic peptide. The Qualified Person (QP) declaration should cover these aspects, as appropriate. In the case of multiple suppliers of the conjugation moiety, for each supplier separate documentation is expected. Peptide-conjugated material from all suppliers should be manufactured and stability data should be generated. The manufacturers of the different components should be included in the dossier. Potential genotoxic impurities resulting from the conjugation moiety should also be addressed.

In numerous development programmes, polymers or other conjugation moieties are coupled to the peptide *via* a chemical linker. The issues mentioned above are also applicable for such chemical linkers, especially for the selection of suitable starting materials and the control of the impurity profile. The critical attributes should be evaluated and a justification for the specification attributes should be provided.

2.10 Medicinal Product Considerations

The quality target product profile (QTPP) relates to quality, safety and efficacy, considering *e.g.* the route of administration, dosage form, bioavailability, strength and stability of a medicinal product containing a synthetic peptide as active substance.

ICH Guidelines ICH Q3B and ICH Q6A are not or only partly applicable to synthetic peptides. The limits for purity as defined in the general monograph of the Ph. Eur. "Substances for Pharmaceutical Use" may also be applicable to the resulting medicinal products. Higher limits should be justified on a case-by-case basis considering the batch analysis history.

Synthetic peptides are included in the scope of ICH Q3D "Guideline for Elemental Impurities" and the requirements laid down in this guideline are applicable for new medicinal products containing synthetic peptides as active substances.[29]

Potential interactions of the peptide with the excipients present in the formulation and also with leachables that could result from packaging materials such as stoppers should be evaluated during pharmaceutical development.

Most of the medicinal product formulations containing synthetic peptides as active substance are for parenteral use. The principles for the choice of sterilisation process for finished products and containers are presented in the form of decision trees in the "Guideline on the Sterilization of the Medicinal Product, Active Substance, Excipient and Primary Container".[40]

Terminal sterilization is preferred to sterilization by filtration and/or aseptic processing because it provides a sterility assurance level (SAL) that it is possible to calculate, validate and control and thus incorporates a safety margin. For aseptic processes, a SAL is not applicable as accidental contamination caused by inadequate technique cannot be reliably eliminated by monitoring, control or validation. Therefore, terminal sterilization provides the highest assurance of sterility and should be used whenever possible. For highly sensitive products such as biological products where terminal sterilization of the medicinal product is not possible, aseptic processing under controlled conditions provides satisfactory quality of the medicinal product.

Aseptic processing cannot be accepted based solely on an increase in impurity levels upon terminal sterilization without further justification. An increased level of impurities does not necessarily preclude terminal sterilization of the medicinal product. The risk induced by the degradation should be balanced with the risk induced with an aseptic manufacturing method also taking in account the posology of the product and the nature of the degradation products. Attempts to find terminal sterilization conditions adjusted to give acceptable impurity levels based on degradation mechanisms of the active substance and the actual bioburden should be described in the quality dossier.

The decision trees are relevant for medicinal products with synthetic peptides as the active substance and applicants intending to use sterile filtration during medicinal product manufacture should have comprehensive data in place on the impurity profile or other CQAs (*e.g.* particle contamination) after treatment at 121 °C for 15 min or other conditions described in the guideline. No change observed in the impurity profile may require a change in the manufacturing process prior to approval. When small changes in the impurity profile are observed, a change in manufacturing process should be considered and a sound justification for sterile filtration should be provided.

In all cases where sterilization is performed by a sterile filtration process, validation data for three medicinal product batches need to be submitted with the application for marketing authorization since sterile filtration is considered a nonstandard process according to the "Guideline on Process Validation for Finished Products."[41]

Additional characteristics for complex finished product dosage forms should be considered on a case-by-case basis.

2.11 Synthetic Peptide Vaccines

Synthetic peptide vaccines are usually composed of 20–30 amino acids containing the specific epitope of an antigen related to infectious and/or chronic diseases including cancers.[42,43] Often they consist of mixtures of different peptide active substances. Whereas the demonstration of sufficient pharmaceutical quality of the drug substance itself is according to what has been described elsewhere in this chapter, the analytical challenges are immense with respect to specificity and selectivity of the applied methods in the mixtures of different peptides.

2.12 Synthetic Peptide Development Programmes Using a Biological Medicinal Product as a European Reference Medicinal Product

The European legislation for medicinal product clearly differentiates between biological medicines ("biologicals") and chemically derived molecules. Biological medicines contain active substances from a biological source, such as living cells or organisms. A biological substance is a substance that is produced by or extracted from a biological source and that needs for its characterization and the determination of its quality a combination of physicochemical and biological testing, together with the production process and its control. The manufacture of biological medicines tends to be more complex than that of chemically derived molecules. Most biological medicines are made by biotechnology, often using sophisticated cell systems and recombinant DNA technology. The quality of biological medicines is tightly controlled and specific guidelines and regulations with additional requirements apply. The level of information that needs to be submitted in the dossier is more extensive than for chemically derived molecules.

An increasing number of biological medicines are "biosimilars" – medicines that are highly similar in all essential aspects to an already approved biological medicine. The EU has pioneered the regulation of biosimilar medicines by establishing a solid framework for their approval and by shaping biosimilar development globally. Since the EU approved the first biosimilar in 2006, healthcare professionals have gained increasing experience with their use. Today, biosimilars are an integral part of the effective biological therapies available in the EU, supported by adequate safeguards protecting patient safety.

The aim of biosimilar development is to demonstrate biosimilarity – high similarity in terms of structure, biological activity and efficacy, safety and immunogenicity profile. Because biosimilars are made in living organisms, there may be some minor differences from the reference medicine. These minor differences are not clinically meaningful, *i.e.* no differences in safety and efficacy are expected. Natural variability is inherent to all biological medicines and strict controls are always in place to ensure that it does not affect the way in which the medicine works or its safety. By demonstrating biosimilarity, a biosimilar can rely on the safety and efficacy experience gained with the reference medicine. This avoids unnecessary repetition of clinical trials already carried out with the reference medicine. Demonstration of biosimilarity relies on comprehensive comparability studies with the reference medicine.[44]

If a biosimilar is highly similar to a reference medicine and has comparable safety and efficacy in one therapeutic indication, safety and efficacy data may be extrapolated to other indications already approved for the reference medicine.

Most of the approved biological medicines contain active substances made of proteins. Some smaller proteins can be produced either by using a biological manufacturing process or through chemical synthesis (*e.g.* solid-phase synthesis). After the end of the data exclusivity and market protection period for the reference medicinal product, applicants for generic or hybrid applications may use both of the manufacturing pathways chemically or by recombinant DNA techniques. It is indisputable, however, that the biosimilar regulatory pathway is not possible for chemically synthesized peptides since these fall outside the definition of a biological substance. The key question is which, if any, additional preclinical and/or clinical data including immunogenicity have to be provided with the respective applications. The legal basis for such so-called "hybrid applications" is Article

10(3) of Directive 2001/83 on the Community code relating to medicinal products for human use. The legal basis for pure generic applications is Article 10(1). The Co-ordination Group for Mutual Recognition and Decentralised Procedures – Human (CMDh) discussed under which legal basis an application for a synthetic peptide can be submitted, claiming equivalence to a reference product containing the same peptide manufactured using recombinant technology and which can be well characterized.[45,46]

It can be assumed that the number of applications where a biological medicinal product is used as a reference product and the active substance is manufactured by solid-phase synthesis will increase in the future. In October 2017, the US Food and Drug Administration (FDA) published the draft guidance "ANDAs for Certain Highly Purified Synthetic Peptide Drug Products That Refer to Listed Drugs of rDNA Origin" and five active substances were explicitly mentioned in this document: glucagon, liraglutide, nesiritide, teriparatide and teduglutide.[47] Other biological medicinal products for which the chemical synthesis pathway may be used in the future are on the European market. The first product on the market in Europe was a teriparatide 20 µg per 80 mL solution for injection and was approved in 2016. A public assessment report is available for this procedure.[48]

As stated by the CMDh, the similarity assessment of the active substances will be part of the assessment. If similarity cannot be proved during assessment, the application will have to be rejected. Currently, no formal European guidance is available on the demonstration of similarity for this type of product. Nevertheless, it is self-evident that analytical comparability testing, comprising physicochemical (structural) and biological (functional) assays and conventional analytical testing, forms the basis of the demonstration of similarity.

A broad panel of analytical methods to demonstrate the similarity between the recombinant reference product and the synthetic version is requested for the side-by-side similarity studies. It is important to note that the reference product used in the comparability studies should be sourced from the European market.

The applicants will need to quantify fully all differences in peptides produced by chemical synthesis and peptides produced by recombinant technology and demonstrate that both products are bioequivalent and comparable. The applicants should consider what analytical tests might be used to confirm comparability and to define and justify, prior to conducting these studies, the acceptance range to conclude comparability. Any observed differences from the reference medicinal product should be evaluated and justified.

The primary structure should investigated by, *e.g.*, MS, peptide mapping with MS/MS, capillary isoelectric focusing (IEF), western blotting, *etc.* Higher-order structures should be addressed by CD analysis, Fourier transform infrared (FTIR) spectroscopy, 2D NMR spectrometry, fluorescence spectroscopy and DSC as appropriate. For certain peptides the oligomeric state should be investigated.

Characterization of purity should be addressed using an orthogonal approach, *i.e.* size-based, charge-based and hydrophobicity-based separation techniques. Aggregation propensity should also be investigated. Differences in the impurity profiles raise the question of whether impurities in the synthetic peptide not present in the biological reference product are qualified. In general, these synthetic peptides have to comply with the requirements of the Ph. Eur. monograph "Substances for Pharmaceutical Use." This monograph allows an identification threshold of 0.5%. However, a full evaluation of the impurity profile also identifying impurities at lower levels than usually required in Europe is of benefit. In general, a limit of quantification (LOQ) of ≤0.1% for HPLC purity testing is required.

Comparative forced degradation studies are also recommended and the suitability of the analytical purity methods to characterize fully the impurity profiles of both products should be demonstrated.

It is the responsibility of the applicant to demonstrate that the HPLC/UPLC purity method is suitable to cover the complete impurity profile of the peptide or whether additional purity testing with a supplementary method is necessary. The presence and potential impacts of different excipients in the respective medicinal products on analytical methods should be addressed.

Process-related impurities from the cell construct [*e.g.* host cell protein (HCP), DNA] or resulting from the manufacturing process (*e.g.* antibiotics and other media components) are not an issue for synthetic peptides in the comparability studies. On the other hand, for synthetic peptides, clearance of reagents and potential genotoxic impurities should be addressed. For residual solvents, the requirements of ICH Q3C(R7) have to be fulfilled.[49]

Cell based functional assays using appropriate cell line(s) should be developed and used in the comparability studies. It depends on the mechanism of action which additional functional assays may be needed to demonstrate similarity (*e.g.* binding kinetics).

The absence of a release specification for the specific biological activity should be appropriately justified, *e.g.* by commercial-scale batch biological activity data and, in addition, by appropriate characterization of higher-order structure by physicochemical testing.

The analytical methods used in the comparability exercise should be suitable, sufficiently qualified and/or validated and sensitive to detect potential differences between both products. In the case that statistical models are used to demonstrate comparability they should be adequately described and justified.

Batches preferably from the commercial process should be used for the side-by-side analyses. The number of batches used in the comparability studies should be adequately justified.

Comparative stability studies can be useful for detecting potential differences in the stability profile of the peptides manufactured either by chemical synthesis or by recombinant techniques. Stability and shelf-life claims cannot be derived from the reference product without their own data.

2.13 Requirements for Clinical Trial Applications

It should be clearly differentiated between the requirements for a dossier for a clinical trial and a marketing authorization dossier. Whereas the latter have to ensure a state-of-the-art quality of a product for wide use in patients, information to be provided for investigational medicinal products (IMPs) should focus on the risk aspects and should consider the nature of the product, the state of development/clinical phase, patient population, nature and severity of the illness and type and duration of the clinical trial itself.

In general, IMPs should be produced in accordance with the principles and the detailed guidelines of GMPs for medicinal products.[50,51]

The most relevant document for the quality of IMPs in Europe is the "Guideline on the Requirements for the Chemical and Pharmaceutical Quality Documentation concerning Investigational Medicinal Products in Clinical Trials" (EMA/CHMP/QWP/545525/2017).[52] Synthetic peptides are explicitly within the scope of this guideline. A second guideline addresses the quality of biologicals IMPs. Synthetic peptides are not within the scope of the "Guideline on the Requirements for Quality Documentation Concerning Biological Investigational Medicinal Products in Clinical Trials" (EMA/CHMP/BWP/534898/2008 rev. 1 corrigendum).[53] Nevertheless, since synthetic peptides are sometimes in the grey area between biologicals and small molecules, some aspects stated in this guideline may also be helpful for sponsors with peptide drug development programmes.

The Investigation Medicinal Product Dossier (IMPD) should include the most up-to-date information relevant to the clinical trial available at time of submission of the clinical trial application.

Information concerning nomenclature, structure (sequence) and general properties should be provided. The name(s) and address(es) and responsibilities of all manufacturer(s), including contractors and each proposed site involved in manufacture and testing, should be provided. A brief summary of the synthesis process, a flowchart of the successive steps including, for each step, the starting materials, intermediates, solvents, catalysts and critical reagents used should be provided.

Materials used in the manufacture of the drug substance (*e.g.* raw materials, starting materials, solvents, reagents, catalysts) should be listed together with a brief summary on the quality and control of any attributes anticipated to be critical. No detailed information on the impurity profile for each protected amino acid derivative used as the starting material is requested; however, a phase-dependent specification for starting materials should be laid down in the IMPD.

Critical steps in the synthesis, tests and acceptance criteria for their control should be briefly summarized. Process validation and/or evaluation are not applicable for drug substances to be used in clinical trials.

It should be documented if the manufacturing process differs significantly from that used for the production of the batches used in the nonclinical studies. In this case, a flowchart of the manufacturing process used for the drug substance used in the nonclinical studies should be presented. Significant changes in the manufacturing process, which may impact on quality, should be discussed (*e.g.* change of route of synthesis).

The structure of chemically defined substances should be established with suitable methodology; relevant data should be provided. For synthetic peptides, a proof for the amino acid sequence is expected for Phase I trials. If present disulfide bridges should be characterized. The submission of the results of MS/MS studies is considered state-of-the-art. For longer peptides, peptide mapping studies may be useful. Higher-order structures should be addressed (*e.g.* by CD). Higher-order structure may be more relevant for the medicinal product section since higher-order structures depend on the composition of the formulation and the concentration of the peptide. Typically, in contrast to biologicals, no bioassay is required in the characterization section of the IMPD. Racemization studies are also usually part of the characterization section.

For impurities in IMPs, a justification that the product is safe for its intended use, considering the anticipated exposure of volunteers and patients, will be required. In contrast to small molecules, for many peptides the impurity profile is much more complex. The limits described for synthetic peptides in the Ph. Eur. monograph "Substances for Pharmaceutical Use" are formally not applicable to clinical trials. Nevertheless, they could serve as a good starting point for future developments, especially considering the identification of impurities for later phases of development. In general, limits proposed for clinical trial materials should be covered by toxicological studies.

Where relevant, aggregation should be addressed, and also a discussion on (potential) mutagenic impurities according to the principles ICH M7(R1) should be provided.

Impurities (*e.g.* degradation products, residual solvents) deriving from the manufacturing process or starting materials relevant to the drug substance used for the clinical trial should be stated. The level of detail necessary depends on the phase of the clinical trial. Absence of routine control for certain solvents/catalysts used in the manufacturing process should be justified.

The specifications, the tests used and their acceptance criteria should be specified for the batch(es) of drug substance(s) used in the clinical trial. Tests for identity, impurities and assay are mandatory. Upper limits, taking safety considerations into account, should be set for the impurities. A limit of "record" or "report results" will not be acceptable for the sum of impurities or overall purity. The limits may need to be reviewed and adjusted during further development. The limits should be supported by the impurity profiles of batches of active substance used in nonclinical and clinical studies. If ICH or Ph. Eur. requirements are met, no further limit justification is expected. The microbiological quality, including bacterial endotoxins for drug substances used in aseptically manufactured products, should be specified. The specification should include acceptance criteria for any relevant residual solvent or catalyst. For certain specification attributes, a limit of "record" or "report results" could be acceptable, especially for early clinical phases. Specifications and acceptance criteria set for previous Phase I or II trials should be reviewed and, where appropriate, adjusted to the current stage of development.

The analytical methods used for the drug substance should be described for all tests included in the specification. For synthetic peptides, orthogonal/additional methods are recommended in order to minimize the risk of coelution of impurities with the main peak or with each other. A single purity method may be sufficient when specificity/selectivity has been demonstrated by validation data and/or forced degradation studies. The development of stability-indicating methods is essential. For some peptides, dimers/multimers/aggregates need to be addressed by additional analytical methods.

For Phase I trials, the suitability of the analytical methods used should be confirmed. The acceptance limits (*e.g.* acceptance limits for the determination of the content of impurities, where relevant) and the parameters (specificity, linearity, range, accuracy, precision, quantification and detection limit, as appropriate) for performing validation of the analytical methods should be presented in a tabulated form.

For Phase II and III clinical trials, the suitability of the analytical methods used should be demonstrated. A tabulated summary of the results of the validation carried out should be provided (*e.g.* results or values found for specificity, linearity, range, accuracy, precision, quantification and detection limit, as appropriate). It is not necessary to provide a full validation report.

In the case of major changes in analytical methods, cross-validation data should be presented, especially for specified unknown impurities identified by their RRT unless otherwise justified. A reanalysis of a preclinical batch with the new method should also be considered, where relevant.

Batch results in a tabulated form or certificate of analysis for batches to be used in the current clinical trial, for batches used in the nonclinical studies and, where needed, for representative batches used in previous clinical trials (*e.g.* in case the comparable quality of batches manufactured by previous processes has to be demonstrated) should be supplied. If data are not available for the batches to be used in the current clinical trial, data for representative batches for each drug substance manufacturer may be submitted instead. The batch number, batch size, manufacturing site, manufacturing date, control methods and the test results should be listed. The manufacturing process used for each batch should be assigned.

A brief justification of the specifications and acceptance criteria for impurities and any other parameters that may be relevant to the performance of the medicinal product should be provided based on safety and toxicity data, and also the methods used for the control of impurities. The solvents and catalysts used in the synthesis should be taken into consideration.

The parameters characterizing the batch of drug substance established as reference standard should be presented, where applicable.

The immediate packaging material used for the drug substance should be stated. If noncompendial materials are used, a description and specifications should be provided.

The stability data available at the respective stage of development should be summarized in tabulated form. Stability data should be provided for batch(es) manufactured according to the representative process (the same/very similar synthesis, comparable batch size) and can be supported by data from batch(es) manufactured by previous processes. The parameters known to be critical for the stability of the drug substance need to be presented. Potential degradation pathways should be described. The retest period should be defined based on the available stability data and should be clearly stated.

The retest period can be extended without a substantial modification submission, if a stability protocol, retest period extension plan and a statement that in case of any significant negative trend the sponsor will inform the competent authority are provided. The stability protocol should cover the maximum planned retest period.

2.14 Scientific Advice and European Initiatives for Early Access Approaches

The EMA and the National Competent Authorities from the Member States provide scientific advice by addressing questions raised by developers, with the objective of facilitating the development and availability of high-quality, effective and acceptably safe medicines, for the benefit of patients. The advice is given in the light of the current scientific knowledge and based on the documentation submitted by developers. Scientific advice is prospective in nature, *i.e.* it focuses on development strategies rather than pre-evaluation of data to support a marketing authorization application. Scientific advice helps to ensure that developers perform the appropriate tests and studies. Scientific advice is not legally binding on authorities or on the medicine developer regarding any future marketing authorization applications for the medicine concerned. In recent years, many developers of peptide therapeutics have asked for scientific advice regarding very different aspects concerning pharmaceutical quality. Resort to EMA or national scientific advice by developers of medicinal products is highly recommended.[54]

Furthermore, the EMA addresses the unique needs of micro-, small- and medium-sized enterprises (SMEs) through the SME office. This dedicated interface has the sole remit of providing regulatory, financial and administrative assistance to small pharmaceutical companies.

PRIME (Priority Medicines) is a scheme launched by the EMA to enhance support for the development of medicines that target an unmet medical need. This voluntary scheme is based on enhanced interaction and early dialogue with developers of promising medicines, to optimize development plans and speed up evaluation so that these medicines can reach patients earlier. Through PRIME, the EMA offers early and proactive support to medicine developers to optimize the generation of robust data on a medicine's benefits and risks and enable accelerated assessment of medicines applications. This will help patients to benefit as early as possible from therapies that may significantly improve their quality of life.[55]

PRIME builds on the existing regulatory framework and tools already available, such as scientific advice and accelerated assessment. This means that developers of a medicine that benefited from PRIME can expect to be eligible for accelerated assessment at the time of application for a marketing authorization.

By engaging with medicine developers early on, PRIME is aimed at improving clinical trial designs so that the data generated are suitable for evaluating a marketing-authorization application. Early dialogue and scientific advice also ensure that patients participate only in trials designed to provide the data necessary for an application, making the best use of limited resources.

A list of products granted eligibility to PRIME is published on the website of the EMA. Several synthetic peptides are already included in this list.

2.15 Conclusion

This chapter has summarized regulatory requirements concerning pharmaceutical quality aspects for synthetic peptides. These include pharmaceutical development activities, the description and control of the manufacturing process, characterization of synthetic peptides, a discussion on the impurity profile including peptide-related and process-related impurities, aspects of the specification for routine release, the description and validation of analytical methods, reference standards, container closure materials and stability aspects.

Disclaimer

The views expressed in this chapter are those of the author and do not necessarily represent those of the BfArM and other European Regulatory Agencies.

References

1. J. L. Lau and M. K. Dunn, *Bioorg. Med. Chem.*, 2018, **26**, 2700–2707.
2. T. Uhlig, T. Kyprianou, F. G. Martinelli, C. A. Oppici, D. Heiligers, D. Hills, X. R. Calvo and P. Verhaert, *EuPa Open Proteomics*, 2014, **4**, 58–69.
3. K. Fosgerau and T. Hoffmann, *Drug Discovery Today*, 2015, **20**, 122–128.
4. European Medicines Agency, European Medicines Regulatory Network, https://www.ema.europa.eu/en/about-us/how-we-work/european-medicines-regulatory-network, accessed March 2019.
5. European Medicines Agency, Quality Guidelines, https://www.ema.europa.eu/en/human-regulatory/research-development/scientific-guidelines/quality-guidelines, accessed March 2019.

6. History, ICH https://www.ich.org/about/history.html, accessed March 2019.
7. ICH Q3A: ImpuritiesTesting: Impurities in New Drug Substances, https://www.ich.org/products/guidelines/quality/quality-single/article/impurities-in-new-drug-substances.html, accessed March 2019.
8. ICH Q3B(R2): Impurities in new drug products, https://www.ich.org/fileadmin/Public_Web_Site/ICH_Products/Guidelines/Quality/Q3B_R2/Step4/Q3B_R2__Guideline.pdf, accessed March 2019.
9. ICH Q6A: Test Procedures and Acceptance Criteria for New Drug Substances and New Drug Products: Chemical Substances Q 6A Specifications, https://www.ich.org/fileadmin/Public_Web_Site/ICH_Products/Guidelines/Quality/Q6A/Step4/Q6Astep4.pdf, accessed March 2019.
10. ICH Q6B: Specifications: Test Procedures and Acceptance Criteria for Biotechnological/Biological Products, https://www.ich.org/products/guidelines/quality/quality-single/article/specifications-test-procedures-and-acceptance-criteria-for-biotechnologicalbiological-products.html, accessed March 2019.
11. ICH Guideline Q11: Development and Manufacture of Drug Substances (chemical entities and biotechnological/biological entities), https://www.ema.europa.eu/en/documents/scientific-guideline/draft-ich-guideline-q11-development-manufacture-drug-substances-chemical-entities-biotechnological/biological-entities_en.pdf, accessed March 2019.
12. ICH Q8(R2): Pharmaceutical Development, https://www.ich.org/fileadmin/Public_Web_Site/ICH_Products/Guidelines/Quality/Q8_R1/Step4/Q8_R2_Guideline.pdf, accessed March 2019.
13. I. Eggen, B. Gregg, H. Rode, A. Swietlow, M. Verlander and A. Szajek, *Pharm. Technol.*, 2014, **38**(3), 48–55.
14. I. Eggen, B. Gregg, H. Rode, A. Swietlow, M. Verlander and A. Y. Szajek, *Pharm. Technol.*, 2014, **38**(4), 48–52.
15. I. Eggen, B. Gregg, H. Rode, A. Swietlow, M. Verlander and A. Szajek, *Pharm. Technol.*, 2014, **38**(5), 52–56.
16. Guideline on Summary of requirements for active substances in the quality part of the dossier CHMP/QWP/297/97 Rev. 1, https://www.ema.europa.eu/en/documents/scientific-guideline/guideline-summary-requirements-active-substances-quality-part-dossier-revision-1_en.pdf, accessed March 2019.
17. Guideline on the chemistry of active substances (EMA/454576/2016), https://www.ema.europa.eu/en/documents/scientific-guideline/guideline-chemistry-active-substances_en.pdf, accessed March 2019.
18. Guideline on Active substance-master-file procedure CHMP/QWP/227/02 Rev. 4 (CHMP/QWP/227/02 Rev 4), https://www.ema.europa.eu/en/documents/report/final-guideline-active-substance-master-file-procedure-revision-4_en.pdf, accessed March 2019.
19. Council of Europe, Knowledge-Database, https://www.edqm.eu/en/knowledge-database, accessed March 2019.
20. Council of Europe, https://www.edqm.eu/, accessed March 2019.

21. Technical guide for the elaboration of monographs on synthetic peptides and recombinant DNA proteins, European Pharmacopoeia, 2nd Revision Edition 2018, European Directorate for the Quality of Medicines & HealthCare (EDQM).
22. The Common Technical Document for the Registration of Pharmaceuticals for Human Use: Quality – M4Q(R1) – Quality Overall Summary of Module 2, Module 3: Quality, https://www.ich.org/products/ctd/ctdsingle/article/m4qr1-quality.html, Accessed March 2019.
23. ICH Q7: Good Manufacturing Practice Guide for Active Pharmaceutical Ingredients, https://www.ich.org/products/guidelines/quality/quality-single/article/good-manufacturing-practice-guide-for-active-pharmaceutical-ingredients.html, Accessed March 2019.
24. ICH Q11: 'Questions & Answers: Selection and Justification of Starting Materials for the Manufacture of Drug Substances', https://www.ich.org/products/guidelines/quality/quality-single/article/questions-answers-selection-and-justification-of-starting-materials-for-the-manufacture-of-drug-s.html, Accessed March 2019.
25. ICH Q9: Quality Risk Management, https://www.ich.org/products/guidelines/quality/quality-single/article/quality-risk-management.html, Accessed March 2019.
26. ICH Q10: Pharmaceutical Quality System, https://www.ich.org/products/guidelines/quality/quality-single/article/pharmaceutical-quality-system.html, Accessed March 2019.
27. M. D'Hondt, N. Bracke, L. Taevernier, B. Gevaert, F. Verbeke, E. Wynendaele and B. De Spiegeleer, *J. Pharm. Biomed. Anal.*, 2014, **101**, 2–30.
28. L. C. Wu, F. Chen, S. L. Lee, A. Raw and L. X. Yu, *Int. J. Pharm.*, 2017, **518**, 320–334.
29. ICH Q3D: Guideline for Elemental Impurities, https://www.ich.org/products/guidelines/quality/quality-single/article/guideline-for-elemental-impurities.html, Accessed March 2019.
30. ICH M7(R1): Assessment and control of DNA reactive (mutagenic) impurities in pharmaceuticals to limit potential carcinogenic risk, https://www.ema.europa.eu/en/ich-m7-assessment-control-dna-reactive-mutagenic-impurities-pharmaceuticals-limit-potential, Accessed March 2019.
31. ICH S6. ICH Guideline S6(R1): Preclinical safety evaluation of biotechnology-derived pharmaceuticals, https://www.ich.org/products/guidelines/safety/safety-single/article/preclinical-safety-evaluation-of-biotechnology-derived-pharmaceuticals.html, Accessed March 2019.
32. V. Thybaud, P. Kasper, Z. Sobol, A. Elhajouji, M. Fellows, M. Guerard, A. M. Lynch, A. Sutter and J. Y. Tanir, *Mutagenesis*, 2016, **31**, 375–384.
33. ICH S2: Guidance on Genotoxicity Testing and Data Interpretation for Pharmaceuticals Intended for Human Use, https://www.ich.org/products/guidelines/safety/safety-single/article/guidance-on-genotoxicity-testing-and-data-interpretation-for-pharmaceuticals-intended-for-human-use.html, Accessed March 2019.

34. S. G. Sawant, M. R. Fielden and K. A. Black, *Regul. Toxicol, Pharm.*, 2014, **70**, 87–97.
35. V. Vergote, C. Burvenich, C. Van de Wiele and B. De Spiegeleer, *J. Pept. Sci.*, 2009, **15**, 697–710.
36. ICH Q2: Validation of Analytical Procedures: Text and Methodology, http://www.ich.org/products/guidelines/quality/quality-single/article/analytical-procedure-development-and-revision-of-q2r1-analytical-validation-copy-1.html, Accessed March 2019.
37. Commission Regulation (EU) No. 10/2011 on plastic materials and articles intended to come into contact with food, http://ec.europa.eu/food/safety/chemical_safety/food_contact_materials/legislation/index_en.htm, Accessed March 2019.
38. Guideline on Plastic Primary Packaging Materials, CPMP/QWP/4359/03, https://www.ema.europa.eu/en/plastic-primary-packaging-materials, Accessed March 2019.
39. ICH Q1A: Stability Testing of New Drug Substances and Drug Products CPMP/ICH/2736/99-ICH (ICH Q1A), https://www.ema.europa.eu/en/ich-q1a-r2-stability-testing-new-drug-substances-drug-products, Accessed March 2019.
40. Guideline on the sterilisation of the medicinal product, active substance, excipient and primary container (EMA/CHMP/CVMP/QWP/BWP/850374/2015), https://www.ema.europa.eu/en/sterilisation-medicinal-product-active-substance-excipient-primary-container, Accessed March 2019.
41. Guideline on Process validation for finished products – information and data to be provided in regulatory submissions (EMA/CHMP/CVMP/QWP/BWP/70278/2012 Rev. 1), https://www.ema.europa.eu/en/process-validation-finished-products-information-data-be-provided-regulatory-submissions, Accessed March 2019.
42. WHO Technical Report Series, Guidelines for the production and quality control of synthetic peptide vaccines; Adopted 1997, TRS No 889, Annex 1, https://www.who.int/biologicals/publications/trs/areas/vaccines/peptide/en/, Accessed March 2019.
43. H. Yang and D. S. Kim, *Adv. Protein Chem. Struct. Biol.*, 2015, **99**, 1–14.
44. Biosimilars in the EU - Information guide for healthcare professionals, European Medicines Agency, 2017.
45. CMDh, Co-ordination Group for Mutual Recognition and Decentralised Procedures – Human, Minutes from the meeting on 22-24 June 2015, available at http://www.hma.eu/fileadmin/dateien/Human_Medicines/CMD_h_/Agendas_and_Minutes/Minutes/2015_06_CMDh_Minutes.pdf, Accessed March 2019.
46. EuropaBio White Paper – Chemically synthesized proteins referencing biological medicinal products, (2018), https://www.europabio.org/healthcare-biotech/publications/chemically-synthesized-proteins-referencing-biological-medicinal, Accessed March 2019.

47. FDA Draft Guidance ANDAs for Certain Highly Purified Synthetic Peptide Drug Products That Refer to Listed Drugs of rDNA Origin addressing synthetic glucagon, liraglutide, nesiritide, teriparatide or teduglutide (October 2017), https://www.fda.gov/downloads/Drugs/GuidanceComplianceRegulatoryInformation/Guidances/UCM578365.pdf, Accessed March 2019.
48. Public Assessment Report (Decentralised Procedure) (2017) https://mri.cts-mrp.eu/Human/Downloads/DE_H_4292_001_PAR.pdf, Accessed March 2019.
49. ICH Q3C(R7): Impurities: Residual Solvents, https://www.ich.org/products/guidelines/quality/quality-single/article/impurities-guideline-for-residual-solvents.html, Accessed March 2019.
50. EU Guidelines to Good Manufacturing Practice: Medicinal Products for Human and Veterinary Use, Annex 13: Investigational Medicinal Products; Brussels, 03 February 2010 ENTR/F/2/AM/an D(2010) 3374, https://ec.europa.eu/health//sites/health/files/files/eudralex/vol-4/2009_06_annex13.pdf, Accessed March 2019.
51. Detailed Commission guidelines on good manufacturing practice for investigational medicinal products for human use, pursuant to the second subparagraph of Article 63(1) of Regulation (EU) No 536/2014; Brussels, 8.12.2017 C(2017) 8179 final, https://ec.europa.eu/health/sites/health/files/files/eudralex/vol-10/guideline_adopted_1_en_act_part1_v3.pdf , Accessed March 2019.
52. Guideline on the requirements for the chemical and pharmaceutical quality documentation concerning investigational medicinal products in clinical trials' (EMA/CHMP/QWP/545525/2017), https://www.ema.europa.eu/documents/scientific-guideline/guideline-requirements-chemical-pharmaceutical-quality-documentation-concerning-investigational_en-0.pdf, Accessed March 2019.
53. Guideline on the requirements for quality documentation concerning biological investigational medicinal products in clinical trials (EMA/CHMP/BWP/534898/2008) rev. 1 corrigendum, https://www.ema.europa.eu/en/requirements-quality-documentation-concerning-biological-investigational-medicinal-products-clinical, Accessed March 2019.
54. R. Pita, F. Ehmann and R. Thürmer in *Biomedical Applications of Functionalized Nanomaterials*, ed. B. Sarmento and J. das Neves, Elsevier, UK, 2018, Chapter 22, 653–679.
55. PRIME: Priority medicines, https://www.ema.europa.eu/en/human-regulatory/research-development/prime-priority-medicines, Accessed March 2019.

CHAPTER 3

Biological and Immunogenicity Evaluation Strategy for Therapeutic Peptides: Chemistry, Manufacturing and Controls Perspective

ARUNAN KALIYAPERUMAL

Lilly Biotechnology Center, Eli Lilly and Company, 10290 Campus Point Drive, San Diego, CA 92121, USA
Email: kaliyaperumal_arunan@lilly.com

3.1 Introduction

Therapeutic peptides have been used for nearly 65 years. Many of the early peptide-based therapeutics initially were purified from animal tissue origins. In recent years, the numbers have increased dramatically, bringing the number close to 70.[1] Additionally, more than 90 peptide-based cancer vaccines are under development.[2] The number of new peptide-based therapies on the market for a wide range of medical indications clearly demonstrates the power of the modality of this class of drugs.[3,4] The new and emerging field of peptide-based vaccines for oncology has opened up a new therapeutic area in which peptides play a critical role. Peptide-based vaccines are more cost-effective than autologous or individualized vaccines. However,

Drug Discovery Series No. 72
Peptide Therapeutics: Strategy and Tactics for Chemistry, Manufacturing and Controls
Edited by Ved Srivastava

Published by the Royal Society of Chemistry, www.rsc.org

they also have a potential drawback because they target only one epitope or a few epitopes of the tumor-associated antigens (TAAs). In general, induction of both antigen-specific cytotoxic lymphocytes (CTLs) and antigen-specific CD4+ helper T cells is necessary for a cancer vaccine to be efficacious. Some polypeptide vaccines (*e.g.* Stimuvax) potentially contain both CD4 and CD8 epitopes. Other approaches to enhance the immunogenicity of a self-antigen are to alter the peptide sequence of TAAs to introduce enhancer agonist epitopes. This increases peptide binding to the major histocompatibility complex (MHC) molecule or the T-cell receptor, resulting in higher levels of T-cell responses and higher avidity T cells.[5–10]

Biological and immunogenicity evaluation is an essential part of peptide-based therapeutic development. A critical part of the evaluation of a peptide-based therapeutic drug indicated for treatment is the accurate determination of its potency. This potency determination is crucial as it pertains to the safety and efficacy of the drug. An optimized cell-based assay is much simpler and more rapid and precise compared with the traditional *in vivo* and/or *ex vivo* assays using live animals or animal tissues.

In the early phase of drug discovery and development, molecules are identified and optimized for their target biological activity, whereas in early process development the evaluation is used to guide the optimization of processes and conditions that result in the best target drug-product profile. Prior to the clinical development stage, potency assays to establish the mechanism of action (MOA) are performed for product release and stability determination.

The structure–activity relationship (SAR) can be measured using various analytical methods such as chromatography (ion-exchange and size-exclusion chromatography) to separate various product species and characterize them using tandem mass spectrometric (MS/MS) peptide mapping and to assess them in various biological assays. This helps in establishing the process and also the best quality attributes for the drug. Additionally, these assays are very useful in other stages of development to assess the impact of process changes and also manufacturing at various geographical locations. During clinical development, robust biological assays are designed and qualified or preferably validated for potency assessment.[11–15]

Simple assay formats are used to assess binding activity, including enzyme-linked immunosorbent assay (ELISA), surface plasmon resonance (SPR) and cell-based fluorescence-activated cell sorting (FACS) analysis for molecules that bind to surface receptors. Functional assays are employed to examine downstream events such as receptor downregulation and receptor phosphorylation or dephosphorylation. For peptides, mechanism-specific assays should be established at the early development stage. For each potency assay, a dose–response curve is optimized so that the concentrations are distributed properly over the complete dose–response curve and adequate coverage along the linear portion of the curve is obtained. Assay parameters are varied appropriately, using design of experiment (DOE) approaches[16–18] to understand the parameters of the interactions and also to pinpoint critical assay parameters, and are optimized for minimal assay variations.[19,20] Appropriate curve-fitting models and statistical tools need to be employed for (*e.g.* goodness of fit) data analysis.

All biologics have the potential to induce an immune response. Immune response or immunogenicity of the therapeutic proteins and peptides remains one of the major risks. Evaluating the immune response to these biological products in terms of the anti-drug antibodies (ADAs) is an important part of the product development and postmarketing monitoring of drug safety.[21] The major causes of immunogenicity to the biological products are still not fully understood, but factors such as aggregation are suspected to trigger an immune response.[22–24] The key challenges are that proteins and peptides can degrade by both chemical and physical degradation pathways. This physical degradation results in the generation of biologics that form aggregates. The ability of proteins to aggregate has been recognized since the early beginnings of protein biochemistry,[25] but that of peptides to a lesser extent. Physical degradation of the biologics can occur during the production and purification stages, but can also result from improper formulation development and storage or handling conditions.[26] Aggregates have been shown to increase the immunogenicity of various therapeutic proteins, which might be explained by their multiple-epitope character and/or by conformational changes of the individual aggregated protein molecules.[27] Insulin aggregates have been shown to induce antibodies against insulin.

There is evidence that the aggregates possess the B-cell antigenic epitopes.[28] The presence of protein aggregates in formulations of human growth hormone correlates with an increased frequency of immune responses.[29] The ADAs can alter the pharmacokinetics, pharmacodynamics, bioavailability and efficacy of the drug.[30] The antibodies against the drug might trigger a range of safety issues from mild to adverse side effects. In the most severe form, it can induce autoimmune disease-like conditions, such as thrombocytopenia[31] and pure red blood cell aplasia,[32] by neutralizing the endogenously secreted proteins. Hence it is essential to monitor the immunogenicity of the therapeutic biologics and characterize the antibodies in terms of binding and neutralizing ability using appropriately validated assays.[33–35]

A risk-based approach strategy has been proposed for antibody testing.[36,37] This approach is based on the level of harm or adversity that an antibody to the therapeutic protein will elicit in the patient. When the risk is greater, the assays must be able to detect the formation of an antibody to the therapeutic protein and/or its native endogenous protein. An endogenous enzyme or a growth factor replacement therapy will be at high risk compared with a monoclonal antibody therapy. The anti-therapeutic protein antibody not only will affect the efficacy of the therapeutic but might also hinder the endogenous protein. The selection of the assay format directly influences the ADA testing and its outcome.

Various methods used have their own advantages and disadvantages. Each of these methods has limitations that might hinder their use in a particular situation. Various formats of the assays that are commonly used for the assessment of the biological activity of the peptide and immunogenicity for both binding and neutralizing antibody are discussed in this chapter.

3.2 Biological Evaluation

3.2.1 Strategy for Assay Development

Potency assays are qualified in the early development stages, optimized in preparation for late-stage clinical testing and validated for their specificity, linearity, precision, robustness and stability (Figure 3.1). In the late stages of drug development, potency testing is integral in determining a drug's specific ability or capacity to elicit a given result. Cell-based potency assays, the central tool used to measure drug efficacy during potency testing, allows the evaluation of how a particular dose and/or lot of a therapeutic will respond in patients. These assays are crucial because they enable us to quantify the biological activity of a peptide, assuring consistency of the identity, purity, strength and stability of the product through all phases of clinical study. Owing to the valuable insight that these assays provide, they are required by regulatory agencies for the release of a drug product under Good Manufacturing Practices (GMPs). Potency assays are one of the most complex and challenging of all bioanalytical methods.[38] To add to this complexity, potency assays often require critical biological reagents to measure the functional effect of the peptides, bringing in additional challenges to securing a source for suitable and consistent reagents to use across all tests. The increased level of regulatory oversight is due to the critical nature of potency testing assays. Any error in estimation of the potency can easily affect the release of the drug for clinical use.

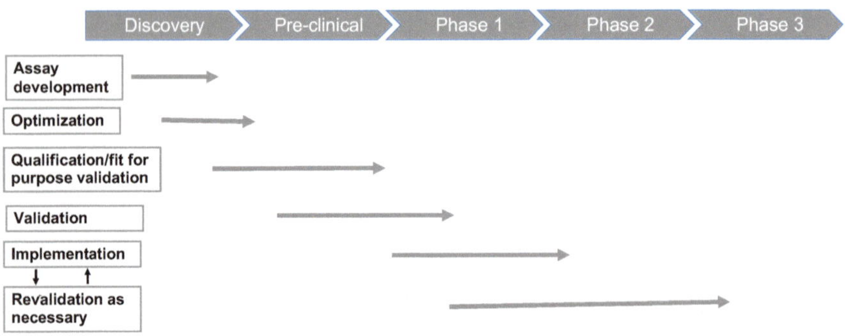

Figure 3.1 Assay development and validation strategy.

3.2.2 Considerations for Potency Assay Design and Validation

It is important to evaluate the approach to assay design, variability and validation. The following should be considered before beginning the potency testing process. The assay design should provide data that will allow the determination of the suitability of the drug for its intended use. It is essential that the assay design must incorporate sufficient replicates for thorough statistical analysis, use sample randomization to reduce bias and include appropriate controls to ensure that the assay is properly executed. Since numerous factors can influence assay variability, rational assay design is an important tool for mitigating variability. The design should essentially reflect the knowledge from each of the factors that could adversely influence reproducibility. The validation feature varies based on the purpose of the assay and where and how the assays are implemented (Figure 3.2, Table 3.1). The use of well-qualified reagents and equipment and well-trained operators helps to moderate variability. In addition, it is imperative that standardized operating procedures and controls are established. Assay validation should be performed on any relevant assay parameters, including, accuracy, precision, specificity, linearity and range, system suitability and robustness (Figure 3.3).

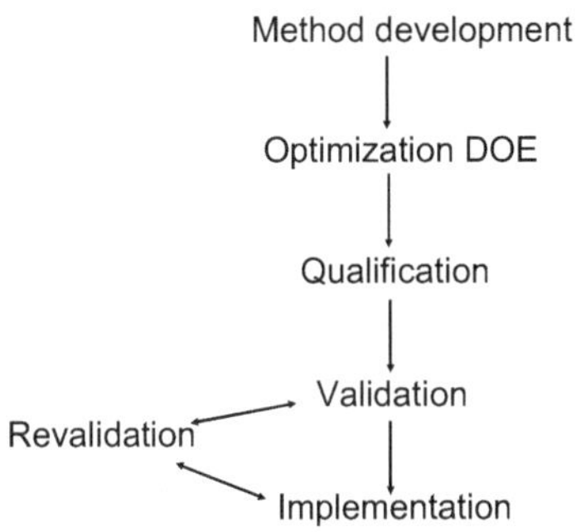

Figure 3.2 Analytical method validation strategy.

Table 3.1 Validation Features of the assays.

Features	Potency	Impurities	
		Quantitative	Limits
Accuracy	Yes	Yes	No
Precision – repeatability	Yes	Yes	No
Precision – intermediate precision	Yes	Yes	No
Specificity	Yes	Yes	Yes
Detection limit	Yes	No	Yes
Limit of quantitation (upper and lower)	Yes	Yes	No
Linearity	Yes	Yes	No
Range	Yes	Yes	No
Robustness	Yes	Yes	No

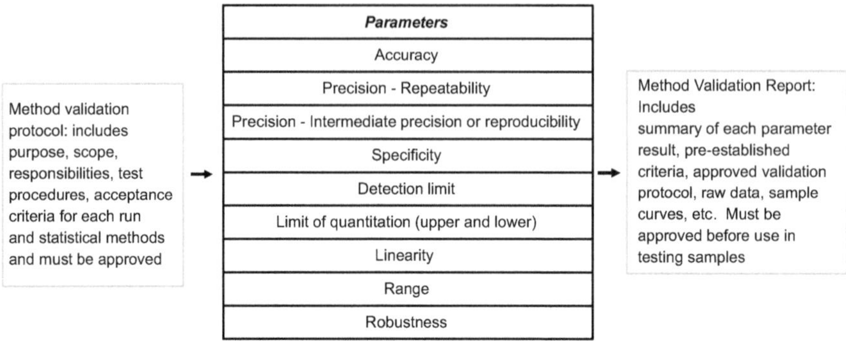

Figure 3.3 Method validation process and parameters.

Calibration curve quality is of critical importance in ligand-binding assays and other quantitative methods, with the accuracy of reported sample results dependent upon the robustness and reproducibility of the assay calibration curve. Although many of the requirements for calibration curves have already been defined in regulatory guidance and lead publications, there are still aspects that have yet to be elucidated or adequately addressed, based on the nature of the peptide and its mechanism of action.[15,39]

3.3 Assay Development and Validation

3.3.1 *Ex Vivo* Bioassays

Tissues or cells from animals or humans can be used to assess the biological activity of the molecules/drug. Peptides such as growth factors can be tested on cells extracted from specific tissues on which they are biologically active and hematopoietic or peripheral human blood cells can be used for a peptide that mimics cytokines. These assays have donor-to-donor variations due to the nature of the tissue or the cells in addition to the extraction process that will also contribute to these variations. Factors such as the reagents (*e.g.* buffers, enzymes and anticoagulants) used in cell extractions process will influence the assay outcome, hence these processes will need controlled collection procedures and qualification of cells before being used in the final assays.[12]

3.3.2 *In Vitro* Bioassays

In vitro assays are primarily carried out using cell lines that respond to the molecules and can be used for the lot release of the therapeutic peptides. These cell lines are mainly derived from tumor cells or immortalized primary cells. They can be dependent on growth factors for their maintenance. A variety of responses as listed in the following can be assessed for the peptides.

3.3.3 Assay Readouts

Assay readouts are very similar in both the biological evaluation and neutralizing antibody testing for cell-based assays. A wide range of readouts is possible and a few that are fairly frequently used are as follows:[40–46]

- cellular responses, *e.g.* proliferation, growth arrest, apoptosis, cytokine release, growth factor-induced angiogenesis;
- signal transduction events of receptor phosphorylation or its downstream signaling components, receptor downregulation, *etc.*;
- gene transcription reporter assays where the transcription of a specific gene and/or associated evaluated using reporters such as green fluoresce protein, enzymes; and
- ligand–receptor-binding assays such as cell-based assays using flow cytometry.

3.3.4 Assay Design

Cell-based bioassay design should reflect the factors that influence the response of cells to the peptide. The response variability is often reflected in parameters such as slope, EC_{50} of the dose–response curve or the ratio of maximum to minimum response. These parameters of the assay, on which potency estimates are established, could result in significant effects on system suitability, precision and accuracy and should be identified and control strategies should be established to minimize their impact on potency estimates.

3.3.5 Cell Line Selection

The development of a cell-based bioassay begins with the selection and/or generation of a cell line. The cell line is a most critical reagent in a cell-based assay. To ensure an adequate and consistent supply of cells for product testing, the cell line(s) used in the assay should be banked. The source of the cell line must have documentation that details the cell line's history from origin to banking and supports its application for commercial use. It is essential that information pertaining to the cell line should be documented during assay development and before banking.[12–14]

3.3.6 Factors Influencing the Assays

Many factors that may affect bioassay performance are common to many cell-based bioassays. To identify these factors, analysts with in-depth knowledge and experience with the cell-based assays should be involved during assay development. Relevant factors include cell type (adherent or nonadherent), cell thawing, plating density and confluence, culture vessels, growth, staging and assay media along with any media requirements.

Additional factors such as incubation conditions (*e.g.* temperature, CO_2, humidity and culture times from thaw), cell harvesting reagents and techniques (*e.g.* for adherent cells, the method of dissociation), cell sorting and cell counting should be monitored and recorded.

3.3.7 Standards and Controls

Appropriate standards and controls should be included in each run of the bioassay to ensure that the results are accurate. Specifically, a reference standard should be included to assess run-to-run variations and to anchor the relative potency results; parallel line analysis is used to ensure appropriate comparison of the sample with the reference standard. Positive and negative control samples along with quality control samples are generated and qualified to provide a specific range of results and system suitability criteria are established. The performance of the assay must be closely monitored over time, *i.e.* control charted, to guarantee that the assay continues to meet the system suitability criteria along with trending.

3.3.8 Design of Experiments and Data Analysis

A systematic approach for bioassay validation using the design of experiments (DoE) that incorporates robustness of critical parameters is provided by the United States Pharmacopeia (USP),[12] in addition to the guidance documents on the development, analysis and validation of biological assays.[12–14] Establishing robustness using DoE approaches has also been reported.[19,20] A DoE-based approach to perform qualification of a bioassay to ensure accuracy, precision, linearity and robustness requires experimental and statistical effort.[47–49] An approach incorporating critical assay parameters provides a more realistic representation of a routine assay during testing. To enable an investigational new drug (IND) application, a DoE approach after initial development of a cell-based bioassay was performed to qualify the bioassay comprehensively for accuracy, precision, linearity and robustness using a number of normal operating ranges specified in the assay development.[18]

3.3.9 Assay Validation

Validation of a method is the confirmation by testing various parameters and establishment of unbiased evidence that specific requirements for the intended use have been achieved. This is imperative as it defines whether it will yield reliable results. The intended use of a method must be specified before any time-consuming and expensive validation experiments are performed.

3.3.9.1 Accuracy

Accuracy is a measure of how close a measured value is to the actual value for the analyte. Known concentrations are measured multiple times and the analytical assay may report different values (*e.g.* concentrations). Validated assays are mostly expected to have an accuracy of $\leq\pm 15\%$ at all concentrations in the assay range. The accuracy can be determined for each sample, then the mean accuracy can be reported. Accuracy is calculated as a percentage from the difference in measured concentrations and actual concentrations.

3.3.9.2 Robustness

Robustness or ruggedness is the ability of a method to remain unaffected by small variations in parameters such as column, pH, temperature, ionic strength, CO_2, plate lots/source in the method. The robustness should be investigated as part of the method development and the results should be reflected in the method validation protocol before other validation parameters are investigated. The reason for this is that a validation is linked to the method validation protocol and changes in the latter might demand a new validation to be performed.

3.3.9.3 Precision

Precision is defined as the closeness of agreement between independent test results obtained under specified conditions. There are three different types of precision depending on the stipulated conditions, namely repeatability, intermediate precision and reproducibility. Repeatability is the variability observed when many of the factors present in the conduct of the experiment, *e.g.* laboratory, technician, days, instrument, reagent lot, are held constant and the time between the measurements is kept to a minimum. To evaluate the reproducibility conditions, all factors are varied and measurements are carried out over several days. In the case of intermediate precision, all factors except laboratory are varied, and for clarity the factors changed are indicated in the validation report. Repeatability is sometimes called within-run or within-day precision and intermediate precision is also known as between-run or between-day repeatability.

3.3.9.4 Limits of Detection and Quantification

The limit of detection (LOD) is the lowest amount of analyte that can be detected, but not necessarily quantitated as an exact value. The LOD is a low concentration that is statistically distinguishable from the background or negative control but is not sufficiently precise or accurate to be quantitated, *e.g.* the LOD can be calculated based on the standard deviation (SD) of the response (S_y) of the curve and the slope of the calibration curve (S) at levels approximating the LOD according to the equation $\mathrm{LOD} = 3.3(S_y/S)$.

The working range for a method is defined by the lower limit of quantification (LLOQ) and upper limit of quantification (ULOQ). For the LLOQ, there is more than one definition and these are classified as either determined based on the signals from the instrument or calculated concentrations from samples. In the case of an instrument-based signal, a number of blank samples are analyzed and the average and SD of the signal are calculated. The ULOQ is the highest concentration on the calibration curve that can be determined in a given analytical assay with the required precision and accuracy. The highest concentration is selected arbitrarily by the bioanalytical scientist. If concentrations are observed above the ULOQ, the samples can be diluted and tested within the calibration range.

3.3.9.5 *Linearity and Range*

In most bioassay analyses, the concentration–response curve is not a straight line, but may approximate a straight line over a limited range of concentrations. A linear-response model may be assessed to determine if it is justified for the data in hand. Difference testing methods for assessing linearity face the same problems as do standard methods for difference testing of parallelism. A subset of the concentrations measured in the assay will be selected in order to derive a linear concentration–response curve. The subset can often be identified graphically by using only concentrations in the linear range; this will result in parallel straight lines for the range of relative potencies expected during routine use of the assay. The assay fails parallelism tests when the potency measurements produce assay response values outside the linear range and in those cases repeat assays will be required. The repeat assays together with the valid assays may generate a biased estimate of potency because of the selective process of repeating assays when the response is at the extremes of the concentration–response curve and should be avoided. The range is defined as the concentrations of analyte or assay values between the low and high limits of quantitation where linearity, accuracy and precision are acceptable. The range of an analytical procedure is the interval between the upper and lower concentrations of analyte in the sample for which it has been demonstrated that the analytical procedure has a suitable level of precision, accuracy and linearity. For comparing the reference standards and the test sample, a parallelism or nonsimilarity analysis can be used. Theoretically, parallelism and dilution linearity are similar. The major difference is that in dilution linearity experiments the samples are spiked with the analyte at such a high concentration that after dilution the effect of the sample matrix is likely to be negligible. For parallelism, on the other hand, no spiking is allowed, and only samples with high endogenous concentrations of the analyte must be used. However, the concentrations must be lower than the ULOQ. The goal of investigating parallelism is to ascertain that the binding characteristic of the analyte is the same as that of the reference standard.

3.3.9.6 *Specificity*

Specificity is the ability to assess unequivocally the target analyte in the presence of components that may be expected to be present. Typically, these might include impurities, degradants, formulation, *etc.* This definition has the following implications: selectivity can be defined as "the ability of the bioanalytical method to measure and differentiate the analytes in the presence of components that may be expected to be present." The terms "selectivity" and "specificity" are often used interchangeably but their significances are different. Selectivity is something that can be graded whereas specificity is an absolute characteristic. Specificity can be considered as the ultimate selectivity. Different validation parameters of the specificity have to be defined prior to the validation, such as that the specificity includes any cross-reactivity to a shorter form of the peptide. For example, if the analyte is a peptide of a specific length, do slightly longer or shorter peptides also give rise to a signal in the assay? Investigations to address the degree of interference by spiking samples with substances identified to interfere have to be conducted.

3.4 Specific Assays of Peptide-based Vaccines

Most peptide-based vaccines in clinical trials target cancer antigens, such as differentiation-associated antigens or certain oncofetal antigens (CEA, MUC-1). These vaccines were able to induce antigen-specific T-cell responses.[50] The results observed were encouraging in a randomized Phase III trial involving patients with stage IV or locally advanced stage III cutaneous melanoma.[51] The peptide in Montanide ISA-51 adjuvant plus IL-2 demonstrated a statistically significant improvement in overall clinical response and resulted in longer progression-free survival. This was one of the first Phase III trials to demonstrate a clinical benefit for a peptide vaccine in melanoma.

Anti-tumor immunity in humans has been associated with the presence of T cells directed at cancer neoantigens, a class of human leukocyte antigen (HLA)-bound peptides that arise from tumor-specific mutations.[52] They are highly immunogenic because they are not present in normal tissues and hence bypass central thymic tolerance. A combination of 20 neoantigens has been demonstrated in a clinical study.[53]

Systematic discovery and evaluation are feasible with the recent availability of massively parallel gene sequencing for the detection of all coding mutations within tumors and of machine learning approaches to predict reliably those mutated peptides with high-affinity binding of autologous human HLA molecules.

The neoantigens are usually identified using whole-exome sequencing (WES) of matched tumor- and normal-cell DNA from individual patients. Identified somatic mutations are confirmed by examining the expression of mutated alleles by RNA sequencing of the tumor. The predicted mutated peptides were selected based on the likelihood of binding autologous HLA-A or HLA-B proteins of the patient-targeting neoantigen.[54–56]

3.4.1 HLA Binding Assays

3.4.1.1 HLA Class I: Cell-based HLA Class I Peptide-binding Assay

The peptides that are identified and synthesized for vaccines need to be qualified/validated for binding to appropriate HLA to be effective in triggering the immune response in humans. Many available humanized animal models (*e.g.* mouse carrying human immune system components or hematopoietic cells) can be used; however, for an initial screen and validation a binding assay is more appropriate. The synthetic peptides are assessed for their ability to bind competitively the HLA class I generally in B-cell lines that express them. The cells undergo acid treatment to remove the naturally HLA-bound peptides from the cells. The acid treatment is a crucial and critical step. The pH of the buffer and the temperature, and also the duration of the treatment, are critical. After washing, cell aliquots are added to the incubation plate, which is subsequently incubated for 24 h at 4 °C in the dark to allow the assembly of the new HLA–peptide complexes. Test peptides are dissolved and diluted in a titration plate. The concentrations of the test peptides to be used differ based on the nature of the peptide. Lower concentrations can easily be tested, but sometimes the use of higher concentrations might cause unexpected problems due to solubility and therefore should be avoided. Following preparation of the titration plate, an incubation plate containing the fluorescent reference peptide and controls is prepared, then samples from the titration plate are pipetted into the corresponding wells of the incubation plate. Cells are added to these plates and after incubation the cells are washed and analyzed by flow cytometry. The percentage fluorescence intensity inhibition compared with cells with only reference peptide added are used to determine the IC_{50} of the added peptide.[57–59]

3.4.1.2 HLA Class II: HLA Class II Peptide-binding Assay

Relative binding affinities are determined by a competitive binding assay.[60–62] Briefly, 10 nM soluble HLA-DR protein is incubated with 10 nM biotinylated benchmark peptide and 30 nM HLA-DM (as required) in the presence of various amounts of unlabeled competitor peptides for 2–4 days (depending on the affinity of the benchmark peptide) at 37 °C. Bound biotin–peptide is detected using a solid-phase immunoassay and Eu^{2+}-labeled streptavidin. IC_{50} values are obtained from the curve fit of the binding data and converted to K_D values using the Cheng–Prusoff equation, in which the K_D of the biopeptide is defined on the basis of a saturation binding study. This approach is not the most precise for assessing peptide affinity because of its indirect nature, but it is more efficient to perform than direct binding assays and less demanding in terms of reagents than SPR or isothermal titration calorimetry (ITC) and it provides a measure of affinity that is sufficient for a rapid screening.[63]

3.5 Immunogenicity Evaluation

3.5.1 Antibody Testing Strategy

ADA testing involves a two-tier approach.[21,64] Initially, the sample is tested for the presence of the ADAs using a binding assay followed by the ability of those antibodies specifically to neutralize the drug. Additional, but optional, characterization of the immune response in terms of isotype of the antibody in addition to affinities has also been reported.[37] A representative antibody testing strategy is given in Figure 3.4. Samples for antibody testing are initially analyzed using a screening immunoassay. Samples that yield a signal below the immunoassay's cut point for antibody detection are negative. Samples that give a signal equivalent to or higher than the immunoassay's cut point are tested in a confirmatory immunoassay. The samples that are positive are further tested in a cell-based or competitive ligand-binding assay for presence of neutralizing antibodies. Under certain circumstances, other characterization immunoassay testing may also be performed on samples that test positive in the confirmatory immunoassay. These tests may include determination of the isotype, IgG subclass, dissociation rates, affinity and relative concentration of the antibodies.[83,84]

3.5.2 Risk-based Antibody Testing

Antibody testing should consider the severity of the consequences of an antibody response to the therapeutic protein. Factors such as biological function or cross-reactivity to endogenous proteins and health status should be taken into consideration while making decisions on the assay format. The chosen format should detect the clinically relevant antibodies. An example of high risk is an endogenous version of the protein (*e.g.* erythropoietin) used as replacement therapy and used repetitively in life-threatening situations. A nonendogenous, redundant or nonreplacement therapy to be used as single-dose treatment will fall into a lower risk category.

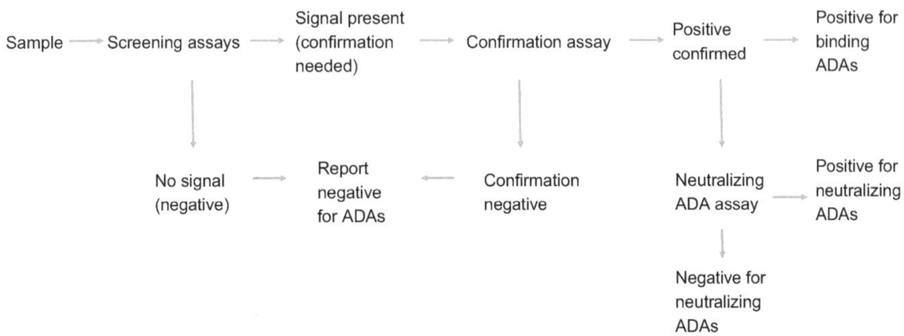

Figure 3.4 Immunogenicity testing strategy.

3.5.3 Binding Antibody Testing

Binding antibody testing commonly uses three different platforms. Binding immunoassays such as direct binding, sandwich and bridging formats are extensively used.[65–68] The second platform is based on mass-related immunoassay using MS or SPR.[35,64] The mass-based technique has been reported to detect more of the lower affinity antibodies to the drug. The third and more recent technology is bead-based cytometric detection of antibodies in serum using flow cytometry.[65,69] In all the assay platforms, it is essential to choose the format that detects the ADA with appropriate sensitivity. Table 3.2 describe the various forms of assays used for the detection of binding ADAs. The details of the assays are described below.

3.5.4 Sandwich Immunoassay

In a sandwich immunoassay, the drug is coated or captured onto a plate directly, by passive adsorption or using a biotin/avidin-tagged drug (Table 3.2, first row), followed by addition of test sample serum that may contain the ADA. The bound antibody is revealed by using a secondary antibody labeled with either an enzyme or a chemiluminescence-responsive tag (*e.g.* ruthenium). A variation of the sandwich immunoassay in which the drug is coated in limited quantity followed by test serum addition is the

Table 3.2 Schematic representation of assay formats.

No.	Procedure	Format[a]
1	Simple immunoassay (*e.g.* sandwich ELISA)	
2	Bridging immunoassays (*e.g.* ELISA, chemiluminescence)	
3	Mass related (SPR)	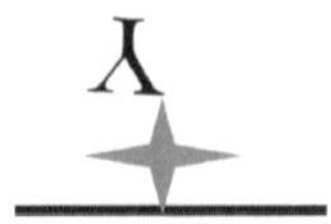
4	Cytometry based	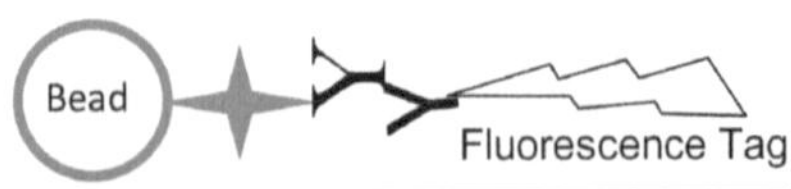

[a]★, Detection agent (enzyme or chemiluminescence tag); ⅄, anti-drug antibody; Y, secondary detection antibody (labeled); ✦, drug.

bridge format (Table 3.2, second row). In this format, a plate is coated with the drug and exposed to the test serum. The difference from other sandwich assays is that the labeled drug is used to form a bridge. Hence this requires that one of the sites in the antibody be available for binding to the labeled drug. This assay format is most versatile and has been widely used for the detection of ADAs. The ease of use and the reliable cost-effective readout technology give optimal sensitivity and a quick turnaround time. The limitation of this technique is that the repeated washing steps necessary after the addition of each of the regents reduce the detectability of relevant low-affinity ADAs. Also, in certain disease populations, serum results in a higher background owing to nonspecific binding.

3.5.5 Mass-related Immunoassay

In a mass-related immunoassay, the binding of antibody to the drug remains similar to that in the other immunoassays but detection is based on the change in mass due to the formation of an immune complex and is performed using either SPR or MS.

3.5.6 SPR

The drug is either coated onto the surface directly or captured using avidin/biotin chemistry (Table 3.2, third row). The immune complex formed by interaction with the serum ADA is revealed by changes in the light characteristics. This technique can be used without any modification or labeling of the drug. It has been shown to detect low-affinity relevant antibodies.[70] The widely used SPR technique has a higher turnaround time and usually has lower sensitivity; however, this is offset by its ability to detect lower affinity antibodies and to achieve more detailed characterization on the affinity and dissociation rates.[64]

3.5.7 Mass Spectrometry

Coupled liquid chromatography and mass spectrometry (LC-MS) has emerged as a useful technique for the detection and quantitation of ADAs with high specificity and selectivity and a wide dynamic range. This method captures ADAs and/or a drug–ADA complex using the biotinylated drug immobilized on streptavidin magnetic beads and separated from human plasma samples using a magnet. The ADAs are eluted from the beads and subjected to trypsin digestion followed by LC-MS detection of specific universal peptides for each ADA isotype. The LC-MS data are analyzed using a cut-point and calibration curve. This method has the advantages of measuring the ADAs in the presence of high drug levels. Additionally, LC-MS can be used in a multiplexing mode to measure the concentration of the drug in the sample along with semiquantitative measurement of ADAs.

3.5.8 Cytometry-based Immunoassays

This technique uses the same basic principles of ADA interactions with a drug to form immune complexes. Detection is based on fluorescently labeled reagents. Chemically modified flow cytometry beads are coated with drug or captured onto the beads using avidin/biotin chemistry followed by the secondary revealing with antibodies that are fluorescently labeled or in a bridging format with the fluorescently labeled drug.[65] Multiplexing of the assay can be easily achieved using this technique to detect cross-reactivity of the antibodies with other related molecules.[69,71] The limitations of this technique are similar to those of the sandwich and bridging immunoassays, *i.e.* the need for conjugation of the reagents and also repeated washing of the beads.

3.5.9 Acid Dissociation-based Assay for Drug Interference

All assays mentioned above measure the ADAs in serum from subjects who have been exposed to the drug. If the sample contains free drug from a prior exposure, it might interfere with the detection of ADAs. Various techniques to reduce the effect of free drugs have been used. The removal of the free drug increases the sensitivity of the assay. Partial dissociation of high-affinity ADAs might remove the drug–ADA complexes. The acid dissociation technique[72–74] has the ability to detect the ADAs by dissociating the immune complexes. Figure 3.5 shows a schematic view of the technique. The serum sample containing the ADAs and the drug–immune complexes are dissociated by subjecting them to lower pH (pH 3.0). The samples are then incubated for a specified amount of time, usually 15 min–2 h, then the pH is adjusted to neutral (pH 7.0) while adding a 10–100 molar excess of the labeled drug and the mixture is incubated on a plate coated with the drug. The bridged ADA is revealed using the bound labeled drug. This acid dissociation technique greatly improves the detection of the ADAs in the presence of as much as a 50 molar excess of the drug over the antibody without affecting the sensitivity of the assay. The limitations are that not all ADAs may be detectable by this technique; especially the very high-affinity antibodies may not be dissociated at the pH levels that are safe for treating the serum samples. Moreover, the assay involves washing steps that might affect the detection of low-affinity antibodies.

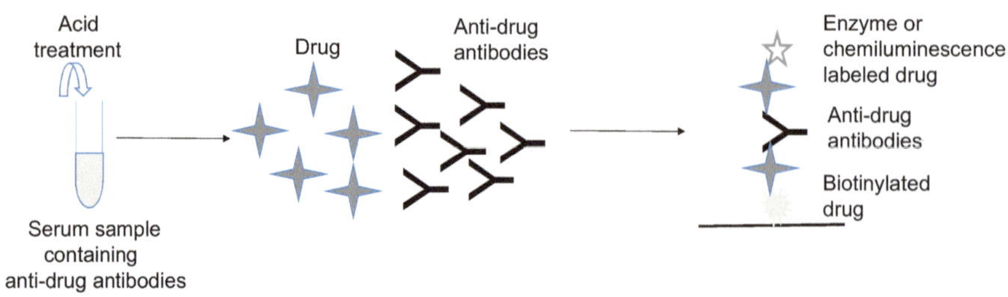

Figure 3.5 Acid dissociation bridging assay.

3.5.10 Considerations for Binding Antibody Immunoassays

All the assays that depend on coating onto plates or a solid surface have the potential to mask the epitopes. The conjugation techniques used in the sandwich and bridging formats can also mask or modify the epitopes owing to chemical conjugation at a specific site. Moreover, the isotype-specific secondary antibody reagents used to reveal the drug specific antibody in direct and sandwich-type assays bind to different classes to varying extents, resulting in a wide range of signals. To minimize the variation of this signal, a standardized mixture of these reagents can be prepared and used throughout the study. The bridging format relies on at least one site being available for the labeled drug to bind to and make the bridge between the plate-bound and the labeled drug. The bridging can be mediated through a nonantibody-mediated agent in this assay. This issue of lack of antibody specificity could be solved using protein G and L that binds and depletes the antibodies in the test serum.

Sensitivity and affinity play an important role in all of these assays. The assays used in clinical testing must detect the lowest possible levels of the antibody. The use of well-qualified reagents and combinations of the various formats provides a more complete analysis of ADAs. For example, a mass-related immunoassay such as SPR in combination with a bridging immunoassay will provide the ability to detect lower and higher affinity ADAs. The bridging assay format will provide better sensitivity for higher affinity antibodies whereas the lower affinity antibodies are best detected using SPR.

3.6 Neutralizing Anti-drug Antibody Testing

Neutralizing antibody assays are used once a sample has been found to contain ADAs. Two major formats, cell-based (Section 3.6.1) and non-cell-based receptor or ligand-binding/inhibition assays (Section 3.6.5) are used in the detection of neutralizing antibodies.

3.6.1 Cell-based Neutralizing Antibody Assay

A cell-based assay provides the most appropriate model for the detection of neutralizing antibodies. Detection is based on the ability of the ADAs to eliminate or reduce the biological activity of the drug.

3.6.2 Cell Line Selection

The most important aspect of the cell line selected to be used in the assay should be that it is representative and is close to the mechanism of action of the drug. Second, the cell line should also be tolerant to the presence of serum and work over a wide range of serum concentrations. If an appropriate cell line is not available, cell lines can be engineered with appropriate receptors to read the responses to the drug. The cell lines should be well characterized in terms of responses to drug, serum concentration and culture for a period of time before being used in the assay.

3.6.3 Assay Endpoints for Cell-based Assays

The various commonly used endpoints of cell-based assays is proliferation, cytokine secretion, phosphorylation and specific gene expression analysis.[75–80] The assay endpoints can be an early or late biological response to the drug or its inhibition of the ligand that interacts with the cell. Examples of early time-point responses are binding of the drug to the cell surface receptor, its internalization or early signaling events such as phosphorylation. Table 3.3 lists the most commonly used endpoints. The drug-induced cell growth can be monitored by cell proliferation assessment. This can be measured by incorporation of radioactively labeled thymidine into the DNA of the growing cells. Alternatively, the intracelullar level of cyclic adenosine monophosphate (cAMP) (Table 3.3, first row) can be used as a readout for the cell viability. Protein expression resulting from the action of the drug or ligand on the cells can be measured as readout, *e.g.* cytokine induction (Table 3.3, second row). When the expression of a protein or other signals are difficult to measure owing to very low expression, the messenger RNA expression can be measured using a polymerase chain reaction (PCR) or branched DNA technique (Table 3.3, third row). This technique greatly amplifies the signals and provides higher sensitivity. The response to the ligand or the drug to its receptor can be measured using signaling events, *e.g.* phosphorylation of proteins (Table 3.3, fourth row). These responses are relatively fast and can be measured much faster than other cellular responses, usually in less than 30 min. The relatively small window of exposure of the test serum to the cell reduces the deleterious effects of serum on the cells. When there is no specific receptor-expressing cell line available, the appropriate receptors can be engineered along with a reporter gene (*e.g.* luciferase) (Table 3.3, fifth row). As these cell lines are engineered, they can be made to satisfy very specific needs, *e.g.* tolerant of testing serum.

Table 3.3 Readouts commonly used in cell-based neutralizing antibody assays.

No.	Method used for the measurement	Potential mechanism of action	Readout endpoints
1	Thymidine uptake, cAMP measurement	Cell growth or death	Proliferation
2	ELISA, chemiluminescence	Induces or alters protein expression	Protein expression
3	Branched DNA technology	Affects mRNA expression	mRNA expression
4	ELISA, chemiluminescence	Phosphorylates or signal results in phosphorylation of proteins	Phosphorylation
5	Luciferase, β-galactosidase	Affects or alters reporter gene expression	Reporter gene expression

3.6.4 Cell-based Assay Formats

These assays fall into two categories, direct and indirect assays, which are described in detail in Sections 3.6.4.1 and 3.6.4.2, respectively. Schematic representations of these assays are given in Figures 3.6 and 3.7. The assay format depends on the mechanism of action of the drug. The interaction of the serum with the cell line also dictates the format of the assay.

3.6.4.1 Direct Assay

A direct assay such as a growth factor drug interacting with its receptor on the cell surface results in proliferation of the cells. The ADA will bind to the growth factor resulting in inhibition of cell proliferation (Figure 3.6).[75] Typically, the direct assay format is used for proteins or drugs that act directly on the system, such as hormones or cytokines. The cells are exposed to the drug and the effector function such as proliferation or cell death is read by measuring the [^{3}H]thymidine uptake or intracellular cAMP.[79] Owing to prolonged interactions of the drug, receptors on the cell line and the serum ADA, there is a possibility of the serum having effects on the assay in the direct assay format. To monitor these effects, appropriate "controls" in the assay, such as serum directly exposed to the cell line in the absence of any drug and cell line controls, should be included. Cells along with the drug and the positive control will provide a good system suitability test for the assay. Any unanticipated or harmful effects of the matrix on the cell can be monitored using these controls to provide a robust assay. Monitoring of the signal in response to the ligand will provide the viability of the cells and the extent of response over the background will provide the consistency of the response and the potency of the drug product used in the assay.

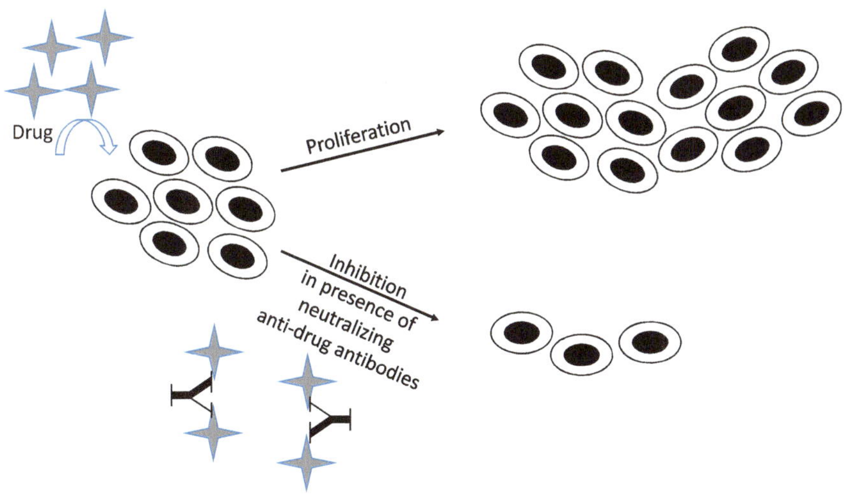

Figure 3.6 Direct assay.

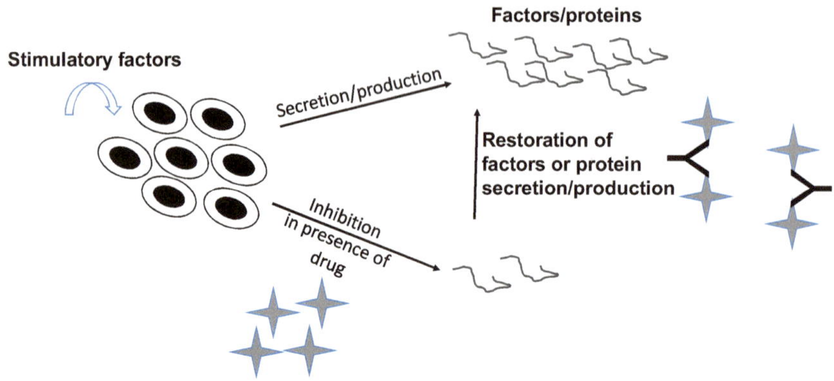

Figure 3.7 Indirect assay.

3.6.4.2 Indirect Assay

In an indirect assay, typically the drug is an antibody against the growth factor, ligand or its receptors. Products such as monoclonal antibodies, soluble receptor and receptor antagonists fall in this category. Addition of the drug inhibits the ligand-induced response, *e.g.* cytokine secretion; the ADA will bind and neutralize the drug, thereby restoring the ligand-induced response (Figure 3.7). Indirect assays are usually used with monoclonal antibody drugs and soluble receptors. The drug acts on the target of interest, which has the biological effector function. When the drug is added to the ligand, the effector function such as production of cytokine is blocked. The secreted cytokine is measured by an immunoassay. The neutralizing ADAs restore the function of the ligand by inactivating the drug. It is necessary to include appropriate controls to monitor the function and potency of these reagents or molecules in each assay.

3.6.5 Considerations for Ligand-binding Antibody Bioassays

A cell-based assay design should incorporate the following details. The cell line selected should be representative of the mechanism of action of the drug. The format can be decided based on the biological action of the drug on the selected cell line. Once the design has been finalized, the assay should be optimized for the correct concentration of the drug to be used in the assay, including culture conditions such as cell density and incubation time, which depend on the assay endpoints for each of the assays.

3.6.5.1 Receptor Binding Assay

The receptor binding assay format can be used in cases where the effector cell lines are not available with appropriate specificities and if there are issues with the serum, *e.g.* certain disease sera are not well tolerated by the

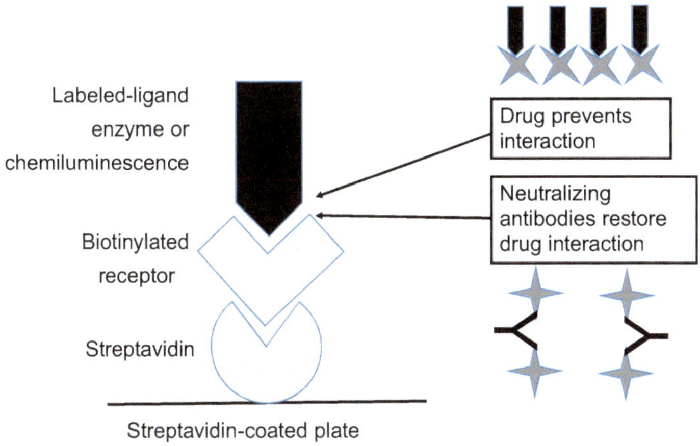

Figure 3.8 Ligand-binding assays.

cells. A schematic representation of the assay is given in Figure 3.8. In this example, the drug is an antibody against the ligand. The biotinylated receptor is bound to avidin-coated plates. The ruthenylated (RU)-ligand binds to a biotinylated receptor in the absence of drug. Ruthenium is detected by electrochemiluminescence. The drug blocks the binding of the (RU)-ligand to the biotin receptor. ADA neutralizes the drug and the (RU)-ligand binds to the biotinylated receptor. Alternatively, the receptor for the ligand can be coated onto a plate. The use of avidin/biotin capture chemistry gives the advantage of the assay being carried out in the solution phase. As these assays do not involve cells, the effect of serum is minimal and higher concentrations of serum can be used in the assay. The nonspecific binding can be lowered by additional washing of the plates and the signal can be enhanced by using various signal amplification methods.

The limitations include that the receptor for the drug needs to be well characterized and available in a purified form. When multiple isoforms of the receptors are present, the effects on each of them must be evaluated separately. Even though the receptor protein can bind the ligand, it may not have the same affinity for the cell surface receptor. Chemical modifications of the receptors for use in the assay might affect their biological activities and their stability must be evaluated in the assay and during storage. The assay-specific controls for the evaluation of these parameters must be included in each of the assays.

3.7 Specific Assays for Peptide-based Vaccines

3.7.1 Enzyme-linked Immunospot (ELISpot) Assay

The main application of the ELISpot assay is in monitoring of immune responses in both humans and animals. The cytokine ELISpot assay is

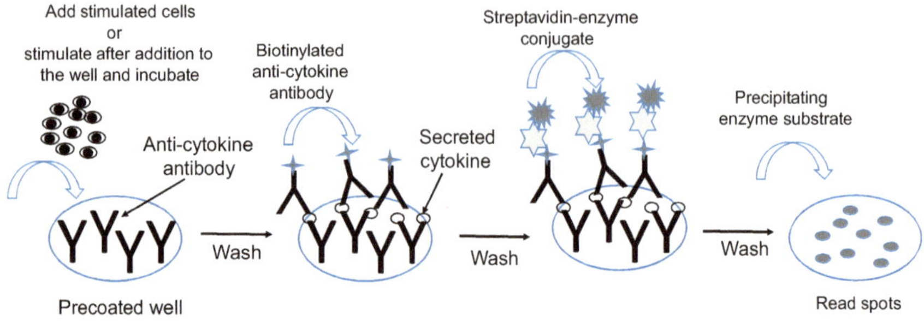

Figure 3.9 ELISpot assay.

based on the enzyme-linked immunosorbent assay (ELISA) method and is designed to estimate the number of cytokine-secreting cells. This is an extremely sensitive assay and therefore useful for detecting low frequencies (one cell in 200 000 total cells) of cytokine-secreting cells. The cytokine released in response to an antigen can be traced to a single cell, hence the responder cell frequencies can be estimated. Cells can be stimulated with the respective peptides in the presence of appropriate antigen-processing cells and in the anti-cytokine antibody-coated plate or stimulated in a separate plate and then transferred to the precoated plate (Figure 3.9); when the assay is complete, the plates can be analyzed on a plate reader.

More recently, the assay has been adapted utilizing fluorochrome-conjugated detection antibodies, thereby allowing the simultaneous detection of multiple distinct cytokines and subsequent T-cell subpopulation analysis. Typically, these are carried out in 96-well format membrane plates instead of plastic-bottomed plates. Briefly, the appropriate anti-cytokine antibody is coated onto the membrane followed by the addition of cells and stimulated using the antigens of interest or stimulated in a separate plate and added to the antibody-coated plates and incubated overnight at 37 °C in the presence of 5% CO_2 in an incubator. The plates are washed and the secondary anti-cytokine antibody conjugated to an enzyme or fluorescence probe are added and the plates are further incubated. The plates are subsequently washed and the secreted cytokines are revealed as spots using a precipitatable enzyme substrate or read directly in the ELISpot reader as images. The number of spots correspond to the number of cells that secreted that specific cytokine. The frequency of the cells can be calculated from the total number of cells added to the well and the number of spots observed for the appropriate cytokine on the membrane.

3.7.2 Surface and Intracellular Staining

Intracellular cytokine staining combined with flow cytometry is one of a number of assays designed to assess T-cell immune responses. It has the specific advantage of permitting the simultaneous assessment of multiple phenotypic, differentiation and functional parameters pertaining to responding T cells, most notably the expression of multiple effector cytokines. These attributes make the technique particularly suitable for the assessment of T-cell immune responses induced by vaccines. Depending on the nature of a given vaccine and the test setting, there are approaches that may be taken at different stages of the assay that are more suitable.[81]

For flow cytometry, cells are stimulated with the antigen of interest and then 1 $\mu g\,mL^{-1}$ of brefeldin A is added to retain the cytokine in the cytoplasm. After incubation for 4–6 h at 37 °C with 5% CO_2, these cells are stained for surface markers (*e.g.* CD4+ or CD8+) using antibodies against the surface markers of interest in separate tubes and incubated for 30 min at 4 °C in the dark. For intracellular staining (*e.g.* IFN-γ, IL-4) after fixation and permeabilization, cells are stained with suitable antibodies and incubated for 30 min at 4 °C in the dark. The stained cells are then washed and analyzed using a flow cytometer.[82]

3.8 Conclusion

Assays for estimation of potency, immunoassays and bioassays for the detection of ADAs provide valuable tools for establishing the appropriate functional activity and safety evaluation of therapeutic peptides. These assays are very complex because they involve various cell lines and biological molecules used as therapeutics. The key to achieving robust and good-quality assays is to choose the right format and design and develop them keeping in mind the various parameters involved. The assay design should include the appropriate format, cell line, endpoint, control antibodies and assay controls, while avoiding any interferences. Once the assay has been designed, it should be optimized for drug product concentration, assay matrix, response range, assay configurations of plates and controls for reasonable sensitivity. All assays have to be validated for sensitivity, precision, robustness, interference from any matrix, stability and ruggedness before use related to either release of the product or clinical evaluation of ADAs.

References

1. R. Lax and C. Meenan, Innovations in Pharmaceutical Technology, *Biopharma*, 2012, **42**, 54.
2. A. C. L. Tan, A. Goubier and H. E. Kohrt, *J. Immunother. Cancer.*, 2015, (3), 48.
3. K. Kaur, I. Singh, P. Kaur and R. Kaur, *Asian. J. Res. Biol. Pharm. Sci.*, 2015, **3**(3), 75.
4. S. S. Usmani, G. Bedi, J. S. Samuel, S. Singh, S. Kalra, P. Kumar, A. A. Ahuja, M. Sharma, A. Gautam and G. P. S. Raghava, *PLoS One*, 2017, **12**(7), e0181748.

5. C. Guo, M. H. Manjili, J. R. Subjeck, D. Sarkar, P. B. Fisher and X.-Y. Wang, *Adv. Cancer Res.*, 2013, **119**, 421.
6. A. H. Dzutsev, I. M. Belyakov, D. V. Isakov, D. H. Margulies and J. A. Berzofsky, *Int. Immunol.*, 2007, **19**, 497.
7. J. W. Hodge, M. Chakraborty, C. Kudo-Saito, C. T. Garnett and J. Schlom, *J. Immunol.*, 2005, **174**, 5994.
8. Y. Hou, B. Kavanagh and L. Fong, *J. Immunol.*, 2008, **180**, 1526.
9. K. R. Jordan, R. H. McMahan, C. B. Kemmler, J. W. Kappler and J. E. Slansky, *Proc. Natl. Acad. Sci. U. S. A.*, 2010, **107**, 4652.
10. S. A. Rosenberg, J. C. Yang, D. J. Schwartzentruber, P. Hwu, F. M. Marincola, S. L. Topalian, N. P. Restifo, M. E. Dudley, S. L. Schwarz, P. J. Spiess, J. R. Wunderlich, M. R. Parkhurst, Y. Kawakami, C. A. Seipp, J. H. Einhorn and D. E. White, *Nat. Med.*, 1998, **4**, 321.
11. Analytical Procedures and Methods Validation for Drugs and Biologics, Guidance for Industry, FDA, July 2015 Pharmaceutical Quality/CMC. https://www.fda.gov/downloads/drugs/guidances/ucm386366.pdf.
12. USP <1032> Design and Development of Biological Assays.
13. USP <1033> Biological Assay Validation.
14. USP <1034> Analysis of Biological Assays.
15. M. Azadeh, B. Gorovits, J. Kamerud, S. MacMannis, A. Safavi, J. Sailstad and P. Sondag, *AAPS J.*, 2018, **20**(22), DOI: 10.1208/s12248-017-0159-4.
16. R. Wilson, *Bioanal. Assay*, 2018, **10**(10), 737.
17. T. A. Little, *Bio. Pharm. Int.*, 2014, **27**(3).
18. T. Coffey, M. Grevenkamp, A. Wilson and M. Hu, *BioProcess Int.*, 2013, **11**(6), 42.
19. X. C. Chen, l. Zhou, S. Gupta and F. Civoli, *J. Immunol. Methods*, 2012, **376**(1–2), 32.
20. L. Kutle, N. Pavlović, M. Dorotić, I. Zadro, M. Kapustić and B. Halassy, *Vaccine*, 2010, **28**(33), 5497.
21. M. Wadhwa, C. Bird, P. Dilger, R. Gaines-Das and R. Thorpe, *J. Immunol. Methods*, 2003, **278**, 1.
22. A. S. Rosenberg, *AAPS J.*, 2006, **8**, E501–E507.
23. H. Schellekens, *Biotechnol. Annu. Rev.*, 2008, **14**, 191.
24. M. Kessler, D. Goldsmith and H. Schellekens, *Nephrol., Dial., Transplant.*, 2006, **21**(Suppl 5), v9–v12.
25. S. J. Shire, M. Cromwell and J. Liu, *AAPS J.*, 2006, **8**, E729–E730.
26. E. M. Cromwell, E. Hilario and F. Jacobson, *AAPS J.*, 2006, **8**, E572–E579.
27. S. Hermeling, D. J. A. Crommelin, H. Schellekens and W. Jiskoot, *Pharm. Res.*, 2004, **21**, 897.
28. G. Schernthaner, *Diabetes Care*, 1993, **16**(Suppl 3), 155.
29. W. V. Moore and P. Leppert, *J. Clin. Endocrinol. Metab.*, 1980, **51**, 691.
30. A. L. Hjelm Skog, M. Wadhwa, M. Hassan, B. Gharizadeh, C. Bird, P. Ragnhammar, R. Thorpe and H. Mellstedt, *Clin. Cancer Res.*, 2001, **7**, 1163.

31. J. Li, C. Yang, X. Yuping, A. Bertino, J. Glaspy and M. Roberts, *Blood*, 2001, **98**, 3241.
32. N. Casadevall, J. Nataf, B. Viron, A. Kolta, J. J. Kiladjian, P. Martin-Dupont, P. Michaud, T. Papo, V. Ugo, I. Teyssandier, B. Varet and P. Mayeux, *N. Engl. J. Med.*, 2002, **346**, 469.
33. A. R. Mire-Sluis, Y. C. Barrett, D. Devanarayan, E. Koren, H. Liu, M. Maia, T. Parish, G. Scott, G. Shankar, E. Shores, S. J. Swanson, G. Taniguchu, D. Wierda and L. A. Zuckerman, *J. Immunol. Methods*, 2004, **289**, 1.
34. D. Geng, G. Shankar, A. Schantz, M. Rajadhyaksha, H. Davis and C. Wagner, *J. Pharm. Biomed. Anal.*, 2005, **39**, 364.
35. S. L. Klakamp, H. Lu, M. Tabrizi, C. Funelas, L. K. Roskos and D. Coleman, *Anal. Chem.*, 2007, **79**(21), 8176.
36. S. Gupta, S. R. Indelicato, V. Jethwa, T. Kawabata, M. Kelly, A. R. Mire-Sluis, S. M. Richards, B. Rup, E. Shore, S. J. Swanson and E. Wakshull, *J. Immunol. Methods*, 2007, **321**, 1.
37. E. Koren, H. M. Smith, E. Shores, E. Shankar, D. Finco-Kent, B. Rup, Y.-C. Barrett, V. Devanarayan, B. Gorovits, S. Gupt, T. Parish, V. Quarmby, M. Moxness, S. J. Swanson, G. Taniguchi, L. A. Zuckerman, C. C. Stebbins and A. R. Mire-Sluis, *J. Immunol. Methods*, 2008, **333**, 1.
38. R. D. Schrock, *BioProcessing J., Trends & Developments in BioProcess Technology*, 2012, **11**(3), 4.
39. E. Bortolotto, R. Rousseau, B. Teodorescu, A. Wielant and G. Debauve, *Bioprocess Int.*, 2015, **13**(6), 26.
40. X. Wei, S. J. Swanson and S. Gupta, *J. Immunol. Methods*, 2004, **293**, 115.
41. M. D. Sadick, A. Intintoli, V. Quarmby, A. McCoy, E. Canova-Davis and V. Ling, *J. Pharm. Biomed. Anal.*, 1999, **19**, 883.
42. S. Sakamoto, M. Yokoyama, K. Prakash, J. Tsuruha, S. Masamoto, R. H. Getzenberg and Y. Kakehi, *J. Biomol. Screen.*, 2003, **8**, 701.
43. K. Wu, C. Yang, M. A. Davoli, M. D'Amico, A. Li, K. Cveklova, Z. Kozmik, M. P. Lisanti, R. G. Russell, A. Cvekl and R. G. Pestell, *J. Biol. Chem.*, 2003, **278**, 51673.
44. I. Wala, S. J. Swanson and S. Jing, *J. Pharm. Biomed. Anal.*, 2007, **45**, 583.
45. Y. Yua, C. Piddingtonb, D. Fitzpatrickb, B. Twomeyb, R. Xu and S. J. Swanson, *J. Immunol. Methods*, 2006, **316**(1–2), 8.
46. R. Camacho-Sandovala, E. N. Sosa-Grandea, E. González-Gonzáleza, A. Tenorio-Calvoa, C. A. López-Moralesa, M. Velasco-Velázquezb, L. Pavón-Romeroc, S. M. Pérez-Tapiaa and E. Medina-Riveroa, *J. Pharm. Biomed. Anal.*, 2018, **155**, 235–240.
47. Z. J. Cheng, Proceedings of the 103rd Annual Meeting of the AACR, Mar 31–Apr 4, Chicago, IL. AACR Cancer Res. 2012, 72(8), supplement abstract #2840.
48. X. Wei, D. S. Grill, A. C. Heatherington, S. J. Steven, J. Swanson and S. Shalini Gupta, *J. Pharm. Biomed. Anal.*, 2007, **43**(2), 666.

49. H. Gazzano-Santoro, A. Chen, B. Casto, H. Chu, E. Gilkerson, V. Mukkua, E. Canova-Davise and C. Kottsb, *J. Pharm. Biomed. Anal.*, 1999, **21**(5), 945.
50. L. Buonaguro, A. Petrizzo, M. L. Tornesello and F. M. Buonaguro, *Clin. Vaccine Immunol.*, 2011, **18**, 23.
51. D. J. Schwartzentruber, D. H. Lawson, J. M. Richards, R. M. Conry, D. M. Miller, J. Treisman, F. Gailani, L. Riley, K. Conlon, B. Pockaj, K. L. Kendra, R. L. White, R. Gonzalez, T. M. Kuzel, B. Curti, P. D. Leming, E. D. Whitman, J. Balkissoon, D. S. Reintgen, H. Kaufman, F. M. Marincola, M. J. Merino, S. A. Rosenberg, P. Choyke, D. Vena and P. Hwu, *N. Engl. J. Med.*, 2011, **364**, 2119.
52. T. N. Schumacher and R. D. Schreiber, *Science*, 2015, **348**, 69.
53. P. A. Ott, Z. Hu, D. B. Keskin, S. A. Shukla, J. Sun, D. J. Bozym, W. Zhang, A. Luoma, A. Giobbie-Hurder, L. Peter, C. Chen, O. Olive, T. A. Carter, S. Li, D. J. Lieb, T. Eisenhaure, E. Gjini, J. Stevens, W. J. Lane, I. Javeri, K. Nellaiappan, A. M. Salazar, H. Daley, M. Seaman, E. I. Buchbinder, C. H. Yoon, M. Harden, N. Lennon, S. Gabriel, S. J. Rodig, D. H. Barouch, J. C. Aster, G. Getz, K. Wucherpfennig, D. Neuberg, J. Ritz, E. S. Lander, E. F. Fritsch, N. Hacohen and C. J. Wu, *Nature*, 2017, **547**, 217.
54. O. J. Finn, *Nat. Rev. Immunol.*, 2018, DOI: 10.1038/nri.2017.140.
55. B. M. Carreno, V. Magrini, M. Becker-Hapak, S. Kaabinejadian, J. Hundal, A. A. Petti, A. Ly, W.-R. Lie, W. H. Hildebrand, E. R. Mardis and G. P. Linette, *Science*, 2015, **348**(6236), 803.
56. J. G. Abelin, B. D. Keskin, S. Sarkizova, C. R. Hartigan, W. Zhang, J. Sidney, J. Stevens, W. Lane, G. L. Zhang, T. M. Eisenhaure, K. R. Clauser, N. Hacohen, M. S. Rooney, S. A. Carr and J. C. Wu, *Immunity*, 2017, **46**, 315.
57. J. H. Kessler, B. Mommaas, T. Mutis, I. Huijbers, D. Vissers, W. E. Benckhuijsen, G. M. Th. Schreuder, R. Offringa, E. Goulmy, C. J. M. Melief, S. H. van der Burg and J. W. Drijfhout, *Human. Immunol*, 2003, **64**, 245.
58. S. H. van der Burg, E. Ras, J. W. Drijfhout, W. E. Benckhuijsen, A. J. Bremers, C. J. M. Melief and W. M. Kast, *Human. Immunol.*, 1995, **44**, 189.
59. J. H. Kessler, W. E. Benckhuijsen, T. Mutis, C. J. M. Melief, S. H. van der Burg and J. W. Drijfhout, *Curr. Protoc. Immunol.*, 2004, 18.12.1–18.12.15.
60. P. E. Jensen, J. C. Moore and A. E. Lukacher, *J. Immunol. Methods*, 1998, **215**(1–2), 71.
61. Z. Zavala-Ruiz, I. Strug, M. W. Anderson, J. Gorski and L. J. Stern, *Chem. Biol.*, 2004, **11**(10), 1395.
62. A. Ferrante and J. Gorski, *J. Immunol.*, 2007, **178**(11), 7181–7189.
63. R. Salvat, L. Moise, C. Bailey-Kellogg and K. E. Griswold, *J. Vis. Exp.*, 2014, (85), e51308.
64. S. J. Swanson, *Nephron. Clin. Pract.*, 2004, **96**, c88.

65. J. Ferbas, J. Thomas, J. Hodgson, A. Gaur, N. Casadevall and S. J. Swanson, *Clin. Vaccine Immunol.*, 2007, **14**(9), 1165.
66. L. Neyer, J. Hiller, K. Gish, S. Keller and I. Caras, *J. Immunol. Methods*, 2006, **315**(1–2), 80.
67. M. Brickelmaier, P. S. Hochman, R. Baciu, B. Chao, J. H. Cuervo and A. Whitty, *J. Immunol. Methods*, 1999, **227**(1–2), 121.
68. J. O. Koskinen, J. Vaarno, R. Vainionpää, N. J. Meltola and A. E. Soini, *J. Immunol. Methods*, 2006, **309**(1–2), 11.
69. M. K. Park, D. E. Briles and M. H. Nahm, *Clin. Diagn. Lab. Immunol.*, 2000, **7**(3), 486.
70. J. A. Lofgren, S. Dhandapani, J. J. Pennucci, C. M. Abbott, D. T. Mytych, A. Kaliyaperumal, S. J. Swanson and M. C. Mullenix, Comparing ELISA and surface plasmon resonance for assessing clinical immunogenicity of panitumumab, *J. Immunol.*, 2007, **178**(11), 7467.
71. E. Morgan, R. Varro, H. Sepulveda, J. A. Ember, J. Apgar, J. Wilson, L. Lowe, R. Chen, L. Shivraj, A. Agadir, R. Campos, D. Ernst and A. Gaur, *Clin. Immunol.*, 2004, **110**, 252.
72. A. Patton, M. Mullenix, S. J. Swanson and E. Koren, An acid dissociation bridging ELISA for detection of anti-drug antibodies in the presence of drug, *J. Immunol. Methods*, 2005, **304**, 189.
73. J. S. Bourdage, C. A. Cook, D. L. Farrington, J. S. Chain and R. J. Konrad, An Affinity Capture Elution (ACE) assay for detection of ADA to monoclonal antibody therapeutics in the presence of high levels of drug, *J. Immunol. Methods*, 2007, **327**(1–2), 10–17.
74. J. A. Lofgren, I. Wala, E. Koren, S. J. Swanson and S. Jing, *J. Immunol. Methods*, 2006, **308**(1–2), 101.
75. X. Wei, S. J. Swanson and S. Gupta, *J. Immunol. Methods*, 2004, **293**, 115.
76. M. D. Sadick, A. Intintoli, V. Quarmby, A. McCoy, E. Canova-Davis and V. Ling, *J. Pharm. Biomed. Anal.*, 1999, **19**, 883.
77. S. Sakamoto, M. Yokoyama, K. Prakash, J. Tsuruha, S. Masamoto, R. H. Getzenberg and Y. Kakehi, *J. Biomol. Screen.*, 2003, **8**, 701.
78. K. Wu, C. Yang, M. A. Davoli, M. D'Amico, A. Li, K. Cveklova, Z. Kozmik, M. P. Lisanti, R. G. Russell, A. Cvekl and R. G. Pestell, *J. Biol. Chem.*, 2003, **278**, 51673.
79. I. Wala, S. J. Swanson and S. Jing, *J. Pharm. Biomed. Anal.*, 2007, **45**, 583.
80. Y. Yua, C. Piddingtonb, D. Fitzpatrick, B. Twomeyb, R. Xu and S. J. Swanson, *J. Immunol. Methods*, 2006, **316**(1–2), 8.
81. S. G. Smith, K. Smits, S. A. Joosten, K. E. van Meijgaarden, I. Satti, H. A. Fletcher, N. Caccamo, F. Dieli, F. Mascart, H. McShane, H. M. Dockrell and T. H. M. Ottenhoff, *PLoS One*, 2015, DOI: 10.1371/journal.pone.0138042.
82. H. Ghaffari-Nazari, J. Tavakkol-Afshari, M. R. Jaafari, S. Tahaghoghi-Hajghorbani, E. Masoumi and S. A. Jalali, *PLoS One*, 2015, DOI: 10.1371/journal.pone.0142563.

83. Assay Development and Validation for Immunogenicity Testing of Therapeutic Protein Products, Guidance for Industry Draft guidance, April 2016, Pharmaceutical Quality/CMC, Revision 1. https://www.fda.gov/downloads/Drugs/Guidances/UCM192750.pdf.
84. Guideline on Immunogenicity assessment of therapeutic proteins, May 2017, EMEA/CHMP/BMWP/14327/2006 Rev 1. https://www.ema.europa.eu/documents/scientific-guideline/guideline-iwmmunogenicity-assessment-therapeutic-proteins-revision-1_en.pdf.

CHAPTER 4

Chemistry, Manufacturing and Controls: Active Pharmaceutical Ingredient and Drug Product

PREM RAMIYA

Cerus Corporation, 2550 Stanwell Drive, Concord, CA 94520, USA
Email: ramiyahp@yahoo.com

4.1 Introduction

Chemistry, manufacturing and controls (CMC) provides guidance on how to advance new molecules from preclinical to Phase I through commercialization and launch of the product. This is further divided into guidance for active pharmaceutical ingredients (APIs) and drug products (DPs) or final dosage form for delivery. This chapter introduces the guidance and industry practices for APIs and DPs.

As shown in Figure 4.1, drug development involves the identification of the target molecule through discovery and preclinical studies to identify the characteristics that will enable an organization to sponsor a molecule as a "new clinical entity" (NCE) for further development. The drug development process, in general, takes 10–15 years and, if the NCE passes all milestones/hurdles, it becomes a product in the market. CMC governs the drug development process and aids in identifying the activities required for a successful outcome with regulatory bodies around the world.

Drug Discovery Series No. 72
Peptide Therapeutics: Strategy and Tactics for Chemistry, Manufacturing and Controls
Edited by Ved Srivastava

Published by the Royal Society of Chemistry, www.rsc.org

Figure 4.1 Drug development process.

4.2 Drug Substance or Active Pharmaceutical Ingredient (API)

Patient safety is critical for any drug that is administered in a clinical/commercial setting, so it is imperative that the drug be manufactured under controlled conditions and maintain its integrity throughout the course of clinical development. Once the NCE has been identified, the molecule will have to be manufactured under current good manufacturing practice (cGMP) conditions in-house or outsourced depending on the sponsor's ability to perform these in-house or not. Outsourcing is common among most of the smaller biotechnology companies and large pharma companies outsource activities depending on the available resources or for a secondary supply requirement. Selection of the contract manufacturer is an important activity and careful due diligence is required to select the right supplier. Evaluate two or three suppliers for their past performance and examine recommendations from current customers who are using those suppliers, technical capability, quality and regulatory history prior to working with one of them. Dual sourcing is always beneficial but not necessary during the early phase of development. One of the most important aspects of manufacturing under cGMP conditions is that the technical leader works with the quality function to implement the necessary controls in the manufacture of the NCE. This includes setting up the specification for the drug substance, in-process tests/controls and purification/crystallization as is required to meet the purity, strength and quality requirements. Quality control and quality assurance personnel support these activities to make sure that the API is manufactured under cGMP conditions and they perform site visits/audits to monitor the controls in place to prevent any contamination during the manufacture of the NCE.

Analytical methods required to perform in-process tests and release tests must be established in the early phase of development and qualified as the NCE progresses through various development stages from preclinical to Phase I to Phase II, *etc.* Based on the complexity of the NCE, the analytical methods vary. For example, for small molecules, identity [by ^{1}H NMR spectroscopy, Fourier transform infrared (FT-IR) spectroscopy or high-performance liquid chromatography (HPLC) retention time or equivalent], purity and impurities determined by HPLC can be used, whereas for larger molecules such as peptides, one needs to add more analytical tools to characterize the molecule, such as using two HPLC methods [reversed-phase HPLC (RP-HPLC) for purity/impurities and size-exclusion chromatography (SEC)-HPLC to determine dimer formation], Edman degradation to determine the sequencing of the peptide and amino acid analysis to determine the composition of the peptide. It is necessary to choose orthogonal methods to measure the purity of the drug substance, whether it is a small molecule or peptides, due to coeluting peaks in HPLC or because other methods such as ^{1}H NMR spectroscopy do not have sufficient resolution to provide a clear identity/purity of the drug substance. Recently, regulators have been requesting orthogonal methods for identity and purity determinations on a regular basis. Depending on the length of the peptide molecule, the analytical tools employed may vary. The steps involved in manufacture and adequate controls are described in the Common Technical Document (CTD),[1] but a summary for Phase I is provided in the following sections for novice readers and as a refresher for experienced personnel.

4.2.1 Description of Active Pharmaceutical Ingredient (API)

Describe the chemical, physical and biological characteristics of the NCE or API.

4.2.2 Name and Address of the Manufacturer

State the manufacturer's name and address and any regulatory identification such as FEI (Facility Establishment Identification) provided by the FDA when the manufacturing facility was established.

4.2.3 Method of Preparation of the Drug Substance

A description of the preparation of the drug substance and a flow diagram of the manufacturing process with in-process control should be narrated in this section.

4.2.4 Control of the Drug Substance

In this section, in-process tests, analytical methods, specifications for the drug substance, justification of specifications, batch analysis and qualification/validation of analytical procedures should be summarized. Control of starting materials, methods to characterize raw materials and stability of raw materials should be considered when providing the data in this section.

4.2.5 Characterization of the Drug Substance

Elucidation of the structure by orthogonal analytical techniques, solid-state characteristics and impurity identification should be summarized in this section.

4.2.6 Reference Standards

A reference standard has to be certified and verified by analytical techniques to be of the highest purity so that it can be used as a reference for purity and biological assays. Additional characterization is required for certifying the reference standard such as by differential scanning calorimetry (DSC), thermogravimetric analysis (TGA) or X-ray powder diffraction (XRPD) in addition to the specifications set for the drug substance.

4.2.7 Container Closure System

Specify what type of primary and secondary containers will be utilized for the storage and shipment of the drug substance. Shipping validation should be carried out to determine the logistics and shipping configuration.

4.2.8 Stability of the Drug Substance

The stability of the drug substance should be determined under recommended storage conditions for a period of 2–3 years along with accelerated conditions for 6 months at the minimum.

4.3 Drug Product

When an NCE is selected for development, a drug delivery strategy should be established based on the preclinical model and dosing regimen. Around 85–90% of drugs that come to market have an oral dosage formulation but for peptides and oligonucleotides the preference has been an intravenous (i.v.) formulation. In those instances, the final drug product presentation is lyophilized in vials and reconstituted for delivery to subjects/patients. If the API is stable in solution, such as in phosphate-buffered saline or other buffers, then a sterile single-use vial can be manufactured for clinical use. As discussed in the API section, the sponsor must decide to outsource or manufacture the drug product in-house depending on their resources and available capacity. The majority of smaller biotech companies tend to outsource the drug product manufacturing activities. Selection of an appropriate contract manufacturer should be evaluated in detail considering their past performance/experience, technical expertise in the area, capacity, resources for new projects, *etc.*, prior to selecting the most suitable contract manufacturer. A team consisting of quality personnel with technical expertise able to evaluate a contract manufacturer's capability should visit and meet the company's management and personnel who will be performing the work prior to partnering for such an activity. Selection of a contract manufacturer is a critical activity that will set the tone for the manufacture of the drug product on time and with high quality.

The manufacturing process should be established in the early stages of development and further optimized as the program goes through development. The following sections provide a summary of the information required for the drug product.

4.3.1 Composition of the Drug Product

Describe all the components and the composition of the investigational drug product, including novel excipients and buffers. The quality of the excipients (grade, source) should be provided. The quantitative composition of the drug product should also be provided.

4.3.2 Excipients

If the drug product contains novel excipients, additional manufacturing information needs to be provided.

4.3.3 Name and Address of the Manufacturer

The name and address and FEI number for the drug product manufacturer are required.

4.3.4 Pharmaceutical Development and Manufacturing Process for the Drug Product

Provide details on how the formulation was developed, stability of the formulations evaluated and the one chosen for further development and used in preclinical and clinical studies. The formulation generally contains buffers, excipients and inactive ingredients that will stabilize the API so that the drug product is stable for a period of time (normally 2–3 years depending on the stability of the formulated drug product) prior to use. Provide information on the manufacturing process, in-process tests and container closure integrity. If the drug product is an injectable product, describe how it is sterilized (for example, aseptic manufacturing or terminal sterilization).

4.3.5 Specifications and Analytical Methods for the Drug Product

Describe the analytical methods used for determination of purity, assay, impurities, particulate matter, endotoxins, sterility and excipients (if applicable). Assign specifications based on the batch history and supporting data.

4.3.6 Container Closure for the Drug Product

Choose an appropriate container closure for the drug product based on how the drug product will be delivered in the clinic or commercially. For oral drug products, one can compress them into tablets or fill them in capsules after formulation and then these can be packaged in blister packs. For an injectable product (as is the case for most peptides), it is filled in sterile single-use vials, either as a liquid fill or lyophilized as a powder for reconstitution or other modes of delivery using devices such as Kwikpen or Flexpen, which provide dosing accuracy in addition to glycemic control for patients with type II diabetes.

4.3.7 Preparation of Placebo (If Required)

Describe the scheme and manufacturing procedure and release criteria for the placebo (if one is required for the clinical study).

4.3.8 Labels

Submit a mock-up or representative copy of the proposed labeling that will be provided to clinical investigators for the study. These labels should carry a cautionary statement as required by 21 CFR 312.6(a). The statement reads "Caution: New Drug – Limited by Federal law to investigational use only."

4.3.9 Stability Study

A stability study should be carried out to establish the shelf-life of the drug product based on preliminary data collected in preclinical studies. Recommended storage conditions (1–3 years), and also accelerated storage conditions (minimum of 6 months), should be studied to establish the shelf-life of the drug product.

4.4 Phase II CMC for the Drug Substance

As the NCE progresses to the next phase of development (Phase II), additional information should be gathered on the new molecule and furnished to clinical investigators and regulatory authorities. This is summarized below for drug substance and drug product.

4.4.1 Description of Active Pharmaceutical Ingredient (API)

Provide the nomenclature, chemical structure and general properties such as solubility.

4.4.2 Name and Address of the Manufacturer

Add or delete if there are any changes to the manufacturer of the API.

4.4.3 Method of Manufacture of the Drug Substance

Describe any changes to the manufacturing process of the drug substance and provide an updated flow diagram of the manufacturing process. Add, if any, new in-process tests that had been implemented since the earlier manufacturing process. Provide stereochemical control during the manufacture of the API from starting materials, chiral intermediates and any impurities (such as epimers) generated. It is important to characterize any impurities above a certain threshold based on the class of molecule under development. For a peptide, a smaller amount of chiral impurities can alter its biological properties. as shown by Finder *et al.*,[2] who observed a faster aggregation rate of recombinant amyloid-β peptide (Aβ1–42) compared with its synthetic counterpart.

4.4.4 Control of the Drug Substance

Provide the structures of the starting materials (if not provided earlier), source, grade of the material, updated methods for characterizing raw materials and updated stability of the raw materials. Any changes with regard to in-process tests, analytical methods, specifications for the drug substance, justification of specifications, batch analysis and updated validation of analytical procedures (custom) and verification of compendial methods should be provided. Polymorphs can influence the solubility and/or bioavailability of the API when it is formulated, so polymorphs should be identified early in the development program and the correct polymorph should be selected for further processing into the drug product.

4.4.5 Characterization of the Drug Substance

Add any additional characterization studies that had been carried out, such as particle size analysis and DSC/TGA, and other related physical properties that can influence the drug product formulation.

4.4.6 Reference Standards

Preferably, international or national standards should be utilized if available (such as USP reference standards). A reference/working standard (if custom prepared, as in most cases) has to be certified and verified by analytical techniques to be of the highest purity so that it can be used as a reference for purity and biological assays. Additional characterization may be required for certifying the reference standard using solid-state techniques such as DSC, TGA, XRPD and SEC-HPLC (in the case of peptides for identifying dimers/agglomerates) in addition to meeting the specifications set for the drug substance.

4.4.7 Container Closure System

Shipping validation should be carried out to determine the logistics and shipping configuration. Continue verification of the shipping containers as the drug substance is shipped from the drug substance manufacturer to drug product manufacturer.

4.4.8 Stability of the Drug Substance

Provide updated data on the stability of the drug substance and continue placing manufactured lots in stability studies as the program advances to

Phase II. Trend data to establish the shelf-life and the use of predictive software are encouraged [*e.g.* JMP (jump), Minitab].

4.5 Phase II CMC for the Drug Product

As mentioned previously, most companies tend to outsource manufacturing activities to contract manufacturers depending on the nature of the molecule. As the program matures, the CMC package for NCE needs to be aligned well with the regulatory requirements and program requirements. These are summarized below.

4.5.1 Composition of the Drug Product/Batch Formula

Update any changes to the components and composition of the investigational drug product, including novel excipients and buffers. Provide any changes (if any) to the quantitative composition of the drug product. Provide a reason for the changes and justify them.

4.5.2 Control of Excipients

List the grade of excipients that are compendial (USP, NF, *etc.*) and if the source is of synthetic, human or animal origin. If there are any noncompendial excipients, describe the flow diagram for manufacture, the process for manufacturing, specifications, analytical methods of characterization, *etc.*

4.5.3 Name and Address of the Manufacturer

State if there are any additions or deletions from the manufacturer listed under Phase I.

4.5.4 Pharmaceutical Development and Manufacturing Process for the Drug Product

Provide an updated flow chart of the manufacturing process and intermediates that lead to the final drug product. The description of the manufacturing process can include step-by-step unit operations as they occur. Additional information may include the equipment utilized, controls during the manufacturing process, whether there are novel formulations involved (such as liposomal, microspheres, implants, *etc.*) and any reprocessing procedures and controls. The QTPP (quality target product profile) should be defined for the NCE. The QTPP is a summary of the quality characteristics of the drug product that ideally will be achieved to ensure the desired quality, safety and efficacy.

4.5.5 Specifications and Analytical Methods for the Drug Product

List any changes from Phase I to the specifications set for the drug product. Describe details of all analytical procedures and method validations that were completed for noncompendial methods. Provide data on the particle size distribution and polymorphic forms if appropriate for the drug substance. Describe microbiological attributes and procedures established to control endotoxin limit and preservative tests if a preservative is used in the formulated drug product. Provide test results and analytical data from batch release for the manufactured batches.

4.5.6 Container Closure for the Drug Product

Determine the container closure integrity if the product is in a sterile single-use glass vial (either liquid fill or lyophilized powder) *via* a dye ingress or microbial ingress test method. Establish a packaging and labeling procedure consistent with the regulatory requirements in the region where the clinical studies are conducted. Shipping validations also need to be performed to determine the best way to move the drug product from the manufacturing location to the distribution center and then to clinical sites.

4.5.7 Stability Study

Provide updated stability data for the batches manufactured, list the tests performed at each time point and provide stability specifications for the drug product (if different from release criteria) and analytical procedures utilized. Establish/predict the shelf-life based on actual stability data and use of statistical software such as Minitab or JMP.

4.6 Phase III CMC for the Drug Substance

Any changes that occurred during the course of development should be summarized and furnished to regulatory authorities in an annual update or information amendment based on the nature of the changes. Any major changes such as new suppliers or changes in the specification should be provided in an information amendment whereas information such as data on new/additional batches of API can be included in an annual update as summarized below.

4.6.1 Description of Active Pharmaceutical Ingredient (API)

Provide information, if not submitted previously, on solubility, partition coefficient, dissociation constant (pK_a), hygroscopic properties, polymorphs, particle size, melting point, optical rotation, stereochemistry and biological activity.

4.6.2 Name and Address of the Manufacturer

List all contract manufacturers, including the manufacturer of the drug substance, release testing laboratory (if different from the manufacturer) and stability testing laboratory (if different from the drug substance manufacturer).

4.6.3 Method of Manufacture of the Drug Substance

Describe the final manufacturing process of the drug substance and update the flow diagram of the manufacturing process with all in-process controls and final isolation of the drug substance. Provide the batch size range, the ratio of reagents, solvents (if different from the previously submitted process), operating conditions, control of critical steps and intermediates and control of crystalline forms (if relevant). Describe reprocessing procedures and controls if these do not affect the overall impurity or safety profile of the final drug substance. Critical process parameters should be evaluated to provide adequate control of the manufacturing process and to that effect one can utilize DOE (design of experiments) and QbD (quality by design) principles to guide the sponsor. QbD is a systematic approach to development that begins with predefined objectives and emphasizes product and process understanding and process control based on sound science and quality risk management.

4.6.4 Control of the Drug Substance

List any new reagents, solvents and biological raw materials, if not provided earlier. Control critical steps in the manufacturing process to ensure reaction completion, identity, purity and changes in critical controls reported during Phase II. Provide analytical procedures and acceptance criteria for any intermediates that are isolated.

4.6.5 Characterization of the Drug Substance

Details of elemental analysis, molecular weight determination, NMR (^{1}H and ^{13}C), IR and UV spectroscopy and mass spectrometry and, if applicable, single-crystal X-ray diffraction analysis should be provided if not submitted previously.

4.6.6 Reference Standards

If there is no international/national standard available, then the sponsor has to generate a primary reference standard during Phase III studies. Describe the analytical procedures utilized to characterize the primary reference standard and any additional updates from Phase II on the characterization of the primary reference standard and source. One can continue to use the working standard from Phase II or generate a new one for lot release activities provided that they are characterized in a similar fashion.

4.6.7 Container Closure System

List any changes in the type of primary and secondary containers that will be utilized for the drug substance storage and shipment.

4.6.8 Stability of the Drug Substance

Provide updated stability data and any changes in the stability specification if it is different from the release specification. The stability of the drug substance should be determined under recommended storage conditions for a period of 2–3 years along with accelerated conditions for 6 months at the minimum.

4.7 Phase III CMC for the Drug Product

At the end of Phase II, discussions with regulatory authorities should be conducted to gain a regulatory pathway and insight into and guidance on the next phase of development. In addition, the Phase III clinical design and endpoints for the clinical study (SPA – special protocol assessment) should be discussed and agreed with the regulatory body so that there is a clear regulatory pathway for approval. For accelerated approval, the discussions with the regulatory body should occur in Phase II. No major changes are allowed in the manufacture of the drug product as the chemistry is locked at the end of Phase II. Any changes that occurred during development should be summarized and furnished to the regulatory authorities in an annual update or information amendment based on the nature of the changes. Any major changes such as new suppliers or changes in the specification should be provided in an information amendment whereas information such as data on new/additional batches of API can be included in the annual update.

4.7.1 Composition of the Drug Product

Describe all components and composition of the investigational drug product, including novel excipients and buffers. The quality of the excipients (grade, source) should be provided. The quantitative composition of the drug product should also be provided.

4.7.2 Excipients

If the drug product contains novel excipients, additional manufacturing information needs to be provided.

4.7.3 Name and Address of the Manufacturer

The name and address and FEI number of the drug product manufacturer are required, if not provided earlier.

4.7.4 Pharmaceutical Development and Manufacturing Process for the Drug Product

Details on how the formulation was developed and the stability of the formulations evaluated and the one chosen for further development and used in preclinical and clinical studies should be provided. A formulation generally contains buffers, excipients and inactive ingredients that will stabilize the API so that the drug product is stable for a period of time (normally 2–3 years depending on the stability of the formulated drug product) prior to use. Provide information on the manufacturing process, in-process tests and container closure integrity. If the drug product is an injectable product, describe how it is sterilized (*e.g.* aseptic manufacture or terminal sterilization). QbD principles should be applied to manage effectively the pharmaceutical development and understanding of the manufacturing process for the drug product.

4.7.5 Specifications and Analytical Methods for the Drug Product

Describe the validation of analytical methods for the determination of purity, assay, impurities, particulate matter, endotoxins, sterility and excipients (if applicable). Assign specifications based on the batch history and supporting data.

4.7.6 Container Closure for the Drug Product

Describe any changes to the container closure of the final dosage form and whether there are any additional presentations such as a different strength (*e.g.* regular Tylenol® contains 325 mg of acetaminophen whereas extra-strength Tylenol® contains 600 mg of acetaminophen) based on the commercial marketing strategy. Provide extractable and leachable data as needed to support the container closure for the NCE.

4.7.7 Preparation of Placebo (If Required)

Describe the scheme and manufacturing procedure and release criteria for the placebo (if one is required for the clinical study).

4.7.8 Labels

Submit any changes to the label and provide justification. As mentioned before, the labels should carry a cautionary statement as required by 21 CFR 312.6(a). The statement reads "Caution: New Drug – Limited by Federal law to investigational use only."

4.8 Registration and Process Validation

As the sponsor gears up to complete the Phase III clinical trials, for an NDA there is a requirement to demonstrate that the manufacturing can deliver the product to the commercial marketplace and the product is stable for a minimum of 12 months. Therefore, the sponsor will manufacture "registration/validation batches" based on the current manufacturing process and establish 12-month stability for the drug substance and 12-month stability data for drug product prior to submission of the NDA. A minimum of three batches is necessary to justify that the process is robust and reproducible and delivers the product with the expected purity and quality. These batches can be used for commercial sale once the product has been approved. Commercial labeling and kitting of the final product should be presented in the NDA so that the regulatory authorities can provide feedback on the proposed labeling and any restrictions based on the drug profile.

4.9 Risk Management

There are several risks when developing an NCE for commercialization, especially if the sponsor is outsourcing all manufacturing and testing activities. As in most industries, mergers and acquisitions are common in the pharmaceutical industry and some of the manufacturers are purchased by others whose business model could change, such that they are no longer available. In the best interests of the development of the candidate entity, it is better to have a qualified additional supplier even if it is not regularly producing your drug substance or drug product. It certainly adds cost and effort to establish a dual supply chain for the drug substance and drug product, but it is worth the effort as in many cases such incidents have occurred. Critical quality attributes (CQAs) should be identified for the manufacturing process (such as polymorphs), for the performance of the drug (dissolution/bioavailability) and stability (water content, light protection, impurity control). A process FMEA (failure mode effect and analysis) should be prepared considering all critical parameters and an RPN (risk priority number) assigned based on severity, occurrence and detection for all steps involved in the manufacturing process. To reduce the risk of failure, the RPN should be kept low. Logistics for drug distribution should be in place early in the development and fully established prior to filing the NDA so that the sponsor is ready for commercial distribution after regulatory approval.

4.9.1 Quality and Regulatory Risk

As most companies outsource all activities, the quality unit should audit on a regular basis so that the manufacturers maintain their GMP status and whether there are any regulatory risks in utilizing their services or facility for manufacturing clinical and eventual commercial materials from them. Any deviations should be thoroughly investigated by technical and quality personnel to understand and identify the root cause and a potential remedy for the cause. Out-of-specification incidents could occur during testing of the drug substance and drug product and these should be investigated and any root cause from those investigations should be addressed so that they do not occur again.

4.10 Conclusion

Drug development is a long and arduous process but it is worth its weight in gold if one can treat or slow an ailment or cure a disease. This chapter and ref. 3–6 provide an overview of CMC for the clinical development of NCEs and eventual commercialization.

References

1. M4 Organization of the Common Technical Document for the Registration of Pharmaceuticals for Human Use Guidance for Industry, October 2017 https://www.fda.gov/downloads/drugs/guidancecomplianceregulatoryinformation/guidances/ucm073257.pdf.
2. V. H. Finder, I. Vodopivec, R. M. Nitsch and R. Glockshuber, The recombinant amyloid-β peptide Aβ1–42 aggregates faster and is more neurotoxic than synthetic Aβ1–42, *J. Mol. Biol.*, 2010, **396**, 9–18.
3. Content and Format of INDs for Phase I Studies of Drugs, Including Well-characterized Therapeutic Biotechnology-Derived Products, https://www.fda.gov/downloads/Drugs/.../Guidances/ucm074980.pdf.
4. INDs for Phase 2 and Phase 3 Studies, Chemistry, Manufacturing and Controls Information, https://www.fda.gov/downloads/Drugs/Guidances/ucm070567.pdf.
5. IND Meetings for Human Drugs and Biologics, https://www.fda.gov/downloads/drugs/guidancecomplianceregulatoryinformation/guidances/ucm070568.pdf.
6. ICH, Q8, Q9, Q10., http://www.ich.org/products/guidelines/quality/training-programme-for-q8q9q10.html.

CHAPTER 5

Peptide Manufacturing Methods and Challenges

SATYANARAYANA KOTA

Peptides Laboratory, Natco Research Centre, Natco Pharma Limited, B-13, Industrial Estate, Sanathnagar, Hyderabad 500018, India
Email: dr_ksn@natcopharma.co.in

5.1 Introduction

Peptides are a unique class of pharmaceutical compounds that are molecularly positioned between small molecules and proteins. Peptides have been the most studied class of biopolymers and the subject matter of a large number of publications, and will continue to be so in the years ahead.[1] Emil Fischer christened peptide chemistry in 1901 by first preparing the unprotected synthetic peptide glycylglycine by hydrolysis of glycine di-ketopiperazine.[2] Subsequently, many scientists have contributed in various ways towards progress in synthetic methodologies for peptides. Some of these landmark contributions include azide coupling by Curtius,[3] *N*-carboxy anhydrides by Leuchs,[4] the benzyloxycarbonyl (Cbz) protecting group by Bergmann and Zervas,[5] a complete synthesis of vasopressin by du Vigneaud *et al.*,[6] solid-phase synthesis by Merrifield[7] and *tert*-butoxycarbonyl (Boc) and 9-fluoromethoxycarbonyl (Fmoc) protecting groups by Carpino.[8] The progress in peptide synthesis from the early days to the present is phenomenal. Merrifield, in the early part of his career, began with the task of preparing a pentapeptide, Ser-His-Leu-Val-Lue at Rockefeller University. In his own words, "my overall yield was 7% and it took 11 months to synthesize."[9]

Drug Discovery Series No. 72
Peptide Therapeutics: Strategy and Tactics for Chemistry, Manufacturing and Controls
Edited by Ved Srivastava

Published by the Royal Society of Chemistry, www.rsc.org

In the present world, the same pentapeptide can be synthesized in an hour, with much higher purity than that achieved by Merrifield.[10]

The pharmaceutical industry's interest in peptides as drug candidates, based on their potency, enormous diversity of functions, high degree of specificity and general lack of toxicity, has existed since their general structure was first elucidated, *i.e.* more than 100 years ago. However, the initial development of this unique class of drugs was limited to the extraction of naturally occurring peptide hormones, such as insulin and adrenocorticotropic hormone (ACTH). The situation changed dramatically around the 1950s, when du Vigneaud reported the synthesis of two peptide hormones, oxytocin and vasopressin, demonstrating that peptides could be manufactured by chemical synthesis. Their short half-lives complicated the early development of synthetic peptides for therapeutic use together with a lack of oral bioavailability, their relatively complex manufacturing methods, which we now refer to as "solution- or liquid-phase peptide synthesis" (LPPS), high costs and the lack of sensitive and discriminating analytical methods for their characterization. However, Merrifield's solid-phase peptide synthesis (SPPS) method, which was reported in 1963, revolutionized the field, especially since the method could be automated, leading to an explosion in peptide research. The introduction of high-performance liquid chromatography (HPLC) around 1970 led to improved analytical methods. The use of SPPS for manufacturing purposes was hampered by the requirement for liquid hydrogen fluoride (HF) during the resin cleavage step in Merrifield's original strategy and the method had scale-up limitations. Consequently, the pharmaceutical industry did not consider peptides to be ideal drugs having commercial viability and, during the 1980s and early 1990s, strived hard to simulate their activity through the development of small-molecule peptide mimetics, which was unsuccessful. Meanwhile, the introduction of new protecting groups and SPPS resins, which eliminated the need for liquid HF, facilitated the scale-up of manufacturing processes.[11]

Enfuvirtide (Fuzeon®), a 36 all-natural amino acid residue peptide used for inhibition of entry of human immunodeficiency virus (HIV) into the CD4 cell, was the first long peptide to be produced in metric ton quantities. This development demonstrated that complex peptides could be manufactured economically by SPPS on a large (metric ton) scale. Although the product was not commercially as successful as it was projected to be during its development, it changed the landscape of peptide manufacture and instilled confidence in the peptide community and the pharmaceutical industry that peptides could be manufactured on a large scale.[12] Prior to enfuvirtide, Fmoc-protected amino acids, the building blocks used for peptide synthesis, were very expensive and not commercially available in large quantities for use in peptide manufacture. After the commercial success of enfuvirtide, the availability of Fmoc-protected amino acids increased and the cost decreased significantly.[13] In a way, the successful large-scale production of enfuvirtide rekindled the interest of the pharmaceutical industry in the development of peptides for therapeutic applications.

The medical definition of the peptide is "a molecule consisting of two or more amino acids in a chain covalently linked with an amide bond and smaller than proteins." However, the general understanding in the classical literature is that a peptide is a polymer of 2–50 amino acids.[14] There is no clear scientific consensus on the criteria that distinguish proteins from peptides, including the exact size at which a chain(s) of amino acids becomes a protein. The US Food and Drug Administration (FDA), in its guidance for industry, distinguishes proteins from peptides based solely on size (*i.e.* number of amino acids) and has defined peptides as "any alpha-amino acid polymer with a specific defined sequence that is 40 amino acids or less in size and made by chemical synthetic methods," while proteins are defined as "any alpha-amino acid polymer with a specific defined sequence that is greater than 40 amino acids in size." In the same guidance, the FDA also amended the statutory definition of a protein to exclude synthetic polypeptides and defines chemically synthesized polypeptides as "any alpha-amino acid polymer that is made entirely by chemical synthesis and is less than 100 amino acids in size." Unless peptides meet the statutory definition of "biological products" (*e.g.* peptide vaccines), they are not considered "biological products" and thus regulated under the Federal Food, Drug and Cosmetic (FD&C) Act, in contrast to "proteins," which are regulated under the Public Health Service Act.[15] The European Medicines Agency (EMA) does not distinguish peptides or proteins based on the size, mode of synthesis, extraction from their natural sources or whether recombinant or synthetically produced. Peptides are treated as biological entities by the EMA.[16]

The strategy for the larger scale production of peptides under good manufacturing practice (GMP) is largely determined by the size and characteristics of the peptide. There are a variety of techniques such as LPPS, SPPS, hybrid synthesis and recombinant DNA methods for the commercial manufacture of pharmaceutical-grade peptides. The downstream processing steps such as purification and isolation are critical process steps for generating highly purified active peptide pharmaceutical ingredients (APPIs) that meet stringent regulatory requirements. Until 2000, peptides used to be manufactured mostly by LPPS and these peptides generally had short chain lengths. After the invention of enfuvirtide, the method of manufacture shifted to solid-phase and hybrid synthesis. Some of the FDA-approved peptides with their use, company and method of synthesis are listed in Table 5.1.

This chapter focuses on the synthesis of peptides by solution-phase, solid-phase and hybrid techniques, highlighting the challenges concerning current GMP (cGMP). In the production of peptides, downstream chromatographic purification and product isolation become critical and sometimes are more important than the synthetic processes *per se*. A knowledge of the sources of related peptide impurities and their control is essential for the development of an efficient manufacturing process. Hence this chapter touches on these two aspects also.

Table 5.1 List of some FDA-approved peptides. Source: https://www.accessdata.fda.gov/scripts/cder/ob/ (as on 17 Sep 2018).

No.	Peptide name	Brand name	Company	Length (amino acid residues)	Method of synthesis[a]	Application	Year of FDA approval
1	Vasopressin tannate	Pitressin Tannate	Parke Davis	9	SPS	To narrow blood vessels	1982
2	Lypressin	Diapid	Novartis	9	SPS	Nephrogenic diabetes insipidus	1982
3	Desmopressin	DDAVP	Ferring	9	SPS	Diabetes insipidus	1984
4	Calcitonin human	Cibacalcin	Novartis	32	SPS	Osteoporosis	1986
5	Teriparatide	Forteo	Eli Lilly	34	rDNA	Osteoporosis	1987
6	Goserelin	Zoladex	Tersera	10	SPPS	Breast and prostate cancer	1989
7	Cyclosporin	Sandimmune	Novartis	11	E	Immunosuppressant	1990
8	Histrelin	Supprelin La	Endo	9	SPS	Prostate cancer	1991
9	Calcitonin salmon	Miacalcin	Mylan	32	SPS	Postmenopausal osteoporosis	1991
10	Leuprolide acetate	Lupron Depot	Abbvie	9	SPPS	Prostate cancer	1985
11	Bacitracin	Neosporin	Casper Pharma	11	E	Antibiotic	1982
12	Eptifibatide	Integrilin	Schering	6	SPS	Antiplatelet drug	1998
13	Octreotide	Sandostatin	Novartis	8	SPS	Neuroendocrine tumours	1988
14	Bivalirudin	Angiomax	The Medicines	20	SPS	*Percutaneous transluminal coronary angioplasty (PTCA)*	2000
15	Cetrorelix	Cetrorelix	EMD Serono	10	SPS	Gonadotropin-releasing hormone (GNRH) antagonist	2000
16	Triptorelin	Trelstar	Allergan Sales	10	SPPS	Prostate cancer	2000
17	Nesiritide	Natrecor	Scios	32	rDNA	Congestive heart failure	2001
18	Glatiramer	Copaxone	Teva	Polymer	SPS	Multiple sclerosis	2002

19	Enfuvirtide	Fuzeon	Hoffmann La Roche	36	Hybrid	HIV-1 infection	2003
20	Exenatide	Byetta	Astra Zeneca	39	SPPS	Type 2 diabetes	2005
21	Pramlintide	Symlin	AstraZeneca	37	SPPS	type 1 and 2 diabetes	2007
22	Lanreotide	Somatuline Depot	Ipsen Pharma	8	SPPS	Neuroendocrine tumours	2007
23	Degarelix	Firmagon	Ferring	10	SPS	Prostate cancer	2008
24	Tesamorelin	Egrifta	Theratechnologies	44	SPPS	HIV-associated lipodystrophy	2010
25	Liraglutide	Victoza	Novo Nordisk	31	rDNA	Type 2 diabetes	2010
26	Lucinactant	Surfaxin	Wind Tree	21	rDNA	Respiratory distress syndrome (RDS)	2012
27	Peginesatide	Omontys	Takeda	21	SPPS	Anaemia	2012
28	Pasireotide	Signifor	Novartis	6	SPPS	Cushing's disease	2012
29	Teduglutide	Gattex	NPS	33	rDNA	Adults with short bowel syndrome	2012
30	Albiglutide	Tanzeum	GlaxoSmithKline	645	rDNA	Type 2 diabetes	2014
31	Linaclotide	Linzess	Allergan Sales	14	SPPS	Irritable bowel syndrome with constipation	2017
32	Semaglutide	Ozempic	Novo Nordisk	31	SPPS	Type 2 diabetes	2017
33	Plecanatide	Trulance	Synergy Pharma	16	SPPS	Chronic idiopathic constipation	2017
34	Etelcalcetide	Parsabiv	Kai Pharma	8	SPPS	Hyperparathyroidism	2017
35	Abaloparatide	Tymlos	Radius Health	34	Hybrid	Osteoporosis	2017
36	Angiotensin II	Giapreza	La Jolla Pharma	8	SPPS	Increased blood pressure in adults with septic shock	2017

[a]rDNA, recombinant DNA; Hybrid, combination of solution-phase and solid-phase synthesis; SPPS, solid-phase peptide synthesis; SPS, solution-phase synthesis; E, extraction from natural sources.

5.2 Solution-phase Peptide Synthesis

Homogeneous solution-phase peptide synthesis is the method of choice for manufacturing small peptides (6–10 amino acid residues); most of the approved peptide pharmaceuticals in use today are manufactured using solution-phase methodology. The method has been shown to be useful for the synthesis of very long peptides and even small proteins. The longest peptide manufactured for commercial use by this method is calcitonin, containing 32 amino acid residues. The solution-phase method is still used to produce small-to-medium length peptides in quantities of up to hundreds of kilograms or even metric tons per year. Most peptides approved before 2000 were produced by this method. The purities of the crude peptides produced by solution-phase methods are generally better than those obtained by the solid-phase method and usually require fewer chromatographic steps for purification. Another advantage of solution-phase synthesis is that peptides can be produced in regular small-molecule cGMP pharmaceutical facilities, provided that the peptides are noncytotoxic and not of the "high potency" category.

Homogeneous LPPSs involves reversible protection of the carboxylic acid group of the C-terminal amino acid and the amino group of the N-terminal amino acid and activation of the free carboxyl group of the N-terminal amino acid, followed by step-by-step amide coupling. Isolation, purification and characterization of the peptide are performed at each step. This technique allows good quality control and it is a time-efficient process, yet not laborious. When a large-scale synthesis is planned, one must give considerable thought to the choice of synthesis method, *i.e.* linear or convergent synthesis, the selection of segments, their assembly and the orthogonal protecting groups used.[17]

5.2.1 Linear *Versus* Convergent Synthesis

In the manufacturing strategy involving linear synthesis, amino acids are coupled in succession until the entire sequence has been assembled. In convergent synthesis, small peptides, usually with 2–5 amino acid fragments (or segments), are assembled to give a final sequence. The size of the peptide usually determines the strategy and peptides up to five amino acids in length are generally manufactured by a linear strategy, whereas for longer peptides a convergent strategy is more suitable. In the convergent strategy, overall yields are normally better than those realized by linear synthesis. Additionally, since quality monitoring is involved for every fragment, the quality of the final peptide is usually very high. If a convergent strategy is planned, the selection of fragments is crucial and the following factors should be taken into consideration. Fragments should end with glycine or proline at the C-terminus as the risk of racemization is minimal with these two residues; if this is not possible, the second option is alanine or arginine.[18] Generally, one should restrict the fragment size to no longer than five amino acids. Before embarking on a large-scale convergent synthesis, it is prudent to test several possible fragment strategies on a smaller scale by solid-phase methods, then couple them in the solution phase. Using this approach, the development process is accelerated and gives good information about the extent of racemization and also provides a hint about the solubility of the segments.

5.2.2 Protecting Groups for Solution-phase Synthesis

5.2.2.1 α-Amino Group Protection

A number of different α-amino protecting groups are used in peptide synthesis, However, some of them are used only in industrial manufacturing. An excellent review of the various α-amino protecting groups that are used in peptide synthesis can be found in *Synthesis of Peptides and Peptidomimetics*.[19] For large-scale solution-phase synthesis, the most frequently used protecting groups are the *tert*-butoxycarbonyl (Boc) and benzyloxycarbonyl (Z) groups.[20] The Fmoc group is less preferred, because of the reactivity of the dibenzofulvene by-product and its removal post-deprotection. The Boc group is removed by acids and for large-scale production the most common reagents are neat trifluoroacetic acid (TFA) or TFA in dichloromethane or HCl–EtOAc mixtures. The latter method has the advantage of isolating the cleaved peptides as solid hydrochloride salts rather than oils. These isolated salts can be conveniently purified. The Z group is removed by reduction, usually by catalytic hydrogenolysis with H_2 and also by reduction with sodium in liquid ammonia. In the past, HBr in acetic acid (AcOH) was also frequently used for this purpose. However, on a large scale, the peptides precipitate as hydrobromide salts, which are hygroscopic and difficult to isolate.

5.2.2.2 α-Carboxyl Group Protection

A number of different α-carboxyl protecting groups are available for the peptide chemist to choose from, but some are suitable only for large-scale manufacturing processes. An excellent review of different α-carboxyl protecting groups that are used in peptide synthesis can be found in *Synthesis of Peptides and Peptidomimetics*.[21] The most suitable carboxyl protecting group for large-scale solution-phase synthesis is the free acid itself. However, this limits the choice of coupling agents to an active ester as the only method. For peptides that occur as C-terminal amides, protection is not required. A benzyl or *tert*-butyl ester is the most preferred carboxyl protecting group. These groups are removed in the same manner as the Z and Boc groups. Aliphatic esters such as methyl or ethyl are also used for carboxyl protection. Saponification, employed using NaOH or KOH, is the method of choice for the cleavage of these groups. This methodology has a high propensity for racemization and in such cases hindered amino acids such as proline are preferred. However, the problem with proline is its very slow or sometimes even zero reactivity. The advantage of simple esters such as methyl or ethyl is that they can be converted into hydrazides by treatment with hydrazine, which can then be used directly for coupling *via* the azide method.

5.2.2.3 Side-chain Protection

A detailed discussion of various side-chain protecting groups used in peptide synthesis is available in *Synthesis of Peptides and Peptidomimetics*.[22] In general, it is desirable to use the minimal side-chain protection approach for large-scale synthesis, as this will minimize the number of steps. However, the disadvantage of the minimal side-chain protection approach is that the peptide generally develops a hydrophilic character, which often complicates extraction of the peptide during work-up procedures. The δ- and ε-amino functions of ornithine and lysine must be protected and, as for the α-amino function, Z and Boc groups are recommended. The carboxyl functions of aspartic and glutamic acids must also be protected, preferably by the same esters as those mentioned above for α-carboxyl protection, *i.e.* benzyl or *tert*-butyl esters. The thiol function on cysteine may be protected by an acetamidomethyl (Acm) group or as a benzyl sulfide; the former group has the advantage of removal followed by concomitant disulfide formation with iodine in a single step. The latter is preferably cleaved by reduction with sodium in liquid ammonia. Trityl (Trt) is another possible protecting group and the combination of Acm and Trt groups is useful in facilitating site-directed cyclization.

5.2.3 Coupling Techniques

5.2.3.1 Coupling of Individual Amino Acids

Peptide chemists have a variety of coupling methods at their disposal today.[23] However, the most commonly used reagents in industrial-scale syntheses are mixed carboxylic–carbonic anhydrides, carbodiimides such as dicyclohexylcarbodiimide (DCC), water-soluble carbodiimide (WSCDI) and uronium reagents such as *O*-(benzotriazol-1-yl)-*N*,*N*,*N*′,*N*′-tetramethyluronium tetrafluoroborate (TBTU) and *O*-(benzotriazol-1-yl)-*N*,*N*,*N*′,*N*′-tetramethyluronium hexafluorophosphate (HBTU). In addition to these methods, commercially available preactivated amino acid derivatives, such as active esters, *e.g.* hydroxysuccinimide (HOSu) and *p*-nitrophenol (HONp) esters and *N*-carboxyanhydrides (NCAs), may be employed. These derivatives have the added benefits of not generating any by-products from the activating agent and also of being compatible with an unprotected C-terminal amino acid residue in the amino component. The latter is also the case for the azide coupling method. For the carbodiimides and uronium reagents, C-terminal protection is mandatory.

5.2.3.2 Coupling of Segments

When coupling segments, the risk for racemization is increased, owing to the possibility of oxazolone formation. However, several methods are available for this type of coupling.[24] The most common is the carbodiimide method (*e.g.* with DCC or WSCDI) with an added auxiliary nucleophile such as 1-hydroxybenzotriazole (HOBt) or 1-hydroxyazabenzotriazole (HOAt). In recent times, uronium reagents such as TBTU and HBTU have become popular owing to increased commercial availability. The mixed anhydride method, using isobutyl chloroformate, with or without an added auxiliary nucleophile, is also frequently used. The azide method, owing to potential low racemization, is also preferred in segment coupling. However, the azide anion produced as a by-product raises safety concerns at the manufacturing site.

5.2.4 Challenges in Manufacturing Peptides by Solution-phase Methods

The processes developed in the laboratory using the synthetic strategies discussed above cannot be adapted to manufacturing setups "as is." Many synthetic strategies and unit process operations such as distillation, filtration and drying operations routinely carried out in the laboratory cannot be carried out in the same fashion at the manufacturing scale.[17] A development scientist must therefore take many technical, commercial and safety aspects into consideration during the product development process for easy transfer of the technology. Some of these aspects are discussed below.

5.2.4.1 Technical Challenges

In any multistep process, the number of steps should be kept to a minimum in order to increase the overall yield. The greater the number of steps, the greater are the time required and the labour involved. Therefore, attempts should be made to decrease the number of steps. Typically, the synthesis of long peptides is performed by segment condensations, *i.e.* by the convergent strategy in industry. Convergent synthesis is not only more suitable and convenient but also advantageous from a safety aspect. From the quality perspective, convergent synthesis offers the advantage of efficient in-process controls, involving isolation, purification and characterization of each fragment and fixed quality specifications. This helps in determining the root cause in the event of batch failures. Identifying the root cause of failures and designing effective corrective and preventive action (CAPA) are key components of GMP manufacturing strategies.

In cases of failure, it is easier to resynthesize a shorter segment in a convergent synthesis than a longer sequence in a linear synthesis. Peptides of up to five amino acids in length are usually conveniently assembled and isolated by conventional methods, using extractions and precipitations.

However, from a practical point of view, it may be advantageous to use even smaller segments of two or three amino acids in length. The reason for this is that it is often easier to obtain purer segments by precipitation or even crystallization. Second, the use of smaller segments may shorten the production time because of the possibility of using parallel production equipment.

The use of extreme reaction conditions should be avoided in large-scale manufacturing processes. Reaction temperatures of >120 or <-10 °C need specialized and expensive equipment. Hence the development scientist should double-check such requirements in the process before transferring the technology. Other extreme reaction conditions, such as high pressure, very dry conditions, very long reaction times, highly exothermic reactions or the use of very specialized equipment, are also not routine in the manufacturing process. Reaction volumes and work-up volumes should be thoroughly optimized. High reaction volumes and work-up volumes not only need more solvent but also increase disposal volumes and costs. Higher reaction and work-up volumes necessitate a larger reactor size and result in lower productivity.

The development scientist should be selective in using solvents. Residual solvents in pharmaceuticals are considered as impurities. Harmonized Guideline Q3C(R6) of the International Council on Harmonisation of Technical Requirements for Registration of Pharmaceuticals for Human Use (ICH) classifies organic solvents into three types and specifies the allowable limits to be present in the final pharmaceutical products.[25] These limits are fixed based on their tolerable daily intake (TDI), the basis for this toxicity of the solvent. The most preferred solvents are environmentally friendly class 3 solvents such as acetone, ethyl acetate, diethyl ether, pentane, ethanol, tetrahydrofuran, formic acid and acetic acid. These solvents have higher permitted residual solvent limits in the final product, thus demanding less time and labour during manufacturing operations. Class 2 solvents include chloroform, acetonitrile, toluene, methanol and hexane. These solvents have higher residual solvent limits than class 3 solvents and their use should be limited. Class 1 solvents such as carbon tetrachloride, dichloroethane and benzene are highly toxic and should be completely avoided in manufacturing processes. Table 5.2 lists solvents with allowable limits in pharmaceuticals.

Purification of the final product by chromatography is demanding in terms of time and resources on a multikilogram scale. With the currently available technology, it is not possible to scale up in a linear manner. The Purification section (Section 5.5) discusses more challenges involved in the purification of peptides.

5.2.4.2 Commercial Aspects

Several different protected and activated amino acid derivatives and coupling agents are now available commercially. However, sourcing these

Table 5.2 ICH classification of solvents with allowable limits in pharmaceuticals.

Class of solvents	Solvent (maximum allowable limit/ppm)		
Class 1 (solvents to be avoided)	Benzene (2)	1,2-Dichloroethane (5)	1,1,1-Trichloroethane (1500)
	Carbon tetrachloride	1,1-Dichloroethene (8)	
Class 2 (solvents to be limited)	Acetonitrile (410)	1,4-Dioxane (380)	Pyridine (200)
	Chloroform (60)	Ethylene glycol (620)	Tetrahydrofuran (720)
	Cyclohexane (3880)	Formamide (220)	Toluene (890)
	Dichloromethane (600)	Hexane (290)	Xylene (2170)
	N,*N*-Dimethylacetamide (1090)	Methanol (3000)	
	N,*N*-Dimethylformamide (880)	Methyl butyl ketone (50)	
Class 3 (solvents with low toxic potential)	Acetic acid (5000)	Ethyl acetate (5000)	Methyl ethyl ketone (5000)
	Acetone (5000)	Formic acid (5000)	Methyl isobutyl ketone (5000)
	Anisole (5000)	Heptane (5000)	
	Ethanol (5000)	Isopropyl acetate (5000)	

raw materials from reliable vendors who are capable of maintaining the same quality material year after year is a major challenge in the manufacture of peptides. For these reasons, the manufacturers need to build up a good relationship with the vendors by engaging them as partners rather than just suppliers. Vendors' support becomes very critical during the regulatory review and approval process of New Drug Applications/Abbreviated New Drug Applications (NDAs/ANDAs). The regulators generally would like to know all of the details of the starting materials and for this the vendor's support is needed throughout the review process.

A number of protecting groups and activated amino acids are available commercially. However, owing to the high cost and the limited availability of good suppliers, one must aim at minimum protection strategies. For example, the use of side-chain protected arginine derivatives, such as the expensive 4-methoxy-2,3,6-trimethylbenzenesulfonyl (Mtr) or 2,2,4,6,7-pentamethyldihydrobenzofuran-5-sulfonyl (Pbf) derivatives, may be preferred from a chemical point of view, but because of their high cost, the inexpensive HCl salt or nitro derivative are the preferred alternatives. Similar considerations can affect the choice of coupling additives, such as HOBt and HOAt. Even though the latter is currently the preferred reagent, especially for demanding couplings, its cost is prohibitively high on a large scale and therefore the less expensive reagent (HOBt) is typically used.

5.2.4.3 *Safety Aspects*

While designing the process, the development scientist should also keep safety and environmental considerations in mind. The reagents, chemicals and solvents used in the industrial processes must not only cheap be and commercially available but also be safe to use for both humans and the environment. Some of these considerations are as follows:

- *Solvents:* Diethyl ether is a routinely used solvent in the laboratory. However, owing to its low flash point, it is highly prone to fire and/or explosion and hence should generally be avoided in manufacturing processes. Similarly, dichloromethane is an ozone-depleting solvent restricted by regulatory agencies.
- *Reagents:* The popular benzotriazol-1-yloxytris(dimethylamino)phosphonium hexafluorophosphate (BOP) coupling reagent is to be avoided owing to the presence of hexamethylphosphoramide (HMPA), which is a carcinogen. The use of HF as a cleavage reagent in commercial manufacturing is now obsolete owing to its hazardous and corrosive nature.
- The manufacture of peptides involves huge quantities of solvents and chemicals. Typically, the production of 1 kg of peptides generates more than 1000 kg of waste, and much more in solid-phase synthesis.

5.3 Solid-phase Methods

Solid-phase peptide synthesis was invented by Bruce Merrifield in 1963, for which he was awarded the Nobel Prize in Chemistry in 1984.[7] It was one of the greatest disruptive innovations in the field of chemistry up to the present day. Since the start of experimental chemistry in the seventh century, chemical syntheses were carried out in a homogeneous solution phase in regular glassware. Rarely the heterogeneous phase was used, in cases such as catalytic hydrogenation. In many ways, SPPS can be considered as the first technology at the laboratory scale in the field of chemistry. It was the ingenuity of Merrifield to think of a heterogeneous solid phase for application in the synthesis of complex organic molecules such as peptides. As solid-phase synthesis represented a paradigm shift, there was huge resistance to acceptance by the chemical community conversant with regular solution-phase synthesis.

SPPS offers the great advantage of carrying out all reactions – amino acid couplings, deprotection, peptide elongation – in one pot. The unreacted reagents and by-products are removed by repeated washings. The desired peptide is removed from the solid support by treatment with a cocktail containing a strong acid. The purification and isolation are performed as a final step after the desired length of peptide has been built. The reaction cycles are short compared with solution-phase synthesis, a factor that allows for shorter manufacturing times. For many years, the solid-phase synthesis approach was thought to be unsuitable for the large-scale manufacture of peptides for reasons that include (i) high process costs and lack of scalability, (ii) lack of adequate in-process controls during the assembly of the protected peptide on the resin support used for the synthesis and (iii) the generally low purity of the final products.

Because of these perceived drawbacks, SPPS was not used much in commercial manufacturing until about 2000. Advances in protecting groups, the invention of the base-labile Fmoc amino protecting group by Carpino in 1970 and the development of high-performance liquid chromatographic (HPLC) purification and analysis in the 1980s created confidence and hope that SPPS could be a method for manufacturing peptides. Based on the protecting groups used for synthesis, SPPS can be classified into two strategies: Boc/Bzl (benzyl) and Fmoc/^{t}Bu. The Boc/Bzl method was adopted by Merrifield at the time of the invention of SPPS technology.[7] In this strategy, α-amino groups are protected by Boc and the carboxylic acid is protected as a benzyl ester. Here temporary cleavage of Boc is effected with TFA and final permanent cleavage from the resin is achieved with HF. The major drawback of the Boc/Bzl method is the handling of corrosive acids such as TFA and HF. This process used to be the technology followed for solid-phase synthesis until the introduction of the Fmoc group, and it is now obsolete. In the Fmoc/^{t}Bu strategy,[26] the temporary protecting group is Fmoc, which is cleaved with a mild base such as piperidine, and final cleavage is performed with TFA. Currently, solid-phase synthesis in the industry uses only the Fmoc/^{t}Bu strategy, hence further discussion in this chapter is limited to this approach. In this section, the methodology of SPPS and challenges in commercial manufacture are discussed. A typical SPPS strategy involves the following steps: (i) attachment of the first protected amino acid derivative to the polymeric carrier (resin); (ii) cleavage of the temporary protecting group; (iii) coupling of the remaining amino acid derivatives into the desired peptide sequence; (iv) multiple washes of the resin after each step; (v) detachment of the peptide (either protected or deprotected) from the polymeric carrier; (vi) cleavage of the temporary and permanent protecting groups on the resin or in solution; (vii) isolation of the crude peptide; (viii) purification of the peptide; and (ix) isolation of the final product.

5.3.1 Solid Supports for Synthesis

A general review of the synthesis of peptides on supports can be found in *Synthesis of Peptides and Peptidomimetics*.[28a] Polystyrene is the most common core resin in SPPS; other core matrices such as polyacrylate, polyacrylamide and poly(ethylene glycol) are also commercially available, but their use is limited to research laboratories. Polystyrene is hydrophobic and is soluble in hydrophobic nonpolar solvents. Crosslinking of polystyrene chains with 1–2% divinylbenzene (DVB) yields crosslinked polystyrene, which is insoluble in nonpolar organic solvents. Although the crosslinked polystyrene resins are insoluble in organic solvents, they become solvated and swell in solvents such as dimethylformamide (DMF) and dichloromethane (DCM). A 1 g amount of 1% DVB-crosslinked resin swells to 4–6 times its original volume in DCM. In contrast, 1 g of 2% DVB-crosslinked resin swells to only 2–4 times its original volume in DCM. The swelling factor is important in solid-phase synthesis since the reaction kinetics are diffusion controlled. Consequently, a resin that swells more will have a higher diffusion rate of reagents into the core of the matrix, resulting in shorter reaction times and better chemical conversions.

5.3.1.1 Bead Size

Polystyrene beads are available in sizes ranging from <1 to 750 μm in diameter. The reaction kinetics are generally faster using smaller beads owing to the higher surface area-to-volume ratio. In practice, however, beads that are too small can lead to extended filtration times. Beads in the range 35–150 μm in diameter offer a good balance of reaction kinetics *versus* reliability. Bead size is commonly reported as the Tyler mesh size, which is inversely proportional to the nominal diameter. Two commonly available resin sizes are 100–200 and 200–400 mesh (75–150 and 35–75 μm, respectively). In an industrial setup 100–200 mesh is commonly employed.

5.3.1.2 Linkers

Merrifield's original solid support was 2% DVB-crosslinked polystyrene, which was derivatized by chloromethylation. This is not very suitable as it required harsh reagents such as HBr–AcOH or HF for cleavage of the peptide from the resin. To make synthesis friendlier, linkers were introduced. The purpose of the linker is to provide a linkage between the synthetic peptide chain and the solid support. By manipulating the structure of the linker, resins ranging from extremely acid labile to base labile can be prepared. Based on the type of linker, it possible to produce a C-terminal ending with acid or amide or any similar functional groups. Most linkers are designed to release peptide acids or amides upon treatment with TFA. A large variety of resins are commercially available and some of the commonly used resins are listed in Table 5.3.

5.3.2 Cleavage of Temporary Protecting Groups

Cleavage of temporary, *i.e.* N^{α}, protecting groups during the peptide assembly stage is an important step in all peptide synthetic processes and the orthogonality principle should be applied. In the case of the Fmoc/^{t}Bu strategy, the N^{α} protecting groups are generally cleaved with a 5–20% solution of piperidine in DMF.[27]

5.3.3 Coupling Reagents

As in the case of solution-phase synthesis (as discussed earlier),[23] a wide range of coupling reagents and methods are also available for use in the solid-phase approach. The choice is often influenced both by the type of the peptide sequence and by the ease/difficulty faced in couplings. Carbodiimides, *i.e.* *N,N*-dicyclohexylcarbodiimide (DCC) and *N,N*-diisopropylcarbodiimide (DIC), in the presence of 1-hydroxybenzotriazole are often the preferred reagents for this purpose, owing to their low cost compared with alternative reagents. Industrially, DIC is preferred over DCC as it gives

Table 5.3 List of commonly used resins in solid-phase peptide synthesis.

Resin	Core polymer	Linker structure	Final product	Cleavage
Merrifield	Polystyrene crosslinked with 1.0% DVB		Acid	Acid labile
Wang	Polystyrene crosslinked with 1.0% DVB	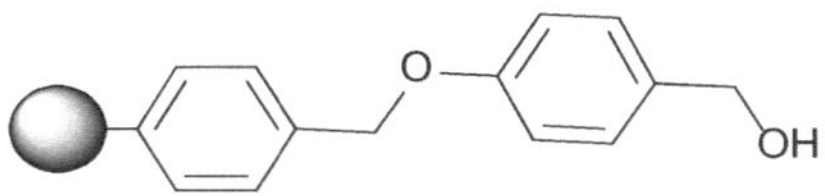	Acid	Acid labile
HMPA AM resin	Aminomethyl-polystyrene crosslinked with 1.0% DVB	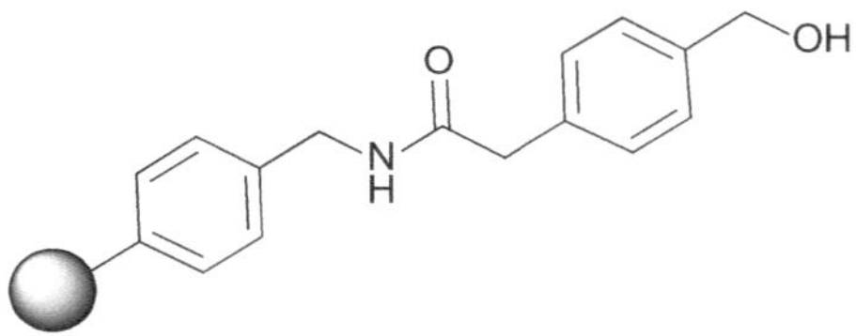	Acid	Acid labile
Tanta gel MB PHB	Poly(ethylene glycol) grafted on polystyrene crosslinked with 1.0% DVB	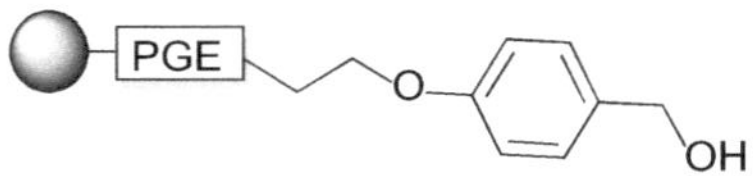	Acid	Acid labile
2-CTC	Polystyrene crosslinked with 1.0% DVB	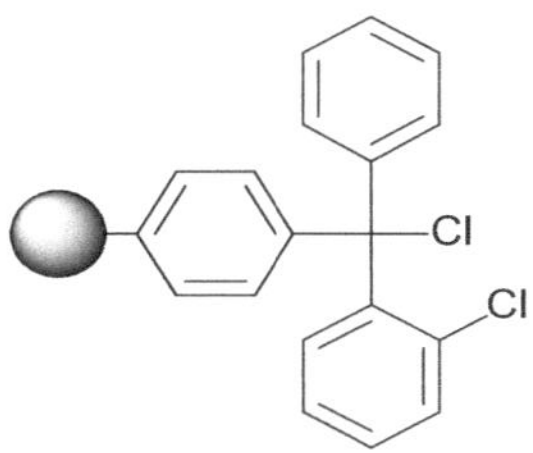	Acid	Hyper acid labile

Table 5.3 (*Continued*)

Resin	Core polymer	Linker structure	Final product	Cleavage
HMPB AM resin	Aminomethyl-polystyrene crosslinked with 1.0% DVB	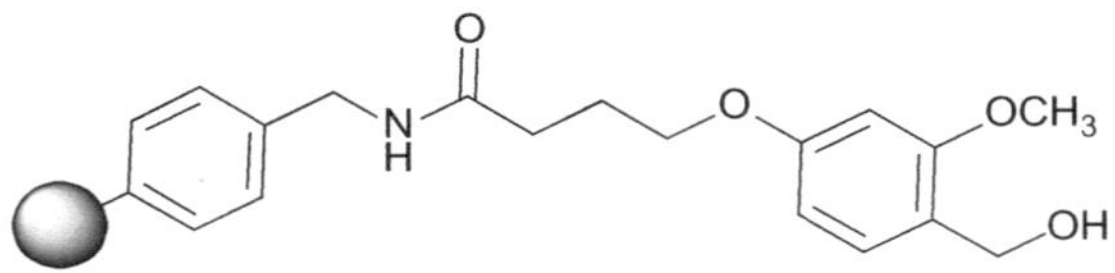	Acid	Hyper acid labile
Rink acid	Hydroxymethyl-polystyrene crosslinked with 1.0% DVB	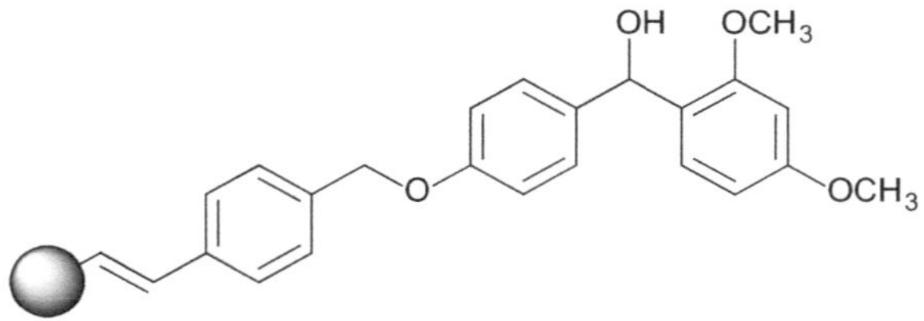	Acid	Hyper acid labile
SASRIN resin	Polystyrene crosslinked with 1.0% DVB	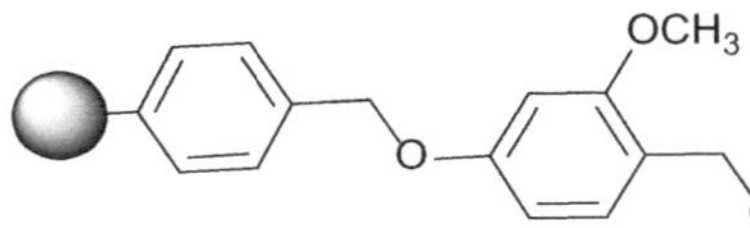	Acid	Hyper acid labile
Rink amide resin	Polystyrene crosslinked with 1.0% DVB	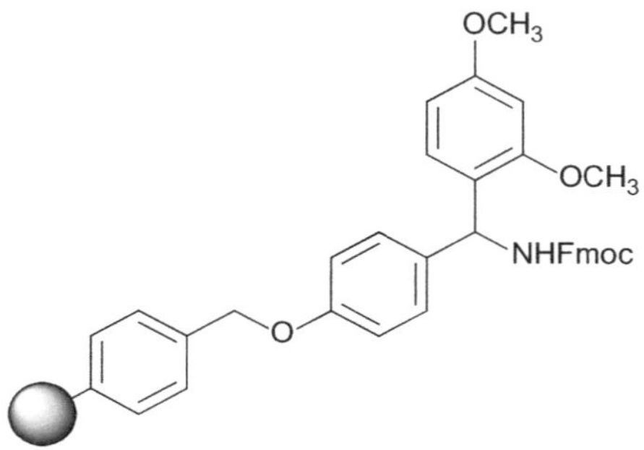	Amide	Acid labile

Rink amide AM resin	Aminomethyl-polystyrene crosslinked with 1.0% DVB	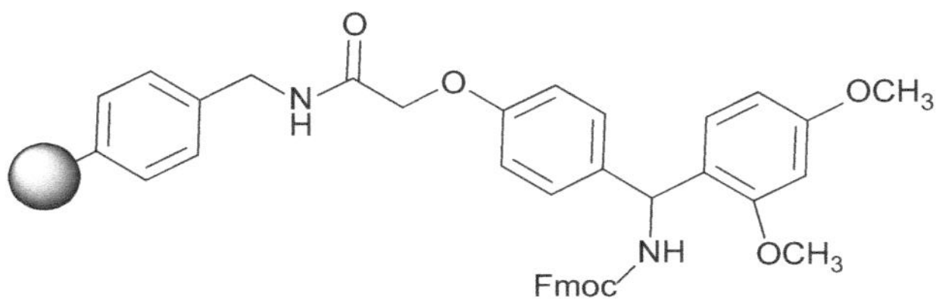	Amide	Acid labile
PAL resin	Polystyrene crosslinked with 1.0% DVB	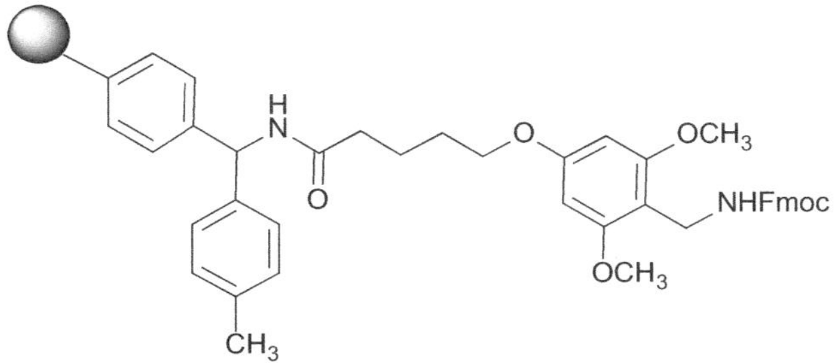	Amide	Acid labile
Rink amide MBHA resin	Polystyrene crosslinked with 1.0% DVB	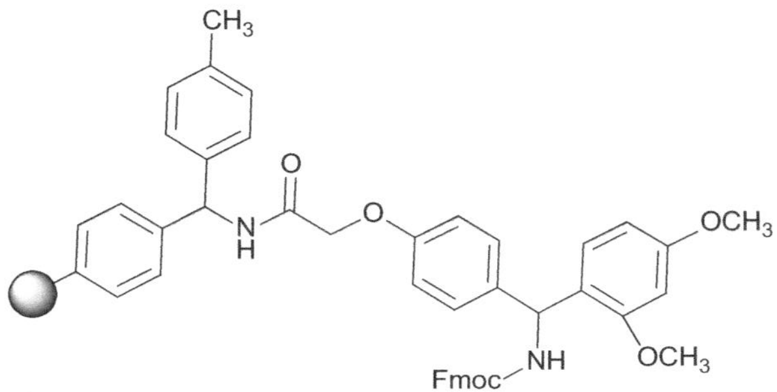	Amide	Acid labile

Table 5.3 (*Continued*)

Resin	Core polymer	Linker structure	Final product	Cleavage
Tantagel MB RAM	Polystyrene crosslinked with 1.0% DVB	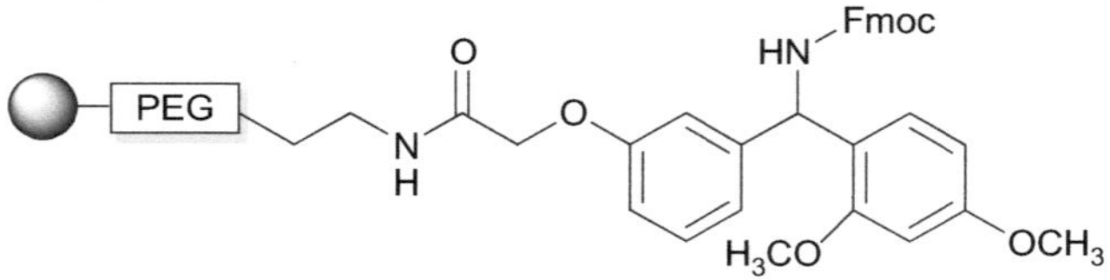	Amide	Acid labile
Sieber resin	Polystyrene crosslinked with 1.0% DVB	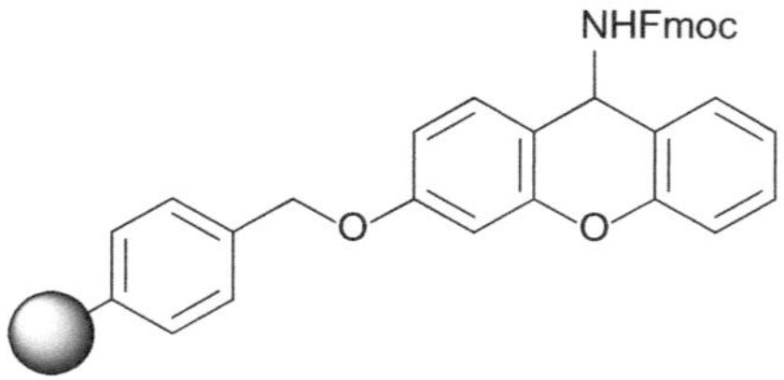	Amide	Hyper acid labile
MBA AM	Aminomethyl-polystyrene crosslinked with 1.0% DVB	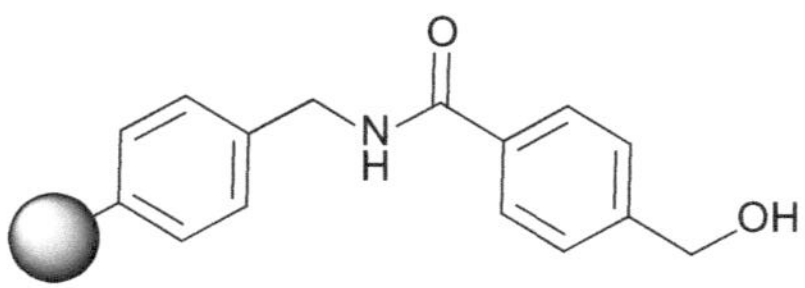	Acids, amides, esters based on the nucleophile of the cleavage reagent	Base labile

soluble urea. Other coupling agents such as TBTU and HBTU are also used industrially.

5.3.4 Cleavage/Deprotection

In most solid-phase strategies, the side-chain deprotection step and detachment from the resin are carried out in a single step. The method for detachment of the peptide from the matrix is determined by both the C-terminal modification of the peptide, if any, and the synthetic strategy employed. Neat TFA is the reagent of choice for this purpose. In the case of peptide fragments with retention of side-chain protecting groups, hyper acid-labile resins, such as 2-chlorotrityl, HMPB AM, SASRIN and Sieber resins, and 1–5% TFA in DCM are used.

5.3.5 In-process Monitoring

A weakness of the solid-phase method in general, especially from the point of view of large-scale cGMP manufacturing, is the lack of proper quantitative in-process control during the peptide assembly stage. Since the growing peptide chain is bound to the insoluble carrier during all steps, *i.e.* during deprotections and couplings, the possibility of performing a reliable analytical evaluation of the process, in terms of peptide properties and application of state-of-the-art analytical chemistry, is very limited. The methods used for monitoring the completeness of cleavage of temporary protecting groups and completeness of the coupling reactions are visual qualitative tests, namely (i) the ninhydrin (Kaiser) test,[28b] (ii) the chloranil test (suitable for proline),[29] (iii) the picric acid test,[30] (iv) the Bromophenol Blue (3′,3″,5′,5″-tetrabromophenolsulfonphthalein) test[31] and (v) the pyridine hydrobromide test.[32]

The sensitivity of the individual tests can vary and depends considerably on the peptide sequence, owing to the difference in the accessibility and reactivity of the amino groups bound to the polymer. It should be emphasized that, for secondary amines, the ninhydrin test can provide false results and, therefore, the chloranil test is recommended. Direct monitoring of the peptide assembly can be achieved through the detachment of the peptide from small samples of the peptide–resin using TFA and subjecting it to HPLC analysis. It is usually safe to carry out two visual qualitative tests serving as in-process control tests. Regulatory agencies recommend the use of two tests rather than a single test for reaction monitoring.[33] The Kaiser test is the most commonly used primary test, followed by a second test such as the chloranil or Bromophenol Blue test.

5.3.6 Purification/Isolation

The purification of the final product is one of the most critical steps in the solid-phase synthesis process. The choice of methods is dependent on the nature and complexity of the contaminants in the crude product. Because no intermediate purifications are performed in the solid-phase method, the crude products often contain complex mixtures of contaminants, such as deletion sequences, truncated sequences and diastereomers, which can pose a considerable challenge in purification. Whereas preparative reversed-phase HPLC (RP-HPLC) is often used on a laboratory scale, a combination of two orthogonal methods, such as ion-exchange chromatography (IEX) and RP-HPLC, is often employed on a large scale. However, it is important to reiterate that, since purification steps are often demanding on time and resources, minimization of the extent of side products through careful optimization of all synthetic steps, and also the cleavage step, can be critical to the overall success of a large-scale solid-phase process.

5.3.7 Challenges in Manufacturing by Solid-phase Methods

The challenges for economic and environmentally friendly processes are multipronged. Some of the challenges discussed for solution-phase synthesis also apply to solid-phase processes. The additional challenges are (i) sourcing of raw materials such as resins and Fmoc-protected amino acids and (ii) purification.

5.3.7.1 Sourcing Challenges

The quality of peptides depends to a great extent on the quality of the resins used. Unfortunately, the number of commercial suppliers of these resins is limited. Generally, the batch-to-batch reproducibility of resins has been unsatisfactory with most suppliers. The quality of the resin cannot be fully guaranteed by simple quality control tests such as particle size, swelling and substitution. Hence it is generally advisable to carry out a performance test for every new lot of resin even from the same manufacturer.

The quality of protected amino acids dramatically affects the purity, impurity profile and yield of the peptide. It is important that the quality of these raw materials is adequately controlled for commercial manufacture. Since protected amino acids form part of the final molecule, their quality should be controlled by setting proper specifications. The specifications must include assay, related impurities such as dipeptides, free amino acids and β-alanine and chiral purity. If the protected amino acids are sourced from different suppliers, their major quality attributes should be identical.

It is common to use a large excess of up to two or more equivalents of the protected amino acid to drive the reactions to completion and increase the purity of the products. This practice is unacceptable from an economic standpoint in large-scale processes and therefore reagents and reactants should be used in as close to stoichiometric amounts as possible, in order to minimize raw material costs.

5.3.7.2 Purification Challenges

In solid-phase synthesis, the intermediates are not purified and not controlled by quality checks. Therefore, the crude peptide is often a complex mixture of contaminants such as deletion sequences, insertion and truncation products and diastereomers. These contaminants pose significant problems during purification since the relative retention times of their peaks are often very similar to that of the main peak. Removal of such contaminants is therefore very difficult, leading to substantial reductions in yields and a corresponding cost escalation. It is therefore critical to use a sensitive in-process method for monitoring coupling reactions and that techniques such as end-capping and acetylation be performed in order to minimize the formation of deletion sequences. Similarly, the coupling reagent should be chosen carefully so as minimize the possibility of racemization. If the purification difficulties are encountered at an early stage in development, it is prudent to reconsider the synthetic approach rather than search for more efficient purification.

5.4 Hybrid Methods

In this approach, a large peptide is split into a number of fragments that can be easily synthesized by solid-phase synthesis. These fragments are cleaved from the solid support with their protecting groups intact and are assembled in solution-phase synthesis. New-generation chlorotrityl chloride (CTC) resin is commonly used for synthesizing fully protected peptide segments by the solid-phase technique. These fully protected segments are subsequently assembled by solid- or solution-phase methods.[34] Commercial production of metric ton quantities and FDA approval of enfuvirtide gave a fillip to hybrid synthesis.[12b] The hybrid approach is particularly attractive for the manufacture of large molecules since it combines the benefits of both the solid- and solution-phase methods.

The advantage of a hybrid approach is that the solid-phase synthesis of fragments can be developed and scaled up relatively rapidly. One can also avoid many of the solubility problems often encountered in solution-phase syntheses of relatively long segments. Production cycle times are short compared with solution-phase methodologies and yields and purities are often higher. The other advantage is that the quality of fragments can be controlled and hence the quality of the final peptide will be superior to that obtained by a complete solid-phase method. Through optimization of the sequences of the fragments, especially based on their solubility characteristics, the final stages of the process, *i.e.* the fragment assembly and deprotection, can be scaled up conveniently by conventional solution-phase methodologies. Thus, the hybrid synthesis possesses the advantages of both the solid- and solution-phase methods, namely rapid synthesis of fragments with high purities by the solid-phase method and full monitoring of coupling reactions, with isolation, purification and characterization of intermediates, by solution-phase methodology.

5.5 Purification of Peptides

Peptides are produced by solution-phase synthesis, solid-phase synthesis, hybrid synthesis, extraction from natural sources and recombinant DNA technology. The crude peptides, irrespective of the method of preparation, are generally not pure enough to be used as pharmaceuticals and they need extensive and laborious purification. Chemically prepared peptides contain many impurities and the physicochemical properties of these impurities are very similar to those of the product of interest. Hence the classical purification techniques such as precipitation and crystallization that are applied to small organic molecules are not suitable for peptides. Crystallization is rarely attempted for the final purification of peptides except for small peptides with fewer than five amino acids.[17,35]

Peptides are generally purified by countercurrent distribution and chromatographic techniques such as gel permeation chromatography (GPC), IEX and RP-HPLC. Countercurrent distribution was used for the purification of many of the early peptide drug substances, but it has now been replaced by more powerful techniques, such as those based on preparative RP-HPLC.[17] The regulations stipulate that impurities in peptides should be controlled to a level of $<0.1\%$, which made the industry pay more attention to robust purification methodologies.[36] The ease of purification of peptides by chromatography depends on the quality of the crude peptide, type of peptide (hydrophilic, hydrophobic and conjugated) and the chain length of the peptide. Generally, peptides with fewer than 10 amino acid residues do not pose many difficulties and can be purified by one or two chromatographic steps. For peptides with >20 amino acid residues, conjugation poses problems and multiple chromatographic steps are required. The purification strategy for peptides varies from case to case and depends on the manufacturing technology used for the preparation, *i.e.* solid phase, liquid phase or rDNA.

5.5.1 Types of Chromatographic Techniques

Proteins and peptides are purified using chromatographic techniques that separate them according to differences in their specific properties, such as charge (ion-exchange chromatography, IEX), size (gel permeation chromatography, GPC), hydrophobic interaction (reversed-phase chromatography, RPC) and biorecognition (affinity chromatography, AC). However, peptides are usually purified by IEX, GPC and RPC techniques.

5.5.1.1 Ion-exchange Chromatography (IEX)

The stationary phase in IEX is either positively or negatively charged and separation is based on a reversible interaction between the charged peptide and oppositely charged chromatographic medium.[37] Depending on the net differences in the charges of the solutes, separation takes place. The method involves loading a column with the sample and then the conditions are altered so that the bound substances are eluted differentially. The elution is usually performed by increasing the salt (usually NaCl) concentration or changing the pH. The target peptide is concentrated during binding and collected in a purified and concentrated form.

The resins employed in IEX are of four types, namely strong cation-, strong anion-, weak cation- and weak anion-exchange resins. IEX is generally used in the capture and intermediate stages of purification of the peptides. It is the method choice in the purification of peptides prepared by solution and recombinant technology. The main advantages of IEX are high resolution and high capacity.

5.5.1.2 Gel Permeation or Size-exclusion Chromatography (GPC or SEC)

Gel permeation media separate molecules according to differences in molecular size. In this type of chromatography, molecules are separated based on their molecular weight: higher molecular weight components elute before low molecular weight components.[38] The technique involves loading the sample and then eluting with a suitable mobile phase. The technique can only be used when sample volumes are minimized. Since the buffer composition does not directly affect resolution, the buffer conditions can be varied to suit the sample type or the requirements for the next purification, analysis or storage step.

GPC is generally used at the polishing stages. It cannot be used for water-insoluble peptides. The GPC method was a widely and routinely used method for the purification of peptides before the advent of RPC.

5.5.1.3 Reversed-phase Chromatography (RPC)

RPC is the chromatographic method of choice for peptide purification. The status of therapeutic peptides as an important segment of the pharmaceutical industry owes a considerable debt to RPC-mediated purification and analysis. The remainder of this chapter deals with details of the RPC purification of peptides. RPC separates molecules of differing hydrophobicity based on the reversible interaction between the molecule and the hydrophobic surface of the chromatographic medium.[39] Samples bind as they are loaded onto a column. The conditions are then altered so that the bound substances are eluted differentially. Owing to the nature of the reversed-phase matrices, the binding is usually very strong and requires the use of organic solvents and other additives (ion-pairing agents) for elution. Elution is usually performed by increasing the organic solvent concentration, most commonly acetonitrile. Molecules that are concentrated during the binding process are collected in a purified and concentrated form.

RPC is often used in the final polishing of oligonucleotides and peptides and is ideal for analytical separations, such as peptide mapping. RPC is not recommended for protein purifications if recovery of activity and return to a correct tertiary structure are required, since many proteins are denatured in the presence of organic solvents.

5.5.1.4 *Affinity Chromatography (AC)*

This type of chromatography is specifically used for peptides produced by recombinant technology. AC separates proteins on the basis of a reversible interaction between a protein (or group of proteins) and a specific ligand attached to a chromatographic matrix. AC can be used only when a suitable ligand is available.[40]

The sample is applied under conditions that favour specific binding. The target protein/peptide(s) is specifically and reversibly bound by a complementary binding substance (ligand). Unbound material is washed away and the bound target protein/peptide is recovered by applying conditions that favour desorption. Desorption is performed specifically, using a competitive ligand, or nonspecifically, by changing the pH, ionic strength or polarity. Proteins, which are concentrated during binding, are collected in a purified, concentrated form. The principal advantages of AC are high capacity and high resolution.

5.5.2 Stages in the Purification of Peptides

Generally, it not possible to purify the crude peptide to a pharmaceutical grade in a single chromatographic step. The purity of the crude peptide obtained directly from synthesis varies from a low of about 10% to a high of about 70%. In order to purify crude material of such quality, one must generally carry out multiple chromatographic purification runs to achieve a product meeting present-day regulatory requirements. There are essentially three functional stages in the purification of crude peptide,[41] normally denoted the capture, intermediate and polishing stages. The various functional stages in purification are shown in Figure 5.1. As can be seen, the purity of the product increases from capture to polishing.

5.5.2.1 *Capture Stage*

This is the first step in the purification procedure. The purpose of the capture stage is to isolate, concentrate and stabilize the target molecule from the crude preparation rapidly and with good recovery. At this stage, the sample volume is usually at its highest and the sample may contain a very large number of related and unrelated impurities, including particulates and viscous materials. The main purpose of the capture stage is to remove junk from the product of interest, and sometimes it is also called the junk removal stage. The capture step is not expected to be highly resolving but is

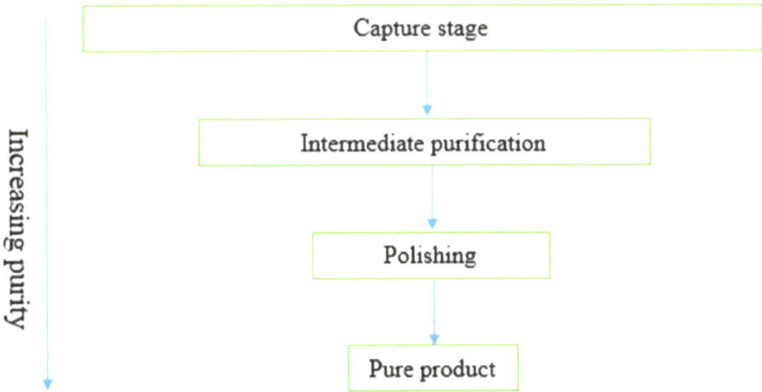

Figure 5.1 Functional stages in the purification of peptides.

required to isolate the molecule of interest from the contaminating substances that are dissimilar to the desired molecule. The capture step may be considered as a group separation rather than as a high-resolution purification. Consequently, time, capacity and recovery are more important than resolution in a successful capture step. AC is commonly used for the capture of peptides produced by recombinant technology. IEX and RPC are the chromatographic techniques of choice for the capture of synthetic peptides produced by both solution- and solid-phase methods.

5.5.2.2 Intermediate Stage

In the intermediate purification stage, the focus is to separate the target molecule from most of the bulk impurities such as other proteins, peptides, nucleic acids, endotoxins and viruses. This technique is mostly used when the peptides are produced by recombinant technology. The critical requirements of this stage are recovery and resolution. The chromatographic technique should ideally be capable of resolving impurities that are often similar to the target molecule regarding functional or structural properties. Depending on the quality of the crude material and final purity requirements, this stage of purification may or may not be required. In the case of synthetic peptides produced by solid-phase synthesis, the intermediate stage is useful in separating some critical closely related impurities, preferably by using RPC.

5.5.2.3 Polishing Stage

Polishing is the final step in the preparation of a product; after this stage of purification, the product is suitable for its intended use. The polishing step is used to remove trace contaminants and impurities. Contaminants at the polishing stage are often very similar to the target molecule. In the case of biomolecules, this step helps in removing dimers, oligomers, aggregates, oxidized amino acids, protease-clipped molecules, deamidated amino acids, *etc.* However, in the case of peptides, its purpose is to remove impurities, concentrate, exchange the counter ion and exchange the solvent to one compatible with lyophilization. The goals of the polishing stage are to produce a peptide with a product purity of almost 100% and with a recovery of about 99%. In the case of biomolecules, it is often performed by SEC to remove the dimers and aggregates. SEC was often the method chosen in the early days of peptide production.

5.5.3 Purification of Peptides by Reversed-phase High-performance Liquid Chromatography (RP-HPLC)

After the Merrifield solid-phase peptide synthesis, the introduction of analytical and preparative HPLC may be considered the most critical event in the field of peptides. Before the introduction of HPLC, the homogeneity of peptides was always in question. Nowadays, HPLC and the higher resolution ultra-high-performance liquid chromatography (UHPLC) are the conventional analytical tools employed for the analysis of peptides. These high-resolution tools demand homogeneous peptides with stringent impurity limits for meeting regulatory guidelines. Owing to these developments, preparative HPLC became an indispensable part of peptide manufacture. Like analytical HPLC, preparative HPLC instrumentation involves a pump–column–detector system. The column can be regarded as the heart of the chromatographic system, and the proper choice of the column is of the utmost importance for successful large-scale purification. The column material and mode of compression (static *versus* dynamic) of the column are crucial factors in the design of a chromatographic system for the industrial-scale purification of peptides. Efficient column packing is another critical factor. Therefore, it is necessary to use sensitive test methods for determination of the number of theoretical plates and symmetry of the packed column. The purification of the product takes place in the column. The factors that influence the purification are the stationary phase and the mobile phase, and a detailed discussion of these two aspects for efficient purification is presented in the following.

The most powerful method for peptide purification is undoubtedly RPC utilizing hydrophobic interactions as the main separation principle. It is characterized by the use of a stationary phase and an aqueous mobile phase containing an organic solvent such as acetonitrile or an alcohol. Various chromatographic media have been used for the large-scale purifications of peptides on reversed-phase media, the most popular being those based on C_4, C_8 and C_{18} alkyl chains attached to a silica surface.

5.5.3.1 Stationary Phase Considerations

The separation mechanism in RPC is based on the hydrophobic binding interaction between the solute molecule in the mobile phase and an immobilized hydrophobic ligand, *i.e.* the stationary phase.[39] Silica was the first polymer to be used as the base matrix for RPC media. Reversed-phase media were originally developed for the purification of small organic molecules and then for the purification of low molecular weight, chemically synthesized peptides. Silica is produced as porous beads that are chemically stable at low pH and in the organic solvents typically used for RPC. The combination of the porosity and physical stability of silica is important since it allows the media to be prepared to have optimal loading capacities and high efficiencies. It is worth noting that, although the selectivity of silica-based media is largely controlled by the properties of the ligand and the mobile phase composition, the different processes for producing silica-based matrices will tend to produce media with different patterns of separation. This is the principal reason for the variation in performance of silica from manufacturer to manufacturer. The chemistry of the silica allows simple derivatization with ligands of various carbon chain lengths. The carbon content and the surface density and distribution of the immobilized ligands can be controlled during the synthesis.

The primary disadvantage of silica as a base matrix for reversed-phase media is its chemical instability in aqueous solutions at higher pH (>7.5). The silica gel matrix can actually dissolve at higher pH and most silica gels are not recommended for prolonged exposure above pH 7.5.

5.5.3.1.1 Ligands. The selectivity of the reversed-phase medium is predominantly a function of the type of ligand grafted to the surface of the silica. Generally, linear hydrocarbon chains (*n*-alkyl groups) are the most popular ligands used in reversed-phase applications. Some typical hydrocarbon ligands are C_4, C_8 and C_{18} alkyl chains, phenyl, biphenyl and phenylhexyl, and many other variants are commercially available from different manufacturers. The chemistry of reversed-phase silica is shown in Figure 5.2. Although a number of ligands are commercially available, for the purification of peptides C_8 and C_{18} ligands are most commonly used. It is not possible to predict theoretically which ligand is most suitable for a particular application, but a good rule of thumb is that the more hydrophobic is the molecule to be purified, the less hydrophobic the ligand needs to be. The more hydrophilic molecules tend to require strongly hydrophobic immobilized ligands in order to achieve sufficient binding for separation.[42] Typically, chemically synthesized peptides are efficiently purified on the more hydrophobic C_8 and C_{18} ligands. Proteins and recombinant peptides, because of their size, behave as hydrophobic molecules and most often bind very strongly to C_8 ligands; they are usually better separated on C_8 ligands. In addition to the ligand structure, the density of the immobilized hydrocarbon ligands on the silica surface also influences the selectivity. Therefore, reproducible chemical derivatization of the silica surface is critical for efficient RPC with consistent batch-to-batch selectivity.

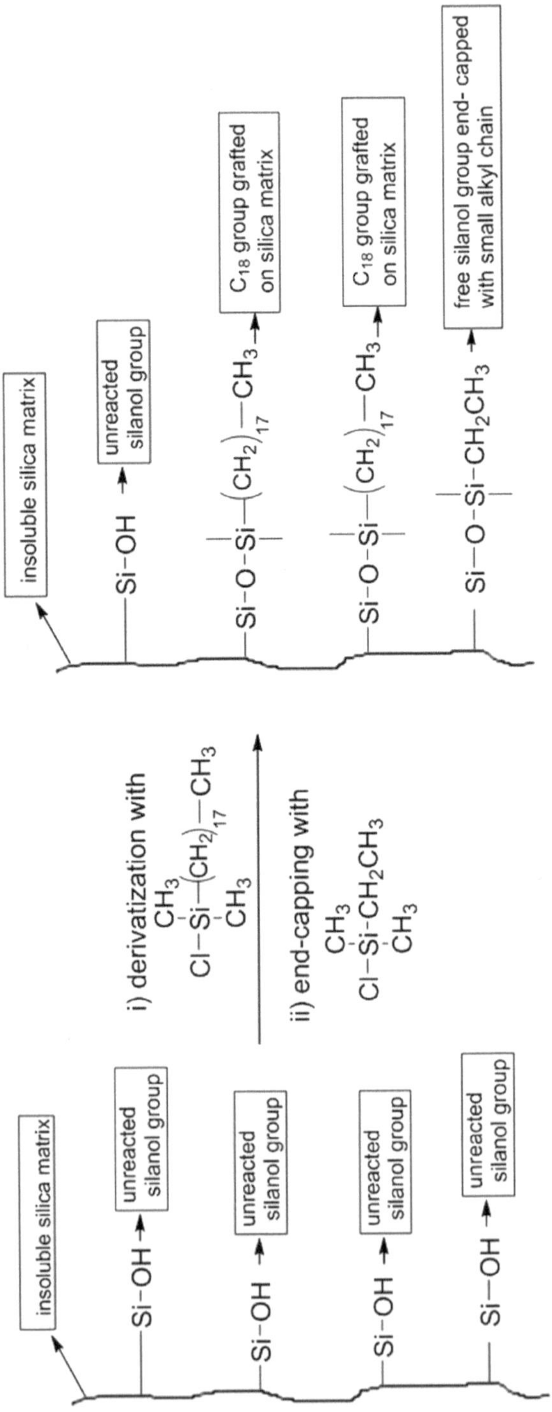

Figure 5.2 Chemistry of reversed-phase silica.

The hydrocarbon ligands are generally coupled to the silica gel *via* silanol groups on the silica surface using chlorotrialkylsilane reagents. Not all of the silanol groups are substituted in this coupling reaction. The C_8 and C_{18} reagents are large and bulky and steric hindrance often prevents complete derivatization of all the available silanol groups. The residual silanol groups are believed to be responsible for the deleterious mixed-mode ion-exchange effects often present during the reversed-phase separation of biomolecules. To reduce these damaging side effects, the residual silanol groups are reacted with smaller alkylsilane reagents such as chlorotrimethylsilane, where the steric effects do not interfere with complete coverage of the silanol groups remaining on the surface of the silica gel. This process is referred to as "end-capping." The extent of end-capping also affects selectivity and therefore reproducibility in the capping step is critical for a well-behaved reversed-phase medium.

5.5.3.1.2 **Particle Size of the Stationary Phase.** The particle size of the silica beads, as measured by their diameter, has an important bearing on the size of the chromatographic bed that can be packed and on the efficiency of separation. Media of larger particle size are generally used for large-scale preparative and process applications owing to their increased capacity and lower pressure requirements at high flow rates. Analytical and small-scale preparative applications are usually performed with 3 and 5 μm beads whereas larger scale preparative applications (pilot and commercial scale) are usually performed with particle sizes >10 μm.

Column efficiency is expressed as the number of theoretical plates and is a dimensionless number based on retention volume and peak width. The greater the number of plates, the better is the resolution between the peaks. Efficiency depends primarily on the physical properties of the media, column and system dimensions. The porosity of reversed-phase beads is a crucial factor in determining the available capacity for solute binding by the medium. Media with a pore size of ~100 Å are used for small molecules and peptides with molecular weight up to 10 000 Da, whereas media with a pore size of 300 Å are used for proteins.

Selection of the right medium for a desired separation is an empirical exercise and should be taken up in the early stages of process development in the laboratory. One should screen media with different chemistries, particle sizes and pore sizes from various vendors. Consideration should be given to a vendor's capability to supply and maintain consistency of quality of material throughout. Since silica is used in the final stages of the manufacturing process, one should also pay attention to the leachable properties of the media.

5.5.3.2 Mobile Phase Considerations

In RPC, the solute binds tightly to the stationary phase. The resolution of components in the solute is dependent on the properties of stationary and mobile phase. The mobile phase contains water, an organic solvent, buffering salts and ion-pairing agents. Elution is effected with a mobile phase containing water and an organic solvent. The factors that need to be considered for efficient elution are the organic modifier, gradient elution and pH.

5.5.3.2.1 Organic Modifier. In RPC, the solutes bound to a hydrophobic stationary phase are desorbed from the reversed-phase medium by adjusting the polarity of the mobile phase so that the bound solute molecules will sequentially desorb and elute from the column. This usually involves decreasing the polarity of the mobile phase by increasing its percentage of organic modifier, which is accomplished by maintaining a high concentration of organic modifier in the final mobile phase (mobile phase B). Generally, the pH of the initial and final mobile phase solutions remains the same. A gradual decrease in mobile phase polarity (increasing mobile phase hydrophobicity) is achieved by an increasing linear gradient from 100% initial mobile phase A containing little or no organic modifier to 100% (or less) mobile phase B containing a higher concentration of organic modifier. The bound solutes desorb from the reversed-phase medium according to their individual hydrophobicities.

There are a large number water-miscible organic solvents that can be used in RPC but only a few are used in industrial practice, namely acetonitrile, methanol, ethanol and rarely 1- and 2-propanol. Their properties which need to be considered for use in an industrial setup are listed in the Table 5.4. The solvents should have a UV cut-off of 210 nm and low viscosity. High-viscosity mobile phases are undesirable since they result in poor mass transport of the solute between the mobile and stationary phases and create a high backpressure. Hence in industrial setups low-viscosity acetonitrile, ethanol and methanol are preferred.

Temperature has a profound effect on RPC, especially for low molecular weight solutes such as short peptides. The viscosity of the mobile phase used in RPC decreases with increase in column temperature. Since mass transport of the solute between the mobile and stationary phases is a diffusion-controlled process, a decrease in solvent viscosity generally leads to more efficient mass transfer and, therefore, higher resolution. Increasing the

Table 5.4 Physicochemical properties of some commonly used solvents in preparative HPLC.

Solvent	Boiling point/°C	Freezing point/°C	UV cut-off/nm	Viscosity/cP at 20 °C
Acetonitrile	82	−45	190	0.36
Ethanol	78	−114	210	1.20
Methanol	65	−97	205	0.60
1-Propanol	98	−126	210	2.26
2-Propanol	82	−89	210	2.30
Water	100	0	<190	1.00

temperature of a reversed-phase column is particularly effective for low molecular weight solutes since they are reasonably stable at elevated temperatures.

5.5.3.2.2 Gradient Elution. In the gradient elution mode, the concentration of organic solvent is lower in the initial mobile phase (mobile phase A) than it is in the final mobile phase (mobile phase B). The gradient, regardless of the absolute range of the percentage of organic modifier, always proceeds from a condition of high polarity (high aqueous content, low concentration of organic modifier) to one of low polarity (lower aqueous content, higher concentration of organic modifier). Gradient elution is the method of choice when performing preparative RPC of peptides. The typical gradients for the preparative RPC of peptides are linear and binary, *i.e.* involving two mobile phases.

Typically, in the exploratory stages of performing a reversed-phase separation of a complex sample, a broad gradient is used for initial screening in order to determine the optimum gradient shape. After the completion of initial screening, the gradient shape may be adjusted to optimize the separation of the desired components. This is usually accomplished by decreasing the gradient slope where the desired component elutes and increasing it before and after. The choice of gradient slope will depend on how closely to the target molecule the impurities elute. Generally, decreasing the gradient slope increases resolution. Gradient slopes are generally reported as the change in the percentage of B per unit time ($\%\mathrm{B\,min^{-1}}$) or per unit volume ($\%\mathrm{B\,mL^{-1}}$).

5.5.3.2.3 pH. The pH of the mobile phase has a profound influence on the way in which a peptide mixture separates. Since peptides have carboxylic and amino functional groups, a lower pH suppresses the ionization of the carboxylic acid and increases the hydrophobicity, which results in late elution, *i.e.* an increased retention time. Similarly, a higher pH increases the ionization of carboxylic groups and thereby increases the polarity (reduces the hydrophobicity) of the compound, which in turn results in early elution, *i.e.* a decreased retention time. This particular property is highly beneficial and is exploited in separating impurities from the compound of interest.

5.5.4 Lyophilization[43]

Peptides after HPLC purification will be in the solubilized form in water and organic solvents such as acetonitrile or methanol. The concentration of peptides is usually low and, since the peptides are thermolabile, their isolation by classical isolation techniques such as precipitation or evaporation of the solvent by distillation poses a challenge. By far the most common method for the isolation of peptides is lyophilization. Prior to lyophilization, there is often a need to concentrate the peptide solution from the previous purification step. Normally, the concentration of a solution containing the peptide is carried out under reduced pressure. As high eluent volumes are a problem associated with large-scale processes, there is a need for high-capacity alternatives to reduce the volume of solvent collected. Such an alternative is reverse osmosis (RO), which provides both a mild and scalable method for removal of low molecular weight salts and organic solvents and is a very rapid means for the concentration of peptide solutions. If RO is used in combination with diafiltration, it is possible to produce peptide solutions with a predetermined concentration of the counter ion. The isolation of peptides by lyophilization not only provides the desired peptide in the form of a solid but also assists in the control of quality attributes such as water content, counter ion and residual solvents. In an industrial setup, the lyophilization step is capital intensive and the size of the lyophilizer often decides the batch size of a product.

5.6 Related Compound (Impurity) Considerations

As described in the previous sections, peptides are manufactured by solution- and solid-phase chemical synthesis and recombinant technologies, hence the related compounds/impurities are specific to the particular technology used for manufacturing. The related compounds in solution-phase synthesis depend on the route of synthesis, hence no generalizations are possible. The related compounds in recombinant technologies will not be similar to those in the chemical synthesis. Currently, the solid-phase method is the mainstay of peptide synthesis, and this section describes related compounds specific to this method and their sources, identification and control strategies.

5.6.1 Types of Related Compounds in Synthetic Peptides

Various types of related compounds are possible during the life cycle of peptides produced by solid-phase synthesis. The origin of these impurities could be from the raw materials used or formed during the manufacturing process or from degradation during storage. The related compounds in peptides manufactured by solid-phase synthesis include deletion sequences, insertion sequences, truncation compounds, diastereomers, substituted compounds and functional group disulfide modifications.[44] In the case of peptide drug products, impurities originating from peptide–excipient interactions are also possible. The process-related impurities with origins and identification/detection methods are listed Table 5.5.

5.6.1.1 Deletion Sequences

Deletion sequences are the most common peptide-related impurities in peptides synthesized by solid-phase synthesis. Deletion sequences lack one or more amino acid residues in the peptide of interest. These sequences

Table 5.5 Origin of and identification method for related impurities in peptides produced by solid-phase synthesis.

Type of impurities	Origin of impurities	Identification method	HPLC detectability
Deletion	Synthesis	LC-MS or LC-MS/MS	+
Insertion	Raw materials or synthesis	LC-MS or LC-MS/MS	+
Truncation	Raw materials or synthesis	LC-MS or LC-MS/MS	+
Diastereomers	Raw materials or synthesis or purification	HPLC spiking with synthesized diastereomeric analogue	±
Substitution	Raw materials	HPLC spiking with synthesized analogue or isolation/AAA	−
Functional group modification	Synthesis or stability	LC-MS or LC-MS/MS	±
Disulfide modification	Synthesis or stability	LC-MS or LC-MS/MS	++

originate from the synthesis due to incomplete removal of the N^{α} protecting group or incomplete coupling.[45]

The qualitative Kaiser colorimetric test is routinely used for checking the completeness of deprotection and monitoring of coupling. However, this qualitative colour test does not guarantee completion of coupling or deprotection. Sequences lacking the first few amino acids at the C-terminal end originate through incomplete coupling of the first amino acid or inefficient capping after the first amino acid coupling. The deletion sequences are more prevalent at sites where difficult-to-couple amino acids are present. The deletion sequence impurities increase as the chain length increases. Deletion peptides at the N-terminal side of a peptide with >20 amino acid residues are a common phenomenon owing to the decrease in coupling efficiencies in the growing peptide chain. Removal of deletion sequence impurities by chromatography is difficult if the missing amino acids are relatively simple, such as alanine or glycine.

The deletion peptides are easily identified by HPLC coupled with a high-resolution mass spectrometer. The deletion peptides show a mass lower by the mass of the missing amino acid residue. The place of deletion can be identified by tandem mass spectrometric (MS/MS) sequence analysis. The deletion peptide impurities can be minimized by properly monitoring amino acid coupling during the coupling stage.

5.6.1.2 *Truncation Sequences*

Truncation sequences are peptides missing one or more amino acid residues at the N- or C-terminus and they are also a type of deletion sequences.

The N-terminal truncated peptide chains originate from inefficient coupling or when capping is used as part of a protocol to prevent deletion sequences.[46] Deletion sequences at the C-terminus are due to incomplete coupling of the first amino acid residue to the solid support or when quenching is part of the synthesis protocol. The structures of truncated sequences are sufficiently different from the target sequence, hence they are removable during the purification stage.

5.6.1.3 *Insertion Sequences*

Other common impurities during SPPS are insertion sequences. Insertion is the presence of one or more amino acid residues in addition to the required amino acids in the target peptide.[47] There are two possible reasons for insertion peptides: the use of excess amino acids to drive the coupling reaction to completion and the impurities in the amino acid starting materials. During SPPS, an excess of Fmoc-protected amino acids is used to ensure maximum coupling. However, excess amino acids can result in an additional amino acid in the desired sequence. The activated excess amino acids acylate not only deprotected amine but also the amide group of the resin-bound peptide. During the next coupling, the transient deprotection will remove the Fmoc protecting group, providing a free reactive amine group. This will further react with the carbonyl moiety of the peptide bond and effectively insert an amino acid into the sequence. The amino acids without a bulky chain adjacent to the amino acids are the most favoured sites for insertion. To prevent insertion sequences, excess activated amino acids must be quenched or extracted before the deprotection step. Removal of insertion sequence impurities is difficult during purification if the inserted amino acids are simple, such as glycine and alanine.

Another major reason for the formation of insertion peptides is dipeptide impurities in the building blocks, *i.e.* the Fmoc-protected amino acids used in the synthesis. Hence it is desirable to have stringent limits for Fmoc and dipeptide impurities in Fmoc-protected amino acids used for the synthesis. The stringent specifications for impurities in the building blocks not only reduce impurities but also improve yields by enhancing the crude peptide purity and the overall purification process of the crude peptide.

Similarly to deletion peptides, insertion peptides can also be identified by HPLC coupled with a high-resolution mass spectrometer. The insertion peptides show an excess mass corresponding to the mass of the inserted amino acids. The place of insertion can be identified by MS/MS sequence analysis.

5.6.1.4 *Diastereomers*

Diastereomeric impurities are another major class of impurities in addition to deletions and insertions in peptide sequences. Diastereomer sequences contain one or more amino acids in the undesired chiral form. These diastereomers have the same mass and same sequence as the desired peptide. Diastereomers are difficult to remove during the purification and also present a great challenge in analytical HPLC separations. The sources and origins of these diastereomeric peptides are epimeric impurities in the Fmoc-protected amino acids (presence of the wrong isomer), racemization during the synthesis because of the coupling conditions, coupling reagents, the propensity of amino acids to racemization, racemization during purification and racemization during storage.[48] The major sources of diastereoisomeric impurities are unwanted isomeric impurities in the building blocks, Fmoc-protected amino acids and racemization during the peptide synthesis. The unwanted amino acid impurities in Fmoc-protected amino acids are directly inserted into the peptide chain and gives rise to diastereomers.

Two mechanisms, namely direct enolization and oxazolone (azlactone) formation, are identified as reasons for diastereomeric impurities in the peptides during synthesis. Both of these mechanisms are base catalysed, hence avoiding a tertiary base during the coupling reduces racemization by the enolization mechanism and coupling from the C- to N-terminus reduces racemization by the oxazolone (azalactone) mechanism.

The most suitable method to determine the chiral purity of a peptide is to acid hydrolyse the peptide and then subject the hydrolysate to HPLC on a chiral column or to derivatize the hydrolysate with chiral reagents such as Marfey's reagent and subject the product to HPLC.[49] Derivatizing the amine functional group of the hydrolysate by trifluoroacetylation and the acid group to C_2–C_4 esters in order to make the amino acids volatile and then subjecting them to gas chromatography (GC)-MS analysis on a chiral column is the most popular methodology.[50] By these methods, one can determine the chiral purity of all of the amino acids present in the peptide.

5.6.1.5 *Substitution Sequences*

Substitution sequences occur when one or more amino acids are substituted by other amino acids, the most common being isoleucine substitution by leucine. This originates through the presence of contaminants in the amino acid derivatives, which can be controlled by setting up appropriate specifications for Fmoc-protected amino acids. Identifying and purifying substitution sequences like diastereomers is equally challenging, especially in the case of isoleucine–leucine substitution.

5.6.1.6 *Modification of Functional Groups and Disulfide*

Amino acid side chains are susceptible to modification either during the synthesis or on storage. Amino acids such as asparagine, aspartic acid and glutamine undergo rearrangement to succinimide and pyroglutamic acid.

Asparagine and glutamine undergo degradation to aspartic and glutamic acid. Primary amides of amino acids such as tryptophan, tyrosine and methionine are prone to electrophilic substitution during acidolysis.

5.7 Conclusion

Peptides are manufactured commercially by liquid-phase peptide synthesis (LPPS) or solid-phase peptide synthesis (SPPS), by hybrid synthesis using both solid- and solution-phase synthesis or by using recombinant DNA technologies. Peptides used to be manufactured by solution-phase synthesis before about 2000 and later mostly by solid-phase or hybrid synthesis. Manufacturing peptides that meet present-day regulatory standards while satisfying commercial viability is a huge challenge. Both the upstream synthesis and downstream purification play a critical role in the commercial success of a product. In order to meet these twin challenges, a development scientist needs to work with diligence and with knowledge and vision of the manufacturing plant from the beginning of process development.

Abbreviations

AC	affinity chromatography
Acm	acetamidomethyl
AcOH	acetic acid
ACTH	adrenocorticotropic hormone
ANDA	Abbreviated New Drug Application
APPI	active peptide pharmaceutical ingredient
Boc	*tert*-butoxycarbonyl
BOP	benzotriazol-1-yloxytris(dimethylamino)phosphonium hexafluorophosphate
Bzl	benzyl
CAPA	corrective and preventive action
Cbz	benzyloxycarbonyl
cGMP	current good manufacturing practice
CTC	chlorotrityl chloride
DCC	*N*,*N*-dicyclohexylcarbodiimide
DCM	dichloromethane
DIC	*N*,*N*-diisopropylcarbodiimide
DMF	dimethylformamide
DVB	divinylbenzene
EMA	European Medicines Agency
FDA	Food and Drug Administration
Fmoc	9-fluoromethoxycarbonyl
GC	gas chromatography
GMP	good manufacturing practice
GPC	gel permeation chromatography

HBTU	*O*-(benzotriazol-1-yl)-*N*,*N*,*N*′,*N*′-tetramethyluronium hexafluorophosphate
HIV	human immunodeficiency virus
HMPA	hexamethylphosphoramide
HOAt	1-hydroxyazabenzotriazole
HOBt	1-hydroxybenzotriazole
HONp	*p*-nitrophenol
HOSu	*N*-hydroxysuccinimide
HPLC	high-performance liquid chromatography
ICH	International Council on Harmonisation of Technical Requirements for Registration of Pharmaceuticals for Human Use (ICH)
IEX	Ion-exchange chromatography
LPPS	liquid-phase peptide synthesis
MS	mass spectrometry
Mtr	4-methoxy-2,3,6-trimethylbenzenesulfonyl
NCA	*N*-carboxy anhydride
NDA	New Drug Application
Pbf	2,2,4,6,7-pentamethyldihydrobenzofuran-5-sufonyl
RO	reverse osmosis
RPC	reversed-phase chromatography
RP-HPLC	reversed-phase high-performance liquid chromatography
SPPS	solid-phase peptide synthesis
TBTU	*O*-(benzotriazol-1-yl)-*N*,*N*,*N*′,*N*′-tetramethyluronium tetrafluoroborate
TDI	tolerable daily intake
TFA	trifluoroacetic acid
^{t}Bu	*tert*-butyl
Trt	trityl (triphenylmethyl)
UHPLC	ultra-high-performance liquid chromatography
WSCDI	water-soluble carbodiimide
Z	benzyloxycarbonyl

Acknowledgements

I sincerely thank Dr Bhujanga Rao Adibatla for review and helpful comments and Dr Nani Babu Palakurthy, Ravi Thammira and Swapna Duddu for their help with the preparation of this chapter.

References

1. (a) O. A. Musaimi, D. A. Shaer, B. G. de la Torre and F. Albericio, *Pharmaceuticals*, 2018, **11**, 42; (b) J. L. Lau and M. K. Dunn, *Bioorg. Med. Chem.*, 2018, **26**, 2700; (c) A. Henninot, J. C. Collins and J. M. Nuss, *J. Med. Chem.*, 2018, **61**, 1382; (d) M. Goodmann, in *Synthesis of Peptides*

and Peptidomimetics, ed. M. Goodmann, A. Felix, L. Moroder and C. Tonolio, Thieme, New York, E22a, 2004, p. 14.
2. E. Fischer, *Chem.-Ztg.*, 1902, **26**, 939.
3. Th. Curtius, *Ber. Dtsch. Chem. Ges.*, 1902, **35**, 3226.
4. H. Leuchs, *Ber. Dtsch. Chem. Ges.*, 1906, **39**, 857.
5. M. Bergmann and L. Zervas, *Ber. Dtsch. Chem. Ges.*, 1932, **65**, 1192.
6. V. du Vigneaud, C. Ressler, J. M. Swan, C. W. Roberts, P. G. Katsoyannis and S. Gordon, *J. Am. Chem. Soc.*, 1953, **75**, 4879.
7. R. B. Merrifield, *J. Am. Chem. Soc.*, 1963, **85**, 2149.
8. (a) L. A. Carpino, *J. Am. Chem. Soc.*, 1957, **79**, 98; (b) L. A. Carpino and G. Y. Han, *J. Am. Chem. Soc.*, 1970, **92**, 5748.
9. A. R. Mitchell, *Biopolymers (Pept Sci)*, 2008, **90**, 175.
10. A. J. Mijalis, D. A. Thomas, M. D. Simon, A. Adamo, R. Beaumont, K. F. Jensen and B. L. Pentelute, *Nat. Chem. Biol.*, 2017, **13**, 464.
11. M. Verlander, *Int. J. Pept. Res. Ther.*, 2007, **13**, 75.
12. (a) B. L. Bray, *Nat. Rev. Drug Discovery*, 2003, **7**, 87; (b) T. Matthews, M. Salgo, M. Greenberg, J. Chung, R. DeMasi and D. Bolognesi, *Nat. Rev. Drug Discovery*, 2004, **3**, 215.
13. (a) T. Bruckdorfer, O. Marder and F. Albericio, *Curr. Pharm. Biotechnol.*, 2004, **5**, 29; (b) J. Bacardit-Carbado, *Int. J. Pept. Res. Ther.*, 2007, **13**, 69; (c) A. N. Thayer, *Chem. Eng. News*, 2011, **89**, 21.
14. Encyclopedia of Britannica https://www.britannica.com/story/what-is-the-difference-between-a-peptide-and-a-protein.
15. Biosimilars: Questions and Answers Regarding Implementation of the Biologics Price Competition and Innovation Act of 2009, Guidance for Industry, April 2015 http://www.fda.gov/Drugs/GuidanceCompliance RegulatoryInformation/Guidances/default.htm.
16. T. Uhlig, T. Kyprianou, F. G. Martinelli, C. A. Oppici, D. Heiligers, D. Hills, X. R. Calvo and P. Verhaert, *EuPa Open Proteomics*, 2014, **4**, 58.
17. L. Andersson, L. Blomberg, M. Flegel, L. Lepsa, B. Nilsson and M. Verlander, *Biopolymers (Pept. Sci.)*, 2000, **55**, 227.
18. M. Bodanszky, *Principles of Peptide Synthesis*, Springer Verlag, Berlin, 1993.
19. J. Podlech, in *Synthesis of Peptides and Peptidomimetics*, ed. M. Goodman, A. Felix, L. Moroder and C. Toniolo, Thieme, New York, E22a, 2004, p. 41.
20. P. Lloyd-Williams, F. Albericio and E. Giralt, *Chemical Approaches to the Synthesis of Peptides and Proteins*, CRC Press, Boca Raton and New York, 1997.
21. H. J. Kohlbau, R. Thurmer and W. Voelter, in *Synthesis of Peptides and Peptidomimetics*, ed. M. Goodman, A. Felix, L. Moroder and C. Toniolo, Thieme, New York, E22a, 2004, p. 193.
22. P. Thamm, W. Kolbeck, H. J. Musiol and L. Moroder, in *Synthesis of Peptides and Peptidomimetics*, ed. M. Goodmann, A. Felix, L. Moroder and C. Tonolio, Thieme, New York, E22a, 2004, p. 315.

23. L. Moroder, in *Synthesis of Peptides and Peptidomimetics*, ed. M. Goodman, A. Felix, L. Moroder and C. Toniolo, Thieme, New York, E22a, 2004, p. 425.
24. Y. Okada, in *Synthesis of Peptides and Peptidomimetics*, ed. M. Goodman, A. Felix, L. Moroder and C. Toniolo, Thieme, New York, E22a, 2004, p. 591.
25. ICH Harmonized Guideline, Impurities: Guideline for Residual Solvents, Q3C(R6), 2016.
26. (a) E. Atherton and R. C. Sheppard, *Solid Phase Peptide Synthesis: A Practical Approach*, IRL Press at Oxford University Press, 1989; (b) W. C. Chan and P. D. White, *Practical approach, Fmoc Solid Phase Peptide Synthesis*, Oxford University Press, New York, 2000.
27. G. B. Fields, *Methods in Molecular Biology, vol. 35, Peptide Synthesis Protocols*, ed. M. W. Pennigton and B. M. Dunn, Humana Press, New Jersey, 1994, p. 17.
28. (a) E. Giralt and F. Albericio, in *Synthesis of Peptides and Peptidomimetics*, ed. M. Goodman, A. Felix, L. Moroder and C. Toniolo, Thieme, New York, E22a, 2004, p. 665; (b) E. Kaiser, R. L. Colescott, C. D. Bossinger and P. Cook, *Anal. Biochem.*, 1970, **34**, 595.
29. T. Vojkovsky, *Pept. Res.*, 1995, **8**, 236.
30. B. F. Gisin, *Anal. Chim. Acta*, 1972, **58**, 248.
31. V. Krchnak and Vagner, *J. Pept. Res.*, 1990, **3**, 182.
32. L. C. Dorman, *Tetrahedron. Lett.*, 1969, **10**, 2319.
33. L. C. Wu, Regulatory Considerations for Peptide Drug Products; 2013 GPhA/FDA CMC Workshop, 2013 http://www.gphaonline.org/media/cms/Larisa_Wu_1.pdf.
34. R. Nyfeler, *Methods in Molecular Biology*, ed. M. W. Pennigton and B. M. Dunn, *Peptide Synthesis Protocols*, Humana Press, New Jersey, 1994, vol. 35, p. 303.
35. L. Andersson and J. Persoson, *Aspects of purification of peptides by preparative chromatography* https://pdfs.semanticscholar.org/4008/e25fc602ca0463fdf46ee9f9718b5d669c93.pdf.
36. US FDA guideline. ANDAs for Certain Highly Purified Synthetic Peptide Drug Products That Refer to Listed Drugs of rDNA Origin Guidance for Industry, 2017.
37. *Ion Exchange Chromatography. Principles and Methods – Handbook*, GE Healthcare. 2016 https://research.fhcrc.org/content/dam/stripe/hahn/methods/biochem/Ion_Exchange_Chromatography_Handbook.pdf.
38. *Gel Filtration. Principles and Methods – Handbook*, Pharmacia LKB Biotechnology, 1999.
39. (a) *Reversed Phase Chromatography. Principles and Methods – Handbook*, Amersham, Biotech, 1999; (b) M. Aguilar, *Methods in Molecular Biology, vol 251, HPLC of Peptides and Proteins, Methods and Protocols*, Humana Press, New Jersey, 2004, 1, p. 3.
40. (a) *Protein and Peptide Purification Technique Selection Guide*, GE Health, 2007; (b) *Affinity Chromatography. Principles and Methods – Handbook*, Amersham Pharmacia Biotech, 1997.

41. *Protein Purification Handbook*, Amersham Pharmacia Biotech AB 1999.
42. Reversed Phase chromatography, LC/GC Chrom academy https://www.chromacademy.com/lms/sco5/Theory_Of_HPLC_Reverse_Phase_Chromatography.pdf.
43. Smart Freeze drying, basic principles, optimum procedures and applications, Christ, 2015 https://www.martinchrist.de/fileadmin/user_upload/christ/04_anwendungen/lyophilization/Smart_FD_Basics_Apps_en.pdf.
44. (a) V. Vergote, C. Burvenich, W. C. Van de and B. Spiegeleera, *J. Pept. Sci.*, 2009, **15**, 697; (b) I. Eggen, B. Gregg, H. Rode, A. Swietlow, M. Verlander and A. Szajek, *Pharm. Technol.*, 2014, **38**(3), 48–55; (c) I. Eggen, B. Gregg, H. Rode, A. Swietlow, M. Verlander and A. Szajek, *Pharm. Technol.*, 2014, **38**(4), 48–52; (d) I. Eggen, B. Gregg, H. Rode, A. Swietlow, M. Verlander and A. Szajek, *Pharm. Technol.*, 2014, **38**(5), 52–56; (e) S. V. Dorpe, M. Verbeken, E. Wynendaele and B. D. Spiegeleer, *J. Bioanal. Biomed.*, 2011, **S6**, 1.
45. (a) V. Sanz-Nebot, F. Benavente, I. Toro and J. Barbosa, *Anal. Chim. Acta*, 2004, **521**, 25; (b) H. L. Ball and P. Mascagni, *Int. J. Pept. Protein Res.*, 1996, **48**, 31; (c) V. Sanz-Nebot, F. Benavente, A. Castillo and J. Barbosa, *J. Chromatogr. A*, 2000, **889**, 119.
46. (a) H. L. Ball and P. Mascagni, *Int. J. Pept. Protein Res.*, 1996, **48**, 31; (b) C. M. Hekman, W. S. DeMond, P. J. Kelley, S. F. Mauch and J. D. Williams, *J. Pharm. Biomed. Anal.*, 1999, **20**, 763.
47. (a) V. Sanz-Nebot, F. Benavente, I. Toro and J. Barbosa, *Anal. Chim. Acta*, 2004, **521**, 25; (b) V. Sanz-Nebot, F. Benavente, A. Castillo and J. Barbosa, *J. Chromatogr. A*, 2000, **889**, 119.
48. (a) M. Bodanszky, *Racemization, Peptides – A Practical Textbook*, ed. M. Bodanszky, Springer, Berlin, 1993, pp. 117–128; (b) R. Steinauer, F. M. F. Chen and N. L. Benoiton, *Int. J. Pept. Protein Res.*, 1989, **34**, 295.
49. (a) A. J. Gordan, T. Hoang, A. Crivici and G. A. Lajoie, *Anal. Biochem.*, 1992, **202**, 210; (b) C. B'Hymer, M. Montes-Bayon and J. A. Caruso, *J. Sep. Sci.*, 2003, **26**, 7.
50. (a) H. Frank, G. J. Nicholson and E. J. Bayer, *Chromatogr. Sci.*, 1977, **15**, 174; (b) K. Nokihara and J. Gerhardt, *Chirality*, 2001, **13**, 431.

CHAPTER 6

Economic and Environmental Factors Affecting the Sustainability of Peptide Therapeutic Manufacturing

E. RODNEY LAX*[a] AND TRISHUL SHAH[b]

[a] Rodney Lax Consulting, Podbielskistrasse 45, 30177 Hannover, Germany; [b] PolyPeptide Laboratories Inc., 365 Maple Avenue, Torrance, CA 90503, USA
*Email: rodney.lax@outlook.com

6.1 Introduction

The terms "peptide therapeutic" and "sustainability" both require qualification. Typically, a "peptide therapeutic" would refer to a polypeptide between two and about 50 amino acids in length, which is either in development or approved as a pharmaceutical for human use. Currently, there are over 70 approved peptide therapeutics marketed worldwide (Table 6.1).

Peptide therapeutics are an attractive class of drugs that, in terms of molecular size, pharmaceutical function and manufacturing costs, build a bridge between small molecules and biologics. The peptide therapeutic drug product industry is thriving and its sustainability does not need to be questioned. As new chemical, recombinant and enzymatic technologies develop and mesh, the scope of this class of drugs is expected to expand in terms of its complexity and diversity. Even a massive cost reduction in the

Drug Discovery Series No. 72
Peptide Therapeutics: Strategy and Tactics for Chemistry, Manufacturing and Controls
Edited by Ved Srivastava

Published by the Royal Society of Chemistry, www.rsc.org

Table 6.1 Approved peptide therapeutics (September 2018). Peptide therapeutics marketed exclusively in Russia[a], China[b], Bosnia and Herzegovina[c] or Australia[d] are indicated.[9]

Peptide	Year	Peptide	Year	Peptide	Year	Peptide	Year
Abeloparatide	2017	Dulaglutide	2014	Leuprolide	1984	Somatostatin	1970s
ACTH (4–10)[a]	2011	Elcatonin	1981	Linaclotide	2012	Taltirelin	2000
ACTH (1–39)	1952	Eledoisin	1970	Liraglutide	2009	Teduglutide	2012
Afamelanotide	2014	Enalapril	1985	Lisinopril	1987	Teriparitide	2002
Alarelin[b]		Enfuvirtide	2003	Lixisenatide	2013	Terlipressin	1978
Alloferon[a]		Eptifibatide	1998	Lucinactant	2012	Tesamorelin	2010
Angiotensin (1–7)	2017	Etelcalcetide	2016	Macimorelin	2017	Tetracosactide	1980
Atosiban	2000	Exenatide	2005	Mifamuritide	2009	Thymodepressin[a]	
Aviptadil	2000	FAR-404[c]	2010	Nafarelin	1990	Thymopentin	1985
Benaglutide[b]	2016	Felypressin	1970s	Neogen[a]		Thymosin-α_1[b]	2009
Bivalirudin	2000	Ganirelix	1990	Nesiritide	2001	Triptorelin acetate	1986
Buserelin	1984	Glatiramer	1996	NOV-002[a]		Triptorelin pamoate	2009
Calcitonin (salmon)	1971	Glucagon	1989	Octreotide	1988	Vapreotide	2005
Calcitonin (human]	1986	GMDP[a]		Ornithine vasopressin[d]	1971	Vasopressin	1962
Carbetocin	2001	Golotimod[a]		Oxytocin	1962	Zinconotide	2004
Carfilzomib	2012	Gonadorelin acetate	1989	Pasireotide	2012		
Carperitide	1995	Goserelin	1987	Plecanatide	2017		
Cetrorelix	1999	Histrelin	1991	Pramlintide	2005		
Degarelix	2008	Icatibant	2008	Romiplostim	2008		
Desmopressin	1972	Lanreotide	2007	Semaglutide	2017		

price of biologics or significant advances in genetic medicine (*e.g.* oligonucleotides, gene editing) would be unlikely to replace peptide therapeutics as a class of drugs. To quote Wieland and Bodanszky, "Nature" created peptides for a purpose.[1]

However, the sustainability of the companies that manufacture the respective active pharmaceutical ingredients (APIs), especially those individual independent contract manufacturing organizations (CMOs) focused on peptide-based drug substances, is less certain. Even though order books through 2018–2019 are full, the last 20 years have seen extensive consolidation among CMOs specializing in this class of API. The number of established good manufacturing practice (GMP)-conforming peptide CMOs, especially those manufacturing peptide new chemical entities (NCEs), that have left this field or that have been acquired is significantly higher than for new organizations entering the peptide therapeutic manufacturing arena. For the near future, most GMP-conforming peptide therapeutic manufacturing will be left in the hands of a very small number of experienced companies. However, one must assume that economic forces will prevail and new CMOs will enter the market.

Because most approved peptide therapeutics are manufactured chemically, this chapter focuses mainly on the sustainability of a relatively small group of CMOs specializing in niche heteropolymer amino acid-based chemistry. We attempt to address the economic factors behind the trend to consolidation and how recent environmental and regional considerations are impacting the cost and sustainability of peptide therapeutic manufacturing today. Manufacturing of peptide therapeutic drug substances may not be limited to chemistry in the future.

6.2 The Allure of Peptide Therapeutics

Most pharmaceutical peptides are analogs or facsimiles of naturally occurring physiological peptides or protein sequences. With >7000 known peptides and over 10 million proteins in the biosphere, the candidate pool is large and diverse.[2] The fact that such molecules have been tuned by evolution over millions of years has resulted in exceptional specificity and high potency (often offsetting the high unit gram cost of manufacture) with a low incidence of unexpected side effects. Because they are metabolized to their component amino acids, they are well tolerated and generally exhibit low toxicity.

Therapeutics based on peptides suffer from several well-recognized weaknesses related to their relatively labile structures, their susceptibility to proteolysis in the gastric tract and their poor uptake through cell membranes (Table 6.2). In the 1980s, a time when peptide therapeutics were starting to find wider approval, these weaknesses were often used as an argument to promote small molecules preferentially as a preferred class of drugs even where peptides seemed advantageous. The need to inject peptide

Table 6.2 Peptide therapeutics SWOT (strengths, weaknesses, opportunities, threats) analysis.

Strengths	Weaknesses
• Tuned by evolution • Exceptional specificity • Low toxicity and side effects • Well tolerated • High potency • Robust class of drugs • Cost of API ≪ biologics • Shorter approval times than for small molecules	• Proteolytic digestion in gut • Poor absorption in gut • Rapid systemic metabolism • Injectable – patient compliance • Labile = cold chain (15% extra cost) • Reconstitution (solubility, viscosity) • Cost of API ≫ small molecules
Opportunities	*Threats*
• Small molecule–biologics bridge • Structural flexibility (*versus* biologics) • LAR (long-acting release) formulations • Alternative delivery options • Conjugates and hybrid structures • Intracellular targets • New technologies for long peptides	• Biologics • Gene-based therapies • Oligonucleotide therapy

therapeutics to bypass both proteolysis and the poor uptake in the gastric tract is associated with patient noncompliance, inconvenience and discomfort. In addition, the relatively labile properties of peptides mandate an expensive cold chain (10–15% additional cost) and often the need to reconstitute a lyophilized drug prior to administration. Many peptides form viscous solutions necessitating either large volumes of highly diluted drug or low-gauge needles.

The landscape around peptides drugs has changed radically in the last 30 years owing to the introduction and overwhelming acceptance of biologics. Even though biologics, which are for the most part proteins (polypeptides), have even more pronounced weaknesses than shorter peptides, they possess the spatial features necessary to address multifocal receptor domains and have opened the doors to treating diseases that small molecules cannot address. Self-injection is no longer seen as a serious impediment to developing polypeptide drugs and indeed biologics have spawned very active research into alternative drug delivery platforms, simultaneously expanding opportunities for peptide-based therapeutics. Many peptide therapeutics, although significantly smaller than biologics, possess a sufficient molecular surface area to address multifocal receptor domains and can target receptors and influence protein–protein interactions that would typically be accessible to biologics, but not to small molecules.

Of the >70 approved peptide therapeutic drugs (in the 2–50 amino acid length range), all but 10, namely adrenocorticotropin (porcine pituitary extract) and the recombinant peptides carperitide, dulaglutide, glucagon, nesiritide, liraglutide, romiplostim, semaglutide, teriparatide and teduglutide are synthesized chemically using either solid-phase peptide synthesis (SPPS) or liquid-phase peptide synthesis (LPPS). Liraglutide and semaglutide are semisynthetic chemical modifications of a recombinant backbone. Dulaglutide and romiplostim are fusion proteins. All approved peptide therapeutics (except the two fusion proteins), irrespective of their current manufacturing process, can demonstrably be assembled synthetically; *vice versa*, some chemically manufactured therapeutics can also be expressed recombinantly (*e.g.* calcitonin). In terms of manufacture, there is no clear delineation line between the domains of chemical and recombinant processes and there is good reason to think that, in the future, an amalgamation of these technologies may occur within CMOs.

Peptide therapeutics are *de facto* significantly more expensive to manufacture than small molecules owing to the number of residues (amino acids) and associated reaction steps. On average, a small molecule drug costs about $1 (US$ used throughout this chapter) per daily dose to produce, with generics typically costing even less.[3]

The price of a peptide API varies significantly depending on the peptide and manufacturer, but the following prices per gram could represent top-end ball-park numbers for peptide APIs containing only canonical amino acids in early development using solid-phase chemistry:

Drug substance	100 g scale	1 kg scale
10-mer	$2000	$600
40-mer	$4000	$1300

The prices *may* decrease significantly (>50%) once a manufacturing process is established and quantities increase (economy of scale). The incorporation of an unusual amino acid or conjugate component could significantly alter the price. While most D-amino acids (Fmoc-D-Ile is an expensive exception) have little influence on the overall price, probably less than 1% of the final price of the API for a single substitution, some exotic amino acids currently used in peptide APIs are not generally commercially available. The latter are not only expensive, they can also be notoriously difficult to source at the development and commercial scale. Conjugate components, such as a poly(ethylene glycol) (PEG), could cost as much as or more than the peptide moiety itself. Table 6.3 gives examples of retail prices of generic and NCE peptide therapeutics.

For most peptide therapeutics, chemical synthesis remains economically more attractive than recombinant expression in addition to offering more flexibility in terms of the peptide sequence by permitting the inclusion of nonproteogenic amino acids and nonamino acid moieties. It is not possible to provide specific data, but an initial GMP recombinant campaign, irrespective of the length or quantity of the peptide, is unlikely to cost less than $2 million. A chemistry-based GMP SPPS campaign for 100 g of a 40-mer would be expected to cost under $500 000.

Table 6.3 Representative pricing (US$ per year) of some peptide therapeutics[a].

Drug	Peptide	Residues	mg per day	US$ per year	Indication
Generics					
Miacalcin	Calcitonin	32	0.03	500	Osteoporosis
Eligard	Leuprolide	9	0.25	5400	Prostate cancer
Sandostatin	Octreotide	8	0.30	8000	Carcinoid tumors, acromegaly
NCEs					
Byetta	Exenatide	39	0.01	3500	Type 2 diabetes
Fuzeon	Enfuvirtide	36	180.00	45 108	Anti-retroviral treatment of HIV-1 infection
Firazyr	Icatibant	10	30.00	98 040	Hereditary angioedema (assumes 3 attacks per year)
Signifor	Pasireotide	6	1.80	165 696	Cushing's syndrome.
Acthar Gel	ACTH (1–39)	39	80.00	204 205	Exacerbations of multiple sclerosis
Gattex	Teduglutide	33	3.80	451 440	Short bowel syndrome

[a]All peptides are synthetic except Acthar Gel, which is prepared from animal tissue, and Gattex, which is recombinant. Calcitonin, octreotide and pasireotide are cyclic. Annual costs are average wholesale prices (AWPs) from GoodRx (www.goodrx.com).

The average cost for developing a recombinant biosimilar is estimated to range from $40 million to $375 million – this compares with a $4 million price for developing a generic small molecule.[4]

Recombinant manufacture is not easily transposable to short peptides. Where it is applicable to longer peptides, the cost structures for recombinant expression and SPPS chemistry are not easily comparable. The chemical manufacture of peptides of ~40 amino acids in length at a 50–100 kg scale is certainly considered to be commercially viable. For larger quantities or longer peptides, fermentation with a genetically modified organism may become the preferred route economically. There are no rigid rules on where the chemical synthesis–recombinant expression switchover point should be. It will depend to a large extent on the availability of in-house technology and intellectual property (IP) considerations. For example, Lilly manufactures teriparatide (human parathyroid hormone), a 34-mer with an annual market in 2017 of less than 2 kg, recombinantly although it can be made easily by SPPS. Conversely, Roche Boulder (now Corden Boulder) developed a manufacturing process for the 36-mer enfuvirtide at the ton scale using a hybrid SPPS approach.

Given the propensity of the industry to move towards longer peptides, it seems likely that the choice for chemical synthesis or recombinant expression will eventually go the full circle as these technologies mesh together for the assembly of small proteins. This will require a third technology to link synthetic and recombinant fragments, one based on enzymatic or chemical ligation.

6.3 Peptide Therapeutics: Inclusion and Exclusion Criteria

To address the sustainability of peptide therapeutic manufacturing properly, we need to define the term "peptide therapeutic."

6.3.1 Defining Peptide Therapeutic

From a scientist's point of view, peptides are linear polymers of amino acids coupled by acid amide bonds. Although some attempt is made at classifying peptides according to their length, the cut-off points between oligopeptides, peptides, polypeptides and proteins are intentionally hazy or nonexistent. Naturally occurring peptides are, after all, a continuum of like molecules ranging from dipeptides with molecular weight (MW) ~200 Da to molecules such as titin, the largest known protein with over 33 000 residues and an MW of up to 3.8 MDa (Figure 6.1).

Typically, shorter peptides are hydrophobic in nature, have no secondary structure and are moderately stable. At the other end of the spectrum, proteins are generally hydrophilic, have conspicuous secondary and tertiary structures and are labile. Ribosome-derived peptides are composed of the

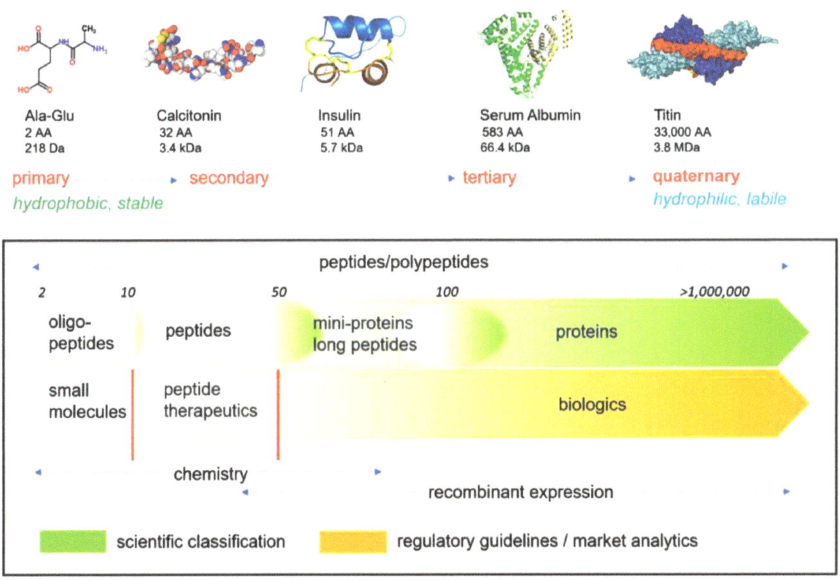

Figure 6.1 Classification of peptides according to sequence length.

20 canonical amino acids. Nonribosomal and synthetic peptides may contain D- and/or exotic amino acids.[5] Most peptide bonds in Nature are between the C-terminal α-carboxy group and the N-terminal α-amino group of L-amino acids, but isopeptide bonds do occur in Nature, in molecules such as glutathione and ubiquitylated proteins.

Regulatory authorities' guidelines are more rigid than those of academic scientists on what constitutes a peptide therapeutic, generally allowing for alignment of peptide-based pharmaceuticals into three categories: small molecule-like peptides, peptide therapeutics and biologics with cut-off points between each category (Figure 6.1). The rationale for this classification is the necessity to provide guidelines for manufacturing molecules of different complexity. That said, the boundary between peptides and biologics has distinct historical, manufacturing technology and subjective bias: traditionally, protein-based medicinal products have been overseen by separate regulatory departments, both in Europe and in the USA. More recently, protein-based biologics have been manufactured almost exclusively by recombinant processes, whereas most approved peptide therapeutics are synthesized chemically.

It is generally accepted that "small molecules" are organic compounds with a molecular weight of less than 900 Da and a size of up to 1 nm. In terms of peptide therapeutics, this corresponds to a peptide of 9–10 amino acids. From a regulatory standpoint, small molecules have more stringent impurity profile requirements, which are set out in the International Council on Harmonisation of Technical Requirements for Registration of Pharmaceuticals for Human Use (ICH) guidelines, than longer peptides. In practice however, the decision to follow ICH guidelines or consider the drug substance to be a longer peptide is made on a case-by-case basis, sometimes on the recommendation of the regulatory authority [especially the US Food and Drug Administration (FDA)] and sometimes by the sponsors based on their previous manufacturing and regulatory experience. Although this is a very rational approach, it does create uncertainty regarding the regulatory expectations for short peptides.

Market analysts also set their own guidelines for what constitutes a peptide therapeutic in order to simplify statistical evaluation of the market. The data collated are a useful indicator of market trends, but the inclusion/exclusion criteria are often questionable. Although dollar numbers are a true indicator of revenues from the approved drugs embraced by their guidelines, the data do not reflect the economic viability of the specialty CMOs, which manufacture a much broader range of peptide products.

Both market analysts and regulatory authorities have a marked penchant for rounding to the nearest full multiple of 10. It can be argued that this predisposition seriously misrepresents the volume of the peptide therapeutic market on the one hand and may lead to additional costs to the manufacture of some peptides on the other. For example, drawing the demarcation line at 50 amino acids excludes most insulin products from the peptide therapeutic market and thereby halves the estimated market value of this class of drugs.

In the context of this chapter, sustainability refers to the long-term viability of CMOs with a manufacturing focus on peptide therapeutics. Since most of these organizations make significant quantities of peptides that do not fall into either regulatory authorities' or marketing analysts' categorizations, we consider that any peptide that can be currently manufactured using synthetic chemistry (including those currently expressed recombinantly) and their conjugates with other molecules is part of the "sustainability" profile.

6.3.2 What Peptides Do CMOs Manufacture?

There is general agreement on the inclusion criteria for peptide therapeutics, but market analysts and regulatory authorities differ on many of the exclusion criteria (Table 6.4). Peptide manufacturers will – depending on capability and capacity – manufacture any peptide irrespective of these inclusion or exclusion criteria. Their job is to generate profitable business using peptide manufacturing technology. So, for example, peptides for

Table 6.4 Inclusion and exclusion criteria for peptide therapeutics.

Market research	Regulatory authorities
Inclusion criteria	*Inclusion criteria*
• Minimum 2 amino acids, amide bond	• Minimum 2 amino acids, amide bond
• No amino acid maximum if manufactured synthetically	• If ≤10 amino acids, might be considered to be small molecule • FDA: ≤99 amino acids if manufactured synthetically
• Conjugates with lipids, proteins, other polymers if peptide component fulfills criteria	• Conjugates with lipids, proteins, other polymers if peptide component fulfills criteria
• Only peptides in clinical trials or approved	• Only peptides in clinical trials or approved
• Human therapeutics	• Human therapeutics
Exclusion criteria	*Exclusion criteria*
• No recombinant peptides >50 amino acids	• FDA: no recombinant peptides >40 amino acids • Ph. Eur.: no recombinant peptides >60 amino acids
• No microbially fermented peptides (antibiotics)	• Peptide antibiotics not specifically excluded
• Only peptides from commercial organizations	• Does not exclude academic or government organizations
• No peptide vaccines	• Therapeutic vaccines – regulatory status not established • Regulatory authorities treat vaccines as biologics
• No peptide diagnostics	• No nontherapeutic peptide diagnostics

veterinary use will be manufactured by these CMOs, as will vaccines, but the revenues from neither will be included in analyses of the peptide therapeutic market. Income from non-GMP manufacture, and also GMP peptides in clinical development, which make up a substantial proportion of the revenue of many CMOs, in some cases 100% of their income, are not included in market analyses.

The absence of vaccines in market analyses of peptide therapeutics deserves additional mention owing to the recent and growing interest in the use of vaccines in personalized medicine. The regulatory and manufacturing challenges are significantly different from those for traditional peptide therapeutics. Personalized peptide vaccines require that, for each patient, several (up to 60) specific peptides would have to be manufactured at the milligram scale, released and administered within a matter of weeks. This requires a completely different regulatory mindset to that for classical drug product development. The time and cost constraints mandate a rapid, automated manufacturing process in which the overall process and not the individual peptide is validated. However, if successful, personalized peptide therapies could open up a significant market akin to that emerging from genomic medicine.

Peptide-related molecules such as peptoids, lantipeptides and depsipeptides are typically included in market analysts' statistics, but 2,5-diketopiperazine (cyclodialanine)-based peptides are not. Microbial fermented peptide antibiotics are not included, but semisynthetic antibiotics such as voclosporin have been (even when the chemical modification does not involve the amino acids *per se*).[6] Microbial fermentation of peptide antibiotics and semisynthetic peptides is typically performed by separate CMOs, partly for historical reasons but also to avoid potential cross-contamination of therapeutic peptides with β-lactams.

Peptide CMOs have a conservative attitude towards novel or unfamiliar manufacturing technology. This is well illustrated by the fact that *tert*-butoxycarbonyl (*t*Boc)-based SPPS chemistry, introduced in the mid-1960s, is still occasionally used to manufacture some approved peptide therapeutics today, even though significantly safer and more scalable 9-fluoromethoxycarbonyl (Fmoc)-based SPPS processes became popular in the early 1980s. This reluctance to change partly reflects an aversion to the risk and the costs (of raw materials) needed for new technologies. However, a crucial factor has been the unwillingness of sponsors to make any modifications to approved processes that would necessitate costly regulatory changes for the drug product. Based on this experience – and short of the introduction of novel economic aqueous chemistry – it is fair to assume that Fmoc-based SPPS will now remain the synthesis method of choice for peptide therapeutics for many decades to come, either as a stand-alone linear SPPS approach or for assembling fragments to be used in hybrid strategies such as SPPS–LPPS hybrid fragment condensation, native chemical ligation (NCL) or chemo-enzymatic peptide synthesis (CEPS).

There are currently no prominent examples of CMOs that offer parallel chemical and/or recombinant technologies for manufacturing approved peptide therapeutics (although Lonza previously provided this option). This may well change in the future if NCL and CEPS are successfully used to ligate synthetic peptides with recombinantly expressed proteins.

There is also a large market for peptides that lie outside the normal definition of therapeutic peptides and that nevertheless may be manufactured by CMOs specializing in peptides. These include cosmeceuticals, nutritional supplements, parenteral nutrition formulas (dipeptides such as Ala-Glu) and over-the-counter (OTC) and online performance enhancers. It is difficult to put numbers on these because peptides are only a small, but expensive, component of each class of substance mentioned. In a recent review of cosmeceutical peptides, the global market for all cosmeceuticals was placed at over $30 billion in 2016; the contribution of peptides to this not stated, but it is growing.[7] In the context of nutraceuticals, most peptide additives are from natural sources, but it should not be forgotten that some existing and emerging artificial sweeteners (*e.g.* aspartame, brazzein) are peptides. Peptide-containing OTC and online performance enhancers lie in a gray area in terms of regulatory and legal control. Another gray area is the status of peptides in cosmetics; the FDA does not recognize cosmeceuticals as a separate class of medicinal products.

There is also a growing market for biomaterials (*e.g.* anti-scaling agents), agricultural products (anti-pest agents) and medical equipment coatings that require significant quantities of peptides. These are all nontherapeutic areas that could help sustain the industry, although the quantities of biomaterials and agricultural products required may be impossible to produce economically using peptide chemistry.

In summary, although marketing reports may provide accurate data based on their own inclusion and exclusion criteria, they do not provide an accurate insight into what peptide CMOs might be manufacturing or how earnings from nontherapeutic peptides contribute to the market. The sustainability of the peptide manufacturing industry is not entirely dependent on revenues from approved peptide therapeutics or those in development.

6.3.3 The Generic and NCE Markets for Peptide Therapeutics

Most pharmaceutical companies keep their generic and ethical drug activities separate. To a certain extent this is also reflected in the generic and NCE API world. CMOs such as Bachem and PolyPeptide have a very strong bias towards peptide NCE manufacture while retaining a strong generic presence. Many other companies, now active in or expected to move into peptide NCE therapeutics as the current NCE providers continue to consolidate, previously had a strong small-molecule or peptide generic background (Almac, Ambiopharm, Asymchem, Aurobindo, BCN, Biocom, Cepia, Corden

Pharmaceuticals, DRL, Hamari, Hemopharm, Jymed, MSN, Neuland, Novotide, Quilo, ShengNuo, Shuangeng, Sun, Teva, USV and Zydus are some examples). The peptide NCE activities of CBL Patras and Senn grew out of their raw material businesses, those of CPC Scientific from its extensive research peptide activities and those of CSBio from its experience in developing peptide synthesizers.

The generic market is limited by the number of off-patent APIs, the current market volume and regional expansion. Dollar opportunities are easy to calculate and once a customer portfolio has been established, output is relatively stable, although profit margins will always be under pressure from new players. There is a limit on how many companies can have a slice of a generic "cake" and be adequately nourished. The competition for popular generics is intense, which results in prices drifting asymptotically towards cost cover levels. As such, the market prices for generic peptides correlate well with the actual manufacturing costs associated with a peptide of a specific length or complexity (which is definitely not the case for peptides sold in the ethical market – see Table 6.3).

In contrast to the generic market, the size of the NCE market is open-ended, but – particularly during development – income can be subject to enormous fluctuations as projects ramp up or fail. Once a peptide NCE or generic has been approved, the sponsor may decide to engage a second CMO as backup or for contingencies. In very rare cases, a third CMO might also be involved.

Recently, the FDA issued a Draft Guidance for the manufacture of peptide therapeutics that have previously been manufactured recombinantly. These include molecules such as liraglutide, teriparatide, glucagon and semaglutide. This has introduced a new class of peptide therapeutics that fall between generics and NCEs. Although the pathway for approval can follow the traditional generic Abbreviated New Drug Application (ANDA) route, there is the added complexity that, according to the Draft Guidance, any impurity above 0.1% that is not present in the Reference Listed Drug (RLD) should be tested for toxicity and immunogenicity. This is a challenge for these synthetic analogs, because the RLD and synthetically manufactured peptide will almost certainly not have the same purity profile. This class could almost be considered to be NCEs, or at least would require that level of expertise from the respective CMO.

6.4 Peptide Therapeutic Market Trends

6.4.1 The Approved Peptide Therapeutic Market

Table 6.1 contains a list of peptide therapeutics approved in the USA and Europe with some additional entries from Australia, China, Japan and Russia. Where the approval date is known, it refers to the first region where it was approved. The list is probably not complete.

Figure 6.2 provides an overview of the number of approvals over the last 20 years. Regression analysis suggests that the current approval rate is close

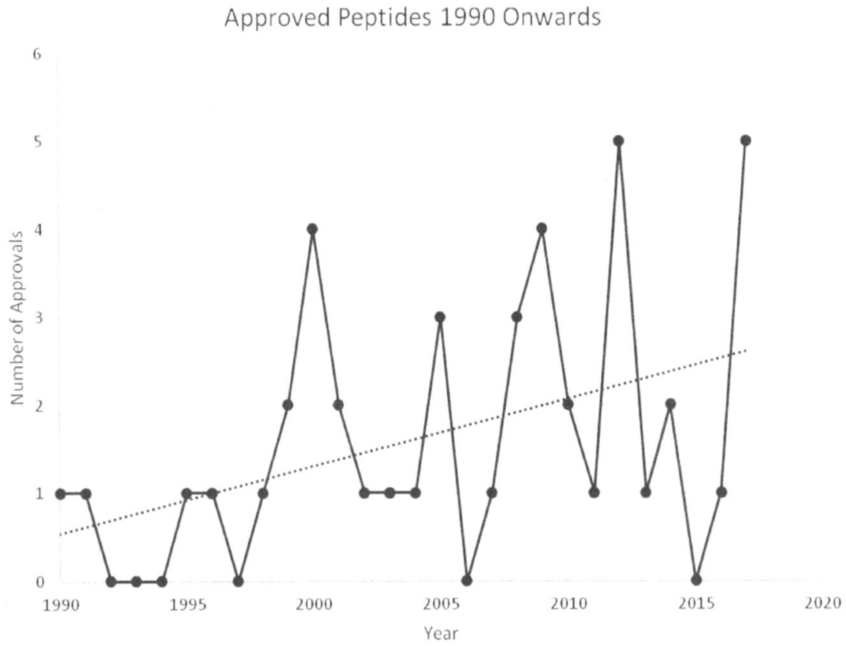

Figure 6.2 Approved peptides since 1990.

to 3, but individual year values can be highly variable. Originally, peptide therapeutics were mainly naturally occurring peptides used for hormone replacement therapy. In the 1980s, analogs [particularly gonadotropin-releasing hormone (GnRH) analogs] gained increasing popularity for the treatment of endocrine cancer. Starting in the late 1990s, many glucagon-like peptide-1 (GLP-1) agonists were and continue to be developed. Today, there is hardly a therapeutic indication that lacks at least one peptide-based pharmaceutical. However, oncology, metabolic diseases and cardiovascular indications dominate.[2,8,9]

Several authors and marketing analysis organizations have provided estimates of the market volume and growth over the years (Figure 6.3). These figures rely on marketing analysts' definitions of peptide therapeutics.[2,8,10,11]

Assuming that the market value for peptide therapeutic drug products today is about $22 billion, it is possible to make some assumptions about the CMO market for approved APIs (drug substances). The contribution of the peptide-based API to the final price of an individual peptide therapeutic drug varies widely, with a highly potent drug substance such as exenatide contributing less than 0.5% and some low-potency or generic peptide APIs contributing over 10%. For generic peptides in a highly competitive market, an average of 7.5% can be used as a yardstick. This would indicate an approved drug substance market of $1.65 billion, of which maybe only 40% (equivalent to $660 million) is contracted out by pharmaceutical sponsors.

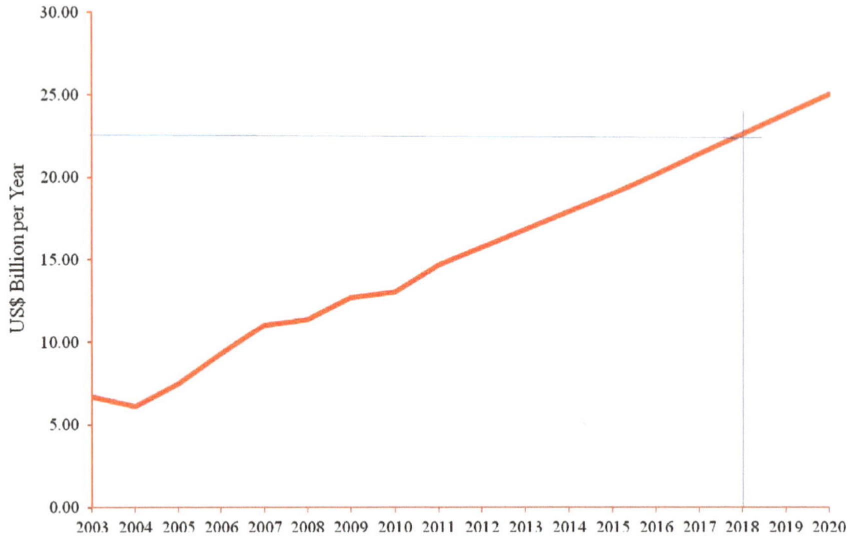

Figure 6.3 The market for peptide therapeutic drug products 2003–2020. Estimated market value today, \$22 billion; peptide therapeutics as a share of all drug products, ~2%; peptide drug product market estimated continuous annual growth (CAG), ~7.6%.
Data derived from ref. 2, 6 and 8.

6.4.2 Peptide Drug Substances in Clinical Trials

There is a healthy pipeline of peptide therapeutics in the clinic. According to Lau and Dunn, there are about 150 candidates in clinical development with 50% of these being in Phase II or III.[9] One of the initial attractions of peptides with their associated low risk of toxicity or unanticipated side effects was an expected higher rate of approval and shorter time to achieve this. This was certainly true of early approvals, probably partly based on the novelty of a new class of therapeutic agents, but as peptides have become mainstream and regulatory requirements have tightened, the time-to-approval advantage has receded. The current time that lapses between the first clinical trial and approval is 9.4 years, which seems to lie between those of small molecules and biologics. However, the data are difficult to interpret accurately as individual classes of drug candidate perform differently at discrete stages of clinical development. Candidates (a) with well-defined endpoints and (b) sponsored by larger pharma companies achieve approval faster.

In total, over 500 peptide candidates have entered clinical trials, with over 70 (14%) being approved. Nine approved peptide drugs have been withdrawn.[9]

The importance of drug discovery and alternative drug delivery formulations is not discussed in this chapter, but both are of essential parts of the current peptide therapeutic landscape. Without a continuous inflow of new candidates and the possibility of ensuring compliant drug self-administration, the future of peptide therapeutics would be severely compromised. The reader is referred to relevant overviews of these topics.

6.5 Commercial Manufacturing Methods for Peptides

6.5.1 Tissue Extraction

Some peptide (and protein) therapeutics, starting with insulin in the 1920s, were originally extracted from animal tissue. ACTH (adrenocorticotropic hormone, 39-mer, 1950s) and salmon calcitonin (32-mer, 1971) are other examples. These were outside the scope of synthetic chemistry when it was introduced in the 1960s. Glucagon was also available as a by-product of insulin production. Apart from adrenocorticotropin, tissue extraction is no longer considered a viable method owing to the potential for prion contamination and the generally low concentrations of the active ingredient in the respective tissue.

6.5.2 Microbial Fermentation

Many currently approved antibiotics are based on nonribosomal peptides and are manufactured by microbial fermentation with or without additional chemical modification. Examples are penicillins, cephalosporins, bacitracin, gramicidin S, polymyxin B, tachyplesin 2, cyclosporin and voclosporin.[12] The microbial fermentation of many antibiotics requires physically separate facilities from those using chemical synthesis because of potential allergic reactions associated with them (especially those containing β-lactam rings).

6.5.3 Chemical Synthesis

A good overall review of peptide chemistry was provided by Sewald and Jakubke.[13]

6.5.3.1 Liquid (Solution)-phase Peptide Synthesis (LPPS)

In its simplest form (*e.g.* for the manufacture of dipeptides), liquid-phase peptide synthesis (LPPS) involves solution-based organic chemistry. As applied to longer peptides, LPPS is characterized by the use of specialized protecting groups to prevent polymerization and some common synthetic strategies. LPPS, which dates back to about 1900 and the work of Fischer and Curtis on the synthesis of glycylglycine and benzoylglycylglycine, respectively, was used to manufacture synthetic peptides until the mid-1960s, when solid-phase peptide synthesis (SPPS) was introduced.[14–16]

Linear and convergent LPPS have been used to manufacture some peptides of significant length and complexity (*e.g.* calcitonin). Nowadays it is preferred only for the synthesis of shorter peptides (*e.g.* GnRH analogs) or peptide fragments, typically with ≤10 amino acids and only when those are needed in large quantities. The upside of LPPS is the scalability of the assembly process. The major downsides are the limitation to short peptides and the number of individual process steps (which need to be monitored).

6.5.3.2 Solid-phase Peptide Synthesis (SPPS)

Solid-phase peptide synthesis (SPPS) was introduced by Merrifield in 1963.[17] By the late 1960s, with *t*Boc (originally introduced by Carpino[18] and McKay and Albertson[19] as a protecting group for LPPS in 1957) as the N-terminal protecting group and hydrogen fluoride (HF) reported in 1965 for final deprotection and cleavage from the peptide resin, SPPS had become the primary methodology for assembling peptides. It can be applied to significantly longer peptides than can LPPS. The intermediates of the multistep assembly process on the resin do not need be isolated, eliminating much of the labor-intensive controls needed in LPPS. Fmoc-based SPPS was introduced initially in 1970 and from the early 1980s onwards has been the predominant technology for manufacturing peptides longer than five amino acids.[20,21] Fmoc chemistry has three significant advantages over *t*Boc chemistry: scalability and independence from the use of both environmentally unacceptable halogenated hydrocarbons (dichloromethane) and hazardous HF.

6.5.3.3 Dual-phase Peptide Synthesis

A number of companies offer Fmoc-based technology using supports that can exist in both the liquid and solid phase (Ajiphase® by Ajimoto; Molecular Hiving™ by Jitsubo; Group-Assisted Purification by GAP Peptides).[22–25] The growing peptide chain is anchored to a hydrophobic support allowing significantly more homogeneous and thus more efficient and controllable reaction conditions than can be obtained using solid supports. Both removal of reagents and post-reaction purification are achieved by addition of polar solvents and precipitation of the peptide on-support.

There are some obvious economic advantages to using this type of technology in large-scale processes for short and medium chain length peptides, but it is still to gain traction as a standard approach.

6.5.3.4 Recombinant Manufacture

A good overall review of recombinant manufacturing technologies (for proteins) was provided by Gupta and Shukla.[26]

Although the recombinant expression of proteins is typically associated with longer polypeptides (proteins) rather than peptides with more than 50 amino acids, the first demonstration of the technology was for the production of somatostatin-(1–14), albeit as part of a fusion protein with β-galactosidase (the somatostatin was cleaved from the protein with cyanogen bromide).[27] The first commercial products manufactured by recombinant expression were human growth hormone and insulin.

Today, over 60 monoclonal antibodies and about 80 other recombinant proteins (mainly enzymes, coagulation factors and immune modulators) are approved.[28,29] Very few peptides with <50 amino acids are manufactured by recombinant expression. Glucagon, liraglutide, nesiritide, semaglutide, teduglutide and teriparatide, which are currently manufactured recombinantly by their innovators, can be synthesized chemically and most generic formulations do indeed contain the chemically synthesized sequence. Nevertheless, recombinant expression as a manufacturing process for peptide therapeutics has penetrated the market, particularly where annual requirements are large.

Recombinant expression systems require proteogenic amino acids as starting materials. Some proteogenic amino acids can be chemically modified *in situ* and it is possible to modify the genetic code and associated molecular apparatus to incorporate nonproteogenic amino acids, but the latter is a major undertaking.[30,31]

Dulaglutide and romiplostim should be considered to be fusion proteins rather than peptide conjugates. Insulin with 51 amino acids is manufactured recombinantly and the insulin market significantly exceeds that of all peptide therapeutics with more than 50 amino acids.

6.5.3.5 *Comparison of Synthetic Chemistry with Recombinant Expression*

Linear Fmoc-based SPPS and recombinant expression represent the two main technologies that are currently employed to assemble therapeutic peptides. To a certain extent the technologies are interchangeable for any particular peptide provided that it only contains proteogenic amino acids. SPPS is significantly more flexible, allowing the incorporation not only of D- and exotic amino acids, but also of non-amino acid residues. It is also more economical at the low kilogram scale. However, SPPS has limitations. Because no chemical reaction goes 100% to completion, SPPS of long peptides is associated with exponentially declining yields and an increasingly complex impurity profile as the sequence length increases that may make a synthetic approach economically untenable. It is generally accepted that a chain length of 50 amino acids is currently the approximate upper economical limit for Fmoc-based peptides that need to be manufactured at a multi-10 kg scale using linear SPPS.

6.5.3.6 *Semisynthetic Synthesis of Peptides*

With respect to peptides, the term "semisynthetic synthesis" generally refers to chemical modification of the products of microbial fermentation, especially the modification of peptide antibiotics in order to improve their therapeutic profile, *e.g.* modification of cyclosporin to form voclosporin or of vancomycin to form telavancin. However, it should also refer to chemical modification of peptides and proteins expressed recombinantly or indeed extracted from tissue.

6.5.3.7 Hybrid Approaches

Hybrid approaches have been, and continue to be, developed with the goal of bypassing the main limitation of SPPS, namely the exponential loss of yield as the chain length increases. This loss can be mitigated by coupling shorter peptides to each other using highly efficient ligation procedures. Hybrid approaches may also be used to overcome difficult couplings that cannot be addressed by linear SPPS and – in some cases – may be the only available route for an individual peptide.

Hybrid strategies require additional one-time development such as reaction optimization, analytical and process validation procedures and hold time studies for each fragment (and also for the final product). After the process has been finalized, the burden of control, release and storage of each fragment and also the final product is an ongoing cost. In this context, the savings gained by an increased overall yield need to be balanced carefully against the additional fragment strategy costs to confirm economic viability. However, there is no doubt that for peptide therapeutics with longer chains (>30 amino acids) that are needed in multi-10 kg quantities, the use of a hybrid strategy is a rational approach and should be evaluated.

6.5.3.7.1 Classical Hybrid Fragment Condensation. By classical hybrid fragment condensation we understand the chemical ligation of peptide fragments assembled by SPPS to each other in solution using LPPS technology. The process uses fully protected (side chain and N-terminal Fmoc) fragments. In principle, the process is analogous to LPPS of a short peptide except that fully protected peptide fragments are used instead of protected amino acids. Bivalirudin, enfuvirtide and exenatide are examples of approved peptides that have been assembled by a hybrid fragment strategy.[32]

The downsides of classical hybrid fragment condensation include the time to develop a suitable strategy (choice of fragments) and the difficulties associated with intermediate purification of fully protected fragments. However, the most commonly cited downside is the need to find coupling points where the fragments have C-terminal glycines or prolines so that racemization can be avoided. Interestingly, neither the process for enfuvirtide nor that for exenatide fulfill this latter requirement.

6.5.3.7.2 Native Chemical Ligation (NCL). Classical NCL requires that one peptide fragment has a C-terminal thioester and the other an N-terminal Cys.[33] The thiol group of the cysteine attacks the C-terminal thioester in a reversible transthioesterification reaction leading to a thioester intermediate, which then undergoes an irreversible intramolecular rearrangement that results in the formation of a peptide bond. This technology, which was introduced by Dawson and Kent in 2009, has been used to assemble a large number of complex peptides and proteins (crambin, 46-mer; human lysozyme, 130-mer; HIV-1 protease, 203-mer; cMyc-Max, 172-mer; erythropoietin, 166-mer; human IL-8, 72-mer; human phospholipase A_2, 124-mer; microprotein S, 116-mer; barnase, 110-mer; lymphotactin, 93-mer; RNAse B, 68-mer; diptericin ε, 82-mer; CCL7, 76-mer).[33,34]

To date, there are no approved peptide therapeutics produced using NCL. This is primarily due to the instability of intermediates and the inability to scale up. It is hoped that these challenges will eventually be overcome.

A further limitation of NCL is the mandatory introduction of an extraneous cysteine footprint at the point of ligation. This can be converted to alanine by desulfurylation.[35] Other thioamino acids have also been used to generate their respective non-thioamino acid counterparts, enabling elimination of the footprint for certain amino acid sequences.

6.5.3.7.3 Chemo-enzymatic Peptide Synthesis (CEPS). The concept of using enzymes to assemble peptides is not new.[36,37] Past strategies involved reverse proteolysis using trypsin and other commonly available proteases. To drive the thermodynamically unfavorable reverse hydrolysis, reactions had to be carried out under anhydrous conditions (*i.e.* in organic solvents). Because longer peptides tend to be hydrophilic, solubility is a problem and most studies were limited to relatively short peptides. No currently approved peptide therapeutics are manufactured using this approach, although some small nontherapeutic peptides (*e.g.* aspartame) and antibiotics are. In this context, it should not be forgotten that the enzymatic conversion of porcine insulin to the human form was a major commercial process from the early 1980s until the late 1990s for Hoechst and it can be assumed that several hundred kilograms were produced using this process.[38–40]

More recently, a number of enzymes have been isolated (*e.g.* butelase) or engineered (sortase, subtiligase, peptiligase, trypsiligase) that can couple peptide fragments to each other in both selective and nonselective manners with respect to the sequences.[41–45] Sortase and butelase require mandatory recognition sequences that are not fully eliminated during ligation and leave a footprint.[46,47] Others, such as subtiligase and peptiligase, can be engineered to recognize most sequences around the coupling points so that footprints can be avoided.[45,48] Critical for commercial manufacture is the ability to work with enzymes that catalyze ligation reactions kinetically (irreversibly), thus eliminating the need for high molar excesses (and cost) of one reaction partner. There are currently no approved peptide therapeutics that are manufactured using CEPS, but appropriate scale-up studies have been performed and the viability of this approach has been confirmed.[49]

CEPS can also provide unique routes for the manufacture of molecules that cannot, or cannot without difficulty, be assembled on a commercial scale using existing chemistry. The assembly of complex macrocyclic peptides is a pertinent example.[50,51]

6.5.4 Downstream Processing

Assembly of the peptide is only one part of the physical manufacturing process. Downstream processing includes deprotection and cleavage from the resin, purification and salt exchange, isolation and, in the case of conjugates, chemical modification and additional processing. Most of these processes involve the use of organic solvents. The downstream process and strategy utilized are dependent on the eventual quantities of the peptide required, the capabilities present at each individual CMO and the critical quality attributes (CQAs) desired by the sponsor.

If the purity specifications are modest, it may be possible to isolate the peptide by precipitation without having to utilize chromatography. However, in most cases reversed-phase chromatography does need to be utilized and, at the commercial scale, this will be performed with sublots. This brings the challenge of hold times; peptide solution sublots can be held for only a fixed amount of time before significant degradation occurs. To avoid this, the peptide sublots may be stored in solution in a cold room or be lyophilized and then reconstituted when all the other sublots are available. The path chosen is highly dependent on the equipment and material flow available at the CMO, and also the stability of the peptide.

Ultrafiltration can also be utilized for removing impurities and organic solvents prior to isolation and/or be used to exchange the peptide solution directly into the formulation buffer, thereby providing a seamless transfer to the formulator where appropriate.

Isolation is typically performed by lyophilization or spray drying or, in suitable cases, by precipitation or crystallization. If the eventual quantities desired are in the multi-100 kg to metric ton range, lyophilization would probably not be the isolation procedure of choice. Precipitation, crystallization or spray drying may be better options. Both precipitation and crystallization would require a substantial amount of development to produce a process that is reproducible and robust – this development could cost several hundred thousand dollars. On the other hand, if equipment is available, spray drying can be developed fairly quickly at a cost equivalent to lyophilization. Spray drying utilizes harsh conditions and its impact on the CQAs has to be monitored closely.

6.6 Economics of Peptide Manufacture

6.6.1 Cost Structure

Attempting to compare the cost structure between CMOs in any detail can be challenging because of the many variables and restricted access to actual costs at CMOs.

A company has to keep close and detailed control of its costs in order to operate efficiently. In simplistic terms, the cost of manufacturing peptide therapeutic APIs at a CMO can be split into direct costs (related directly to production) and indirect costs (management, administration, business development, *etc.*). The latter are considered to be overhead costs, which are spread over all products, which may include nonpeptide entities. Service department costs, such as those arising from quality and regulatory departments, supply chain, *etc.*, might be assigned to either direct or indirect costs depending on the nature of the CMO's business. It is logical to expect a CMO with a single or few large API projects to assign such services directly to a product, whereas a CMO with many smaller projects will view these more as a service to the company rather than being product related. For the purposes of book-keeping, a CMO may assign all overheads to a few profitable peptide projects or, at a multipurpose site, pool all overhead costs irrespective of the source.

Given that small-volume assignments require similar GMP compliance, quality and regulatory support to large-volume projects, CMOs overseeing a few large projects have a significantly lower GMP cost load than those overseeing many small projects.

Geographical location is often considered to be a major cost factor in pharmaceutical manufacture. Wages and associated salary costs probably account for 40–50% of all costs at a peptide CMO in the USA and Europe. Given that the average salary of a chemist in the USA is about $84 000, whereas those in China and India are $27 000 and $10 000, respectively, one would assume that, all other things being equal (*i.e.* equipment and raw material costs), the latter regions would enjoy a massive economic advantage.[52] However, in a niche industry using the services of highly skilled employees, this may be more complex. As more Western-educated chemists and technicians return to their home countries, the salary expectations increase. It is also important to note that manufacturing cost is dependent on the scale of manufacture; at very large scales a main driver is the raw material costs, which are similar worldwide. Additionally, established processes can be automated, resulting in fewer full-time employees and a lower wage bill. The cost of energy is also an important factor and therefore locations with automation and lower energy costs have distinct economic advantages.

6.6.2 Comparative Cost Structure of Different Technologies

The introduction of new technologies or even modest modifications to existing processes for approved peptide therapeutics are often avoided because of regulatory considerations. Exceptions arise when either the quantities of drug substance overwhelm the output of the current manufacturing process, the economic advantages of process modification outweigh the costs of regulatory changes, mandatory raw materials are no longer available or commercially acceptable or environmental laws prevent the use of specific solvents or process aids. However, for APIs in early development, choices can be made and, where appropriate, different processes may be developed in parallel.

6.6.3 Linear SPPS or LPPS?

LPPS is mainly reserved for peptides with fewer than 10 amino acid residues, where the resin can represent a disproportionate cost if SPPS is chosen. Because the reaction intermediates of LPPS are isolated, they are subject to characterization, validation and individual quality control (QC) release. LPPS is associated with significantly higher regulatory costs. In general, much purer crude products can be expected. The economic advantages of LPPS over SPPS are dependent on scale. Figure 6.4 provides some data where both LPPS and SPPS were used to manufacture a five amino acid peptide. The crossover point at which LPPS becomes more economically attractive is around 15 kg. Of course, the use of unusual raw materials would skew such results.

6.6.4 SPPS: Fmoc or *t*Boc chemistry?

*t*Boc chemistry is not scalable owing to the exergonic nature of the HF cleavage step. In addition, the use of methylene chloride is no longer permitted for pharmaceutical API manufacture in some European countries. As a result, *t*Boc chemistry is no longer a real choice for commercial

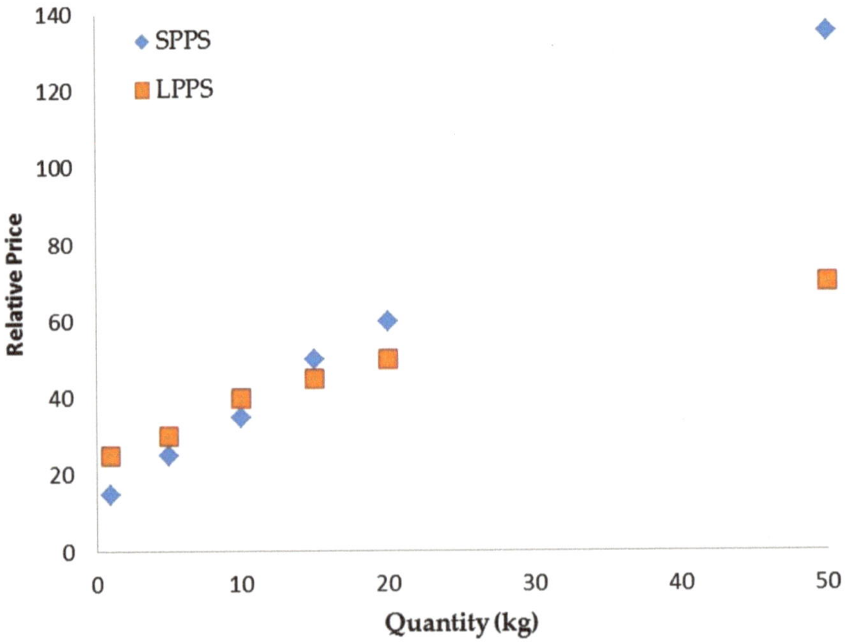

Figure 6.4 Comparison of costs between SPPS and LPPS.

manufacture. As these concerns rise, the use of *t*Boc chemistry becomes less attractive and, as a result, raw materials have become more difficult and more expensive to source.

6.6.5 Linear SPPS or Recombinant Expression?

There is a range of sequences from approximately 30 to 60 amino acid residues where both technologies can be usefully applied when only proteogenic amino acids are present. Those companies that have in-house peptide chemistry, recombinant fermentation, or both, will probably be influenced by their internal options and IP considerations, but for a pharmaceutical company that is using an external CMO, chemistry is the more economical option for early development and clinical trials of an NCE, probably by a factor of 5–10. Moreover, the time required for development and production using the chemical process is significantly shorter than that for a recombinant process, which would be beneficial in the early stages to obtain proof-of-concept information. However, there are often points in development or commercial manufacture where a change of manufacturing technology could become attractive. Such changes should be made very early in development or after regulatory approval.

6.6.6 Linear SPPS or Fragment SPPS Assembly?

Fragment approaches are mainly used to assemble long peptides. By using highly efficient fragment coupling procedures with short peptide fragments generated by SPPS, the low yields associated with linear SPPS can be mitigated. However, all fragment procedures require extensive development (choice and testing of various strategies), additional one-time validation and optimization studies and additional QC release. The choice of utilizing SPPS *versus* a fragment-based approach is dependent on the scale of manufacture. Generally, at scales that warrant batches sizes that exceed 15 kg, fragment synthesis would be preferred. At this crossover point, raw materials are a main cost driver and therefore any additional labor cost necessary for the fragment process becomes diluted and the savings comes from the increased yield.

Very short Fmoc-protected peptide fragments, typically 2–4 amino acids in length, are often used to bridge difficult couplings in peptides or to avoid racemization. These can be viewed as alternative raw materials rather than components of a fragment synthesis. A degree of caution is required when using this approach, especially when unusual building blocks are involved.

Fragments can be very specific to an individual molecule and therefore they may not be produced by multiple vendors or in adequate quantities, resulting in a high cost and lack of availability. This could eventually impact the cost of the manufacturing approach and make a linear SPPS process more attractive.

6.6.7 Linear SPPS or Native Chemical Ligation (NCL)?

NCL technology has been used to assemble long peptides since 1994, but it has not yet found commercial-scale applications.[33] Currently, it must be considered to be a discovery and development tool.

6.6.8 Enzymatic Ligation

Although several enzymes have been shown to assemble peptides efficiently from shorter peptide fragments, most reports focus on milligram-to-gram feasibility studies for new or "difficult to synthesize" peptides or protein conjugates.[45] The Dutch biotech company EnzyPep, which focuses on designing commercial CEPS processes, has reported these for thymosin-α_1, exenatide, liraglutide, semaglutide and the macrocycle MCoT1-II.[50,53,54] In cooperation with the PolyPeptide Group, the manufacture of exenatide at the 100 g scale was successfully performed to demonstrate the scalability of the process.[49] The main advantages of CEPS are dramatically increased yields, better impurity profiles and lower solvent usage. The longer the peptides and the greater the quantities, the more pronounced the advantages become.

6.6.9 Intein Technology

Intein technology would not normally be considered to be a peptide therapeutic manufacturing option, being primarily a technology for assembling proteins on a small scale and as a research tool.[55]

6.7 A Very Short History of Therapeutic Peptide Manufacturing Companies

The market for peptide therapeutic drug substances is growing at about 3–5% per year, driven mainly by increasing volumes of individual peptides and an expanding world market. The number of CMOs and sponsors manufacturing generic peptide drug substances is rising, with several newcomers in the last 5 years and more expected.

The situation is not the same for peptide-based CMOs focused on pharmaceutical NCEs. A brief glance reveals a market with an ever-shrinking number of experienced players. In the mid-1990s there were about 20 experienced CMOs that specialized in pharmaceutical-grade therapeutic peptide NCEs. Today, there are only about five and, what is more alarming, is that of these, only two, namely Bachem and PolyPeptide, are operating under their original name and ownership.

What happened? Some of the original companies from the 1990s simply left peptides to pursue more profitable activities (*e.g.* Avecia) or effectively terminated external contracting (*e.g.* Abbott, Mallinckrodt, Diosynth, NovaBiochem). However, most were acquired by their competitors or other organizations. For example, Bachem (Switzerland) acquired Bachem

California (formerly a completely separate company in spite of the identical name), Peninsula and the American Peptide Company. The PolyPeptide Group acquired the Prague Peptide Institute, NeoMPS (which had previously merged with MPS and was owned by Isochem) and, in 2017, Lonza's peptide group in Belgium (which had previously acquired UCB Bioproducts). Both Bachem and PolyPeptide continue to expand. The Roche Boulder Peptide Division, the Genzyme Peptide Division (which had previously acquired Sygena), Peptisyntha and Orpegen are all now are part of the Corden Pharmaceutical Group that belongs to International Chemical Investors Group (ICIG).

Why did this happen? One answer is obvious: the increasing difficulty of sustaining satisfactory profit. The last 30 years have seen highly significant price attrition while most costs have risen.

6.8 Contributing Factors to Sustaining Profit

6.8.1 Price Development

As peptides have moved from being "innovative" in the 1980s to becoming mainstream drugs, the achievable price target for the APIs has fallen significantly. This is despite the fact that peptides have become increasingly longer and more complex. Although it is not possible or particularly useful to present data to support this for proprietary projects, the evidence for large to modest volume generics is compelling. Figure 6.5 shows how the price of one generic drug substance (salmon calcitonin) has fallen over a 20-year period. The figure also shows how the quoted prices of four other peptide therapeutics have decreased. In general, the quoted and actual sale price of generics with attractive volumes have decreased by approximately 50%. On the other hand (data not shown), generics with low volumes (*e.g.* gonadorelin acetate), which are unattractive for generic API manufacturers, have maintained relatively stable prices.

6.8.2 Materials and Solvents

In contrast to prices, most costs have increased over the same period. Perhaps the most notorious example is acetonitrile, which is used in final purification. As a by-product of acrylonitrile, a solvent that is used extensively in the carpet and car industry, it has been highly sensitive to swings in consumer spending. In 2008 the price soared (Figure 6.6). Although it has returned to a level of about twice what it was prior to 2008, it is a contributor to overall costs. The costs of isopropyl alcohol (IPA), dichloromethane and methanol have also increased, although that of dimethylformamide (DMF) has remained relatively stable.

The only costs that have decreased significantly during the same period are those of the common Fmoc-amino acids, certainly driven in part by the immense requirements of Roche's Fuzeon (enfuvirtide) project. However, in

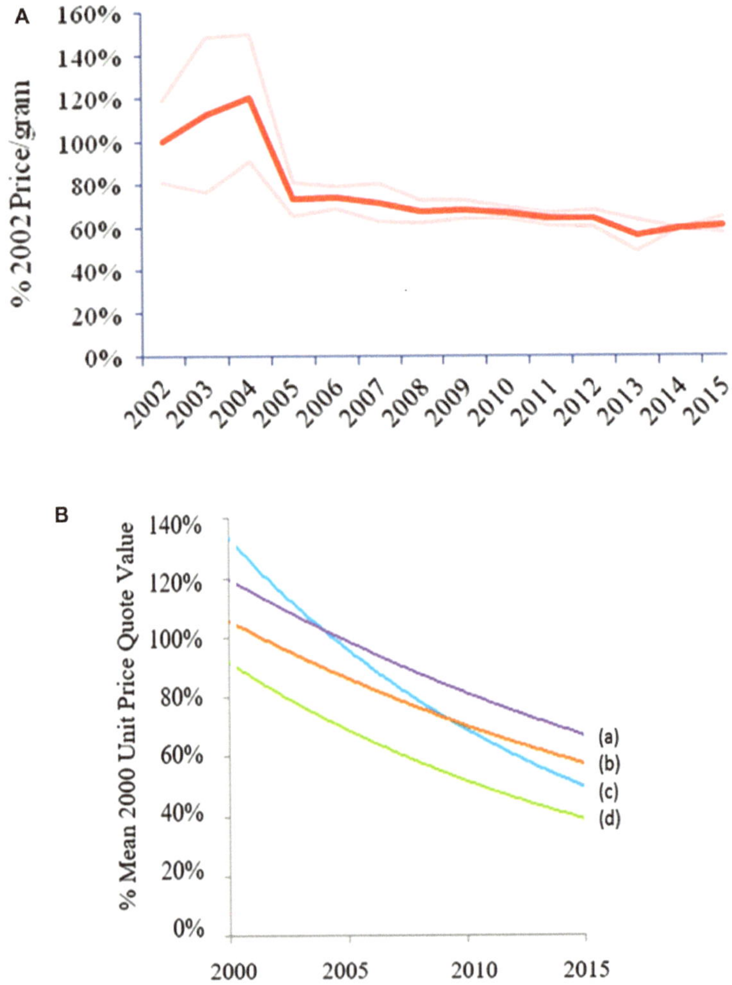

Figure 6.5 Development of peptide sale prices and quotes 2000–2015. (A) Achievable sale price of salmon calcitonin API between 2002 and 2015. Unit price expressed as a percentage of 2002 mean ± SD; data generated only from sales ≥10 g; range = 10 g to >1 kg; $n = \sim 400$). (B) prices on four generic peptides between 2000 and 2015. (a) and (c) are long peptides (>30-mers), (b) and (d) are short peptides (<10-mers). Data are shown as exponential trended lines from a scatter diagram with ~15 points for each generic.
Data provided with permission from the PolyPeptide Group. Data previously presented at the TIDES Conference 2016.

early development when low-kilogram quantities are required, Fmoc-amino acids constitute only a modest part of the overall costs (Figure 6.7) and cannot in any way compensate for the increases seen in the costs of other materials and labor.

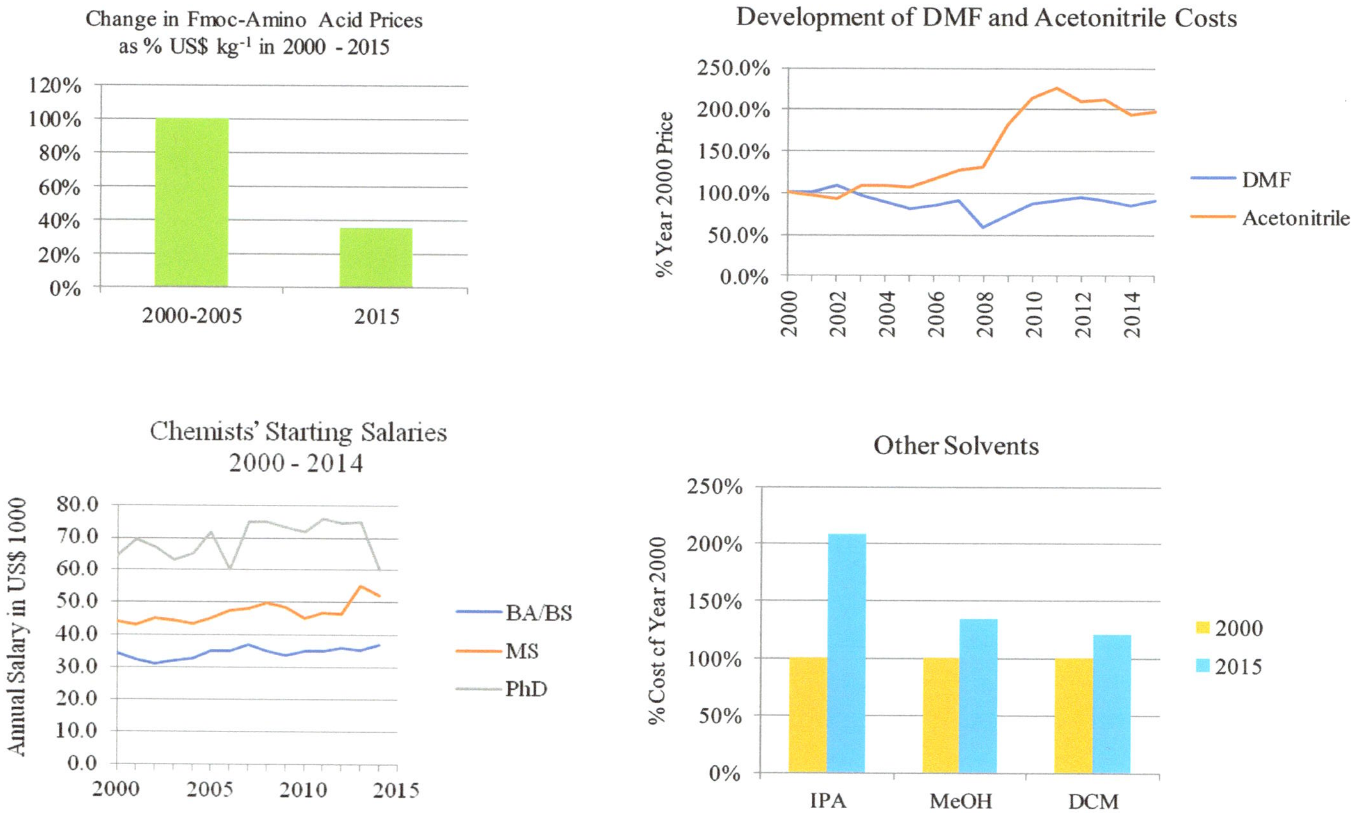

Figure 6.6 Development of manufacturing costs. With the exception of the analysis of chemists' starting salaries (which are from the ACS's Survey of New Graduates in Chemistry and Chemical Engineering, Steve and Clint Marchant, 2015, Data Based Insights, Inc.).
Data provided with permission from the PolyPeptide Group. Data previously presented at the TIDES Conference 2016.

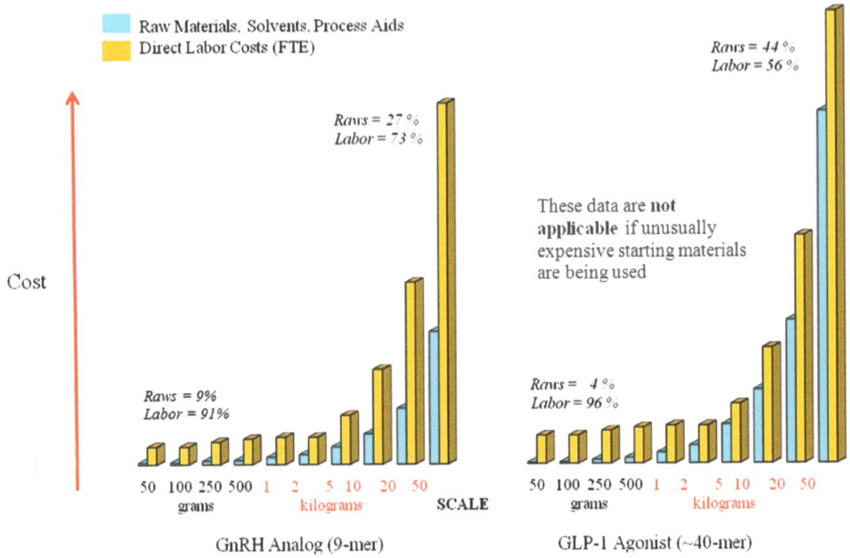

Figure 6.7 Cost split between raw materials and manpower. Analysis was performed in 2012. Analysis assumes capacity restrictions based on current manufacturing scale.

6.8.3 Salary-related Costs

The American Chemical Society (ACS) publishes a report on starting salaries of chemists every 5 years. Chemists' median starting salaries increased by about 10% between 2000 and 2014 when the American Peptide Society last conducted a survey.[56] As an interesting aside, the cost of living between 2000 and 2018 rose by about 45%.[57]

6.8.4 Management and Direction

The inability of many CMOs to build their increasing costs into the final price of their products is often attributed to global competition and the opening of markets to low-cost labor from India and China. However, regional labor costs are not the only cause.

Mismanagement and misdirection are contributing factors. When companies have focused entirely on peptide NCEs without alternative sources of revenue – particularly if none of the peptides were approved drug substances they have been exposed to the periodic ebb and flow of biotech and venture capital cash. Without alternative revenues, CMOs have unrelentingly underbid each other. Once down, realistic prices are hard to re-establish. This situation has been exacerbated in the recent past by some subsidiary CMOs of larger organizations which, bankrolled by their owners and freed of financial risk, were able to pursue sub-cost pricing over many years while maintaining losses that would put independent organizations out of business.

Cost structure has also changed significantly over the last 30 years. Although Quality-by-Design (QbD), risk management and validation of final products are usually invoiced separately to the client, the increased load of day-to-day compliance and regulatory activity and also safety and environmental services remain built into API prices.

Regulatory and quality requirements on raw materials, especially starting materials, have increased significantly over the years. Although these are issues that should be covered by the respective vendors, it is an onus that is still currently falling on the CMOs.

Another often overlooked cost is the time and human resources that are dedicated to CMO–sponsor interaction. Around 30 years ago this was minimal. Indeed, one could solicit significant volume orders for GMP peptides verbally at trade shows. Apart from written confirmation, the next interaction would be 4–5 months later when the product was ready to ship. All this has changed. Nowadays, the proposal process often involves several costly cycles of extensive documentation and sequential downward price discounts. Once the proposal has been accepted, there is very frequent interaction with the CMO or its consultants, often on-site and pulling in many employees. The loss and cost of resources to other projects are significant and they are much more severe for those CMOs handling many small projects than for those handling a few large projects.

There is no question that these interactions during development are useful for both CMO and sponsor (and their agents) because it mandates continuous monitoring of the project's progress and provides the ability to react to unanticipated events. However, it also unavoidably leads to "scope creep" and unforeseen costs. Retaining a cost position for "additional unforeseen development" and also for risk mitigation is critical.

6.8.5 Protection of Intellectual Property

Peptide therapeutics are manufactured by generic processes, such as linear Fmoc-based SPPS, that are in the public domain. The basics of GMP are practiced in all pharmaceutical manufacturing processes. So why is it that some CMOs manufacture peptide therapeutics in higher yield, with higher quality and with better documentation than others? This cannot just be equated solely to higher expenditure on personnel and equipment. A significant factor is experience, in-company know-how and proprietary use of technology. Unfortunately, the peptide manufacturing industry is particularly susceptible to seepage and loss of these assets.

Holding on to experience and know-how is becoming increasingly difficult in a volatile job market where the residence time of employees at any one company or any one region of the world is getting shorter. There is perhaps little that CMOs can do about this except provide training and career incentives to promising employees and exact punitive consequences for breach of confidentiality.

This is, however, only part of the problem. Know-how often flows unintentionally in the form of batch production records (BPRs) from the primary vendor's process *via* the sponsor to a secondary vendor's facsimile process and from there wittingly to unrelated processes at the secondary vendor's facility, from where it starts the cycle anew with a separate sponsor's project. CMOs need to mandate that sponsors take responsibility for protecting their CMOs' know-how.

Another issue, but hardly unique to peptide therapeutics, is how arising IP is shared between sponsor and CMO. Aside from cost considerations, most sponsors choose their CMO on the basis of their peptide manufacturing prowess and regulatory approval history, qualities that were acquired over many years as "spin-off" from third-party projects. While happy to benefit from a CMO's IP accumulated from past projects, some sponsors are very reluctant to allow arising IP to be used by the CMO on future projects. For example, we have seen draft contracts that mandate for perpetuity that the CMO shall never manufacture any API that could potentially compete with the sponsor's drug candidate or other company products or mandate that the CMO may not produce any peptide that could treat the same medical indication. It is important that CMOs insist that arising IP be shared and that the CMO may profit from this, insofar as this does not compete directly with the sponsor's product.

6.9 Environmental Impact of Peptide Manufacturing

Modern analytical techniques allow the detection of APIs and their impurities at low levels. The use of innovative analytical techniques allows us to trace APIs into groundwater and eventually into the lakes and oceans, where they may have toxic effects on aquatic life. Contaminated groundwater or aquatic life is consumed by other animals and humans. This concern and awareness have led to increasingly stringent regulations to control the impact of API manufacturing on the environment. We will now examine how these regulations are affecting the peptide manufacturing industry.

The release of an API into the environment can occur from three primary sources: (a) from normal patient use and excretion of an API and its metabolites into sewer and wastewater treatment systems, (b) from improper disposal of medicines by consumers and (c) from facilities where the API is manufactured, formulated or packaged. Given the chemical nature of peptides and their susceptibility to proteolytic degradation, the toxicity of the peptides themselves is a minor issue. Far more important are the potential environmental issues associated with the reagents and solvents used in their manufacture, but which are not part of the final product.

Regulations extend well beyond the consideration of API release into the environment. Environmental considerations for manufacturing APIs, more specifically peptides, are focused on the overall approach to manufacturing and have resulted in regulations and oversight from a number of different agencies. A holistic overview of the environmental considerations involves looking at a number of different areas:

1. solvent usage – green chemistry;
2. manufacturing facility capabilities;
3. waste disposal;
4. employee safety.

6.9.1 Solvent Usage – Green Chemistry

The type and usage of solvent are critical in the successful manufacture of peptide-based APIs. Solvents used in peptide manufacture include, but are not limited to, DMF, dichloromethane, IPA, acetonitrile, trifluoroacetic acid (TFA), piperidine, methyl *tert*-butyl ether (MTBE) and hexanes. The use and storage of these solvents are highly regulated owing to the potential hazards they pose. Regulations are typically imposed by national authorities and, as such, have regional consequences for peptide manufacturers in terms of both processes and cost. For example, dichloromethane is considered to be carcinogenic and in some countries in Europe environmental agencies have heavily limited its use. Piperidine is regulated by the Drug Enforcement Agency in the USA because it is considered a controlled substance. This requires manufacturing facilities to have secure facilities for storage, with personnel having appropriate security clearance. The process aid anhydrous hydroxybenzotriazole (HOBT) is not permitted in many European countries because of the risk of explosion. Most deprotection and cleavage cocktails contain foul-smelling thiols such as 1,2-ethanedithiol (EDT) and cannot be used without local approval and appropriate scrubbing equipment in place.

In order to comply with some of these regulations and work towards a more sustainable environment, green chemistry has become increasingly popular and to some extent become a mandatory consideration for sponsors outsourcing manufacturing processes, reflected in their mission statements. Green chemistry started to emerge in the 1990s to address problems of chemical pollution and resource depletion. This occurred as a shift in mindset from mandating the reduction of industrial emissions at the "end of the pipe" towards actively preventing pollution through the innovative design of production technologies. In the USA, the Environmental Protection Agency (EPA) has spearheaded green chemistry. The State of California has been in the forefront by passing two laws in 2008 and launching the California Green Chemistry Initiative. California's Department of Toxic Substances has developed regulations to minimize hazardous chemicals and to replace them with safer alternatives. In 2007, the European Union put into place the Registration, Evaluation, Authorization and Restriction of Chemicals (REACH) program, which requires companies to provide data confirming that their products are safe.[58] More recently, regulations in China on pollution and the use of green chemistry have been introduced and are quite stringent.

The practice of green chemistry is guided by a set of principles whose goal is to reduce the environmental and health impacts of chemical production. The 12 principles of green chemistry are given in Table 6.5.[59] If followed, these 12 principles would provide real benefit to CMOs because they

1. encourage the use of new innovative technologies allowing the contract manufacturer to be in the forefront of cutting-edge technology;
2. encourage the development of manufacturing processes that are higher yielding, which in turn would increase the profitability of the manufacturer;
3. lead to reduced waste, resulting in lower costs for hazardous waste disposal;
4. lead to better performance, so that less product is needed to achieve the same function;
5. lead to fewer synthetic steps, allowing faster manufacture of products, increasing the plant's capacity and throughput,
6. ensure compliance with new regulations and minimize costs associated with fines;
7. ensure support from the local community due to the manufacturer's sustainability program;
8. improve overall competitiveness of the chemical manufacturer.

Table 6.5 The 12 principles of green chemistry.

1. It is better to prevent waste than to treat or clean up waste after it is formed
2. Synthetic methods should be designed to maximize the incorporation of all materials used in the process into the final product
3. Wherever practicable, synthetic methodologies should be designed to use and generate substances that possess little or no toxicity to human health and the environment
4. Chemical products should be designed to preserve efficacy of function while reducing toxicity
5. The use of auxiliary substances (*e.g.* solvents, separation agents) should be made unnecessary wherever possible and innocuous when used
6. Energy requirements should be recognized for their environmental and economic impacts and should be minimized. Synthetic methods should be conducted at ambient temperature and pressure
7. A raw material or feedstock should be renewable rather than depleting wherever technically and economically practicable
8. Reduce derivatives – unnecessary derivatization (blocking group, protection/deprotection and temporary modification) should be avoided whenever possible
9. Catalytic reagents (as selective as possible) are superior to stoichiometric reagents
10. Chemical products should be designed so that at the end of their function they do not persist in the environment and break down into innocuous degradation products
11. Analytical methodologies need to be further developed to allow for real-time, in-process monitoring and control prior to the formation of hazardous substances
12. Substances and the form of a substance used in a chemical process should be chosen to minimize potential for chemical accidents, including releases, explosions and fires

With the emphasis on green chemistry, large pharmaceutical companies have started to evaluate alternative solvents to those used currently for peptide manufacture.[60]

6.9.2 Manufacturing Facility Capabilities

The environmental regulations that govern the production of APIs, drugs and excipients have an impact on the capabilities of the manufacturing facility. Depending on the location of the manufacturing site, national and/or state laws and regulations determine which solvents may be utilized in peptide manufacture. The usable solvents are then subject to a second set of regulations imposed by the local municipality and fire department. The local fire department would require permits for the usage of such solvents and, depending on the permit, could limit the amount of a particular solvent that can be utilized in a manufacturing suite. A manufacturing site could apply for additional permits that could increase solvent usage within the facility or in a particular suite, but this comes at an additional cost and also training of personnel in handling the higher level of solvents. The additional costs could have a major impact on the operating costs of the manufacturing facility and the facility's competitiveness in the market place.

Alternatively, to avoid additional permitting and training, solvent usage could be limited by developing processes that are solvent efficient. This presents its own challenges, in that the use of less solvent may result in an API that is deficient in critical quality attributes. If a process can be developed while maintaining the critical quality attribute targets, then not only does a manufacturing facility benefit from less local permitting and more flexibility in its design, but also enjoys cost savings from the use of less solvents.

Another means by which a manufacturing facility can comply with local fire regulations is to design a facility with multiple suites in order to spread the usage of solvents among the different suites, thereby increasing the overall solvent capacity. This, however, is highly dependent on the local municipality and these requirements also mean that reactor size capacity may have to be limited owing to the amount of solvents allowed. Although the immediate reaction may be that the manufacturing facility is not capable of increased capacity, this is not entirely true. Local regulations that limit solvent usage do provide substantial benefits to peptide manufacturing. They improve employee safety by limiting the use of hazardous solvents, reduce fire risks and reduce the impact on the environment. Manufacturing capacity can still be increased and met by sub-batching manufacturing (modular) processes. For example, rather than producing a single batch of peptide resin in a 500 L reactor, the same output can be achieved by producing five batches of peptide resin in five 100 L reactors. This provides the advantage that it reduces the risks of the peptide synthesis. If an entire batch of peptide resin is lost, this could have negative effects on the timelines and cost of the process. If the same negative impact occurs with only one of the sub-lots, the impact is minimized. The use of smaller reactors also provides the benefit of faster manufacturing time of the peptide resin because the draining of a smaller reactor would be faster than that of a larger reactor and the overall coupling time would be reduced. Additionally, the use of smaller reactors allows the use of additional manufacturing tools (*e.g.* heating) that become more difficult to control when larger reactors are used. Furthermore, the use of smaller reactors provides flexibility in the manufacturing environment in that they do not have to be stationary equipment. Rather, they can be assembled and dismantled as needed, allowing the manufacturing facility to be more agile and accommodating to various types of programs. Smaller mobile reactors can be cleaned in separate cleaning areas, reducing the downtime of a manufacturing suite, resulting in greater throughput of the manufacturing facility (which indirectly increases the annual capacity), and they improve utilization, reducing overall costs.

Another consideration that a manufacturing facility needs to take into account is the type of equipment that is utilized, especially if the facility is a multiproduct manufacturing facility.[61] Cross-contamination is always a major concern in multiproduct facilities, particularly for highly potent molecules and at early stages of development when the potency of a molecule may not be known. The use of dedicated equipment reduces concerns about cross-contamination between products. With less risk of cross-contamination, validation of cleaning procedures between batch change-over is simplified because of the lower concern about residual potent product carryover from a different product using the same equipment. With less exhaustive cleaning procedures being necessary, resource utilization is reduced, bringing improved efficiency to the manufacturing process and reduced overall manufacturing costs. Duplicating dedicated equipment improves the production flow: there is less disruption in production planning and less waiting for equipment because it is not otherwise used.

In a multiproduct manufacturing facility, purchasing dedicated equipment for specific products is expensive for the CMO. Therefore, the cost of the dedicated equipment is usually passed on to the sponsor. In addition to the original purchase cost, the equipment has to be maintained on an ongoing basis, with the additional cost of maintenance contracts that are usually borne by the CMO. Over and above expense, the use of dedicated equipment adds logistical challenges. Secondary (backup) equipment is important as mitigation in the event of a mechanical failure to the primary equipment. Storage areas are required for all the dedicated equipment. Maintaining a storage area adds cost and uses valuable real estate that could potentially be used as additional manufacturing space.

Disposable equipment is another approach to dedicated equipment saving on expensive upfront costs and storage costs; however, disposable equipment may not always be feasible for chemical manufacturing because it is made from material that may degrade and/or leach during processing. The impact of disposable equipment on the environment also has to be fully understood.

The use of shared equipment by a CMO greatly reduces the expense of duplicate equipment and multiple maintenance contracts. Having the ability to use the same piece of equipment for various products reduces the amount of equipment that needs to be housed and therefore alleviates this logistical challenge and additional storage expense. It also frees up space to build supplementary revenue-generating manufacturing suites.

The use of shared equipment brings its own set of unique challenges. The manufacturing flow at the CMO may be slowed because a new manufacturing campaign needs to wait for availability of equipment that is already being utilized. The use of shared equipment results in a need for changeover procedures, which incur additional resources, time and costs. The greatest concern with the use of shared equipment is cross-contamination between different products used in the same equipment.

The manufacturing design of the facility is dependent on its geographical location, because this would influence architectural regulations. For example, in earthquake-active areas, it is very unlikely for a manufacturing facility to have multiple levels. In the case of peptide manufacturing, the advantage of gravity addition is lost, but this can be compensated by the use of electric or air pumps on a single floor level.

Naturally, for GMP manufacturing sites, the facility should be designed with material flow where raw materials enter at one end and the final product leaves at the opposite end. The rationale is that the further downstream the process moves, the more controls are in place to ensure there is no contamination from earlier process steps. The facility should also be equipped with airlock systems to prevent contamination from the external environment.

6.9.3 Waste Disposal

Disposal of waste from the manufacturing process is an important consideration in minimizing the environmental impact. It is important for manufacturers not only to have a plan to address disposal of waste from their processes, but also to find creative means to minimize waste production.

Agreements and relationships with local or regional firms that specialize in waste removal and disposal should be set up. Many firms require that organic and aqueous wastes are separated. Therefore, the process design and facility design should be set up to accommodate this requirement. Naturally waste disposal has a cost: the greater the quantity of waste, the higher is the cost of disposal. After removal of the waste from the manufacturing facility, it is either incinerated or, in the case of organic solvents, distilled for other uses. In the latter case, manufacturing sites may benefit from reduced rates from waste disposal companies because the solvent can be used for other purposes. It would be prudent for the manufacturer to have these discussions with the waste disposal firms in advance of constructing or restructuring manufacturing facilities.

In some instances, manufacturing facilities have set up their own incinerators or distillation plants on-site. Although this brings waste disposal under the full control of the manufacturing facility, it does add additional costs and overheads that should be measured. Moreover, the local municipality may have its own reservations about on-site incinerators out of concern for air pollution. The immediate advantage provided by the distillation plant to the manufacturing facility is that solvent newly distilled on-site can be used for the production of other products, thereby reducing the amount of additional solvent required. On the surface this would provide a cost savings to the operation; however, caution should be exercised that the distillation process fully removes all contaminants from the previous product. This would require that analytical testing methods be established and used to prove that the distilled solvents do not introduce other contaminants during the production of a new peptide. With all these additional controls and testing requirements, there may not be an immediate cost benefit to having on-site distillation of solvents; this, however, would have to be evaluated.

In summary, manufacturers could ensure a greater impact on their costs and on the environment by focusing on developing processes that are less solvent intensive and therefore lower in waste production. This would provide the advantage of reduced costs from not only less solvent procurement, but also reduced waste removal costs due to lower volumes. Creative means to reuse solvents in the same manufacturing process would result in lower solvent consumption and lower waste production, *e.g.* in peptide manufacturing some of the later "cleaner" washes during the coupling reactions could be used as initial washes for a later coupling step.

It is the responsibility of the manufacturers to ensure that controls established in their standard operating procedures and facility design are in place to ensure that any leaks of waste do not contaminate the groundwater and eventually the lakes and oceans. Manufacturing facilities should work closely with their local municipalities, engineering firms and waste disposal firms to establish a sustainable waste program.

6.9.4 Employee Safety

Occupational exposure limits (OELs) have existed since 1939, when the National Conference of Governmental Industrial Hygienists published its inaugural acceptable workplace exposure limits. Historically, OELs were based on human experience in the workplace, with limits being based on adverse health effects. As animal testing became more common in the twentieth century, the OELs were set based on the "no observed effect level (NOEL)." If a NOEL was not available, then the lowest observed effect level (LOEL) was used and extrapolated using a factor of 1–10 to determine the NOEL. The OEL is determined using the following equation:[62]

$$\mathrm{OEL} = \text{human breathing rate} \times \frac{\mathrm{NOEL} \times \text{human body weight}}{(\text{safety factor})\, n}$$

During the development and manufacture of an API, it is important to manage risks associated with exposure of employees to the API or its derivatives. Many large pharmaceutical companies have internal committees composed of toxicologists, occupational physicians, industrial hygienists and engineers that set health-based exposure limits for their products.[63] Initially, when a new potential clinical candidate is identified, the toxicologist at the large pharma company will assign an occupational exposure band (OEB). In many cases, toxicological data about the compound are unavailable and therefore a default conservative band is prescribed. Small biotech organizations unfortunately do not have the bandwidth or resources to hire toxicologists and therefore in many cases their molecules are not assigned an OEB. The OEB is a process of assigning chemicals and pharmaceuticals into bands corresponding to a range of exposure concentrations designed to protect workers' health. The bands are assigned based on chemical toxicological potency and the adverse health effects associated with exposure. Many companies and organizations have their own banding system but they all have the same goal – protection of the worker. An example of a banding system is shown in Figure 6.8.[64]

When large pharma and biotech companies contract out the development and manufacturing of their candidate compounds, it is their responsibility to provide as much information available to the contract manufacturer. This information is vital in order for the contract manufacturer to protect its employees from potential adverse effects that the clinical candidate may have. Depending on the OEB of the candidate compound, if assigned, the contract manufacturer may need to utilize specific equipment and laboratories that provide closed systems and appropriate protection for its employees. The use of these specialized equipment and laboratories may impact the cost of the campaign. Therefore, it is important for the sponsor to evaluate properly the equipment being utilized at each contract manufacturer before embarking on a campaign. Most contract manufacturers have their own questionnaires to enable them to obtain information about exposure concerns and appropriate cleaning procedures.

The design of the manufacturing facility is important to ensure that the most efficient processes can be utilized to maximize throughput and capacity and minimize the risk of product contamination, which could be influenced by local laws and regulations.[65] An important aspect of the design

OEB	1	2	3	4	5
OEL (μg/m³)	≥100	10 - 100	1 - 10	0.1 - 1	<0.1
Potential effects of exposure	None to minor	Minor to moderate	Moderate to serious	Serious	Serious
Handling	GMP/PPE	GMP/PPE	Containment	Containment	Containment/ robotics

Figure 6.8 Example of banding system.
Reproduced from ref. 64 with permission.

of the facility is to establish the process flow.[66] Ideally, the facility should have a one-way flow from raw materials to finished product. Appropriate air handling systems should be implanted to allow for air flow in a singular fashion to prevent cross-contamination. In addition to air handling, the facility would have to be designed to accommodate closed system equipment for containment purposes. Depending on the level of containment, the type of project that the manufacturer could accommodate may be impacted. This equipment could include sealed reactors, isolators, laminar flow hoods and local exhaust ventilation. In the case of a potent API, negative air pressure relative to all adjacent areas should be established. The containment controls should not be limited to the manufacturing area, but should also be considered for support services such as development and QC.[67]

In addition to facility engineering controls, the employees should be protected further with personal protective equipment (PPE). This includes gloves, face masks, portable respirators and safety glasses, to name just a few. It is also important that the employees are trained in the use of PPE and in using appropriate procedures for their operation to enhance safety.

6.10 Conclusion

Although the average price of peptides manufactured by independent CMOs has decreased by about 50% over the last three decades, the demands made on these organizations have increased significantly. Apart from the economic pressure to deliver less expensive drug substances in the face of increasing costs, the pressure to develop technology that can address the commercial manufacture of longer and more complex peptides has grown, as has ecological pressure to implement safeguards to protect the environment and employees. None of these pressures is unique to peptide therapeutic manufacture *per se*, but the niche nature of the industry puts it at particular risk.

There is good reason to assume that the attrition in the peptide NCE industry has bottomed out. The nature of peptide manufacturing, the combination of relatively small batch sizes, combined with the need for dedicated space and expertise, make it very unlikely that pharmaceutical companies with sporadic or low-volume peptide NCEs will be able to justify the cost of in-house process development and production. The external contracts awarded by such companies will in themselves support a viable peptide CMO industry. Those CMOs that have a track record of successful regulatory approvals of peptide therapeutic drug substances will have an advantage that will allow them to command premium prices for their services. However, these companies alone do not have the capacity to service the current high demand for peptide manufacture. Other companies will continue to build successful approval histories in the next 5–10 years. The challenges to new CMOs attempting to gain a foothold in the NCE peptide therapeutic CMO market are significant. Nevertheless, it will probably be newcomers that directly or indirectly bring more economic and environmentally disruptive technologies to the conservative world of peptide manufacture.

Despite numerous improvements in analytical technology and downstream processing, there have been no major changes in the basic synthesis technology since the introduction of the Fmoc group in 1970, and this will remain the norm for peptide assembly for the foreseeable future. Despite efforts being made to develop aqueous-based peptide chemistry, results to date have been confined to very short peptides. For long peptides where SPPS yields are very low, the use of fragment approaches such as CEPS and NCL will gain popularity owing to higher yields and lower solvent consumption. Future peptide therapeutics may see the amalgamation of Fmoc-based SPPS and recombinant approaches.

Discrete changes in the way in which Fmoc-based chemistry is used, especially the use of heating, automation and constant-flow processes, will make a large impact on costs and the environmental footprint. Any technologies that reduce organic waste will find intensified usage in the future.

Last, but not least, we need more efficient and rational usage of resources and there is a need to adjust to a changing world.

Acknowledgements

We would like to thank numerous colleagues and friends for their support in putting this chapter together, especially Peter Larsson (PolyPeptide Group) and Marcel Schmidt (EnzyPep BV) for proof-reading and useful suggestions. Thanks are also due to Ved Srivastava for his prompt and helpful responses to our questions.

References

1. T. Wieland and M. Bodanszky, *The World of Peptides. A Brief History of Peptide Chemistry*, Springer Verlag, 1991, p. Preface.
2. K. Fosgerau and T. Hoffman, Peptide therapeutics: current status and future directions, *Drug Discovery Today*, 2014, **20**(1), 122–128.
3. Y. Cohen, Small Molecules: The Silent Majority of Pharmaceutical Pipelines, *Exome*, 2015, https://xconomy.com/boston/2015/11/23/small-molecules-the-silent-majority-of-pharmaceutical-pipelines/.
4. H. Jones, The legal framework applicable to biosimilars in the EU, *Eur. Pharm. Rev.*, 2016, **21**, 12–14.
5. R. Finking and M. A. Marahiel, Biosynthesis of Nonribosomal Peptides, *Annu. Rev. Microbiol.*, 2004, **58**, 453–488.
6. M. S. Butler, A. B. Avril, A. B. Robertson and M. A. Cooper, Natural product and natural product derived drugs in clinical trials, *Nat. Prod. Rep.*, 2014, **31**, 1612–1661.
7. V. V. Pai, P. Bhandari and P. Shukla, Topical peptides as cosmeceuticals, *Indian J. Dermatol. Venereol.*, 2017, **83**, 9–18.

8. A. Kaspar and J. M. Reichert, Future directions for peptide therapeutics development, *Drug Discovery Today*, 2013, **18**, 1–11.
9. J. L. Lau and M. K. Dunn, Therapeutic peptides: Historical perspectives, current development trends and future directions, *Bioorg. Med. Chem.*, 2018, **26**, 2700–2707.
10. Partners, Bionest, "Overview of the Peptide Market. "A growing niche opportunity?"," 2007.
11. S. Ghosh, Peptide therapaeutics market: forecast and analysis 2015–2025, *Chim. Oggi*, 2016, **34**(2), V–VII.
12. R. E. W. Hancock and D. S. Chapple, Peptide Antibiotics, *Antimicrob. Agents Chemother.*, 1999, **43**(6), 1317–1323.
13. N. Sewald and H.-D. Jakubke, *Peptides: Chemistry and Biology*, 2009.
14. E. Fischer and E. Fourneau, Über einige Derivate des Glykokolls, *Ber. Dtsch. Chem. Ges.*, 1901, **14**, 2868–2879.
15. T. Curtius, Über einige neue Hippursäureanalog constituietrte, synthetisch dargestellte Amidosäuren, *J. Prakt. Chem.*, 1882, **26**, 145–208.
16. R. B. Merrifield, Solid Phase Peptide Synthesis. I. The Synthesis of a Tetrapeptide, *J. Am. Chem. Soc.*, 1963, **85**, 2149–2154.
17. R. B. Merrifield, Solid Phase Peptide Synthesis. I. The Synthesis of a Tetrapeptide, *J. Am. Chem. Soc.*, 1963, **85**, 2149.
18. L. A. Carpino, Oxidative reactions of hydrazine. II. Isophthalimides. New protective groups on nitrogen, *J. Am. Chem. Soc.*, 1957, **79**, 98–101.
19. F. C. McKay and N. F. Albertson, New amine-masking groups for peptide synthesis, *J. Am. Chem. Soc.*, 1957, **79**, 4686–4690.
20. L. A. Carpino and J. Han, The 9-fluorenylmethoxycarbonyl amino-protecting group, *J. Org. Chem.*, 1981, **37**, 3404–3409.
21. R. Behrendt, P. White and J. Offer, Advances in Fmoc solid-phase peptide synthesis, *J. Pept. Sci.*, 2016, **22**(1), 4–27.
22. D. Takahashi, T. Inomata and T. Fukui, AJIPHASE®: A Highly Efficient Synthetic Method for One-Pot Peptide Elongation in the Solution Phase by an Fmoc Strategy, *Angew. Chem., Int. Ed. Engl.*, 2017, **56**(27), 7803–7807.
23. Y. Okada, H. Suzuki, T. Nakae, S. Fujita, H. Abe, K. Nagano, T. Yamada, N. Ebata, S. Kim and K. Chiba, Tag-Assisted Liquid-Phase Peptide Synthesis Using Hydrophobic Benzyl Alcohols as Supports, *J. Org. Chem.*, 2013, **78**, 320–327.
24. J. Wu, G. An, S. X. J. Lin, W. Zhou, H. Sun, Y. Pan and G. Li, Solution-phase-peptide synthesis *via* the group assisted purification (GAP) chemistry without using chromatography and recrystallization, *Chem. Commun.*, 2014, **50**, 1259–1261.
25. C. W. Seifert, A. Paniagua, G. A. White, L. Cai and G. Li, GAP Peptide Synthesis through the Design of a GAP Protecting Group: An Fmoc/tBu Synthesis of Thymopentin Free from Polymers, Chromatography and Recrystallization, *Eur. J. Org. Chem.*, 2016, **9**, 1714.
26. S. K. Gupta and P. Shukla, Sophisticated Cloning, Fermentation, and Purification Technologies for an Enhanced Therapeutic Protein Production: A Review, *Front. Pharmacol.*, 2017, **8**, 419.

27. K. Itakura, T. Hirose, R. Crea, A. D. Riggs, H. L. Heyneker, F. Bolivar and H. W. Boyer, Expression in Escherichia coli of a chemically synthesized gene for the hormone somatostatin, *Science*, 1977, **198**(4321), 1056–1063.
28. Animal Cell Technology Industrial Platform (ACTIP), *Monoclonal Antibodies Approved by the EMA and FDA for Therapeutic Use (Status 2017)*, 2017.
29. S. S. S. Usmani, G. Bedi, J. S. Samuel, S. Singh, S. Kalra, P. Kumar, A. A. Ahuja, M. Sharma, A. Gautam and G. P. S. Raghava, THPdb: Database of FDA-approved peptide and protein therapeutics, *PLoS One*, 2017, **12**(7), e0181748.
30. A. R. Bezerra, A. R. Guimarães and M. A. S. Santos, Non-Standard Genetic Codes Define New Concepts for Protein Engineering, *Life*, 2015, **5**(4), 1610–1628.
31. V. Kubyshkin, C. G. Acevedo-Rocha and N. Budisa, On universal coding events in protein biogenesis, *BioSystems*, 2018, **164**, 16–25.
32. B. Bray, Large-scale manufacture of peptide therapeutics by chemical, *Nat. Rev. Drug Discovery*, 2003, **2**, 587–593.
33. P. Dawson, T. W. Muir, I. Clark-Lewis and S. B. Kent, Synthesis of proteins by native chemical ligation, *Science*, 1994, **266**(5186), 776–778.
34. S. B. H. Kent, Total chemical synthesis of proteins, *Chem. Soc. Rev.*, 2009, **38**, 338–351.
35. L. Yan and P. E. Dawson, Synthesis of Peptides and Proteins without Cysteine Residues by Native Chemical Ligation Combined with Desulfurization, *J. Am. Chem. Soc.*, 2001, **123**, 526–533.
36. F. Guzmán, S. Barberis and A. Illanes, Peptide synthesis: Chemical or enzymatic, *Electron. J. Biotechnol.*, 2007, **10**(2), 279–314.
37. F. Bordusa and H.-D. Jakubke, Enzymatic synthesis, *Houben-Weyl E 22a Synthesis of Peptides and Peptidomimetics*, 2001, 642–664.
38. K. Morihara, T. Oka, H. Tsuzuki, Y. Tochino and T. Kanaya, Achromobacter protease I-catalyzed conversion of porcine insulin into human insulin, *Biochem. Biophys. Res. Commun.*, 1980, **92**(2), 396–404.
39. J. Markussen, *Human Insulin by Tryptic Transpeptidation of Porcine Insulin and Biosynthetic Precursors*, 1987.
40. J. Markussen, U. Damgaard, M. Pingel, L. Snel and E. Sorensen, Human insulin (Novo): chemistry and characteristics, *Diabetes Care*, 1983, **6**, 4–8.
41. L. A. Marraffini, A. C. DeDent and O. Schneewind, Sortases and the art of anchoring proteins to the envelopes of gram-positive bacteria, *Microbiol. Mol. Bio. Rev.*, 2006, **70**, 192–221.
42. G. K. T. Nguyen, S. Wang, Y. Qui, X. Hemu, Y. Lian and J. P. Tam, Butelase 1 is an Asx-specific ligase enabling peptide macrocyclization and synthesis, *Nat. Chem. Biol.*, 2014, **10**, 732–738.
43. A. C. Braisted, J. K. Judice and J. A. Wells, Synthesis of proteins by subtiligase, *Methods Enzymol.*, 1997, **289**, 298–313.
44. S. Liebscher, M. Schöpfel, T. Aumüller, A. Sharkhuukhen, A. Pech, E. Höss, C. Parthier, G. Jahreis, M. T. Stubbs and F. Bordusa, N-terminal protein modification by substrate-activated reverse proteolysis, *Angew. Chem., Int. Ed.*, 2014, **53**, 3024–3028.

45. M. Schmidt, A. Toplak, P. J. L. M. Quaedflieg and T. Nuijens, Enzyme-mediated ligation technologies for peptides, *Curr. Opin. Chem. Biol.*, 2017, **38**, 1–7.
46. H. Mao, S. A. Hart, A. Schink and B. A. Pollak, Sortase-mediated protein ligation: a new method for protein engineering, *J. Am. Chem. Soc.*, 2004, **126**, 2670–2671.
47. G. K. T. Nguyen, S. Wang, Y. Qiu, X. Hemu, Y. Lian and J. P. Tam, Butelase 1 is an Asx-specific ligase enabling peptide macrocyclization and synthesis, *Nat. Chem. Biol.*, 2014, **10**, 732–738.
48. A. C. Braisted, J. K. Judice and J. A. Wells, Synthesis of proteins by subtiligase, *Methods Enzymol.*, 1997, **289**, 298–313.
49. J. Pavlas, *Towards Commercial Manufacturing of Therapeutic Peptides Using Chemo-enzymatic Peptide Synthesis (CEPS)*, 2018.
50. M. Schmidt, A. Toplak, P. J. L. Quaedflieg, H. Ippel, G. J. Richelle, T. M. Hackeng, J. H. van Maarseveen and T. Nuijens, Omniligase-1: A Powerful Tool for Peptide Head-to-Tail Cyclization, *Adv. Synth. Catal.*, 2017, **359**(12), 2050–2055.
51. G. J. J. Richelle, M. Schmidt, H. Ippel, T. M. Hackeng, J. H. van Maarseveen, T. Nuijens and P. Timmerman, A One-pot "Triple-C" Multicyclization Methodology for the Synthesis of Highly Constrained Isomerically Pure Tetracyclic Peptides, *ChemBioChem*, 2018, **19**, 1934.
52. ERI, "Salary Expert," 2018.
53. M. Schmidt, A. Toplak, H. J. Rozeboom, H. J. Wijma, P. J. L. M. Quaedflieg, J. H. van Maarseveen, D. B. Janssen and T. Nuijens, Design of a substrate-tailored peptiligase variant for the efficient synthesis of thymosin-α1, *Org. Biomol. Chem.*, 2018, **16**, 609–618.
54. T. Nuijens, E. R. Lax, A. Toplak and M. Schmidt, Efficient chemo-enzymatic peptide synthesis (CEPS) of incretin peptides, 2018.
55. N. H. Shah and T. W. Muir, Inteins: Nature's Gift to Protein Chemists, *Chem. Sci.*, 2014, **5**(1), 446–461.
56. S. Marchant and C. Marchant, *Starting Salaries of Chemists And Chemical Engineers: 2014. Analysis of the American Chemical Society's Survey Of New Graduates in Chemistry and Chemical Engineering*, American Chemical Society, 2015.
57. American Institute for Economic Research, "Cost of Living Calculator," 2018.
58. M. R. Greenberg, *Explaining Risk Analysis, Protecting Health and the Environment*, Routledge, New York, 2017.
59. P. T. Anastas, J. C. Warner and J. C. Green, *Green Chemistry, Theory and Practice*, Oxford University Press, New York, 1998, p. 30.
60. J. Lopez, S. Pletscher, A. Aemissegger, C. Bucher and F. Gallou, N-Butylpyrrolidinone as Alternative Solvent for Solid-Phase Peptide Synthesis, *Org. Process Res. Dev.*, 2018, **22**, 494–503.
61. T. Shah, Considerations for multi-product manufacturing facilities, *Specialty Chem. Mag.*, 2017, **37**, 18–19.

62. R. H. Ku, An overview of setting occupational exposure limits (OELs) for pharmaceuticals, *Chem. Health Saf.*, 2000, 34–37.
63. P. J. Naumann, U. Nigro, L. Bruen and L. Parola, Setting Health-Based Exposure Limits to Support Pharmaceutical Development and Manufacturing, *Contact Pharma*, 2016, https://www.contractpharma.com/issues/2016-05-01/view_features/setting-health-based-exposure-limits-to-support-pharmaceutical-development-and-manufacturing/53331.
64. R. Harris, Outsourcing Formulation Development for High Potency Compounds, *Pharm. Tech. Eur.*, 2016, **28**(13), 8–10.
65. D. Ainsworth, Planning and Designing a Pharmaceutical Facility: A Process Designer's View, *Pharm. Tech. Eur.*, 2005, **17**(9), http://www.pharmtech.com/planning-and-designing-pharmaceutical-facility-process-designers-view.
66. A. A. Liddy, Facility Design: Find the flow, *Pharm. Manage.*, 2012.
67. T. Calkins, Containment of High-Potency Products in a GMP Environment, *BioProcess Intl.*, 2010, 14–18.

CHAPTER 7

A Holistic Quality Control Strategy for Peptide Active Pharmaceutical Ingredients (APIs)

ALEKSANDER SWIETLOW[a] AND ASHER LOWER*[b]

[a] PolyPeptide Laboratories Inc., Torrance, CA 90503, USA; [b] Amgen Inc., Thousand Oaks, CA 91320, USA
*Email: alower@amgen.com

7.1 Introduction to Control Strategies for Peptide APIs

Consistent quality cannot be tested-in. It should be built-in by design.

FDA Guidance

For over a decade, pharmaceutical drug development has been experiencing a paradigm shift. New modalities are pushing the frontiers of science and our imaginations to approach human disease in creative ways, among which synthetic peptides are emerging as a molecular construct commanding significant attention.[1,2] Whether modified to improve pharmaceutical properties or employing unnatural amino acids, the chemical space accessible through simple amide bond formation is astounding.[3–6] For pharmaceutical innovators, this newfound availability of seemingly

Drug Discovery Series No. 72
Peptide Therapeutics: Strategy and Tactics for Chemistry, Manufacturing and Controls
Edited by Ved Srivastava

Published by the Royal Society of Chemistry, www.rsc.org

unlimited complexity comes with the equally large challenge and responsibility to ensure the quality of any new therapeutic.

The approach that has served society adequately historically with respect to ensuring the safety and efficacy of drug products has relied heavily on end-product testing and efforts to rigorously minimize process variability to achieve a consistent product. The rationale is that "Although the sources of variability are not well understood, the process is being tightly controlled and thoroughly tested." Naturally, this approach is neither cost efficient nor elegant as it puts the quality burden unduly on external control factors such as consistency and reproducibility instead of relying on an intrinsically well-designed process. Alternatively, an approach that seeks to build process and product understanding by leveraging science and technology offers to improve outcomes for both patients and innovators by focusing the efforts only on what is important. This approach has been appropriately termed quality-by-design (QbD).[7–10] QbD begins with the end in mind by putting the patient's needs first (patient's safety, product efficacy) and rationally designing the product, process and methods.

QbD seeks to build understanding through a structured, prospective product design and process development framework. Risk-based tools are heavily leveraged to provide a comprehensive development plan for science-based innovation and continual improvement. QbD principles can be applied during early development and should continue throughout the product life cycle. In this paradigm, it is less about knowing the right answers and more about knowing the right questions. The data generated will write the story. With a QbD approach, the innovator, regulators and patients can be confident in the quality of the product because the understanding required to prove the label claims has been derived from a thorough investigation of the important variables and a holistic design of the control strategy. The end goal is a reliable, efficient manufacturing process that produces a high-quality drug with minimal oversight.

This chapter presents a QbD-based approach to the development of a holistic control strategy for peptide active pharmaceutical ingredients (APIs). Guidance is based on the principles outlined in International Council on Harmonisation of Technical Requirements for Registration of Pharmaceuticals for Human Use (ICH) Guidelines ICH Q7–Q11 and examples from commercial and clinical peptide APIs. This discussion is focused on the development of therapeutic peptides manufactured synthetically, although the scientific principles can be applied more generally.

The chapter is organized to allow the reader to reference the available tools and approaches and as appropriate cater the control strategy to the peptide API target. Peptide APIs offer unique challenges owing to their complexity, variety and broad utility.[3,11,12] Manufacturability and quality control (QC) strategies have evolved to a point where synthetic therapeutic peptide candidates containing more than 100 amino acid (AA) residues are being developed.[13,14] The holistic control strategy and process design (PD) approach presented here builds on the experiences and interactions between sponsors and global health regulators and documented in current good manufacturing practice (cGMP) for the pharmaceutical industry. A list of terms is provided in Table 7.1.

Table 7.1 List of terms and definitions used in this chapter.

Term	Definition
Control strategy	A planned set of controls, derived from current product and process understanding, that assures process performance and product quality. The controls can include parameters and attributes related to drug substance and drug product materials and components, facility and equipment operating conditions, in-process controls, finished product specifications and the associated methods and frequency of monitoring and control – ICH Q10
Critical in-process control (IPC)	An in-process control that is a measurement of a CQA(s) (direct or indirect) or product quality indicator with limits that must be met to ensure final product quality
Critical process parameter (CPP)	A process parameter whose variability has an impact on a critical quality attribute and therefore should be monitored or controlled to ensure the process produces the desired quality – ICH Q8 (R2)
Critical raw material (CRM)	CRM includes all GMP starting materials (API starting materials) but may also include other raw materials that have a direct impact on DS quality. An API starting material is a raw material, intermediate or an API that is used in the production of an API and that is incorporated as a significant structural fragment into the structure of the API. An API starting material can be an article of commerce, a material purchased from one or more suppliers under contract or commercial agreement or produced in-house. API starting materials normally have defined chemical properties and structure – ICH Q7
Critical quality attribute (CQA)	A physical, chemical, biological or microbiological property or characteristic that should be within an appropriate limit, range or distribution to ensure the desired product quality – ICH Q8 (R2)
Design of experiments (DoE)	A structured, organized method for determining the relationship between factors affecting a process and the output of that process
Design space	The multidimensional combination and interaction of input variables (*e.g.* material attributes) and process parameters that have been demonstrated to provide assurance of quality. Working within the design space is not considered as a change. Movement out of the design space is considered to be a change and would normally initiate a regulatory post-approval change process. Design space is proposed by the applicant and is subject to regulatory assessment and approval – ICH Q8 (R2)
Drug product (DP)	A finished dosage form, *e.g.* tablet, capsule or solution, that contains a drug substance, generally, but not necessarily, in association with one or more other ingredients – CFR Title 21

Table 7.1 (*Continued*)

Term	Definition
Drug substance (DS)	An active ingredient that is intended to furnish pharmacological activity or other direct effect in the diagnosis, cure, mitigation, treatment or prevention of disease or to affect the structure or any function of the human body but does not include intermediates used in the synthesis of such ingredient – CFR Title 21
In-process control (IPC)	Checks performed during production in order to monitor and, if appropriate, to adjust the process and/or to ensure that the intermediate or API conforms to its specifications – ICH Q7
Knowledge space	Range for a process parameter that is evaluated for process and product impact during process characterization (is expected to be 2–3 times the operating range in most cases). Aka characterized range
Life-cycle management (LCM)	The process of managing the entire life-cycle of a product including its research, design and manufacture, service and disposal
Noncritical IPC	An in-process control that is a measure of process consistency
Noncritical process parameter	A process parameter that has no significant influence on a critical quality attribute. Noncritical process parameters may be controlled within a defined range to ensure process consistency
Normal operating range (NOR)	A range for a process parameter that is listed in the manufacturing description. This is often set based on equipment and/or process capability
Performance indicator	A measurement of a CQA(s) (direct or indirect), product quality indicator or process consistency measurement (*e.g.* a process output)
Process analytical technology (PAT)	A system for designing, analyzing and controlling manufacturing through timely measurements (*i.e.* during processing) of critical quality and performance attributes of raw and in-process materials and processes with the goal of ensuring final product quality. Tools and systems that utilize real-time measurements or rapid measurements during processing, of evolving quality and performance attributes of in-process materials to provide information to ensure optimal processing to produce final product that consistently conforms to established quality and performance standards – ICH Q8 (R2)
Process parameter	A parameter that can be directly controlled. Process parameters are typically thought of as "inputs"
Proven acceptable range (PAR)	A characterized range of a process parameter for which operation within this range, while keeping other parameters constant, will result in producing a material meeting relevant quality criteria – ICH Q8 (R2)
Quality-by-design (QbD)	A systematic approach to development that begins with predefined objectives and emphasizes product and process understanding and process control, based on sound science and quality risk management – ICH Q8 (R2)
Quality risk management (QRM)	A systematic process for the identification, assessment and control of risks to the quality of pharmaceutical products across the product life-cycle – ICH Q9

Table 7.1 (*Continued*)

Term	Definition
Quality target product profile (QTPP)	A prospective summary of the quality characteristics of a drug product that ideally will be achieved to ensure the desired quality, taking into account safety and efficacy of the drug product – ICH Q8 (R2)
Target product profile (TPP)	A format for a summary of a drug development program described in terms of labeling concepts – FDA Guidance for Industry. Includes labeled use, safety and efficacy

7.1.1 Scope and Assumptions

Within the scope of this chapter are peptides made by the following:

- *Chemical synthesis (fully synthetic):* Traditional C-terminally amidated, N-terminally acylated, linear peptides; peptides containing inter-chain disulfide bonds (providing secondary structure); constrained peptides; cyclic peptides; peptide cocktails (immuno- and vaccines). The number of methods for manufacturing peptides of increasing complexity has grown exponentially owing to advances in manufacturing technologies such as solid-phase peptide synthesis (SPPS) and amide-bond facilitating coupling reagents with mild conditions.[15]
- *Hybrid/semisynthetic:* Peptides containing modifications made after main-chain assembly – including conjugation to monoclonal antibodies (mAbs), half-life extension, among other constructs.

Excluded from the discussion are peptides made by

- *Recombinant DNA technologies/extraction from natural sources:* Proteins >100 AAs manufactured *via* cell or cell-free expression (phage display libraries, isolate clones) or fermentation culture.

Since discussion of a holistic synthetic therapeutic peptide CMC control strategy spans the drug development life cycle, phase-appropriate references will be included. For example, the approach taken during candidate selection is significantly different from that following proof-of-efficacy. The requirements for a commercial process control strategy are presented with a focus on development beginning at approximately pivotal material delivery.

- *Toxicology studies to first-in-human (FIH):* Based on limited manufacturing experience, utilizing generic platforms and general guidance for testing and limits. Document process changes and minimal comparability or process evaluation is performed.
- *Phase 1 to Phase 2:* Risk-based tools utilized for commercial process development (CPD) and process characterization (PC) efforts. Product and process-specific data are generated and leveraged. Process evaluation

involves detailed comparison of process, changes, in-process controls (IPCs) and drug substance (DS) release data and intermediate testing results. Statistical analysis typically minimally needs three batches.
- *Phase 3: Process performance qualification (PPQ) to commercial:* Life-cycle management for post-approval changes; continued process verification (CPV) objectives; evaluation of trends in monitored signals, second -source/site changes, changes to presentation/dosing regimen.

Early development will not benefit from a broad accumulation of process and product knowledge and the balance between process control and analytical control will conservatively favor significant amounts of testing. Analytical characterization early on supports the development of process models necessary for the later stages of development. A set of analytical controls for peptide APIs is provided in Section 7.3. Assembling a large analytical data set will assist in building and executing insightful risk assessments (Section 7.4), which are the basis for developing appropriate and robust control strategies.

7.1.2 Enhanced Control Strategy Approach

Process understanding is an expectation of regulatory authorities and the goal of QbD. A process is considered well understood when all critical sources of variability are understood and controlled and process performance and product quality attributes can be consistently predicted. The critical quality attributes (CQAs) are measures of the product that define the safety and efficacy and must be controlled.

A holistic control strategy for a peptide drug product (DP) must ensure its safety and efficacy according to the product's label and must develop an understanding of multiple factors following the process illustrated in Figure 7.1.

Assurance of DS quality is accomplished by purposefully designing the control strategy to meet each CQA. However, it must be noted that a therapeutic target cannot become reality until it can be manufactured reproducibly. Manufacturability is a critical consideration in designing the control strategy, as targets for DS quality attributes must be balanced with current technical capabilities. Assuming that there is a technically feasible route, the therapeutic drug sponsor has a suite of tools at its disposal to control DS quality. The control elements are typically applied in combination and grouped into two main categories: (1) manufacturing/procedural controls based on process design (see Sections 7.5, 7.6 and 7.7) and (2) analytical controls applied at strategic points in the process (see Section 7.3), *e.g.* at the input materials, by in-process testing and release testing.[16–19] See Section 7.2, Figure 7.2 and Section 7.3.

Since it is the desire to achieve as low as possible nonconformance (NC) rates on the finished DS, the design of input controls, procedural controls and IPCs should bear the primary burden in the holistic QC strategy such

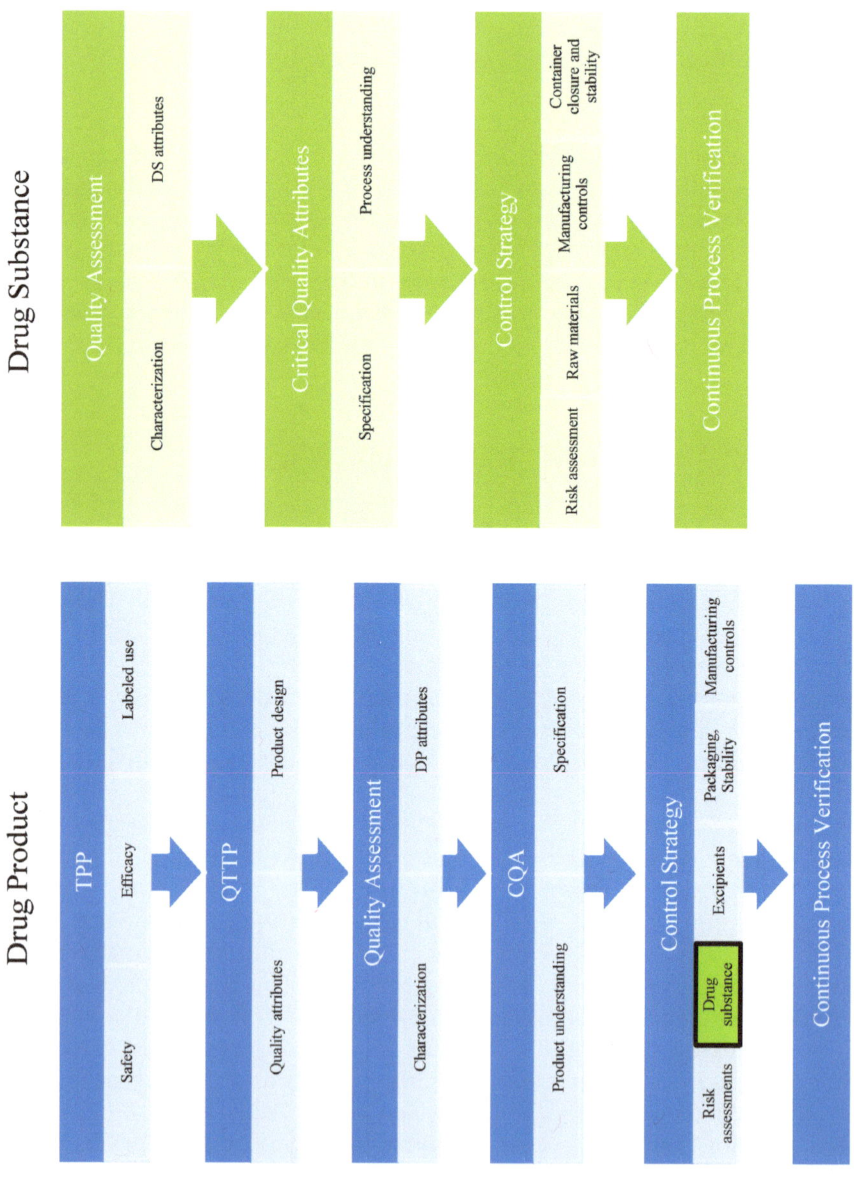

Figure 7.1 A holistic control strategy must be based on thorough product and process understanding.

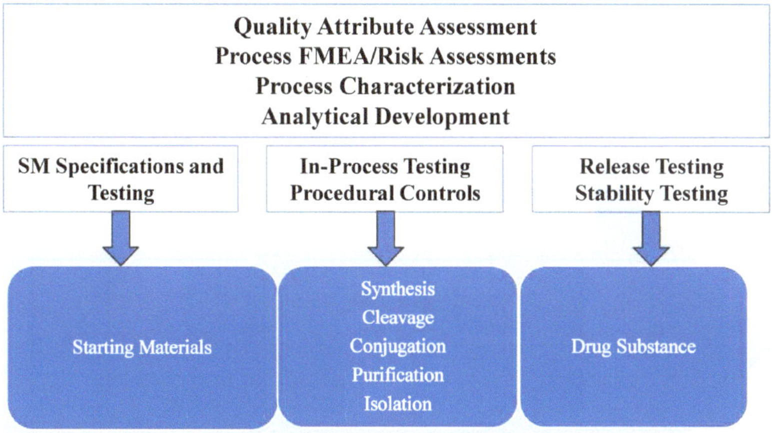

Figure 7.2 Assurance of DS quality is accomplished by purposefully designing the control strategy.

that end-product testing would be only for purposes of confirmation. For therapeutic peptides, control over individual potential and known impurities is primarily achieved by applying three critical levers:

1. tightening of critical raw material specifications;
2. optimization around problematic process conditions;
3. tightening of the purification IPC criteria.

Indeed, a combination of the above options applied thoughtfully and with the appropriate stringency will allow for the finished DS always to meet specification criteria. The trade-off to ultimate stringency is cost and efficiency. By the same token, identifying situations where the control strategy is applied too stringently may be an opportunity for improved efficiency or savings. Early phase development approaches also favor less stringency.

A QbD approach fundamentally begins with the end product in mind. It represents a holistic framework for product planning and execution required to deliver the product which meets the desired, predefined expectations.[8] Recommended steps for QbD implementation and reference to the relevant topics and sections in this chapter are outlined in Table 7.2.

Quality risk management (QRM) as outlined in ICH Q8–Q11 and specifically Q9 indicates that a science-based approach will leverage appropriate risk management tools. A pharmaceutical innovator is expected to have QRM policies that provide methods to assess, control, communicate and review risks. Risk-based tools can play a central role in a strategic and robust product development strategy. Risk-based tools specifically tailored for

Table 7.2 Steps to implement a QbD-based control strategy.

Step	Relevant topic	Section
1. Define the desired product performance and identify the QTPP	Product definition	TPP QTPP DS CQAs Analytical considerations: testing
2. Identify the DS CQAs	Product definition Risk assessment	Product quality risk assessment analytical considerations: characterization Analytical considerations: testing
3. Link process parameters and material inputs to DS CQAs though focused studies	Process design and process characterization	Manufacturing process considerations Route and unit operations Fate and purge studies Planned studies
4. Identify the CPPs and CRM attributes	Risk assessment Process design and process characterization	FMEA Classification of process parameters Raw material risk assessment
5. Identify and control the sources of variability	Risk assessment Process design and process characterization	Manufacturing process considerations Analytical considerations: testing
6. Monitor the process to assure consistent product quality	Risk assessment	FMEA Product quality risk assessment Analytical considerations: testing

constructing a holistic control strategy for therapeutic peptide APIs are presented in Section 7.4.

Process design seeks to achieve an understanding of the impact of material inputs and process parameter variability on DS quality attributes and manufacturing performance indicators. Process characterization occurs during late-stage process development and seeks to optimize the process to meet the DS CQAs consistently and efficiently. Regardless of the stage of development or maturity of the manufacturing process, therapeutic drugs must have a control strategy for each DS CQA.

Early development necessitates a control strategy in which the balance of control elements is weighted heavily on end-product specification tests to assure the consistency and safety of the product. As manufacturing experience and process and product understanding increase during mid-phase development, a stronger emphasis is placed on manufacturing controls. Manufacturing controls refer to the rational arrangement of unit operations, process parameter set points and normal operating ranges (NORs), used to deliver the DS CQAs consistently and efficiently with minimal off-line or redundant testing. Active control using process analytical technology (PAT) by at-line or in-line monitoring can aid in delivering reproducible and robust manufacturing procedures.

Analytical methods to test and quantify drug attributes and measure process performance provide the other required element of a holistic control strategy.[20] Each quality attribute has associated analytical test methods. Additional analytical tools are available for characterization. Section 7.3 provides a comprehensive list of peptide API characterization and test methods.

The integrated control strategy for peptide APIs will incorporate elements of control based on input controls, manufacturing controls, in-process testing and release testing. The control strategy is a result of the above activities and reflects the depth of product and process understanding built through development and manufacturing experience.

7.2 Product Definition

7.2.1 Product TPP

The target product profile (TPP) summarizes the drug label and aids in direct and open communication with regulators. The TPP describes the therapeutic use and aspects of the safety and efficacy of the product. Development of a TPP is foremost a patient-focused exercise that conceptualizes the product usage and then works backwards stepwise to define the requirements of the previous step to ensure consistent delivery of the product. The TPP is a planning tool, which is leveraged beginning in early development. It describes how the end user will utilize the product. It aids in the design of clinical trial strategies.[21]

The TPP is a living document, which evolves throughout development as knowledge of the product increases. Project teams should review the TPP at planned points in the development process to ensure alignment with the program goals. Elements contained within the TPP are as follows:

- indication;
- target patient population and clinical setting;
- route or mode of administration;
- therapeutic entity;
- dose, frequency and duration of therapy;
- clinical safety and efficacy profile, pharmacokinetics and counterindications;
- packaging, storage, stability and expiry.

For parenteral drug products, the delivery device and sterilization are important aspects. Administration will require a needle and syringe, autoinjector device or other combination product.[22]

Since the TPP defines the end use, it guides the development of the DP QTPP and DS CQAs.

7.2.2 Product QTPP

The QTPP describes the quality design criteria for the product and should therefore form the basis for development of the CQAs, critical process parameters (CPPs) and holistic control strategy.

Whereas the TPP is aspirational, the QTPP describes the current state of what is possible. The QTPP provides target quality attribute ranges alongside the observed ranges, with full justification. The QTPP provides an understanding of which aspects are present and necessary to ensure the quality, safety and efficacy of that specific product for the patient. The QTPP must be defined for the DP first; subsequently, the components of the DP, including the active and inactive ingredients, can be further specified. The DS CQAs will match that of the DP in many aspects except for, most notably, physical state, stability and components.

The QTPP translates the TPP into performance-based quality targets and ranges. The QTPP will assign quantitative tests for determining CQAs. The tests and methods can be leveraged for both process design and characterization, and also final DS specifications. The ICH provides recommendations for residual solvents, heavy metals, small-molecule impurities and counter-ion content. Based on indication and administration, product-specific limits for related substances can be proposed based on clinical exposure and process capability. Although peptides are outside the scope of ICH guidance, the general principles still apply.

7.2.3 DS CQAs

The DS CQAs must define the necessary components and limits for downstream DP processing which may impact its safety, efficacy and quality, regardless of level.

The comprehensive list of potential quality attributes to consider should cover all components present in the DS [based on a weight percent (wt/wt%) basis] and all physical definitions of the material, *e.g.* physical properties, identification, assay, individual and total impurities, residual solvents, moisture, microbial limits, *etc.* Table 7.3 presents a list DS quality attributes grouped in general categories and organized to refer to the appropriate control elements. Control elements can be either passive (*e.g.* procedural controls and proper design of the manufacturing process) or active (*e.g.* IPC with limits and potential interventions, intermediate and/or final specification tests). A QbD approach seeks to control the DS quality attributes (QAs) by applying the *right* control elements at the *appropriate* places in the process; therefore, not all DS QAs must be controlled by the final DS specification. The absence of certain tests may be justified based on downstream manufacturing controls or demonstrated capability of upstream steps along with consistent manufacturing history.

Recommendations for control elements linked to impacting steps, in addition to general limits, are provided for each DS QA category in Table 7.3.[19,23] Other limits or additional tests may be justified based on clinical relevance and batch data.

A QA is classified as critical or noncritical based on the severity of harm to a patient. Therefore, QAs are ranked by severity. An attribute's severity does not change as a result of risk management. See Section 7.4.2 for details on DS QA classification.

Table 7.3 Recommended quality attributes for consideration during peptide DS development.

Drug substance quality attribute general categories	Impacting steps/control point(s)	Available control elements	Expected on DS specification? (limits)
Characteristics (physical properties) • Appearance • Solubility • Crystal form • Particle size	Crystallization Drying Lyophilization	Process design	Yes: typically qualitative only, depending on DP process requirements
Identity	DS release testing	Final specification	Yes
Assay (consists >95% on an anhydrous, counter ion-free basis) • Peptide content "as is" • Salt (counter ion content) • Water content	CRMs Reaction Work-up Crystallization Lyophilization	CRM specifications IPC Intermediate or final specifications Procedural control	Yes (95.0 to 105.0% or tighter)
Peptide-related substances (0.05–5% of bulk) • Carry-forward and introduced from RMs (*e.g.* chiral purity) • Impurities: unwanted side reactions and by-products normal to the process • Degradants	Reaction Work-up Crystallization Purification	CRM specifications IPC Intermediate or final specifications Procedural control	Yes: specified, qualified in tox limits Yes: unspecified limited separately
Residual solvents • Organic solvents including TFA and HOAc • Water content	Crystallization Distillation Drying Lyophilization	IPC Intermediate or final specifications Process design	Yes for those used in last 2 steps, ICH Q3C limits (50–20 000 ppm; <2% of bulk)

Table 7.3 (*Continued*)

Drug substance quality attribute general categories	Impacting steps/control point(s)	Available control elements	Expected on DS specification? (limits)
Nonpeptide-related substances • Process reagents • By-products of process reagents	Work-up Crystallization Purification Drying	IPC Process design	No Based on ICH classification and limits – requires justification (100 – 10 000 ppm bulk)
PGTIs, *etc.* (1 ppb–1 ppm of bulk)	Varies	CRM specifications Intermediate or DS specifications Process design	No Requires strong justification
Inorganic impurities (*e.g.* 100–10 000 ppm bulk) • Elemental impurities	Work-up Crystallization Purification	Intermediate or final specifications Process design	Yes for all used, based on ICH Q3D limits
Microbial limits, endotoxin limits	Environmental controls Reaction media Water quality Filtration	IPC Intermediate or final specifications Process design	Yes for parenteral DP

7.2.4 Peptide-related Substances[24]

The European Pharmacopoeia (EP) [General Monographs, Substances for Pharmaceutical Use (2034)] specifies limits for impurities in peptides obtained by chemical synthesis. At the time of writing (2018), the US Pharmacopeia (USP) does not have an equivalent monograph for peptides, whereas peptides as a drug substance category are exempt from ICH Q3A requirements. The USP has drafted ⟨476⟩ "Control of Organic Impurities in Drug Substances and Drug Products," which reiterates the ICH Q3A limits without the exemption for synthetic peptides; however, the chapter is not yet official. None of the above documents set limits for total impurities. Table 7.4 summarizes the EP, ICH and draft USP ⟨476⟩ limits.

Table 7.5 shows an example of a phase-appropriate and technically feasible approach for setting limits that can be used as a starting point for synthetic peptides.[19] However, based on the experience of the authors and US Food and Drug Administration (FDA) presentations at conferences, it is clear that the agency has adopted a case-by-case approach to setting impurity limits. For instance, further tightening of the limits may be required if batch data or fate and purge studies (see Section 7.6.3) indicate that the manufacturing process is capable of consistently delivering material with lower impurity levels. Additional limit adjustments may be desired based on the size and complexity of the molecule. For small peptides (fewer than 10 amino acid residues), the expectation is to set lower impurity limits and that those impurities are well characterized. On the other hand, longer peptides or peptides with added complexity of conjugated groups (for example, PEGylation) may justify wider impurity limits. However, impurities originating from longer peptides may have increased immunogenicity potential, hence *in silico* or *in vitro*

Table 7.4 Summary of EP, ICH and USP (draft) impurity limits (%) for peptide therapeutics.

Threshold	EP General Monograph 2034	ICH Q3A[a] and USP ⟨476⟩[a]
Reporting threshold	0.1	0.05
Identification threshold	0.5	0.10
Qualification threshold	1.0	0.15

[a]Dose ≤2 g per day.

Table 7.5 Example of phase-appropriate impurity limits (%).

	Phase 1 and 2	Phase 3 – market
Reporting threshold	0.1	0.1
Unspecified impurities	≤1	≤0.5
Specified impurities	>1	Based on batch and qualification data
Total impurities	≤5	≤3.0 to ≤5.0[a]

[a]Range adjusted depending on limits of specified impurities.

methods may be required to identify potentially immunogenic epitopes.[25] The immunogenicity of peptide aggregates is discussed in Section 7.3.6.

Impurities should be specified if their level exceeds the indicated threshold. For early-phase development, impurities may be specified on the basis of mass, relative retention time (RRT) or other identifiable properties; for late-phase development, impurities should be specified on the basis of their molecular structure.

7.2.5 Nonpeptide-related Impurities

The properties of these types of impurities usually differ significantly from those of the peptide, allowing for rejection of these impurities. Demonstration of absence in a screening test during development coupled with a strong justification based on process knowledge may be sufficient for not including these in the final DS specification. If detected, then a 0.1% limit is usually applied and a specific method is required.

7.2.6 Mutagenic Impurity Assessment

Peptides and peptide-related impurities are outside the scope of ICH M7 for mutagenicity evaluation. Chemicals used and generated in the production of synthetic peptides are within scope for mutagenicity evaluation (*e.g.* reagents, reaction by-products, monomeric amino acids). Assessment usually involves an *in silico* screen for alerting structures, followed by *in vitro* testing if necessary in order to classify impurities with respect to their mutagenic or carcinogenic potential. The impurity assessment is product specific and references the threshold of toxicological concern (TTC) for the indication and duration of treatment. Specific methods are required. Test methods should take into account the inherently reactive nature of the impurities of concern. However, the very reactivity that renders the impurity a concern is often the same property that will generally ensure its effective removal in the downstream process. Therefore, the demonstration of the absence of the mutagenic impurity in the DS coupled with a strong justification based on chemical fundamentals and process knowledge is sufficient for not including the testing in the final DS specification. The evaluation should be based on assessment of purge factors such as the chemical reactivity, solubility, volatility and ionizability. The overall purge factor is a multiple of purge factors for the individual process steps.[26]

7.2.7 Carryover

Many peptides are produced in multipurpose facilities; therefore, carryover of the residue of a previously manufactured peptide into the next peptide is a potential concern. Although outside a formal control strategy discussion, GMP manufacturing requires that appropriate cleaning and line clearance procedures are implemented.[27]

7.3 Analytical Tools for Characterization and Testing

The analytical characterization data supporting the final API are submitted in Module 3 of the regulatory submission. Module 3 of the Common Technical Document (CTD), Section 3.2.S.3 (Characterization), systematically describes the molecular attributes of the API and the analytical methods used, whereas Section 3.2.S.4 (Control) focuses on routine methods used for the determination of CQAs and results obtained during batch analysis. The purpose of characterization is to confirm unequivocally the structure of the molecule, determine its purity profile and provide a review of its QAs. Advanced methods and instrumentation may be used to this end, whereas validated methods used for control of CQAs are typically a subset of methods used during the characterization studies and may be adapted to meet the requirements of routine QC. These methods focus on the testing of CQAs against established specification limits.

A comprehensive list of well-established analytical methods used for the characterization and control of peptides is presented in Table 7.6. The table highlights the differences in approach between characterization and control and indicates whether the testing is recommended for characterization, control, or both.

As the scope of this chapter does not include detailed descriptions of analytical methods, reference is made to the corresponding USP and EP general chapters or the literature.

7.3.1 Identity

Basic characterization with respect to the identity of a peptide must provide evidence of structure. It is recommended to perform this work on the material that is designated the reference standard as it can subsequently be used as a comparator for other batches of material. The attributes that must be confirmed as part of identity testing are the primary, secondary and tertiary structure (as applicable) and chiral configuration of constituent amino acid residues. The location of disulfide bridges or other intramolecular linkages must be known; if isomers are possible then evidence of correct structure must be provided. Any side-chain modifications, adducts or conjugated groups and their exact locations in the molecule must be demonstrated. Several methods providing evidence of structure, as outlined below, are usually needed depending on the complexity of the molecule. For routine QC analysis, a minimum of two orthogonal methods is required.

7.3.2 Primary Sequence Methods

7.3.2.1 Mass Spectrometry

The evidence of structure requires mass spectrometric (MS) analysis to demonstrate the monoisotopic mass of the molecule and sequence

Table 7.6 Available methods and USP/EP chapters for characterization (Section S.3 of CTD) and routine control (Section S.4 of CTD) of peptide APIs.

	Method	Characterization	Control
Test	USP/EP/Lit.	S.3	S.4
Definition (characters)			
Appearance (physical properties)	Visual inspection EP 5.11	Qualitative description of material. Further characterization, *e.g.* particle size, solubility, color of solution, clarity and opalescence are optional	The statements included in the characters section are qualitative, not to be interpreted in a strict sense and are not requirements
Identification			
(U)HPLC coelution with reference standard	USP ⟨621⟩ EP 2.2.46	One or more methods demonstrating specificity	Method same as for peptide-related substances. Usually the most selective method available
Mass spectrometry	USP ⟨736⟩ EP 2.2.43	High-resolution MS (HR-MS): difference between monoisotopic mass and theoretical mass of 20 ppm or less	A mass accuracy or agreement of ±0.05% from a standard should be sufficient for the identification of large molecules (above *m/z* 2000) when multiply charged ions are employed for the identification test
Amino acid analysis	USP ⟨1052⟩ EP 2.2.56	Several hydrolysis protocols may be used in order to demonstrate presence of most amino acid residues	Hydrolysis protocol should be specified. Typically, 1–2 standard protocols are used. The limits for amino acids that are difficult to recover may be broad or not specified
MS/MS sequencing	USP ⟨736⟩ EP 2.2.43	May be complicated for longer sequences, multiple disulfide bridges, stapled peptides, lactam bridges	A protocol defining minimum sequence coverage is required
NMR spectroscopy	USP ⟨761⟩ EP 2.2.33 EP 2.2.64	^{1}H shift assignments and 2D ^{1}H TOCSY, 2D ^{1}H NOESY, 2D ^{1}H–^{13}C HSQC[55] 2D ^{1}H–^{15}N HSQC,[56] as needed to demonstrate structure	Comparison with reference standard. Not recommended as a routine identity test
Peptide mapping	USP ⟨1055⟩ EP 2.2.55	For sequences where MS/MS does not provide sufficient coverage. Peptide mapping is a comparative procedure because the	Not recommended for synthetic peptides. Commonly used for recombinantly expressed peptides and proteins

		information obtained, when compared with a reference standard or reference material similarly treated, confirms the primary structure of the protein	
N-terminal sequence analysis by Edman degradation	USP ⟨1045⟩	Complicated analysis, especially for sequences that are N-terminally blocked; largely superseded by LC-MS/MS	Not recommended as a routine identity test
Infrared spectroscopy	USP ⟨197⟩ USP ⟨851⟩ EP 2.2.24	Limited use for peptides owing to interference from water vibrations and presence of counter ion (*e.g.* acetic acid)	Not recommended as a routine identity test
Higher order structure	Circular dichroism EP 2.2.41 NMR, FTIR	For investigation of secondary or tertiary structure in aqueous solution	The test is relevant for evaluation of the activity of the formulated API and is typically performed on the drug product
Bio-identity	USP ⟨111⟩ USP ⟨1030⟩ EP 2.7.1	Confirmatory test for longer, complex sequences. Cell based assays or ELISA	Not recommended as a routine identity test. These tests are relevant for evaluation of the activity of the formulated API and are typically performed on the drug product
Assay			
HPLC	USP ⟨621⟩ EP 2.2.46	Part of reference standard characterization	Method the same as for related substances; based on comparison with a reference standard
Amino acid analysis	USP ⟨1052⟩ EP 2.2.56	Part of reference standard characterization	Hydrolysis protocol must be validated, only stable amino acids should be included in the calculation, although the recovery is seldom 100%. Accuracy of individual AAs in the reference standard is critical to overall accuracy.
Nitrogen analysis	USP ⟨1057⟩ Method 7 EP 2.5.33 Method 7 EP 2.5.9	Part of reference standard characterization	Elemental analysis or Kjeldahl analysis or using HPLC with a chemiluminescence nitrogen detector (CLND)

Table 7.6 (*Continued*)

Test	Method USP/EP/Lit.	Characterization S.3	Control S.4
UV spectroscopy	USP ⟨507⟩ Method 1b USP ⟨1057⟩ Method 1 EP 2.5.33 Method 1	Part of reference standard characterization	Only useful for peptide sequences containing Trp, Tyr or Cys. Typically, several of these residues, especially Trp, are required within the sequence to provide a sufficiently high molar absorbance coefficient to ensure accuracy
Quantitative NMR spectroscopy	USP ⟨761⟩, USP ⟨1761⟩ Ref. 57	Part of reference standard characterization	A stable, preferably nonvolatile, internal standard is required
Impurities			
Peptide-related impurities	USP ⟨621⟩ EP 2.2.46	One or more methods to demonstrate separation of peptide-related impurities (generated from starting materials or during manufacturing and degradation products)	Specific method(s) for drug substance; must be validated for both process-related impurities and degradation products; limits for total and individual impurities should be specified
Enantiomeric purity	Chiral amino acid analysis	Screening test for potential stereoisomers	Not necessary if stereoisomers are detected in (U)HPLC; method specificity must be validated
Residual solvents	USP ⟨467⟩ EP 2.2.28 EP 2.4.24	All solvents used in manufacturing, unless justified	May be limited to those solvents used in the final step(s) of the manufacturing process
Elemental impurities	USP ⟨232⟩ USP ⟨233⟩ EP 2.2.58	Screening test to determine the presence of elemental impurities	Requirement if metal catalysts are used in the process; may be required if elemental impurities are detected during screening (ICH Q3D)
Residual TFA	USP ⟨503.1⟩	Only required if TFA is used during the manufacturing process	Required if TFA is used during the manufacturing process; ⟨1065⟩ used for peptides that are not soluble in acid
Residual fluoride	USP ⟨1065⟩ or ion-selective electrode EP 2.2.36	Only required if HF is used during the manufacturing process (Boc chemistry)	Only required if HF used during the manufacturing process (Boc chemistry)

Other residual counter ions	USP ⟨1065⟩	Required if used during the manufacturing process; especially *e.g.* PO_4^{3-}, ClO_4^-	Only required if removal is not adequately demonstrated
Process-related impurities (nonpeptide)	Specific method	Nonpeptide impurities should be screened using suitable methods, *e.g.* NMR, LC-MS or similar	Specific methods for process-related impurities ≥0.1%; potential genotoxic impurities require additional evaluation
Size-exclusion chromatography	EP 2.2.30	Part of a suite of methods to characterize the aggregation properties of the peptide	Not necessary if absence of oligomers was demonstrated in characterization
Specific tests			
Counter ion content	USP ⟨503⟩ EP 2.5.34 USP ⟨503.1⟩ USP ⟨1065⟩ EP 2.2.20	Acetic acid and TFA can be determined with the reversed-phase method or ion-exchange chromatography. Chloride may be determined by argentometric titration	Required
Water content	USP ⟨921⟩ Method Ic EP 2.5.12	Coulometric method most commonly used	Required for solid API
Determination of pH	USP ⟨791⟩ EP 2.2.3	The pH is determined to confirm the buffering properties of the peptide within the established ranges of peptide assay and counter ion content	Not recommended as a routine test
Optical rotation	USP ⟨781⟩ EP 2.2.7	The sensitivity of the method is not sufficient to detect minor levels of racemization	Not recommended as a routine test
Ellman test	ref. 58	Potential orthogonal method to demonstrate the presence of sulfhydryl groups in the peptide	Required if the reduced form(s) of a peptide containing disulfide bonds cannot be determined using the method for related substances
Bioburden	USP ⟨61⟩ USP ⟨62⟩ EP 2.6.12	All lots used for clinical purposes	Requirement if API is used to manufacture a sterile drug product
Bacterial endotoxins	USP ⟨85⟩ EP 2.6.14	All lots used for clinical purposes	Requirement for APIs used in the manufacture of injectable drug products

confirmation by tandem MS (MS/MS or MS^n). Using high-resolution mass spectrometry (HR-MS), the observed monoisotopic mass can usually be established within 20 ppm of the theoretical mass. If multiply charged species are generated, then deconvolution methods are required to convert the observed *m*/*z* values to monoisotopic mass. It is important to recognize that as the overall size (and thus mass) of the peptide increases, the peak corresponding to its monoisotopic mass becomes less abundant in the spectrum. In those cases, where the signal intensity of the monoisotopic mass is less than 10-fold above the background noise, it may be more appropriate to use the average molecular mass as the characterization criterion for identification of the peptide. The less stringent criteria described in USP ⟨736⟩ can be applied to routine QC release testing since QC laboratories typically use less resolving MS instruments.

Sequence coverage should demonstrate the presence of all amino acid residues in their correct order. For linear peptides, obtaining at least one fragment for each amino acid residue in the b- or y-series [when using chemically induced dissociation (CID)] or c- and z-series [when using electron transfer dissociation (ETD)] is usually possible. Data where a fragment during ionization loses a group (*e.g.* H_2O, NH_3) are usually acceptable if the monoisotopic mass confirms the missing group's presence in the intact molecule or orthogonal methods exist confirming the presence of the correct amino acid.

Peptides containing disulfide bridges may require reduction prior to MS/MS analysis; see Section 7.3.3.

7.3.2.2 NMR Spectroscopy

Demonstration of primary structure and atom-specific assignment of chemical shifts follow the procedures developed for proteins at natural abundance isotope distribution using ^1H homonuclear 2D NMR spectroscopy complemented by supporting information from 2D heteronuclear ^{13}C–^1H and ^{15}N–^1H NMR spectra.[28]

Briefly, the first step involves identification of different spin systems, *i.e.* the sets of NMR signals belonging to each amino acid. In this step, 2D correlation spectroscopy (COSY) and 2D total correlation spectroscopy (TOCSY) data are primarily used. COSY provides chemical shift data for protons on adjacent atoms, whereas TOCSY is able to transfer magnetization between all protons connected by adjacent atoms. A comparison of the two sets of spectra makes it possible to identify the protons of the side chains of the individual amino acids. The chemical shifts allow further assignment of specific amino acids.

Many amino acid types display unique spin system patterns whereas others are similar, *e.g.* Asp, Asn, Tyr, Phe, His and Trp – all containing ^{1}HN, ^{1}Hα and two ^{1}Hβ signals in which the chemical shifts can be very similar. Differentiation between these amino acids can be achieved by examining heteronuclear single quantum correlation (HSQC) spectra, in which the spectrum for each proton attached to a heteronucleus, either ^1H–^{13}C or

^{1}H–^{15}N, is evaluated. In HSQC data, His and Trp differ significantly from Phe and Tyr. To connect the different spin systems in a sequential order, the nuclear Overhauser effect spectroscopy (NOESY) experiment is used. Since the neighboring residues are inherently close in space, assignments can be made by the peaks in the NOESY with other spin systems. A second technique to determine which signals arise from protons that are close to each other in space even if they are not bonded is rotating frame Overhauser effect spectroscopy (ROESY). A ROESY spectrum yields through-space correlations *via* spin–spin relaxation.

The rigidity of the structure of different peptides varies significantly and influences the complexity of the assignment process. For very unstructured and dynamic peptides, the chemical shift variation is very small so that most chemical shifts of each amino acid type tend to be very similar ("poor chemical shift dispersion" is observed). These so-called random coil chemical shifts have been extensively studied and using such tabulated values can significantly facilitate rapid identification of the different amino acid types.[29] However, since the degree of spectral overlap is often significant, it may be difficult to resolve the structure of a peptide fully, particularly if the sequence contains several amino acids of the same type. For unstructured peptides containing more than about 25 amino acids or containing repeated stretches of amino acid sequences, a full unambiguous chemical shift assignment may not be possible.

On the other hand, for peptides with a high degree of stable secondary structure or peptides containing rigid structural elements (for example, disulfide bridges), the chemical shift assignment is normally more straightforward as the degree of chemical shift overlap is low.

Once the assignment of a peptide or small protein has been completed, the $^{1}H\alpha$, $^{13}C\alpha$ and $^{13}C\beta$ chemical shifts can be effectively used to determine the secondary structure by comparison with the tabulated random coil chemical shifts.[30]

7.3.2.3 Peptide Mapping and Other Sequencing Methods

Peptide mapping or other sequencing approaches may be necessary when MS/MS or MS^{n} cannot provide full sequence confirmation, mainly due to the high molecular mass of the peptide. The test, based on USP ⟨1055⟩, involves the chemical or enzymatic treatment of a peptide or protein, resulting in the formation of peptide fragments, followed by separation and identification of the resultant (lower molecular weight) fragments in a reproducible manner. Peptide mapping is a comparative procedure because the information obtained, when compared with a reference standard or reference material similarly treated, confirms the primary structure of the protein, is capable of detecting whether alterations in structure have occurred and demonstrates process consistency and genetic stability of the host cells.

However, many pharmaceutically relevant peptides are engineered to make them less prone to degradation by natural enzymes by blocking the

N- or C-terminus or incorporating D- or other unnatural amino acids in the sequence. Disulfide-linked peptides typically need to be reduced (linearized) prior to enzymatic digestion and the resulting sulfhydryl groups blocked with a suitable alkylation reagent, *e.g.* iodoacetamide. Additional sample workup may be necessary if the reagents or enzyme interfere with the final high-performance liquid chromatographic (HPLC) testing.

The historically important methods of Edman degradation (N-terminal sequencing) or C-terminal sequencing utilizing carboxypeptidases[31] have lost most of their relevance today.

7.3.2.4 Amino Acid Analysis

Amino acid analysis (AAA) is a traditional method applied to peptides. Based on USP ⟨1052⟩, it involves digestion of the peptide to its constituent amino acids using a suitable hydrolysis reagent. Quantitation of each amino acid is achieved using either pre- or post-column derivatization. Not all amino acids can be detected or accurately quantitated using this approach. For instance, Gln and Asn will hydrolyze to Glu and Asp, respectively. Recoveries of Trp, Ser and Cys can be low and depend on the hydrolysis protocol.

7.3.3 Secondary and Tertiary Structure

7.3.3.1 Circular Dichroism

Circular dichroism (CD) is considered to be one of the most valuable techniques for characterizing the structure of peptides and proteins in solution. CD refers to the differential absorption of counterclockwise (left-handed, L) and other clockwise (right-handed, R) circularly polarized light by the analyte. From this, complementary structural information can be obtained from a number of spectral regions. In peptides, the chromophores of interest include the peptide bond (absorption below 240 nm), which is used to determine secondary structure, aromatic amino acid side chains (absorption in the range 260–320 nm) and disulfide bonds (weak, broad absorption bands centered around 260 nm). The advantages of CD are that it is a nondestructive, fast and simple technique, consuming small sample quantities. The content of both secondary and tertiary structures can be estimated from CD.[32]

7.3.3.2 FTIR Spectroscopy

Infrared (IR) spectrophotometry is not a useful routine identification tool for a lyophilized DS owing to the similarity of the spectra for many peptides, which tend to be dominated by water vibrations from residual water and carbonyl bands, arising from the peptide backbone and acetate, if present as a counter ion.

However, Fourier transform infrared (FTIR) analysis in solution (using a very short pathlength) of amide bond vibrations can provide information about the peptide structure. Characteristic signals found in the IR spectra of proteins and peptides include the amide I (1600–1700 cm^{-1}) and amide II (1480–1600 cm^{-1}) bands, which arise from the amide bonds that link the amino acids. The absorption of the amide I band is associated with stretching vibrations of the C=O bond of the amide, whereas absorption of the amide II band is associated with bending vibrations of the N–H bond. Since both of these bonds are involved in the hydrogen bonding that takes place between various elements of secondary structure, the locations and shapes of the amide I and amide II bands reveal the secondary structure content of a peptide. However, the underlying component bands of α-helices, β-sheets, β-turns and coils overlap, hence mathematical tools, such as second derivatives of the spectra, are used to resolve the individual band components. A comparison with proteins of known structure is used to determine the secondary structure content using multivariate regression methods[33] or neural network analysis.[34]

7.3.3.3 Cyclic Peptides

Peptides containing disulfide bridges may require reduction prior to MS/MS analysis in order to unfold and linearize the molecule *in situ* using commonly available reducing agents, such as tris(2-carboxyethyl)phosphine (TCEP), dithioerythritol (DTE) or dithiothreitol (DTT). In order to demonstrate the presence of disulfide bonds, an HPLC analysis should be performed to show the change in retention time following application of the reducing agent. In the majority of cases when using reversed-phase HPLC (RP-HPLC) the linear molecule will elute later than the folded one; care must be exercised to ensure a sufficient length of the gradient to elute the peak within the window of the method. HPLC-MS/MS will typically provide sufficient primary sequence information in such cases. If, however, peptide mapping is required then the free sulfhydryl groups need to be blocked in a chemical step, otherwise their presence will likely interfere with the action of the enzyme. The selection of the enzyme should also be driven by the sequence in order to generate shorter peptide fragments of similar length.

Complex folded molecules with multiple disulfide bonds or molecules folded *via* lactam linkages, stapled peptides or containing other unnatural linkers may require either chemical or enzymatic digestion prior to MS/MS in order to confirm their structure. A promising method for characterizing peptide disulfide isomers is MS utilizing electron-transfer dissociation (ETD).[35] For hepcidin, a tightly folded 25-amino acid peptide containing four disulfide bonds, NMR spectroscopy was able to provide the necessary orthogonal complementary information to confirm the disulfide connectivity.[36]

7.3.4 Assay

The major components of a typical peptide powder isolated by lyophilization are the peptide "as is" (including related substances), a counter ion and residual moisture. Methods for assay are listed in Table 7.6; each method has specific precision and accuracy limitations.

While the knowledge of the peptide content "as is" is important for formulation purposes, the recommended method of calculating and reporting assay is as follows:

$$\text{Assay (\%)} = \frac{\text{peptide ``as is'' (\%)} \times [100 - \text{total impurities (\%)}]}{100 - \text{counter ion (\%)} - \text{water (\%)}} \quad (7.1)$$

where peptide "as is" is the total peptide content, including related substances; total impurities includes all related substances and other impurities that may impact the overall assay; the counter ion is *e.g.* acetic acid and water is in the form of residual moisture. The numerator can be simplified if the assay is performed by a method resolving related substances, *e.g.* HPLC, which provides the direct net peptide content.

Since water content has a significant impact on the result, it is important to minimize samples from absorbing/desorbing residual moisture during analysis. The peptide's vapor sorption properties should be characterized to establish safe handling conditions (see Section 7.3.8). Standards used in HPLC analysis should be individually vialed and assayed, to permit direct reconstitution prior to analysis.

Full characterization of the lead lot should be performed with an effort to close mass balance on all components present.

7.3.5 Impurity Analysis

Impurity identification and development of liquid chromatographic (LC) methods targeted towards separation and quantitation of peptide-related substances in the DS are expected to progress throughout the manufacturing development and are an integral part of the development of a successful control strategy. Ultra-high-performance liquid chromatography (UHPLC), utilizing columns packed with sub-2 μm particles is rapidly becoming the standard LC platform. During development, it is recommended to utilize two or more methods using dissimilar (orthogonal) separation mechanisms, in order to maximize the potential for detecting impurities. It is advantageous for impurity identification if the primary method is compatible with MS.[37] It is recommended to synthesize impurities likely to be present in the DS, as that will aid in method development. For example, diastereoisomers, especially those of His and Cys, should be considered as they will not be detected by MS if they coelute with the main peak. See Sections 7.6–7.11 for other potential impurities to consider. Impurities may also be generated *in situ* using forced degradation approaches; see Section 7.6.5.

As the manufacturing process matures, some of the methods may be eliminated in favor of a single method. In those cases where a single method cannot separate all expected impurities, additional methods for the determination of specific impurities may be used.

7.3.6 Aggregation

Aggregation encompasses interactions that result in the self-association of peptide or protein molecules, usually accompanied by a change in their higher order structure. Aggregates include a diverse range of molecular assemblies that can differ in their biochemical and biophysical characteristics. They can range considerably in size, from dimers up to subvisible and visible particles, form precipitates or gels, involve covalent or noncovalent linkages, be ordered or disordered in structure, soluble or insoluble, and their formation can be reversible or irreversible. Aggregation occurs when the peptide or protein is subjected to chemical or physical stress, typically causing partial or full unfolding and subsequent loss of native structure; the nature of the stress factor may define the properties of the aggregate.

Aggregates can elicit an immunogenic response.[38] This response may involve anti-drug antibody (ADA) formation, T-cell activation and innate immune responses. The consequences of an immune reaction to a therapeutic peptide or protein range from the transient appearance of ADAs without any clinical significance to severe life-threatening conditions. Both the FDA and European Medicines Agency (EMA) have published guidance documents for assessing the immunogenicity of therapeutic proteins. The agencies also expect immunogenicity assessments for peptides, especially those that exhibit higher order structure or those that have potential to form covalently bound oligomers.

7.3.6.1 Methods for Assessing Aggregation

7.3.6.1.1 Size-exclusion Chromatography (SEC) (EP 2.2.30). SEC (with UV detection) is the most commonly used method for assessing sample aggregation that is also most suitable for routine testing. The method should be validated using positive controls that confirm its suitability. The difficulty in obtaining an authentic, stable, well-characterized positive control (*i.e.* a sample of aggregated peptide) limits the utility of the method. Although the use of surrogate molecular mass markers may be justified for method validation purposes, it is necessary to demonstrate that aggregates of the peptide being studied are detectable in the method.

7.3.6.1.2 Field-flow Fractionation (FFF). The most common variant, asymmetric flow field-flow fractionation,[39] is an alternative to SEC. Separation is achieved by interactions with an external, perpendicular field, rather than a stationary phase. Samples are subjected to low shear stress and no interactions with a stationary phase occur, which may be an advantage when aggregates are reversible. However, the method has low sensitivity and the sample loading is limited.

7.3.6.1.3 UV Detection (Optical Density). The light scattering (turbidity) in a solution of macromolecules leading to a decrease in transmitted light depends on several factors: refractive index of the analyte relative to the medium, molecular mass, hydrodynamic volume and shape (*e.g.* spheres, rods or coils). Measurement is performed at a wavelength where the absorbance of the molecule is negligible. Although the measurement of optical density is similar to measuring absorbance, owing to the impact of multiple variables the results cannot be easily interpreted.[40]

7.3.6.1.4 Multi-angle Light Scattering (MALS). The drawbacks of optical density measurements can be overcome by measuring the intensity of scattered light at several (at least three) different angles. MALS allows the direct measurement of molar mass.[41] The limitation of the technique is its low sensitivity for low molar mass aggregates (<5 kDa) and the need for a simultaneous mass concentration detector, *e.g.* by refractive index. Separation of different species is also necessary; therefore, SEC-MALS is the preferred method for the characterization of peptide aggregates. Data derived from SEC-MALS can be used to support the development and validation of SEC methods for routine use.

7.3.6.1.5 Sedimentation Velocity Analytical Ultracentrifugation (SV-AUC). Sedimentation velocity, as measured in an analytical ultracentrifuge, is a method for obtaining information about peptide or protein heterogeneity and the state of association or aggregation. Aggregates or different oligomers are detected on the basis of their different sedimentation coefficients. This method can detect aggregates or other minor components at a level below 1% by weight. Sedimentation velocity gives good quantitation of relative amounts of species and will usually give accurate sedimentation coefficients for any aggregates that are seen. Sedimentation velocity is an absolute method. Its calibration is based on fundamental units of length and time, requiring no standard molecules.[42]

Although SV-AUC is sensitive to all forms of oligomers and aggregates, generally only the long-lived aggregates are resolved as separate species (those with lifetimes longer than ~1 h). When a sample exhibits rapidly reversible self-association to form small oligomers, generally these oligomers are not resolved as distinct species, but their presence does shift the average sedimentation coefficient to higher values in a concentration-dependent manner.

The method is also sensitive to differences in conformation. Conformational changes in peptides or protein complexes alter their sedimentation coefficients because they alter the amount of hydrodynamic friction (the frictional coefficient). The frictional coefficient is quite sensitive to the presence of disordered, flexible regions and is also sensitive to the overall shape (extended *versus* compact and globular). Sedimentation coefficients can be measured with high accuracy (±0.5% or better), hence they provide a sensitive and quantitative means to demonstrate comparability of molecular conformation.

7.3.6.1.6 Dynamic Light Scattering (DLS). DLS is the scattering of laser light by small, insoluble particles to gain information on the size distribution of a sample. It has the advantage of being simple to implement and no potentially interfering dye or chemical is required. However, DLS provides the average size of the species present rather than the sizes of the individual aggregates and the technique is also biased towards larger species, which have a greater propensity to scatter light. Samples must be prepared essentially particle and bubble free in order to generate meaningful results.

7.3.6.1.7 Dye Binding Methods. Certain dyes, such as Nile Red, Congo Red and Thioflavin T, exhibit fluorescence upon binding to aggregates. This property has been used to detect and quantitate aggregates of β-amyloid[43] and human calcitonin.[44]

7.3.6.1.8 NMR Spectroscopy. Aggregation can be assessed either by diffusion NMR[45] or by examining the temperature dependence of ^{1}H shifts as a measure of hydrogen bonding.[46] Aggregation of β-amyloid has been characterized by NMR spectroscopy.[47]

7.3.6.1.9 Particles. Concerns about particles in protein formulations have led to the development of several analytical techniques.[48] A method that can be adapted for use with peptide solutions for counting and determining the size of subvisible particles in the low-micrometer range utilizes light obscuration (USP ⟨787⟩, EP 2.9.19). Additional particle characterization can be achieved using microflow imaging (MFI).[49]

7.3.7 Isomerism

Common isomers that require characterization are the following:

1. Leu to Ile substitutions (typical origin – starting materials);
2. Ala to β-Ala substitutions (typical origin – starting materials);
3. Asp to β-Asp substitutions (typical origin – process conditions and stability);
4. L- to D- conversions (typical origin – starting materials and process conditions).

Usually NMR spectroscopy is capable of differentiating and confirming sequence locations for isomers of type 1 and 2. Amino acid analysis is also capable of distinguishing between Leu and Ile and Ala and β-Ala; however, it is not able to establish their positions in the sequence. Owing to the origin of these isomers, it is advantageous to control their levels in starting materials.

The formation of β-Asp from Asp is most common when the next amino acid in the sequence is sterically unhindered, such as Gly or Ser. β-Asp can usually be determined using a suitable HPLC method. Furthermore, evidence of β-Asp formation in the peptide sequence can be inferred from the formation of succinimide intermediates, which are readily detected by both HPLC and MS.

The L- to D-conversion (racemization) of individual amino acid residues in the peptide sequence leads to the formation of stereoisomers. MS cannot distinguish between D- and L-amino acid substitutions and it is not routinely possible to determine absolute chirality using NMR spectroscopy. Although ion-mobility MS/MS is being studied as a possible technique that can distinguish between isomers and potentially provide direct chiral information for amino acids in a peptide,[50] currently the most readily available technique for determining the chirality of amino acid residues in peptides is chiral amino acid analysis (chiral AAA).

Chiral AAA is a multistep procedure in which acid hydrolysis of the peptide is followed by suitable derivatization of the resulting AAs and subsequent determination of the optical isomers of the constituent AAs by chiral gas chromatography with mass spectrometric detection (GC-MS).[51] HPLC-based methods of chiral AAA, which involve pre- or post-column derivatization of AA hydrolyzates, are also available, for example, using Marfey's reagent[52] or numerous other reagents.[53] Regardless of the hydrolysis procedure employed, the method should account for potential racemization during hydrolysis or subsequent derivatization. Unfortunately, both the accuracy and precision of these methods are limited.

An approach that helps address this issue involves the use of deuterated reagents, *e.g.* DCl or D_2O. Substitution of the α-carbon hydrogen with deuterium allows for correction for racemization when using MS-based detection methods. However, the technique has many pitfalls as specific sequences may cause racemization of certain AAs during hydrolysis that cannot be corrected for and lead to false results. For example, a Phe residue adjacent to Cys will racemize during hydrolysis.[54] Another drawback of chiral AAA is that it does not reveal the location of the racemized residue. Furthermore, if a residue is present at several different positions in the sequence, but only one of the positions is racemized, then the results can misleadingly average out the evidence of racemization.

For all the reasons outlined above, chiral AAA is not recommended as a routine test, although it may be acceptable to use it in cases where racemization of a residue is difficult to control through starting materials or process parameters and no other alternative method exists. For example, the impurity D-His^1-exenatide is controlled using chiral AAA (USP Exenatide monograph).

Primarily, however, chiral AAA is recommended as a characterization test to confirm that residues in the sequence are in the correct enantiomeric form. Otherwise, the method may be used to identify residues in the sequence that carry a potential racemization risk, which cannot be controlled by other means. Once such residues have been identified then their corresponding impurities should be synthesized and a suitable HPLC method (or methods) should be developed that are capable of separating those impurities from the target molecule. In general, (U)HPLC is the preferred method for controlling stereoisomer levels. Note that chiral HPLC is not an effective technique for resolving stereoisomers in peptides, mainly owing to size and solubility constraints of the analyte.

7.3.8 Other Characterization Methods

7.3.8.1 UV–Vis Spectrophotometry (EP 2.2.25)

The spectrum of the API in aqueous solution is required. The profile should be consistent with the constituent amino acids of the peptide.

7.3.8.2 X-ray Powder Diffraction (USP ⟨941⟩, EP 2.9.33)

X-ray powder diffraction (XRPD) is used to demonstrate the crystallinity (or lack thereof) of the API. Since most peptides are isolated as lyophilized powders that result in amorphous material, this test is used to confirm the absence of crystal structures due to the absence of peaks in the diffractogram.

7.3.8.3 Dynamic Vapor Sorption (USP ⟨1241⟩, EP 2.9.39)

The amorphous nature of peptide lyophilizates makes them often prone to rapid moisture absorption/desorption. It is therefore useful to determine the moisture sorption isotherm in order to establish the handling conditions under which moisture sorption/desorption is minimized.

7.3.8.4 Thermal Analysis (EP 2.2.34)

Differential Scanning Calorimetry (DSC). DSC provides additional information regarding the morphology of the lyophilized peptide. The presence of a glass transition temperature or a melting point can be used to confirm that the material is amorphous or crystalline, respectively.

Thermogravimetric Analysis (TGA). TGA provides additional information pertaining to volatiles present in the lyophilized peptides. It is typically possible to distinguish between the presence of residual solvents, moisture and decomposition events.

7.4 Risk Assessment

7.4.1 Introduction to the Risk Assessment Process

This section outlines the tools that can be applied to achieve a risk-based approach to peptide API development.

Risk assessment spans the entire drug development timeline and can provide the framework for setting of appropriate control strategies. The three steps in risk assessment are (1) risk identification, (2) risk control/mitigation and (3) risk review.

The risk assessment process begins shortly after definition of the QTPP, with classification of the DS QAs on a spectrum of criticality to determine the likely impact on patient safety. Based on clinical experience, the DS CQAs are defined and appropriate ranges are set. The product quality risk assessment (PQRA) process is covered in Section 7.4.2. The process design strategy which utilizes a risk-based workspace to assess product- and process-specific data is illustrated in Figure 7.3.

Risk assessments are broadly applied throughout the manufacturing industry to address the following:

- *Risk to patient.* An example is a product quality risk assessment, which asks: Are the correct controls in place to ensure the identity, quality, efficacy and safety of the product?
- *Risk to product.* An example is a failure modes and effects analysis (FMEA), which assesses the manufacturing process's ability and likelihood to meet the DS CQAs consistently. FMEA is a quantitative assessment tool using severity, occurrence and detection scores to rank risks and can factor in cost and throughput considerations.

Once the PQRA has defined the targets for DS CQAs, the manufacturing process and testing strategy can be built to ensure consistent delivery to those targets. A systematic framework to organize data and evaluate how input variability affects product quality and manufacturability is useful. A framework based on the FMEA process is presented in Section 7.4.4, which evaluates the process design and control strategy collectively.

7.4.2 Product Quality Risk Assessment (PQRA)

A product quality risk assessment (PQRA) is a general tool used to evaluate the risk of meeting product quality attributes as defined in the QTPP. The assessment is directly tied to patient safety.

The PQRA is a life-cycle document (revisable assessment). The initial PQRA should be completed as early as FIH and then updated at initiation of process design, prior to PPQ and prior to launch, then as necessary for life-cycle management (LCM). The first PQRA is heavily based on prior knowledge (literature and platform experience) and only limited process experience (*e.g.* observed impurities), compared with more in-depth product understanding (improved with stability data) gained later in development. Severity (*e.g.* criticality) scoring is assessed based on immunogenicity, non-immune safety, pharmacokinetics (PK), potency and safety.[59] The highest score is used. Risk is the combination of the severity (impact) score and uncertainty score (confidence measure). Risk ranking and filtering (RRF) is an approved method for related substances and process-related impurities. The marketing authorization (MA) should contain sufficient information to explain and justify the assessment used.

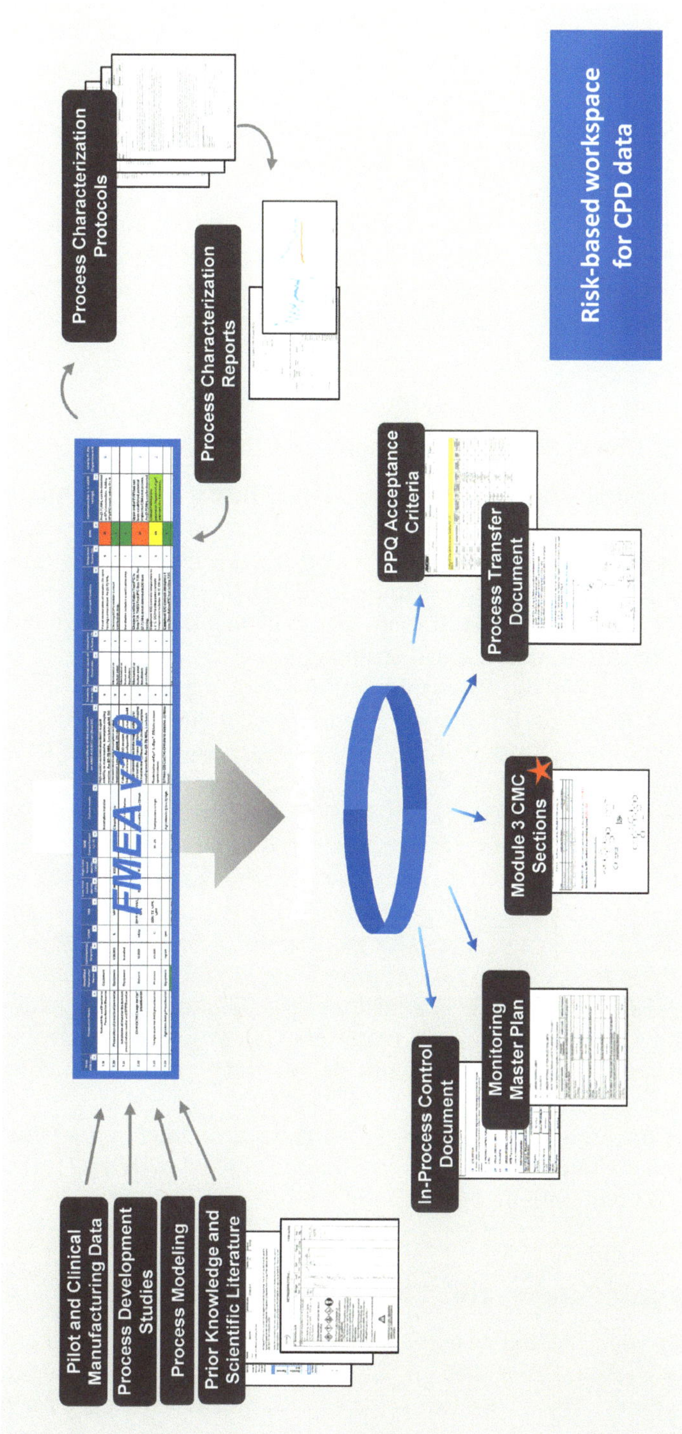

Figure 7.3 Process design strategy that utilizes a risk-based workspace to assess product- and process-specific data.

The DS QAs for peptide APIs should generally follow the guidance given in Section 7.2.3 and Table 7.3. Each QA is scored for severity of the potential impact to product safety and efficacy. The modality, indication, patient population and dosing regimens can affect product quality attribute (PQA) scores. All known and potential impurities should be scored. Revision can be performed as process and product knowledge is gained, *e.g.* as a result of improved method capability.

General DS CQA categories include product purity attributes/product variants; process-related impurities; composition and strength (pH, concentration for solutions; potency for solids); adventitious agents (potential viruses, bioburden, mycoplasma, endotoxin); and other obligatory attributes (appearance, particulates).

7.4.3 High-level Risk Assessment (HLRA)

High-level risk assessments (HLRAs) can be useful tools in early commercial process development. An HLRA can be performed much faster than FMEA but provides similar output in terms of prioritized lists of risks and development focus areas. The output can be used to compare synthetic routes for benefits and liabilities. An HLRA can assist in mapping the strengths and weaknesses of the current control strategy.

HLRAs are typically set up as interaction tables mapping unit operations (see Section 7.7) against DS QAs (see Section 7.2.3). Unit operation (UO) rows should be included chronologically beginning at the GMP sequence start with API starting materials (SMs) listed as the first row. The UOs end with finished DS testing and DS storage. Isolation and testing of intermediates can be included as UO rows. DS QAs are listed as columns. The interaction scoring can be performed in two ways. Approach (1) applies Yes/No scoring wherever there is an impact of a UO on a QA. This is the simplest approach and is used to highlight process steps that require additional development based on their impact on QAs. Approach (2) applies a gradated scoring that has Yes/No for impact and additional clarification for risk. Yes/Green: process data is available and the parameter is considered controlled, and Yes/Red: limited data is available and the attribute may not be controlled. No interaction is gray.

An HLRA process will gravitate the focus towards later steps since these have the greatest impact on DS quality. This is aligned with the QbD process and other risk assessment methods.

7.4.4 Failure Modes and Effects Analysis (FMEA)

A process-focused risk assessment using the FMEA format is recommended during late-stage process design when significant product-specific data become available. The FMEA can serve as a technical workspace for process knowledge management as a way to input all development data generated around a program in one location. The initial FMEA should be drafted after the major elements of the manufacturing process have been reasonably well defined (Phase 1/Phase 2) and the PQAs are understood. Although development may be performed without the benefit of a full FMEA, organization of early development data into a consistent format will prove advantageous for both chemist and analytical scientist. The formal assessment of risks and mitigation plans can be performed at an appropriate time, whereas equal access to data should be made available as early as possible. An FMEA built in the manner described here keeps a real-time account of manufacturing risks and mitigation plans and informs where resources should be placed during process design and process characterization.

Each process parameter, material input and performance indicator is evaluated in the FMEA exercise. Process parameters are scored based on the potential that variability can impact the process, the likelihood of that occurrence and the ability to detect the impact. The result is a risk priority number (RPN), which is calculated by severity (S)×occurrence (O)×detection (D): $S \times O \times D = RPN$. When performed in a standardized, structured manner, the output is a quantitative ranking of risks to product quality and manufacturability. Numerical scoring can contain between three and five levels (*e.g.* 1, 3, 9 and 1, 3, 5, 7, 9), with improved stratification with higher levels. Definitions are adjusted based on weighted scoring. For each category, suggested scoring definitions are provided in Table 7.7.

The severity score is based on the effect of the failure, not the failure itself. The failure mode could be described as temperature too high, which has the effect of increasing impurity levels. Another example could be failure to stir, which has the effect of changing the crystallization behavior and increasing the filtration time. Each effect will have its own severity score. Impacts to product quality will receive higher scores than impacts to process performance (yield, throughput).

The occurrence score is ideally based on process data and manufacturing history. When limited data are available, estimation of the score must be performed.

The detection score accounts for all control elements and their reliability. Control elements include the following:

1. Procedural controls, *e.g.* the manufacturing process description and documentation. Operator training, verification of steps performed, *etc.*
2. Equipment controls, *e.g.* manufacturing capability. Equipment installation and operational qualification will indicate the ability to control set points such as temperature, stirring and mass measurements. The combination of batch data and manufacturing experience will inform the true parameter variability.

Table 7.7 Example severity, occurrence and detection scoring guide for process FMEA risk assessments.

Level	Severity (S)	Occurrence (O)	Detection (D)
1	No impact on product quality or process performance	Remote. No history of failure	Almost certain. Controls are in place and will detect the failure in almost every instance
3	No impact on product quality. Minor impact on process performance (yield, throughput)	Unlikely. Failure could occur at some time. Only isolated incidents observed	High. Controls are in place and will very likely detect the failure
5	No impact on product quality. Or impact on product quality is mitigated by downstream steps Moderate impact on process performance (yield, throughput)	Probable. Failure is probable at some time and has been observed	Moderate. The controls in place have a moderate chance of detecting the failure
7	Minor impact on product quality, which does not result in failure to meet DS or intermediate specifications Major process performance impact (yield, throughput) which may result in deviation and investigation	Likely. Failure is likely and will occur in most circumstances. Repeated failures observed	Remote. Some chance that current controls can detect the failure. Control may be unproven or unreliable
9	Serious impact on product quality. Failure to meet DS or intermediate specifications. Likely NC and batch failure. Unknown impact on DS quality can be given a default S = 9 score	Frequent. Failure is almost inevitable. Consistent failures observed	Current controls cannot detect the failure No controls in place

3. In-process and downstream detection by monitored performance indicators (usually analytical methods). It is useful to include all means to detect the effect of the failure mode in downstream processing steps. Performance indicators are broadly defined and may include measurements such as processing time, yield and quality targets. IPCs are used to ensure that purity and impurity targets are met. Detection elements include testing of intermediates and final DS release testing. The robustness of the test is included in the detection score to weigh the accuracy, specificity and reproducibility

of the assay. A redundancy of controls will contribute to a low detection score, whereas the absence of controls will result in a high detection score.

Subject matter experts from manufacturing, quality, process, analytical and regulatory areas should be involved in the development of the FMEA. Inputs should include QTPP; DS CQAs; technical information found in the investigational new drug (IND) filing; manufacturing IPCs and specifications for process intermediates; manufacturing batch records; analytical methods; development reports; platform knowledge; and relevant scientific literature. Execution of an FMEA is best accomplished by prepopulating a template with as much information as possible and then scoring (severity, occurrence, detection ratings) should be performed within a team setting. The gradient scoring system can be broken down into high/medium/low risks based on the team assessment. The team will then define mitigation plans for the highest risk scores and select which medium risks (typically business risks) should be part of the plan. The risk assessment, mitigation plans, protocols and summary reports should be documented for future reference.

It is often the case that each process step will have a characterization plan in addition to general studies such as impurity fate and purge, challenge studies and process hold-time definitions. The FMEA risk assessment is a life-cycle document that captures the current state of process and product understanding. It includes all generated process data [knowledge space ranges, set points and acceptable ranges (ARs)], information describing the impact of variability and likelihood of the effect, in addition to the means to detect and control excursions. As such, the FMEA should be updated as new information becomes available.

7.4.4.1 Template Organization

Columns are labeled as shown in Table 7.8. The table consist of series of rows and columns. Chronological organization of the manufacturing process is listed in the rows. An individual row is included for each process parameter, material attribute, performance indicator, IPC and intermediate release test. An example abbreviated FMEA template for an SPPS coupling unit operation is shown in Table 7.8.

7.4.5 API or GMP Starting Materials

The designation of API starting materials and their specifications are a critical element of the holistic control strategy as it often has a direct impact on DS quality. This is especially true for peptides, where the sheer number of chemical steps often results in the accumulation of many low-level impurities.

Evaluation of starting materials and reagents should be part of a QbD development plan and begin early for peptides. Risk assessments can be applied to help identify and mitigate supply chain and product quality

Table 7.8 Example FMEA template for the reaction unit operation.

Parameter name	Classification	Suggestions for FMEA team	Commercial set point	NOR (range target ± X)	UOM	Reference data
Solvent charge	**Non-Critical**				kg kg^{-1} L kg^{-1}	List reports documenting range-finding experiments and representative runs at the commercial set point
Solvent quality – purity by GC (or other	Specification test	Indicate if multiple vendors and/or manufacturing	List actual specification	NOR is not especially relevant : may	%	List specification document (if it exists) and

Low limit tested during CPD	High limit tested during CPD	AR	Failure mode	Potential effects of the excursion on final DS	(S)	(O)	Current controls	(D)	RPN	Risk level
Typically low end is determined by solubility of reagents/ intermediates	Typically high end comes from early development		Continuous variable Amount is: (1) low, (2) high	Indicate the impact of 2–3 × NOR solvent undercharge on solubility of SMs and reagents, effect on kinetics, reaction conversion at normal endpoint and ability to meet reaction completion IPC criteria (if an IPC exists). Mixing and heat transfer may become important at the more concentrated end. Describe any increased impurity formation due to undercharge and indicate if effects can be mitigated in real time or if downstream processing steps reduce the impact. Indicate how downstream crystallization, distillation and work-up steps are affected by solvent variability. If relevant, indicate impact at CPD range limits and if the AR extends to the entire characterized range of if an edge-of-failure has been found. If an undercharge results in IPC or spec. failure, S = 9. If it only impacts yield, S = 7 or 5. If the CPD range does not cover 2–3 × NOR, document impact as "unknown" (*e.g.* a knowledge gap) and S = 9 until data can be generated	5	3	Add all control elements here. Includes procedural, IPCs and spec. testing Typical language: Solvent charge is controlled to ± xx.x kg (*e.g.* 5%) and document-ed in the batch record. Reaction completion IPC can detect a slower reaction rate. Indicate if a downstream distillation, crystallization or work-up step can remove additio-nal solvent. Indicate where this specific solvent is detected and controlled downstream (all the way to DS release testing)	1	15	Low
Ideally the low end extends further than	Typically high end is near 100%		For specifica-tion tests the failure mode	Typical language is: Within specification	5	1	Incoming raw material specifications.	1	5	Low

Table 7.8 (*Continued*)

Parameter name	Classification	Suggestions for FMEA team	Commercial set point	NOR (range target ± X)	UOM	Reference data
method as appropriate) Pull data from both program experience and from industry knowledge		sites employ different specifications, *e.g.* spec. is ≥99.5% at xx, spec. is ≥99.9% at xx	test acceptance criteria Typically ≥99.9	add typical observed range here		representative runs at the commercial set point.
Critical raw material or API charge Use proper name but may include common name or abbreviation as intended for filing	**Non-Critical**		Typically 1.00 equiv. for limiting reagent. (3) sfs	NOR is typically ≤ 1% of set-point for critical charges.	Use units of equiv. or kg kg^{-1}	List reports documenting range-finding experiments and representative runs at the commercial set point

...ow limit tested ...uring CPD	High limit tested during CPD	AR	Failure mode	Potential effects of the excursion on final DS	(S)	(O)	Current controls	(D)	RPN	Risk level
the spec. limit, but this is optional			is the purity exactly meets the specification limit	range, no known impact on DS CQAs. Outside range, unknown impact on DS purity profile. Depends on impurity and level If no data exist at or near the spec. limit, default to S = 9. Knowledge about the typical impurities in the specific solvent will direct the impact assessment. Specified (known) impurities, as well as other specified attributes are broken out as separate rows			Or accepted based on vendor CoA. Solvent is not retested prior to introduction			
...ypical range is 0.90 equiv.	Typical range is 1.10 equiv.		Continuous variable, typical failure modes are: Amount is (1) low, (2) high	(1) Indicate the impact of 2–3 × NOR undercharge on kinetics, reaction conversion at normal endpoint and ability to meet reaction completion IPC criteria (if an IPC exists). Describe any increased impurity formation due to undercharge and indicate if effects can be mitigated in real-time or if downstream processing steps reduce the impact. If relevant, indicate impact on CPD range limits and if the AR extends to the entire characterized range of if an edge-of-failure has been found. If an undercharge results in IPC or spec. failure, S = 9. If it only impacts yield, S = 7 or 5. If the CPD range does not cover 2–3 × NOR, document	5	1	Add all control elements here. Includes procedural, IPCs and spec. testing Typical language: CRM or API charge is controlled to ± 0.xx kg (approx. xx%) and is documented and verified in the MBR. Reaction completion IPC can detect a slower reaction rate, which would trigger xx (*e.g.* extended reaction time). If an increased impurity formation is known, indicate where impurity xx is detected downstream (*e.g.* reaction completion IPC, wet-cake IPC, release testing) and if it is or is not controlled adequately by the acceptance criteria (or	3	15	Low

Table 7.8 (*Continued*)

Parameter name	Classification	Suggestions for FMEA team	Commercial set point	NOR (range target ± X)	UOM	Reference data
Reagent charge	**Non-Critical**	Example of reaction with excess reagent.	Typical: 1.10 equiv. (0.xx kg kg^{-1}) Check sfs based on NOR and process impact of variability (2) sfs is probably acceptable (1) sfs may be O.K., depending on	1.08 –1.12 equiv.	equiv. or kg kg^{-1}	CPD studies; PC and DoE

ow limit tested uring CPD	High limit tested during CPD	AR	Failure mode	Potential effects of the excursion on final DS	(S)	(O)	Current controls	(D)	RPN	Risk level
				impact as "unknown" (*e.g.* a knowledge gap) and S = 9 until data can be generated			procedural controls). Indicate if, in downstream step xx, a reprocess (*e.g.* recrystallization) is triggered if intermediate purity is <xx.x%. Also indicate if the resulting step xx impurity xx is completely rejected in downstream step xx D = 1 if multiple or redundant checks are in place (*e.g.* procedural controls, IPC controls and release testing) or if the current controls are considered strong. FIO tests which are not actionable do not reduce the detection score. Extent of method validation can impact the detection score: Qualified but not validated methods may be D = 3 or 5. Procedural control alone by one operator with MBR documentation may be D = 3 or 5			
9	1.3		Amount is: (1) low, (2) high	1. Potential for incomplete reaction with higher levels of residual SM. Typical language: Minimal impact within 2 × NOR at low-end which would narrowly pass reaction completion IPC.	9	1	Include procedural controls: reagent charge is controlled to ± 5% and documented in the MBR. Charge is verified by two operators Include any necessary	3	27	Low

Table 7.8 (*Continued*)

Parameter name	Classification	Suggestions for FMEA team	Commercial set point	NOR (range target ± X)	UOM	Reference data
			excess and downstream clearance			
Reaction temperature	**Non-Critical**	Robust data set typically necessary. Kinetic studies are useful. Temp is more likely to be a CPP than most other parameters.	Set point should be well balanced within AR	Reflects process capability for parameter control. Around ambient ±1–2 °C is typical.	°C	CPD, PC and/or DoE studies

ow limit tested during CPD	High limit tested during CPD	AR	Failure mode	Potential effects of the excursion on final DS	(S)	(O)	Current controls	(D)	RPN	Risk level
				Indicate if edge-of-failure is known. Indicate if side reactions or different by-products are generated as a result of an undercharge. Discuss fate of residual SM. Based on complete balanced chemical equation, indicate fate of reagent by-products inherent to the process. Also discuss if there are specific impurities in the reagent which require tracking (*e.g.* PGTIs) 2. Indicate if side reactions or by-products are generated as a result of an overcharge. Potential to form elevated level of impurity xx. If the reagent is used in excess, discuss removal of excess reagent and by-products in downstream process. Some minimal data set is typically required to justify absence of testing in final DS. Note if reagent or by-products are PGTIs or other mutagenic concerns			discussion on fate of reagents and by-products based on balanced chemical equation and expected kinetics. Typical language: Residual SM is detected in reaction completion IPC, dry product analysis (FIO) and downstream processing. Impurity xx is detected at reaction completion IPC and API release to ≤xx.x%. The resulting step xx impurity xx is completely rejected in step xx			
			Temperature is: (1) low, (2) high	1. Reduced reaction rate and potential for initially failing IPC result for reaction completion which would trigger extended	5	1	Temperature is controlled to ±x °C based on equipment IQ/OQ. Temperature is documented in MBR. Reaction	1	5	Low

Table 7.8 (*Continued*)

Parameter name	Classification	Suggestions for FMEA team	Commercial set point	NOR (range target ± X)	UOM	Reference data
Reaction time	**Non-Critical**				min	
Reaction completion IPC – typically residual SM (limiting reagent) by HPLC analysis is measured	**Non-Critical**	List test Method xx.	Limit *e.g.* ≤0.5	Not especially relevant for IPCs. Can list typical value here.	%	Method developmen report, CPD and PC studies, DoE data

risks associated with API SMs. The risk assessments can include the following:

- *Business risk/supplier risk:* Consider capacity, geography, natural disaster recovery. Single- or multiple-sourcing opportunities. Supplier quality systems, service and support capabilities.
- *Material risk:* Safety, toxicity, immunogenicity, carcinogenicity, physical stability, handling, cleaning.

ow limit tested uring CPD	High limit tested during CPD	AR	Failure mode	Potential effects of the excursion on final DS	(S)	(O)	Current controls	(D)	RPN	Risk level
				reaction time. No known impact on DS CQAs 2. Increased reaction rate and potential for elevated impurity levels – mapping and DOE indicated increase in yield and no effect on purity			completion IPC for reaction completion. Total reaction time is documented in the MBR			
			Step time is: (1) short, (2) long	None known	1	1	Step time is documented in the MBR	1	1	Low
:0.05	Include routine data and spiking in parentheses <0.10 (*e.g.* 2.2% spike)		For example: >0.5% residual SM	Describe decision if case of failed IPC result. Typical approach is: Batch is held at xx ± 5 °C and resampled one more time at xx h or multiple samples can be taken up to xx h. Decision required if IPC does not pass. (likely NC and investigation). Include full downstream impact including process rejection capability or if downstream IPC or spec. failure would occur. Typical statement: Residual SM is approximately xx% rejected during crystallization	9	3	IPC control point for reaction completion. Validated method. Acceptance criteria has been proven. Residual SM can be purged during isolation. API release testing can detect elevated impurity levels	1	27	Low

- *Process impact risk:* Assessment of quality attributes: purity, assay, physical properties.

Integrated control strategy development is most concerned with the process impact risk assessment. It is recommended for commercial processes that multiple sources of API SMs are evaluated through the entire process. This provides a comprehensive evaluation of known and potential impurities in the raw materials and bolsters the control strategy. It increases the knowledge around raw material quality variability and allows a more accurate estimation of the process impact.

7.5 Process Design, Parameters and Performance Indicators

Process design refers to the inclusive development package used to define how to manufacture the finished pharmaceutical drug product according to the TPP. The focus in this chapter is on the active DS and delivering a manufacturing process that ensures that each DS CQA is met.

7.5.1 Important Concepts and Tools for Process Design

The relationship between control space, design space and knowledge space for process parameters is illustrated in Figure 7.4. The description provided to regulatory agencies in MA 3.2.S.2.2 is binding and should contain the information necessary and important for the manufacture and control of the drug substance. Each step of the synthesis should be described in detail. This includes amounts of reactants, solvents and reagents, reaction conditions, yields, in-process controls and circumstances for repeating certain steps of the process. Recommendations for process parameters and performance indicators are given in the Route and Unit Operation Sections 7.7.6–7.7.11. It is useful to align all parameters' proper name, unit of measure and significant figures to the manufacturing description found in MA 3.2.S.2.2 early in the development history and maintain consistency throughout data collection and documentation.

Module 3 Section S.2.2 may be described in terms of acceptable ranges (ARs) if the applicant has shown sufficient process understanding to demonstrate the impact of this operational flexibility on DS CQAs and process performance, wherein implementation of changes to the control space is allowed without prior approval.[60]

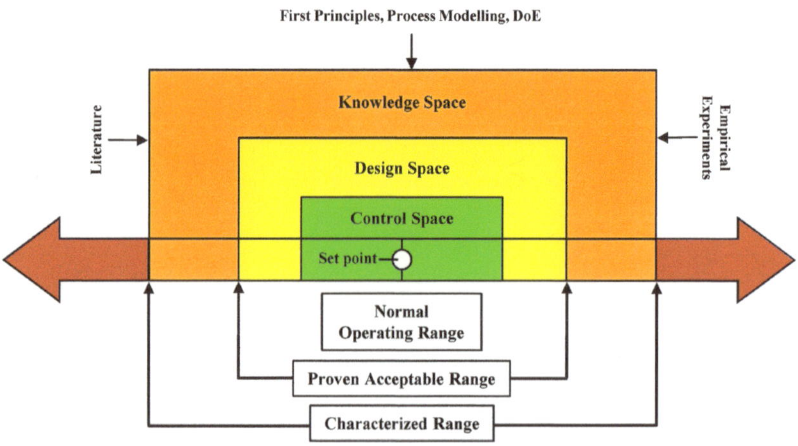

Figure 7.4 Relationship between control space, design space and knowledge space for process parameters.

The EMA has provided a Q&A Guidance: "Working within the approved proven acceptable range is not considered as a change to the marketing authorization dossier. Changes to the target value within the registered PAR can be managed under the company's Pharmaceutical Quality System without regulatory action. Consequently, there is no specific need to include a target set point within the registered proven acceptable range (PAR), but if included no variation will be required when changed."

7.5.2 Process Parameters and Performance Indicators

Process parameters are physical terms that can be directly adjusted and controlled (*e.g.* process inputs), typically seen described in the process description and manufacturing batch procedure. Process parameters can be either categorical or continuous variables. Process parameter examples include thermodynamic/kinetic terms such as concentration, temperature and time, mass transfer terms used to describe mixing and addition rates and amounts of components based on measured weight, volume, *etc.*

A diagram that represents process parameter distribution for a continuous variable is shown in Figure 7.5. The set point, normal operating range (NOR), acceptable range (AR) and characterized range (CR) are labeled. The NOR is the observed all-cause variability in a parameter. The NOR for a parameter will be affected by both the equipment and process operation and is scale and site specific. Parameters describing mass and heat transfer will be affected by scale and equipment power.

Procedural controls refer to the collective set of operational instructions used to define the manufacturing process unambiguously. The level of detail

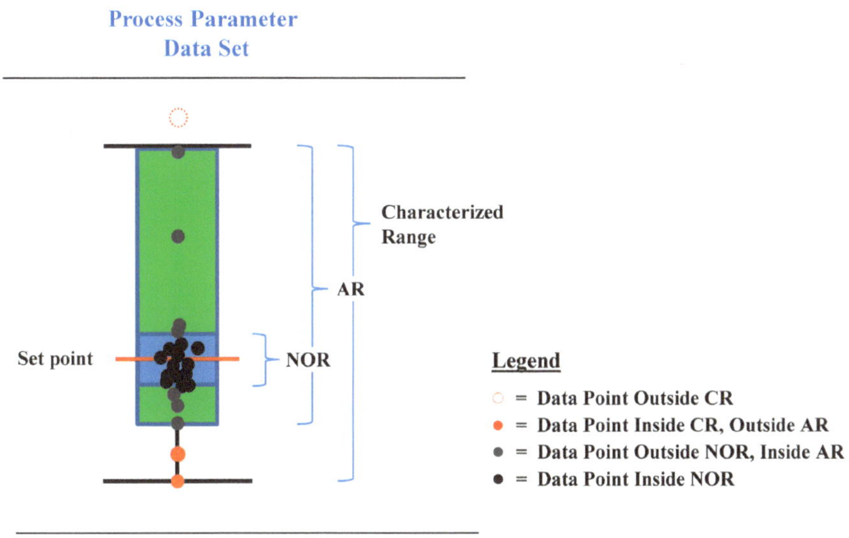

Figure 7.5 Process parameter data set visualization showing important qualitative relationships.

in the process parameter description should enable the process to be understood and reproduced by a suitably trained scientific audience.

Performance indicators (PIs) are measures of process performance that can be measured but only indirectly controlled. Typical PIs are yield, purity and step time and are measured outcomes of the previous manufacturing actions and inputs. Although PIs such as step yield and throughput are business metrics and do not directly measure DS CQAs, it is necessary to consider their impact on the overall manufacturability. For instance, an extremely long filtration step could render the process inoperable without any impact to CQAs.

Although PIs are predominately passive control elements, a subset of PIs can be used to control the process actively and the term in-process controls (IPCs) is used. Different actions can be taken depending on the IPC result. IPCs allow for adjustments to the process to ensure that quality or performance targets are met (typically that some limit is not exceeded) during manufacture. Active adjustments to the process *via* in-line or off-line measurements are currently necessary for all peptide API manufacturing and are considered state-of-the-art for many important unit operations. For example, reaction steps often employ an in-line or off-line assay based on concentration of the limiting component to determine the reaction endpoint (suitability for forward processing); the drying endpoint is determined by similar means and analytical methods. Notably for peptide manufacturing, pooling decisions during chromatographic steps are often defined as critical IPCs since the analysis and actions taken directly influence DS CQAs. On the other hand, extraction and filtration steps are often performed without the need for IPCs. Critical IPCs are measures of product quality or process performance, which must be controlled within a certain range.

A process that relies purely on manufacturing (procedural) controls to ensure product quality may be achievable in the future or applicable for situations where high failure rates are acceptable (*e.g.* a low-cost/industrial-type process), but the current cost–benefit situation for APIs argues for extensive analytical characterization and testing during development, followed by a gradual reduction in the necessary analytical control elements as product and process understanding increases and the integrated control strategy becomes concrete.

7.5.3 Design of Experiments (DoE)

First-principle models should be used whenever they reasonably predict process performance and product quality. This is usually true for thermodynamics and reaction kinetics, where first-principle models are fairly accurate. When multiple variables are interacting and simple models do not provide adequate predictive capability, the use of statistically derived mathematical models may prove useful.

Design of experiments (DoE) can be used to link input variability to process performance and product quality measures. The DoE design purposefully varies input parameters in a multidimensional interaction space and measures the impact on product and process. Statistical analysis of the data allows for quantitative evaluation of the correlation. In addition to primary interactions, secondary and tertiary interactions can also be identified using DoE more readily than by univariate experimentation, which might overlook the impact of secondary interactions. The output of DoE experimentation is often plotted as a multidimensional response surface. The minima of the response surface often indicate the process optimums. The purpose of surface analysis is to indicate the direction where process parameter set points should be adjusted to provide improved performance.

DoE can be used for screening experiments, response surface modeling, regression modeling and comparative experiments. Screening experiments are used to identify which input parameters, selected from a broad list, affect process performance and product quality. In screening, the number of levels should be lower and apply fractional factorial, half-factorial and Plackett–Burman designs. The parameters that show interaction can be further explored in follow-up experiments, while the parameters with no correlation will be screened out.

Practical ranges should be applied for input factors. High and low limits that are not achievable should not be included (*e.g.* equipment capacity, solvent boiling point). In some cases, the initial DoE experiment will not give the process minimum (optimal conditions) but will show that further experimentation is required. In this case, the use of an initial "screening DoE" to identify which parameters have some interaction is useful. In the screening DoE, the most likely CPPs are varied over a wide range. The use of logarithmic scaling is recommended (*e.g.* 0.1, 1.0, 10%), as opposed to linear scaling (*e.g.* 1.0, 2.0, 3.0%), as this is most likely to identify the important interactions. The follow-up "characterization DoE" can refine the list of parameters and study the impact over a more reasonable range. Selection of the number of factors and levels for the DoE will depend on the type of optimization sought.

DoEs are typically executed in the context of individual unit operations. The inputs to the DoE (variables) need to be well defined. The outcomes (process and analytical measurements) need to have sufficient resolution and relevance to product quality and manufacturability.

For optimization, full factorial, central composite, Box–Behnken design for response surface modeling, simplex mixture and D-optimal design approaches are available.

7.6 Planned Studies During Process Design and Process Characterization

7.6.1 Define Synthetic Route

Although there is no prescriptive approach to developing a manufacturing process for peptide APIs, retrosynthetic analysis will usually identify a few primary options for investigation.[15] The paper exercise should begin by identifying any unusual constructs such as conjugation of the peptide, unnatural residues or other fully synthetic portions of the drug substance. A convergent process that divides the DS into equally complex portions will provide the best economy for manufacture. Identifying the likely chemical bond disconnects and the resulting late-stage intermediates is an important input criterion when building a rationale for pursuing one route over another. Isolated intermediates in the process should be stable and well characterized. Depending on the functional group sensitivity, the order of assembly of the late-stage fragments will be impacted.

Among the arguably infinite ways to construct a chemical target, a scientifically justified approach will balance consistent product quality with manufacturability aspects. The only limits are creativity and the resource and time commitments required to explore alternative synthetic routes. It is for those reasons that laboratory evaluation typically compares one, two or at most three routes during early development. That being said, the manufacture of therapeutic peptides is typically performed using the sequence outlined in Figure 7.6.

The advantages and disadvantages of LPPS *versus* SPPS approaches should be evaluated in the context of the innovator's resources and future program objectives. For initial clinical development, the advantage of a platform SPPS approach is clear.[61] In the rare cases where metric ton-scale manufacture is required, developing an optimized LPPS process is recommended. There is considerably more flexibility in the development of LPPS chemistry as opposed to the constraints imposed by SPPS. However, since there is no one fit-for-purpose system for LPPS, each molecule will require a unique impurity control approach to maximize the benefits.[15]

Peptides with sequences of >60 AAs typically require some form of fragment condensation.[58–62] Convergency is achieved by selecting appropriate disconnects in the main chain that can be selectively coupled. Consider selecting an amino acid at the C-terminus that is not prone to epimerization. Additionally, individual fragments should have good solubility.

Native chemical ligation (NCL)[62] and chemo-enzymatic synthesis[63,64] are options that have frequently been applied with success.[65,66] A sequence-related assessment is required for these approaches. There are many options for streamlining and improving the convergency in the manufacturing process to assemble peptides. Using dipeptides in SPPS is an option, especially for sequences with Gly–Gly and where pseudoprolines can be employed. Resins with variable linker motifs including preloaded initial (C-terminal) amino acids are available and can offer a major advantage for long sequences. Unnatural AAs and other synthetic fragments that must be incorporated within the structure (for instance, to facilitate later NCL) will require special consideration. Many advanced intermediates may be suitable for API SM justification given that the control strategy is clear and robust for those individual fragments. Impurities formed in earlier steps of the process to make the amino acid derivatives (AADs) may be suitably removed by

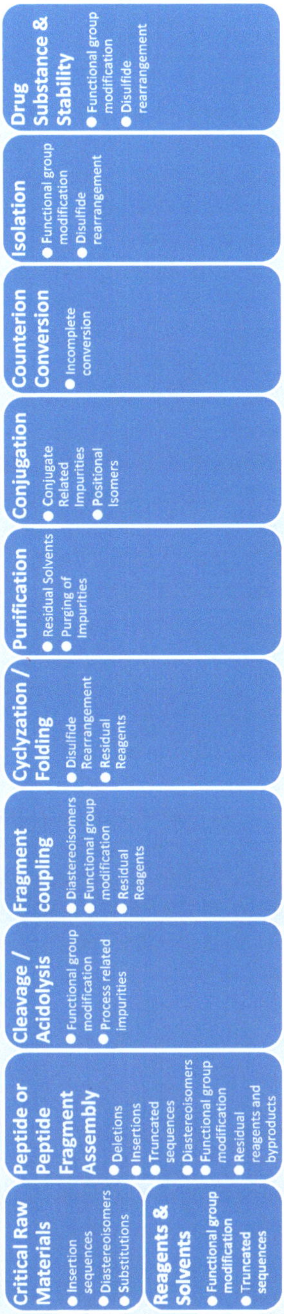

Figure 7.6 Synthetic peptides usually follow the manufacturing flow shown. Typical impurities are listed per step.

downstream process steps and capability demonstrated by appropriate testing, in which cases stringent purity control may be less important.

7.6.2 Selection, Characterization and Justification of Starting Materials: ICH Q7 and Q11

Selection of API SMs is a decision heavily influenced by the synthetic route and ability of the process to control the quality and consistency of the outputs. SM manufacturers should provide detailed information on the route of manufacture, controls in place, testing strategy and history. This will support the scientific justification of API SM selections. Assessment of multiple commercially available sources for SMs is recommended, since different routes of manufacture may be employed and API SM specifications should cover all routes.

The innovator should have sufficient visibility into the API SM synthetic route to demonstrate that impurities in the SMs are suitably identified and controlled. Impurities that may persist in downstream manufacture and affect the DS quality is led prominently by related substances (both structurally related and functional group similarities), metals and potential genotoxic impurities (pGTIs). Structurally related substances are problematic because they often have similar solubility to the desired API and therefore persist in downstream operations that are specifically intended to remove impurities. Functional group similarity is most commonly observed to impact downstream processing because the impurity reacts in the same or a similar manner to the desired API. In addition to depleting the available reagents and thereby diminishing productivity, for structurally similar compounds, the products may persist for multiple steps and are more likely impact DS quality. If impurities are formed in the final step of SM manufacture and that impurity can impact DS CQAs, the step likely will need to be included.

Analytical methodology that assesses a wide arrange of possible impurities can be applied; see Section 7.7.2 on SM specifications. For impurities that can come from SMs and persist in the DS, it may be appropriate to control impurities in the SM. Detection of impurities is more powerful in the early steps since it comprises analysis of less complex mixtures, hence more specific methods are available. It allows a more rational process design with the focus on preventing impurities (at the source) rather than removing them later. It also improves the justification of controls since it can be demonstrated (with data) that certain concerns are not relevant.

A specific development work package addressing SMs is recommended. The topics to be covered are (1) analytical characterization and (2) suitability for the desired sequence.

For analytical characterization, the focus is on the identification and quantification of related substances since these are likely to have the highest criticality. A method that can accurately and specifically identify the known and likely impurities in the SM will allow for comparison between sources and aide identification of DS impurities where a mechanistic evaluation is necessary. Chiral and achiral HPLC and thin-layer chromatographic (TLC) methods are frequently used. Assay and/or methods to determine the content of inorganics and volatiles should be used during development and may be used for commercial control. AADs should control for carboxylic acid impurities since these will directly activate and react with the peptide chain and the reaction is often kinetically faster than for the protected AADs. A specification for free acetic acid is frequently set in the SMs at ppm levels. Note that based on the relative difference in molecular weight, even ppm levels of a carboxylic acid contaminant can be super-stoichiometric in the reaction. Example conversions that convert impurity wt% to mol% can be found in Figure 7.7. The calculation assumes that all of the impurity present in the AAD reacts and therefore gives the maximum impact to DS impurity. Calculation of the impact of trace impurities introduced from reagents, solvents and raw materials can also be performed using the equations shown.

7.6.3 Fate and Purge of Impurities, By-products and Reagents

The limits for impurities in the final DS should be as low as reasonably possible, all things being equal. However, the cost–benefit curve for maximizing the purity of the DS argues that as the API purity approaches 100.00% the cost would similarly approach infinity. A suitable balance between cost and quality is therefore required.

Control of impurities will naturally improve as process and product understanding increases during development.

7.6.3.1 Methods

Demonstration of impurity clearance is typically done by spiking, by generation of stressed reaction streams or by intentionally skipping certain manufacturing steps. Demonstration of clearance of related substance impurities and process reagents and by-products to acceptable levels is required for a holistic control strategy. Acceptable levels depend on the

Equivalents method for Reagents	$impurity\ molar\ max = (reagent\ equiv.)\dfrac{(MW\ reagent)}{(MW\ impurity)}(Spec\ limit)$
Wt ratio method for solids / solvents	$impurity\ molar\ max = (reagent\ wt\ ratio)\dfrac{(MW\ SM)}{(MW\ impurity)}(Spec\ limit)$
Volumes method for solvents	$impurity\ molar\ max = \dfrac{(solvent\ volumes)}{(solvent\ density)}\dfrac{(MW\ SM)}{(MW\ impurity)}(Spec\ limit)$

Figure 7.7 Calculation of maximum process impact based on impurity level in reagents, SMs and solvents.

impurity class and impact on product safety and efficacy. Analytical methods must be capable of detecting to the required levels.[20]

A process flow diagram is necessary to visualize the introduction and removal of impurities. Complete, balanced chemical equations are also required. A mass balance of all process streams (both product and waste streams) should be performed as part of the characterization but can also be performed routinely. If an HLRA was performed in the manner described in Section 7.4.3, the DS CQAs and impacting process steps have already been arranged. A table that shows observed impurity levels at each step of the process will identify the critical control points. One of the goals of process design is to organize the unit operations in a manner that orthogonally removes various classes of impurities through the process.[20]

Related substances are the most common impurities and arise from numerous sources.[24] For related substances, it is useful to build an impurity rejection table as illustrated in Figure 7.8. Impurities that are not well rejected are of the most concern and will require the most attention to prevent and control their presence.

7.6.4 Classification of Process Parameters

"A Critical Process Parameter (CPP) is one whose variability has an impact on a Critical Quality Attribute (CQA) and should be monitored or controlled within a defined range to ensure the process produces the desired quality." – ICH Q8 (R2).

Ideally, a well-designed manufacturing process and integrated control strategy would have no CPPs; however, this is an unreasonable expectation for therapeutic peptide APIs, owing to both their complexity and the time constraints placed on development. Risk-based approaches can leverage scientific understanding and process data as a framework to identify potential CPPs even early in development so that sufficient controls can be developed.

Classification of process parameters as critical or noncritical occurs during process characterization (Phase 2/Phase 3), where process parameters are intentionally and purposefully varied to determine the impacts on DS CQAs and process performance. Finding the edge-of-failure for each parameter is often not necessary to demonstrate control of the process. However, understanding the impact of NOR variability is a reasonable expectation from the control strategy. If the impact to a DS CQA is significant across the characterized range of a process parameter, a more detailed analysis is warranted, potentially involving additional experimentation. Selection of a reasonable range to characterize can be based on the perceived ability to control the parameter and the observed normal variability.[9,67] NOR may be well understood or merely estimated, therefore typically a more conservative 2–3×NOR assessment is adopted. Figure 7.9 provides examples of qualitatively evaluating a process parameter's criticality. Note that the last two examples represent gaps in process knowledge where the classification has not yet been determined.

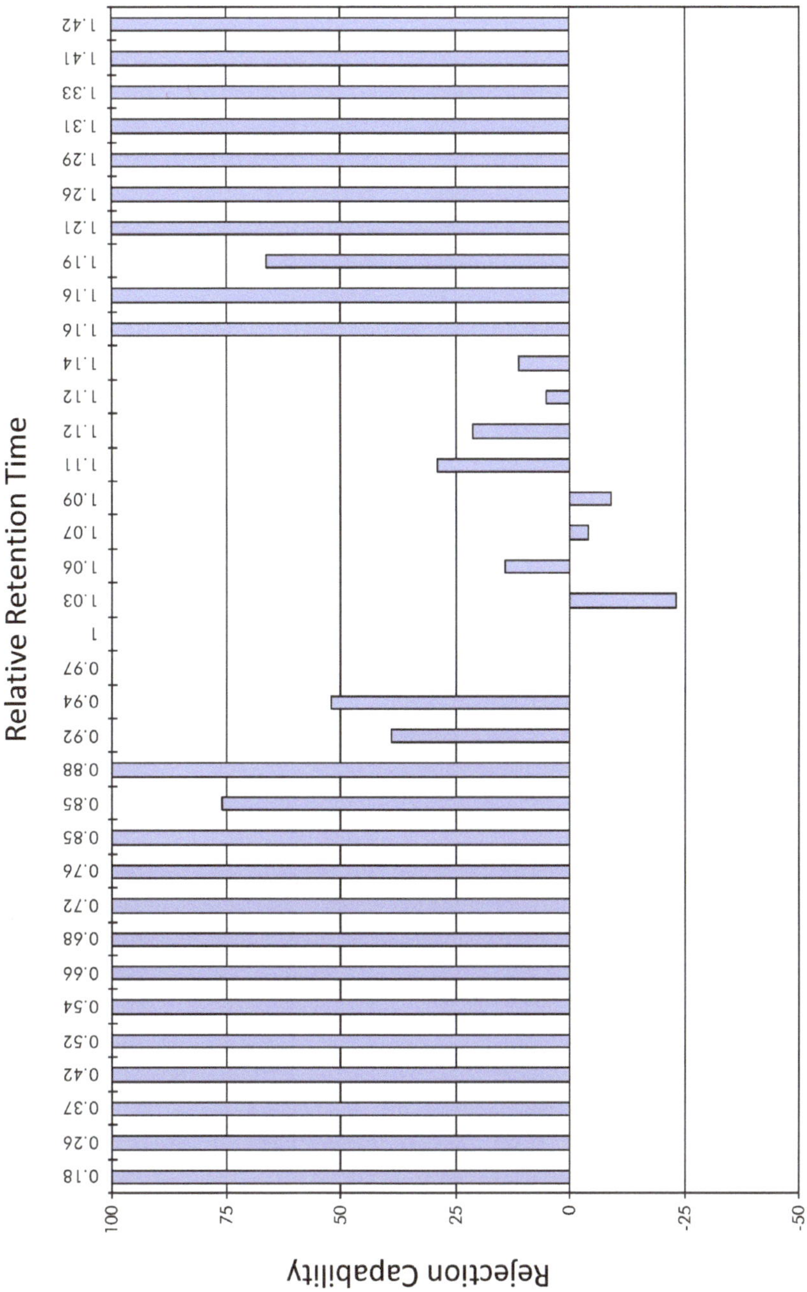

Figure 7.8 Example peptide API showing HPLC purification rejection capability is highly dependent on impurity RRT.

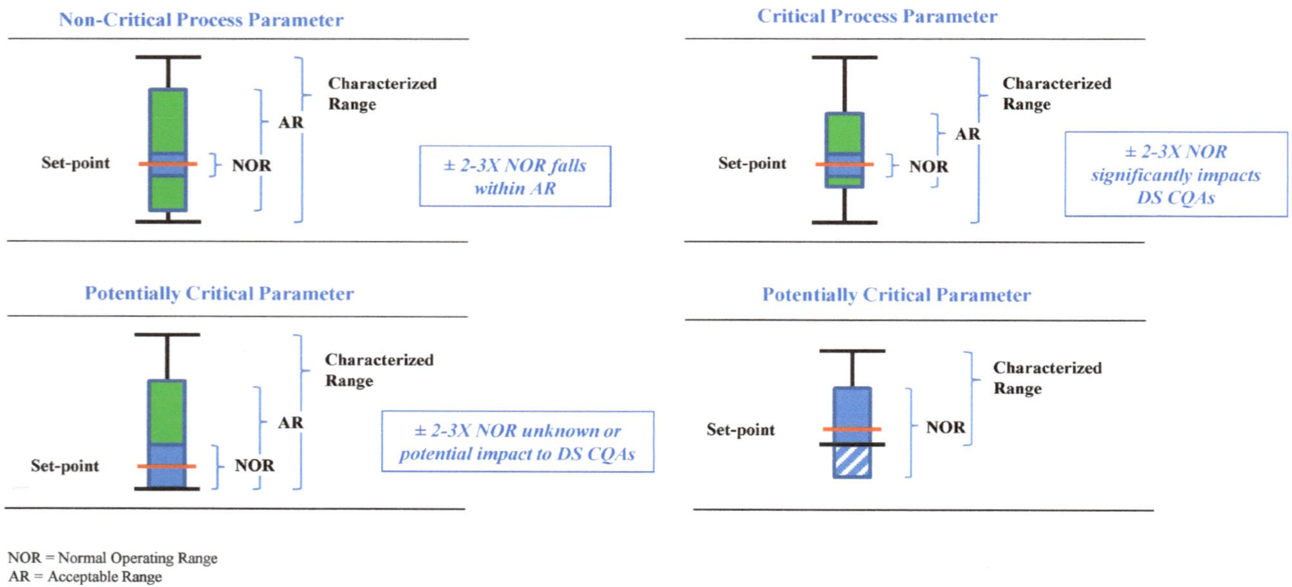

Figure 7.9 Use of routine process capability for parameter classification.

To justify the approach for parameter classification to regulatory agencies, the control strategy should be applied consistently. One method to ensure a structured approach to classification introduces the concept of an impact ratio. Data across the range (no less than 2–3×NOR from the process set point) is used to calculate the impact ratio. If the impact ratio exceeds a predefined threshold (20% is often used), the parameter is classified as critical. Downstream process capability is assumed when evaluating the limit.

The impact ratio is computed as the difference between the CQA limit and the CQA value at the process set point divided by the difference between the CQA limit and the CQA value at the parameter limit (typically 2–3×NOR is used), as shown in the equation

$$\text{impact ratio} = \frac{\text{CQA value at process parameter limit} - \text{CQA value at process parameter set point}}{\text{CQA limit} - \text{CQA value at the process parameter}} \quad (7.2)$$

This approach has also been referred to as the Z-test for impact practicality from a statistical approach. This provides flexibility by allowing a wider operating space and larger parameter variability, providing that the impact on DS CQAs is far away from the CQA limit. For use in the calculation, the CQA limit can be derived from fate-and-purge studies, specification acceptance criteria or IPC limits, historical manufacturing data, DS or DP specifications or prior knowledge.

A flat response surface over the characterized range results in NPPs (noncritical process parameters). Changes to the CQA limit will have a direct impact on the CPP classification. In cases where the CQA limit is lowered, a reassessment of the classification for all parameters is necessary to determine if any new CPPs arise. Naturally, if a CQA limit is raised, some parameters may no longer be classified as critical (Figure 7.10). An example that demonstrates process parameter classification using an impact ratio is provided in Section 7.8.3.

7.6.5 Stress Testing

Stress testing[68] of the drug substance is performed to help identify likely degradation products that can be used to establish degradation pathways and to validate the stability-indicating power of the analytical methods used. Stress testing is carried out on a single batch of drug substance and should include the impact of the following:

- temperature in increments of 10 °C above the accelerated testing temperature (planned for stability testing);
- humidity where appropriate (*e.g.* 75% RH or greater, at or above the planned stability accelerated temperature);
- oxidation;

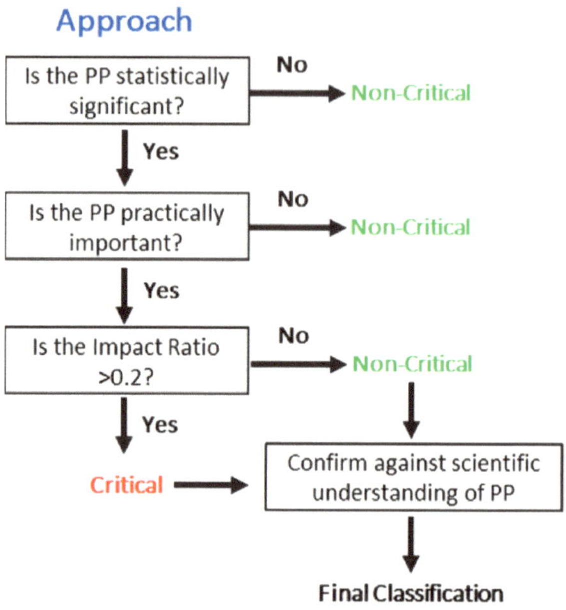

Figure 7.10 Flow chart for classification of process parameters as critical or noncritical.

- photolysis (as a solid powder and/or in solution);
- hydrolysis across a wide range of pH values when in solution;
- conditions expected to be encountered during manufacturing, but with increased temperature or extended time;
- reducing conditions.

Studies supporting method validation (demonstrating method specificity) should include mass balance and peak purity studies to investigate any peaks potentially coeluting with the main peak. The samples are analyzed with the analytical method(s) used for purity determination of the drug substance. Where possible, these studies are preferably performed with MS detection in addition to UV analysis for degradant identification. If the primary method(s) used for purity determination are not compatible with MS then a secondary MS-compatible method can be used for identification of the degradants.

The conditions are chosen to generate between 5 and 20% total degradation to ensure that only primary degradation occurs and that secondary degradation is limited.

Early-phase (preclinical and Phase 1/2) studies are limited to the evaluation of the most common stress factors, at the minimum the temperature effect on the isolated peptide and a limited hydrolysis study that can simulate a holding time effect during manufacture. Suitable testing should include conditions expected to be encountered during manufacture such as the peptide in the solution used prior to the final lyophilization of the DS or on the final pools obtained during purification, involving increased temperature or extended time. Additional conditions may be added depending on the known properties of the peptide being studied. For example, peptides containing easily oxidizable residues, such as Met and Cys, should include an oxidation-forced degradation condition.

Phase 3 studies should encompass the complete set of degradation conditions outlined in Table 7.9. Reduction is the only condition exempt from the 5–20% degradation level requirement as it may be challenging to achieve. It is also the only condition not specifically required by ICH guidelines, although it is informative to perform the test, especially if reducible bonds (primarily disulfide bonds) exist in the peptide. Major degradants should be identified at this stage; for technical reasons, it is useful to set a threshold of 0.3% (as determined by LC-UV) for identification by mass. At 0.5% and higher levels, the expectation is to determine the structure of the degradant.

7.6.6 Stability Studies

The purpose of stability studies is to provide evidence of how the quality of a drug substance changes over a period of time under the influence of factors such as temperature and humidity. These studies form the basis for assigning retest dates and storage conditions. The requirements of formal stability studies are established in ICH guidelines and the studies are conducted following a written protocol. For most peptides isolated as lyophilized powders, the preferred long-term stability storage condition is -20 ± 5 °C.

Accelerated conditions should be included in the study, typically 25 ± 2 °C, 60 ± 5% RH. Although this exceeds the requirements for accelerated conditions, it is recommended to represent excursions from the recommended storage conditions, such as removal of the product for sampling at room temperature and shipment excursions.

In addition, it is of value to study the drug substance stability under the condition of 5 ± 3 °C, which is considered accelerated with respect to -20 ± 5 °C, in order to evaluate and justify lesser excursions from the recommended storage conditions. Furthermore, the data from such a study

Table 7.9 Recommended stress test conditions.

No.	Stress factor	Condition
1	Temperature	40–75 °C up to 7 days
2	Simulated manufacturing	Depending on process
3	Humidity	Open dish 75% RH/25 °C or higher up to 7 days
4	Oxidation	0.1–0.3% H_2O_2 up to 2 days
5	Reduction	Solution of reducing agent TCEP, DTT or DTE
6	Acid	Low pH 1–2 up to 7 days
7	Base	High pH 8–10 up to 7 days
8	Light	1.2 and 3.6 million lux hours (per ICH Q1B) as powder and/or solution

(if appropriately extended) can provide support for long-term storage under 5 ± 3 °C. Long-term stability testing is conducted at time intervals of 0, 3, 6, 9, 12, 18 and 24 months and annually thereafter. Accelerated stability is conducted at 0, 1, 3 and 6 months. It is recommended to prealiquot samples for the stability testing at the start of the study in order to minimize environmental exposure of bulk material during stability pulls.

The minimum testing required is appearance, purity by HPLC/UPLC and periodic verification of moisture content (for powders). Additional testing may involve assay, if a suitable method is available. Further testing is driven by requirements of the DP.

7.7 Peptide-specific Route and Unit Operation Discussion

7.7.1 Raw Materials

Raw materials encompass the starting materials, resins, reagents and solvents used in the manufacture of an API. Of these, starting materials, which are defined as a raw material, an intermediate or an API that is used in the production of an API and that is incorporated as a significant structural fragment into the structure of the API (definition from ICH Q7), are generally considered the most critical because their quality attributes directly impact the quality attributes of the API. Examples of starting materials are derivatized amino acids, di-, tri- and longer protected peptide fragments and pseudoprolines. If SPPS is used, then the second most critical raw material is the resin used to anchor the peptide chain. Reagents and solvents impact the CQAs of the API to a lesser extent provided that they are free of impurities that may directly react with peptide intermediates during synthesis. In all cases a supplier questionnaire is mandatory in order to understand the presence of potential impurities in materials purchased. This includes, but is not limited to, the content of elemental impurities and mutagenic impurities.

7.7.2 Starting Materials

The amino acid derivatives used as building blocks of the peptide have a major impact on the quality of the crude API, as commonly encountered impurities will typically become incorporated as part of the peptide structure and subsequently lead to impurities in the API. Depending on the peptide, these impurities may be difficult to remove during the purification stage, leading to low overall API purity or significant losses in product yield. It is therefore important to establish tight limits for impurities in the starting materials.

The rational basis for setting impurity limits in starting materials is the specification limit for individual impurities in the API. Generally, the limit for any unspecified impurity in the starting materials should be set below the limit of unspecified impurities in the API, with a safety margin to allow for process variability. Impurities in starting materials known to contribute to impurities in the API should be controlled at levels based on process understanding. For example, the level of D-amino acid residues in a peptide is determined by the level of D-enantiomer in the starting material plus the margin for racemization during activation and coupling less the impurity rejection capability of the purification.

Other impurities in starting materials that may contribute to impurities in the peptide API are residual solvents in starting materials, such as alcohols, esters and carboxylic acids that may participate in coupling reactions.

7.7.3 Resin (SPPS)

The resin is usually considered a critical raw material. The impact of resin selection in SPPS should be understood, as the appearance of certain impurities is specific to the type of resin used. For example, Wang resins often give rise to an impurity characterized by a mass increase of 106 Da in peptides containing Trp residues due to alkylation of Trp by the hydroxymethylphenyl linker.[69] Currently, no generic methods exist that can fully characterize a resin, despite efforts in the area.[70,71] The minimum resin testing requirements are substitution, swelling and bead size.

7.7.4 Starting Materials for Conjugation Constructs

If the peptide is to be conjugated, for instance to a synthetic polymer [*e.g.* poly(ethylene glycol) (PEG)], a polymer obtained from natural sources [*e.g.* poly(sialic acid) (PSA)] or a recombinantly expressed biologic [*e.g.* a monoclonal antibody (mAb)], then each of these materials must be treated as a starting material and adequately characterized. The minimum testing requirements are identity, purity, polydispersity (for any polymers) and assay of available reactive groups that will be utilized for conjugation. In addition, if linkers are used then those should also be characterized with respect to identity, purity and assay. For low molecular weight compounds, the expectation is the full characterization with respect to impurities.

7.7.5 Reagents and Solvents

Minimum requirements for control of reagents and solvents are appearance, identity and purity. Potential impurities that may react during peptide synthesis should be considered. For example, *N*,*N*-dimethylformamide (DMF) may degrade, yielding traces of formic acid and dimethylamine, both of which may lead to impurities in peptide synthesis.

It is useful to reference American Chemical Society (ACS) standards or other industrial standards when developing specifications for reagents and solvents. Pharmacopeial requirements, on the other hand, should not be applied, as they are designed for use in finished dosage forms.

To ease the testing burden of frequently purchased reagents and solvents, it is recommended to implement reduced testing programs per ICH Q7 and the ICH Q7 Q&A of June 2015.

7.7.6 Unit Operations

A list of commonly used unit operations (UOs) for peptide synthesis is given in Table 7.10. The purpose of the UO (example goals) should indicate what impact it has on the DS CQAs and process performance and align directly with the integrated control strategy. The selection and organization of UOs defines the manufacturing process and available control strategies. Considerable creativity can be exercised in the organization and execution of

Table 7.10 Commonly used unit operations in peptide manufacturing.

Unit operation	Abbreviation	Primary goal
Distillation	DST	Concentration, solvent swap, removal of volatile solvents and impurities. Process efficiency
Reaction	RXN	Chemical transformation (covalent bond-forming/-breaking) step. The desired outcome is a list of expectations for both quality and manufacturability (PIs and IPCs)
Crystallization (or precipitation)	XTL	Control of DS or API CQAs including rejection of all known and potential impurities to acceptable levels based on acceptance criteria set in the DS specification
Drying	DRY	Removal of volatile solvents and impurities, Physical properties control, homogenization
Liquid–liquid separation (*e.g.* extraction)	LLS	Removal of reagents, by-products, solvent or inorganics. Process efficiency
Solid liquid – separation (*e.g.* filtration)	SLS	Either capture of material (isolation of solid API from slurry) or removal of insoluble components (catalyst or resin removal)
Chromatography (reversed-phase, ion-exchange, *etc.*)	PUR	Separation of organic components based on differential retention (adsorption and desorption) from selected stationary phase; followed by fraction selection and pooling to achieve net purification of the API stream
Packaging and storage	PKG	Sampling of bulk solid intermediate (API) or final DS for release testing, followed by discharge and secure packaging and movement to storage
Lyophilization	LYO	Removal of water, volatile reagents, by-products and solvent

UOs. New technologies and scientific innovation are continuously creating additional UOs and increasing the capabilities of others. Determining the organization of UOs is performed during Phase 1/Phase 2 (process design) including the start of the GMP sequence. Optimization of the UO for the product is typically performed during Phase 3 (process characterization) and is used to place commercial acceptance criteria appropriately on the control strategy. Certain UOs benefit from development of scale-down models during development.

7.7.7 Impurities and Control in Liquid-phase Peptide Synthesis (LPPS)

The impurity control approach heavily leveraged in LPPS seeks to identify precipitation and washing conditions to remove impurities. This is accomplished by extensive screening of solvents or solvent combinations looking for candidates with high solubility for the starting materials, reagents and by-products while also exhibiting low solubility for the desired product. Crystallization development is beyond the scope of this chapter. It is not uncommon for a crystallization approach to target a final purity of ≥99.5%, with individual impurities controlled to ≤0.1% and all with product losses of ≤5%. The targets for purity and impurities using a precipitation process (uncontrolled crystallization) are lower because thermodynamics are not contributing to the purity upgrade. Losses to the mother liquors can be reduced to <1%. By comparison, although there is technically no product lost during SPPS, there is also no control over related substance impurities, as these are carried forward to the final cleavage step in SPPS.

7.7.8 Impurities and Control in Solid-phase Peptide Synthesis (SPPS)

SPPS is generally used for linear peptides of <30 AAs. The length of the peptide affects synthesis as aggregation becomes a problem at around 30–40 AAs. The yields of the desired product drop precipitously in this range. However, assembly of sequences of 60 AAs is possible. SPPS is considered a reliable platform to assemble linear peptides rapidly with minimal development necessary.[61,72] SPPS can be amenable to large-scale production as equipment and SMs are readily available. SPPS is usually not the throughput constraint. Purification and isolation are usually the bottlenecks to large-scale production since they are often reliant on specialized equipment, *e.g.* direct axial compression (DAC) columns and lyophilizers (Table 7.11). The approach taken by most innovators is to generate clinical material using SPPS, while exploring if alternative routes of manufacture are feasible and superior in terms of throughput and possibly quality. SPPS can permit scaling to a few kilograms of DS, which is usually sufficient for Phase 1 and 2 studies.

Table 7.11 SPPS process parameters and performance indicators.

Potential CPPs and IPCs	Range to be studied	Performance Indicator	Comments
Deprotection			Multiple cycles are performed until IPC meets criteria. The limit should reasonably tight to avoid deletion impurities. Problematic residues may require cleavage and HPLC/NMR analysis to obtain sensitivity. Efficiency decreases as sequence length increases
• Amine based used[a]		UV determination	
• Concentration (equivalents)	5–20% w/w		
• Deprotection time	15 min – 2+ h		
• No. of deprotection cycles	$N = 1, 2, 3$		
• Mass transfer (mixing)	Agitation or mixing		
Activation/preactivation			Solution phase allows for direct assessment of conversion, concentration and impurities
• Solvent(s)[a]		IR	AR conditions for AADs prone to stereochemical erosion should be established. Degradation of activated AADs can be accelerated in the presence of resin
• Coupling reagents[a]	1.1–3 equiv.	HPLC	
• Base used[a]		NMR	
• Stoichiometry (equivalents)			
• Temperature	Cryogenic or room temperature		
• Time	Minutes to hours (for stability)		
Coupling			Two independent IPCs are recommended to monitor conversion. The sensitivity of Kaiser and TNBS tests is between 0.1 and 0.5% and cannot completely prevent deletion sequences. Automatic capping can reduce deletions by replacing with the capped impurity sequence
• Stoichiometry (equivalents or concentration)	1.1 to 3+ equiv.	Kaiser	
• Temperature	0–60 °C	TNBS	
• Time	From hours to days	Chloranil	
• Mixing			
• No. of coupling cycles			
• Recoupling conditions	Recoupling is triggered by IPC result		Reaction kinetics decrease as peptide-on-resin length increases

Capping			Terminal amino group is capped in final step
• Solvent(s)		Kaiser	Automatic capping during synthesis can reduced the level of deletion sequences. The truncated capped sequence is easier to remove in subsequent steps
• Reagent(s)	Acetic anhydride and triethylamine	TNBS	
• Time		Chloranil	
• Temperature			
Washing			Diffusion rates depend on the solid support and sequence and must be studied
• Solvent selection[a]	DMF, ACN, IPA	UV/IR	
• Washing time	Minutes to hours	Resin volume	Incomplete removal of excess AAD or capping reagent can lead to insertion sequences or truncated capped products. Insertion impurities are difficult to remove
• No. of wash cycles	$N = 1, 2, 3$		
	Agitation or mixing		
Drying			
• Temperature	Ambient to elevated temperature	HS-GC	
• Pressure	Reduced pressure	LOD	
• Time	Hours to days	Karl Fischer	
• Mixing	Intermittent		

[a]Categorical variable.

7.7.8.1 General Impurities in SPPS[73]

- For all Fmoc-AADs:
  - Fmoc-β-Ala impurities in AADs.[74] Leads to *endo*-β-Ala impurities.
  - Free amino acid (des-Fmoc): leads to double-hit impurities; *endo*-AA.
  - Enantiomeric impurities: lead to point diastereomers.
  - Acetic acid: leads to an acetylated, capped sequence.
- Specific Fmoc-AADs:
  - Asparagine and glutamine. Dehydration of unprotected asparagine can occur.[75]
  - Cystine. Racemization during activation or coupling is common.
  - Arginine to lysine conversion. Results in des-guanidinium.
- From process
  - Deletion sequences: incomplete coupling, incomplete Fmoc deprotection.
  - Point diastereomers: from racemization, particularly during AAD activation or long coupling cycles.
  - Acetylated, capped sequences can result from carry-over of capping reagents to the next cycle.

7.7.8.2 Selection of Resin and Linker for SPPS

C-Terminal carboxylic acids and amides are possible based on the solid support (resin) and primary linkage to the first AA (linker). Preformed linkers facilitate the correct C-terminal product and deprotection conditions.[76] Resin attributes that affect the process include loading (available reactive functionality per gram of resin), particle size (affecting diffusion of reagents into the solid matrix) and swelling volume (as a function of solvent). These attributes should be considered during characterization for their impact on coupling cycles. Coupling reagents such as hydroxybenzotriazole (HOBt) can affect resin swelling. HOBt and ethyl cyano(hydroxyimino)acetate (Oxyma) are preferentially dissolved on the resin, which affects the concentration of these reagents in solution. The selection of peptide coupling agents has exploded, offering many choices to develop conditions that are optimal for the sequence.[77]

7.7.9 Impurities and Control in Cleavage/Deprotection/Isolation

Following main-chain assembly, cleavage of the peptide from the resin and removal of semipermanent protecting groups is accomplished under acidic conditions. Several standard cocktails and conditions are available.[78]

Since there are simultaneous reactions occurring during cleavage, there are a large number of potential side reactions, by-products and degradants that must be controlled. Balanced chemical equations should be used to ensure that the necessary stoichiometric components are present in the cocktail solution. Standard cocktails provide a composition based on volume ratios, but each peptide will have a different molecular weight and the relative stoichiometry of reagents that each standard cocktail presents is therefore different. Reactive by-products should be quenched or scavenged. Quenching and scavenging reagents should be present at appropriate concentrations to reduce side reactions. Liberated functional groups should ideally be stable to the cleavage conditions (free thiols, amino, hydroxyl and terminal amide functions can be particularly problematic). Orthogonal stepwise deprotection steps may be necessary to prevent premature folding or to control disulfide bond formation.

Mass transfer may be an important consideration for scale-up since cleavage chemistry is occurring on the solid support. Optimization of the cocktail solution often is not performed until the PC phase. Identification of initial conditions should follow reactivity principles and employ a standard cocktail screen. An analytical method that can determine purity and assay (such as HPLC) should be used to track the product concentration and total impurities as a function of time for several cocktail variants. The extent of initial development on the cleavage step is based on both quality and business factors. There should be sufficient data to confirm that impurity formation is controlled to reasonable levels (<5% for any individual impurity) and within downstream purification capability (typical targets are >80% crude purity). Particular attention should be given to close-eluting impurities since these have the largest impact on purification performance. If close-eluting impurities are formed during the cleavage step, it is appropriate to perform process characterization work during early development since the step will have a direct impact on DS CQAs. Process risk assessments should identify this as a high risk. Impurities, by-products and degradants that are well rejected during crude peptide isolation or downstream purification do not have a direct impact on DS CQAs and present a business risk (cost/throughput/yield) and are typically not addressed until PC and scale-up to commercial batch size. The use of an internal reference amino acid allows for the quantitation of cleavage efficiency. Table 7.12 provides examples of potential issues based on the amino acid protecting group in the sequence.

Table 7.12 Examples of potential issues based on the amino acid protecting group in the sequence.

Amino acid	Cleavage conditions considerations
Arginine (R)	2,2,4,6,7-pentamethyldihydrobenzofuran-5-sulfonyl group (Pbf) is preferred since it is cleaved readily and easily scavenged
Histidine (H)	Acetylation of the imidazole ring could occur
Methionine(M)	Susceptible to alkylation and oxidation
Tryptophan (W)	Susceptible to oxidation and alkylation by unscavenged electrophiles
Tyrosine (Y)	Alkylation by electrophiles or back-reaction with the resin, especially for C-terminal Tyr

7.7.10 Impurities and Control of Side-chain Modifications and Fragment Coupling

7.7.10.1 Control of Cystine- and Methionine-containing Peptides

Unwanted oxidation pathways should be understood as they lead to impurities that may be reversible. Some sulfur chemistry is better performed on a solid support prior to cleavage. Many peptides contain one or more well-defined intramolecular disulfide bonds between cysteine residues. Although it may be possible to install disulfide bonds during main-chain assembly, typically this is done following cleavage and deprotection. Orthogonal thiol protecting groups can be employed to control further the disulfide bond-forming step(s). This is particularly useful when more than one disulfide bond is present. Intermolecular disulfide bond-forming steps are less common but necessary for many sequences. This can be accomplished by direct oxidation in the presence of the two components or by preactivation of one sulfide.

Control of disulfide impurities is particularly perilous as scrambling can occur in many situations, if not properly understood. Fortunately, misfolded or otherwise incorrect products can easily be detected by HPLC and other analytical methods, since the properties are typically vastly different chromatographically. Additionally, the redox process is reversible and under equilibrium in many cases, making optimization efforts more likely to yield suitable conditions.

7.7.10.2 Side-chain Modification

Glycosylation, PEGylation and tagging are becoming increasingly common for peptide therapeutic targets. It should be noted that any advanced intermediates or SM intended to be incorporated into the final product must have suitable controls over the identity, purity and assay; see Section 7.7.1, Raw Materials. It is the expectation of regulators that the manufacturing process for the appended side-chain is thoroughly understood and controlled.

Fragment coupling by native chemical ligation (NCL) is a powerful technique for late-stage assembly.[6,14,62,79] The disconnection for fragment coupling in solution should be at a location where α-racemization is not likely. Cysteine and histidine are poor choices; glycine and proline are ideal.

7.7.11 Impurities and Control in Packaging and Storage

Following final isolation (typically lyophilization), packaging and storage of the API and final DS are necessary. A holistic control strategy includes considerations of the primary container, secondary container and container closure system. A study of extractables should be performed using HPLC-MS or GC-MS. During unloading and packaging, the room conditions, temperature, humidity and protection from light should be considered. Stability studies are required for the DS. It is recommended that intermediates have a stability program that will define the lifetime of the intermediate and retest period. This will support manufacturing at multiple sites and shipping between locations, as necessary.

7.8 Integrated Control Strategy

The therapeutic peptide innovator should summarize the DS process understanding and control strategy in a series of controlled documents during development and throughout the product's life cycle. At the time of MA authoring, at the minimum the documentation should consist of the following:

- TPP, QTPP, DS CQAs and DS commercial specifications;
- product quality risk assessment and process risk assessment;
- analytical methods, method development, qualification and validation reports;
- manufacturing procedure for commercial-scale process and laboratory and scale-down equivalents;
- process development and process performance qualification protocols and results;
- UO summary reports or step summary reports from process characterization;
- scale-up and campaign reports;
- comprehensive process design and DS control strategy summary report.

7.8.1 Drug Substance Specifications

The holistic control strategy must ensure that each DS CQA can be met. Quality is demonstrated through a combination of procedural controls, in-process tests and end-product specification tests. Consistency and robustness are the result of a well-understood manufacturing process balanced with a scientifically justified analytical testing plan.

The DS specifications are derived from the DS CQAs. Ranges and limits are based on toxicological qualification, process and analytical capability, batch analysis, process characterization data and stability.

An appropriately designed process and integrated control strategy will demonstrate robustness across the entire design space. Understanding the impact of both process and material input variability on DS CQAs is fundamental to designing and controlling DS manufacturing processes. Understanding the edge-of-failure for the design space can also identify latent quality and manufacturing optimizations. Design space filings are limited and have been approved for only a few steps of peptide manufacture. Individual UO design spaces could be filed. The quality risk management (QRM) and risk assessment process ensures that the appropriate level of control is in place to meet each DS CQA consistently.

7.8.2 Process Capability Assessment (6σ Analysis)

Process capability assessment (6σ analysis) is a method to weigh the manufacturing process's ability to deliver to the CQA limits. Although 6σ process capability is a worthy goal for pharmaceutical manufacturing, given that IPCs can be applied at will anywhere in the process to ensure suitable control, the focus for pharmaceutical development should be on *appropriately weighting the available control elements* to ensure that both an economical and robust commercial process is achieved. A flow chart that aids in minimization of redundant testing of DS QAs based on improving the manufacturing process design is provided in Figure 7.11.

7.8.3 Case Study: Holistic Control Strategy for DS CQA: Des-amido Content

By means of a case study, this section outlines the development of a holistic control strategy for a specific peptide DS impurity using the principles from this chapter. With the focus on process understanding built on sound science and data, a direct, purposeful design of the process and controls was implemented. The integrated control strategy required elements of risk assessment, analytical and process characterization and application of manufacturing and analytical controls.

7.8.3.1 Elements of the Control Strategy

- framework focused on process understanding using sound science and data;
- analytical methods used to identify, track and control the impurity;
- methods used to identify where the impurity is introduced and where it is removed;
- high-level and detailed risk assessments;
- classification of critical process parameters.

This case study covers all major elements of the control strategy for a related substance impurity common in synthetic peptides: the hydrolysis degradant deamidation (des-amido). The control strategy elements are presented by topic, in a retrospective fashion as opposed to following the development timeline. Chronologically, the des-amido impurity was identified early during preclinical development, as hydrolysis was both likely and known for the peptide in this case study (eight amino acid sequence abbreviated as AA_{1-5}–Arg–Ala–Arg–NH_2 to show terminal residues and identify the location of modification).

Note that these principles can be applied for any impurity including those identified late in development since integration of process understanding should be a continuous process.

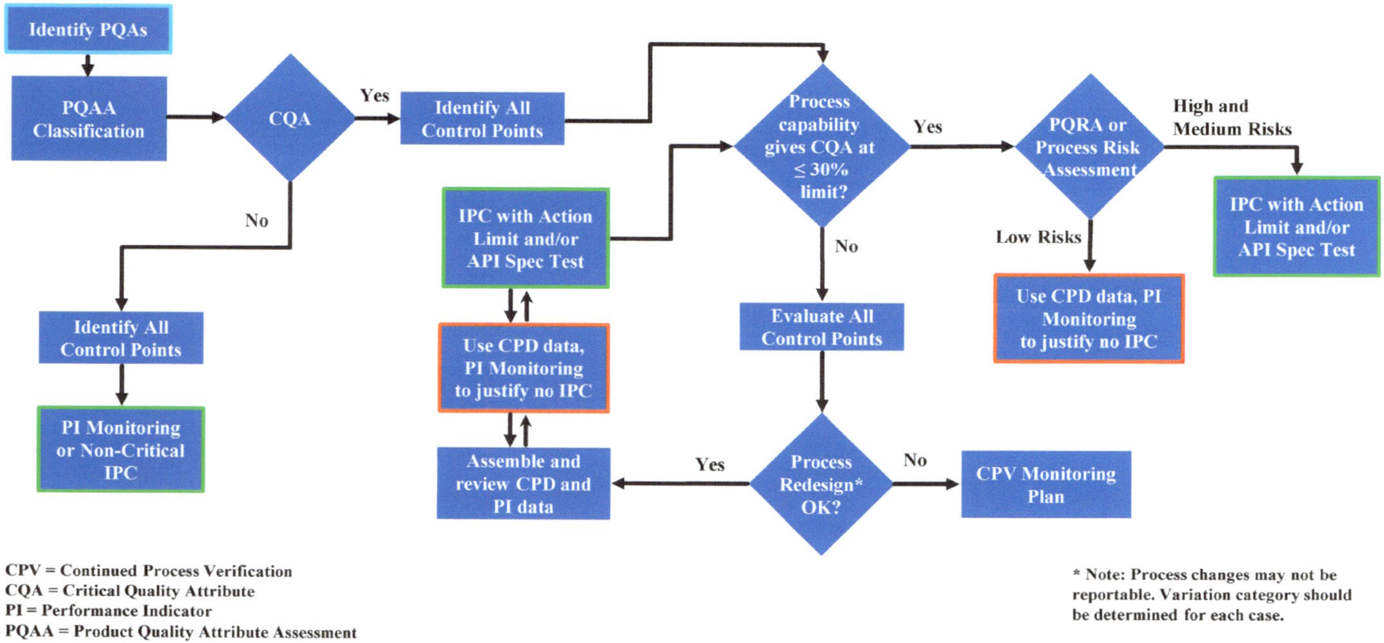

Figure 7.11 Flow chart for shifting redundant IPC testing to manufacturing controls.

7.8.3.2 Background

The peptide DS in this case study contains a small, highly charged main chain and a cystine-bridged side chain. Preclinical design at the TPP stage preferred modifications to the native peptide sequence which lead to the clinical candidate having C-terminal amidation and N-terminal acetylation of the main chain, the rationale being that improvements to the DS stability will translate into improved pharmacokinetic properties. The DS proved fairly stable both as a powder and in solution but, as expected, in the presence of water hydrolysis of the C-terminal amide can and does occur, resulting in the free carboxylic acid impurity (des-amido). Fortunately, the des-amido impurity was fairly well resolved from the main peak (HPLC RRT = 1.09) and easily monitored throughout development.

Risk assessments were used to identify the process steps that impacted the des-amido content. Des-amido formation was possible during DS storage and under process conditions where water is present. In this example, the impurity was not introduced from raw materials, but could be formed at several points in the process. Paper evaluation identified that problematic process conditions were present during cleavage (TFA-mediated release of the linear main chain from the resin following SPPS), purification, salt exchange, lyophilization and packaging and storage. Reduction of the des-amido impurity was only achieved during purification. For each impacting step, an analytical method capable of identifying and quantifying the impurity level was developed. In general during early development, detection may not be as sophisticated and certain impurities including deamidation may only be observed at the DS stage.

In this example, several options were available to control the des-amido impurity. From the process design perspective, the purification employed chromatographic principles, allowing for removal through appropriate fractionation and pooling criteria. Even so, the downstream process included UOs with the potential to increase the des-amido level (including storage). Additional controls were necessary to meet DS and ultimately DP release specifications. From the analytical perspective, good detection of the des-amido impurity allowed for process characterization around problematic steps, with the added safeguard of routine detection at DS release using the HPLC purity method. These elements of process understanding integrated to form the holistic des-amido QC strategy.

7.8.3.3 Analytical Methods Used to Detect and Control the Des-amido Impurity

In general, related substance impurities and degradants (including des-amido) should be detected during DS characterization, stress testing and/or stability studies by the routine HPLC methods. Analytical development including HPLC-MS/MS and orthogonal methods should provide assurance of impurity detection. Once identification has been performed using MS/MS or related techniques, authentic standards of known and likely impurities should be synthesized to facilitate further work. For all specified impurities, gram quantities will be needed for analytical method development and validation. Significant material will also be necessary for fate and purge studies. In this arena, suitable scale-down models are particularly efficient.

In this example, the DS release analytical purity method (HPLC) could detect the des-amido impurity and be used to control its level. The limit was established based on the DP specification limit, considering the potential increase during DS storage. Working back in the process from the DS specification limit, the IPC limit at the key purification stage needed to be lower to account for the potential increase in the subsequent steps leading to DS release. The purification IPC method was therefore developed with des-amido impurity detection in mind.

7.8.3.4 Risk Assessments Applied to Des-amido Impurity Control

In early development, the PQRA (see Section 7.4.2) identified the des-amido impurity as a DS CQA. Detection and control of the impurity were required to ensure patient safety. Therefore, the impurity was considered to be a specified related substance in the toxicology batch, FIH and Phase I clinical studies with limit of ≤2.0% by HPLC at DS release.

An HLRA (see Section 7.4.3) was used to identify likely process control points and determine the risk of meeting the DS specification. As the only known rejection point, the purification was leveraged to limit the impurity to ≤0.5% by the pooling IPC. In that way, a 1.5% increase could be tolerated between the purification and DS release. However, given the lack of traceability and absence of detailed information on the impurity through the salt exchange, lyophilization and packaging and storage steps, the HLRA still identified this as a high risk. The HLRA highlighted areas of suboptimal process design and noted the potential that this CQA may not be robustly controlled by the process. The HLRA assessment identified this as a business risk (*i.e.* that a batch of DS may not meet specifications) and the decision to act to mitigate that risk was optional, given that a routine analytical DS release method was in place. Note the difference between a process-focused and a patient-focused risk assessment. The patient-focused PQRA did not identify a high risk since the impurity is controlled through the DS specification and the method is validated for that purpose.

Later, as the product matured through the clinic and the prospect of a commercially viable product emerged, interest in improving the process robustness and efficiency heightened. Applying the key levers introduced in Section 7.1.3 for controlling peptide-related substance impurities, only *optimization around problematic process conditions* and *tightening of the purification pooling IPC criteria* were useful mitigation options. Indeed, even tightening the purification pooling criteria to ≤0.05% was not sufficient to eliminate the risk fully since the downstream increase was not well understood. This was the correct time to perform a detailed FMEA.

The FMEA (see Section 7.4.4) identified numerous high- and medium-risk process parameters directly related to the control of des-amido formation. Some of the risks are exemplified below:

- *During cleavage:* The impacts of temperature, reaction time and reagent stoichiometry on des-amido formation were not well understood.
- *During chromatography:* Acceptance criteria for the des-amido impurity in the purification pooling IPC may not be set appropriately. This resulted in risks in two ways. In the case that the des-amido content does not increase downstream of the purification, the IPC limit is set too stringently (at the cost of yield) – which was scored as a medium risk. In the case that the des-amido content increased more than 1.5% downstream of the purification (resulting in a DS specification failure), this results in a high risk.
- *During salt exchange:* The impacts of temperature, time and concentration on des-amido formation were not well understood.

Mitigation plans were developed that focused on developing a better understanding of the impact of process parameter variability on des-amido formation throughout the process. This consisted primarily of generating laboratory-scale data in univariate and DoE studies. Representative data from the cleavage UO are given in Table 7.13. Based on improved process understanding, the control strategy was further refined and the FMEA was updated. Ultimately, the integrated control strategy would consist of a combination of procedural controls, critical process parameters (CPPs) and IPCs. Additionally, the des-amido impurity required assessment in the DS as part of release testing and stability studies.

7.8.3.5 Manufacturing Control: Critical Process Parameter Classification

Facilitated by risk assessments, process parameters impacting des-amido formation were identified for several UOs. Specifically, the FMEA assessment of the cleavage step identified potential CPPs in reaction temperature, reaction time and reagent stoichiometry. For the temperature impact, at the low-end a reduced reaction rate was observed and at the high-end there is the potential for elevated des-amido formation.

Table 7.13 Effect of temperature variations on impurity levels.[a]

Experiment	15 °C	20 °C	25 °C	35 °C
Purity of desired product/%	78.49	89.97	89.98	90.55
Des-amido Impurity level/%	0.28	0.22	0.21	0.45
Incomplete cleavage impurity level/%	12.70	0.19	n.d.	n.d.

[a]n.d. = not detected.

To cover a reasonable range of temperatures and to establish an AR, laboratory-scale experiments at 15 °C (minimum), 25 °C (center point) and 35 °C (maximum) were performed. The results were compared with those of a fourth experiment performed at the set-point condition (20 °C); see Table 7.13.

Both a reduced yield and increased level of a reaction intermediate indicated that the minimum point of 15 °C was not sufficient for the desired outcome and 15 °C was considered outside the AR. At the maximum tested temperature of 35 °C, the amount of des-amido impurity was increased compared with the center-point conditions. The maximum allowable level of des-amido impurity was based on downstream rejection capabilities and the final DS specification, with an upper limit of the in-process stream set at $\leq$0.6%. At the process set point of 25 °C with an NOR of $\pm$3 °C, the level of des-amido impurity was 0.21%. However, the level of des-amido impurity was sensitive to variations in temperature and increased to 0.45% at 35 °C. The impact ratio calculation for the effect temperature on des-amido impurity is carried out with the equation

$$\text{impact ratio} = \frac{0.45 - 0.21}{0.6 - 0.21} = \frac{0.24}{0.39} = 0.6 \tag{7.3}$$

In this case, the impact of temperature variability at approximately 3×NOR caused an increase in des-amido content that was greater than 20% of the remaining safety buffer between the process set point and CQA limit (Figure 7.12). Exceeding this threshold meant that temperature was classified as a CPP. This parameter was monitored and appropriately controlled to maintain consistent product quality.

Overall, the characterized range of 15–35 °C was considered the knowledge space. The commercial set point was 25 °C with a NOR of $\pm$3 °C and an AR of 20–35 °C. Note that operating at 3×NOR at the high end (35 °C) provided material of acceptable quality and was considered within the AR although it was not desirable for manufacturability reasons. The influence of temperature

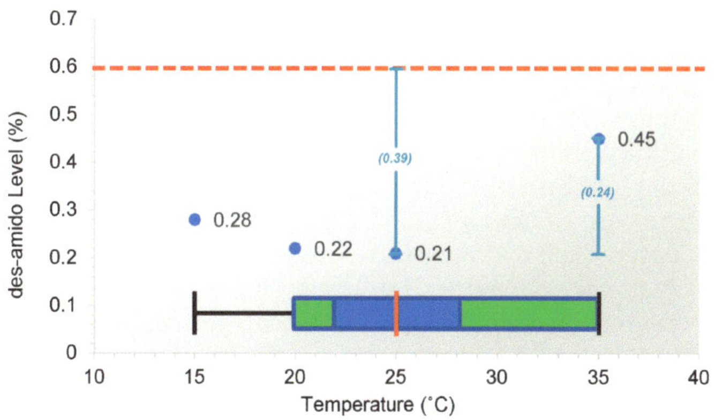

Figure 7.12 Impact of cleavage temperature on des-amido level.

on the CQA des-amido was clearly demonstrated and the impact ratio method identified the need for enhanced control and routine monitoring.

7.8.3.6 Des-amido Integrated Control Strategy

From the beginning, establishing control of des-amido levels in the drug product was clearly understood based on the TPP discussion. As a known degradant, the related-substance des-amido impurity was a DS CQA and also identified on the DP QTPP. It was controlled at the DP stage by an appropriate limit at release and through packaging and storage conditions maintained throughout the intended shelf-life of the product. A rational limit at the DS stage was established based on an understanding of the potential for downstream increase and needed detection controls.

The key control elements at the DS stage included foremost appropriate design of process conditions to limit formation (upper limits for the hold time and temperature were established for each impacting step) and detection and control though purposely placed in-process analytical controls (IPC at purification) and release criteria. Monitoring of the des-amido level in DS, DP and stability studies indicated that the level was well controlled by the combined control elements and below the specification limits for both the DS and DP. Given that the controls were purposely designed based on a process understanding developed though sound science and data, confidence in the robustness of the holistic control strategy for the des-amido impurity was established and accepted by global regulatory agencies.

7.9 Conclusion

There is no doubt that it is a challenging endeavor to develop a holistic control strategy for peptide APIs. The trends in the pharma industry are clear: (1) increasingly complex drug substance targets are being imagined, synthesized and introduced to the clinic, (2) shorter clinical development timelines are a reality and expected and (3) improvements in analytical technologies and methods capable of scrutinizing product quality, and all these are applying pressure on the process design scientist. To meet these challenges, a strategy for product development must be clear from the beginning. After decades of effort, the scientific and regulatory community has consolidated the learnings (both good and bad) into a strategic framework that puts patient safety and product efficacy at the forefront, namely the quality-by-design (QbD) concept. The key components of QbD include beginning with the end in mind, driving a patient-focused design down into development and building a control strategy based on sound scientific understanding of the product and process. These concepts have been outlined in guidance documents from the ICH and clarified by EMA and FDA white papers. It is the hope of the authors that the concepts and guidance provided here can prove beneficial to the peptide API community and further enrich the efforts to develop life-changing medicines and advance the frontiers of science.

References

1. J. L. Lau and M. K. Dunn, *Bioorg. Med. Chem.*, 2018, **26**, 2700–2707.
2. S. B. Kent, *Chem. Soc. Rev.*, 2009, **38**, 338–351.
3. T. Uhlig, T. Kyprianou, F. G. Martinelli, C. A. Oppici, D. Heiligers, D. Hills, X. R. Calvo and P. Verhaert, *EuPa Open Proteomics*, 2014, **4**, 58–69.
4. A. Henninot, J. C. Collins and J. M. Nuss, *J. Med. Chem.*, 2018, **61**, 1382–1414.
5. T. J. Harmand, C. E. Murar and J. W. Bode, *Curr. Opin. Chem. Biol.*, 2014, **22**, 115–121.
6. L. Raibaut, N. Ollivier and O. Melnyk, *Chem. Soc. Rev.*, 2012, **41**, 7001–7015.
7. M. A. T. Kannissery Pramod, N. A Charoo, S. H. Ansari and J. Ali, *Int. J. Pharm. Invest.*, 2016, **6**, 129–138.
8. L. Zhang and S. Mao, *Asian J. Pharm. Sci.*, 2017, **12**, 1–8.
9. T. Garcia, G. Cook and R. Nosal, *J. Pharm. Innovation*, 2008, **3**, 60–68.
10. P. C. Collins, *AIChE J.*, 2018, **64**, 1502–1510.
11. A. A. Kaspar and J. M. Reichert, *Drug Discovery Today*, 2013, **18**, 807–817.
12. K. Fosgerau and T. Hoffmann, *Drug Discovery Today*, 2015, **20**, 122–128.
13. L. R. Malins and R. J. Payne, *Curr. Opin. Chem. Biol.*, 2014, **22**, 70–78.
14. V. Agouridas, O. El Mahdi, M. Cargoet and O. Melnyk, *Bioorg. Med. Chem.*, 2017, **25**, 4938–4945.
15. F. Guzman, S. Barberis and A. Illanes, *Electron. J. Biotechnol.*, 2007, **10**, 279–314.
16. I. Eggen, M. Verlander, A. Swietlow, H. Rode, A. Szajek and B. Gregg, *PharmaTech*, 2014, **38**, 1–5.
17. A. Szajek, I. Eggen, H. Rode, A. Swietlow, M. Verlander, A. Y. Szajek and B. Gregg, *PharmaTech*, 2014, **38**, 6–9.
18. I. Eggen, B. Gregg, H. Rode, A. Swietlow, M. Verlander and A. Y. Szajek, *PharmaTech*, 2014, **38**, 10–14.
19. V. Vergote, C. Burvenich, C. Van de Wiele and B. De Spiegeleer, *J. Pept. Sci.*, 2009, **15**, 697–710.
20. M. D. Argentine, P. K. Owens and B. A. Olsen, *Adv. Drug Delivery Rev.*, 2007, **59**, 12–28.
21. U. S. D. o. H. a. H. Services, F. a. D. Administration and C. f. D. E. a. R. (CDER), 2007.
22. W. J. Lambert, *AAPS PharmSciTech*, 2010, **11**, 1476–1481.
23. A. Swietlow and R. Lax, *Chim. Oggi; Chem. Today*, 2004, 22–24.
24. M. D'Hondt, N. Bracke, L. Taevernier, B. Gevaert, F. Verbeke, E. Wynendaele and B. De Spiegeleer, *J. Pharm. Biomed. Anal.*, 2014, **101**, 2–30.
25. J. A. McMurry, S. Kimball, J. H. Lee, D. Rivera, W. Martin, D. B. Weiner, M. Kutzler, D. R. Sherman, H. Kornfeld and A. S. D. Groot, *Curr. Mol. Med.*, 2007, 7.
26. A. Teasdale, D. Elder, S.-J. Chang, S. Wang, R. Thompson, N. Benz and I. H. Sanchez Flores, *Org. Process Res. Dev.*, 2013, **17**, 221–230.
27. E. guideline, 2014.

28. J. Cavanagh, W. J. Fairbrother, A. G. P. III, M. Rance and N. J. Skelton, *Protein NMR Spectroscopy – Principles and Practice*, Academic Press, 2nd edn, 2006.
29. D. S. Wishart, C. G. Bigam, A. Holm, R. S. Hodges and B. D. Sykes, *J. Biomol. NMR*, 1995, **5**, 67–81.
30. Y. Wang and O. Jardetzky, *Protein Sci.*, 2002, **11**, 852–861.
31. A.-C. B. Valentina Bonetto, H. Jornvall and R. Sillard, *J. Protein Chem.*, 1997, **16**, 371–374.
32. G. Böhm, R. Muhr and R. Jaenicke, *Protein Eng., Des. Sel.*, 1992, **5**, 191–195.
33. K. E. Wilcox, E. W. Blanch and A. J. Doig, *Biochemistry*, 2016, **55**, 3794–3802.
34. J. A. Hering, P. R. Innocent and P. I. Haris, *Spectroscopy*, 2002, **16**, 53–69.
35. K. L. Durand, L. Tan, C. A. Stinson, C. B. Love-Nkansah, X. Ma and Y. Xia, *J. Am. Soc. Mass Spectrom*, 2017, **28**, 1099–1108.
36. J. B. Jordan, L. Poppe, M. Haniu, T. Arvedson, R. Syed, V. Li, H. Kohno, H. Kim, P. D. Schnier, T. S. Harvey, L. P. Miranda, J. Cheetham and B. J. Sasu, *J. Biol. Chem.*, 2009, **284**, 24155–24167.
37. K. Zeng, I. Geerlof-Vidavisky, A. Gucinski, X. Jiang and M. T. Boyne, 2nd, *AAPS J.*, 2015, **17**, 643–651.
38. K. D. Ratanji, J. P. Derrick, R. J. Dearman and I. Kimber, *J. Immunotoxicol.*, 2014, **11**, 99–109.
39. M. U. S. Mats Leeman and L. Nilsson, *LCGC Europe*, 2015, **28**, 642–651.
40. R. D. Camerini-Otero and L. A. Day, *Biopolymers*, 1978, **17**, 2241–2249.
41. P. J. Wyatt, *Analyttca Chtmrca Acta*, 1993, **272**, 1–40.
42. P. H. Brown and P. Schuck, *Biophys. J.*, 2006, **90**, 4651–4661.
43. C. G. Taylor, G. Meisl, M. H. Horrocks, H. Zetterberg, T. P. J. Knowles and D. Klenerman, *Anal. Chem.*, 2018, **90**, 10385–10393.
44. B. Demeule, R. Gurny and T. Arvinte, *Int. J. Pharm.*, 2007, **329**, 37–45.
45. Y. Cohen, L. Avram and L. Frish, *Angew. Chem., Int. Ed. Engl.*, 2005, **44**, 520–554.
46. N. J. Baxter and M. P. Williamson, *J. Biomol. NMR*, 1997, **9**, 359–369.
47. S. N. B. Reif, *Biochemistry*, 2005, **44**, 1444–1452.
48. S. Zolls, R. Tantipolphan, M. Wiggenhorn, G. Winter, W. Jiskoot, W. Friess and A. Hawe, *J. Pharm. Sci.*, 2012, **101**, 914–935.
49. D. K. S. Peter Oma and David King, *Pharmacopeial Forum*, 2010, **36**, 311–320.
50. C. Jia, C. B. Lietz, Q. Yu and L. Li, *Anal. Chem.*, 2014, **86**, 2972–2981.
51. V. Schurig, *J. Chromatogr. B: Anal. Technol. Biomed. Life Sci.*, 2011, **879**, 3122–3140.
52. R. Bhushan and H. Bruckner, *Amino Acids*, 2004, **27**, 231–247.
53. I. Ilisz, R. Berkecz and A. Peter, *J. Pharm. Biomed. Anal.*, 2008, **47**, 1–15.
54. P. C. Charles Brown, C. Eckers, W. Neville, G. Okafo and P. Camilleri, *Anal. Commun.*, 1996, **33**, 433–436.
55. S. N. Takeshi Saito, M. Kinoshita, T. Ihara, S. Kinugasa, A. Nomura and T. Maeda, *Metrologia*, 2004, **41**, 213.

56. L. E. K. Ouwen Zhang, J. P. Olivier and J. D. Forman-Kay, *J. Biomol. NMR*, 1994, **4**, 845–858.
57. M. Schmidt, A. Toplak, H. J. Rozeboom, H. J. Wijma, P. Quaedflieg, J. H. van Maarseveen, D. B. Janssen and T. Nuijens, *Org. Biomol. Chem.*, 2018, **16**, 609–618.
58. P. W. Riddles, R. L. Blakeley and B. Zerner, *Methods Enzymol.*, 1983, **91**, 49–60.
59. Committee for Medicinal Products for Human Use (CHMP), Public Assessment Report for Perjeta, Procedure No. EMEA/H/C/002547/0000, 2012.
60. J. Lepore and J. Spavins, *J. Pharm. Innovation*, 2008, **3**, 79–87.
61. M. Verlander, *Int. J. Pept. Res. Ther.*, 2007, **13**, 75–82.
62. D. Bang, B. L. Pentelute and S. B. Kent, *Angew. Chem., Int. Ed. Engl.*, 2006, **45**, 3985–3988.
63. K. Y. a. K. Numata, *Molecules*, 2014, 13755–13774.
64. M. Schmidt, A. Toplak, P. J. Quaedflieg and T. Nuijens, *Curr. Opin. Chem. Biol.*, 2017, **38**, 1–7.
65. T. Nuijens, A. Toplak, P. J. L. M. Quaedflieg, J. Drenth, B. Wu and D. B. Janssen, *Adv. Synth. Catal.*, 2016, **358**, 4041–4048.
66. M. Schmidt, A. Toplak, P. J. L. M. Quaedflieg, H. Ippel, G. J. J. Richelle, T. M. Hackeng, J. H. van Maarseveen and T. Nuijens, *Adv. Synth. Catal.*, 2017, **359**, 2050–2055.
67. S. S. Kevin, D. Seibert, J. D. Mitchell, K. L Griffiths and B. McGarvey, *J. Pharm. Innovation*, 2008, 105–112.
68. M. Blessy, R. D. Patel, P. N. Prajapati and Y. K. Agrawal, *J. Pharm. Anal.*, 2014, **4**, 159–165.
69. V. Castro, H. Rodriguez and F. Albericio, *Org. Lett.*, 2013, **13**, 246–249.
70. K. D. J. Andrew and R. Vaino, *J. Comb. Chem.*, 2000, **2**, 579–596.
71. D. Rentsch, C. Stähelin, M. Obkircher, R. Hany, M. Simeunovic, D. Samson, G. Loidl and F. Dick, *ACS Comb. Sci.*, 2012, **14**, 613–620.
72. J.-A. F. Muriel Amblard, Jean Martinez and Gilles Subra, *Mol. Biotechnol.*, 2006, **33**.
73. S. V. D. Mathieu Verbeken, *J. Bioanal. Biomed.*, 2012, **s6**, 1–15.
74. E. Hlebowicz, A. J. Andersen, L. Andersson and B. A. Moss, *J. Peptide Res.*, 2005, **65**, 90–97.
75. R. L. Noble and G. B. Fields, *J. Peptide Protein Res.*, 1990, **35**, 161–214.
76. J. A. Moss, *Curr. Protoc. Protein Sci.*, 2005, **7**(18), 1–19.
77. F. A. Ayman El-Faham, *Chem. Rev.*, 2011, **111**, 6557–6602.
78. P. W. W. Chan, *Fmoc Solid Phase Peptide Synthesis: A Practical Approach*, Oxford University Press, 1999.
79. J. B. Blanco-Canosa and P. E. Dawson, *Angew. Chem., Int. Ed. Engl.*, 2008, **47**, 6851–6855.

CHAPTER 8

Higher-order Structure Characterization and Comparability Assessments for Peptides

RENATA VARGA* AND ANUJA RANE

Teva Pharmaceuticals Inc., Analytical Sciences and Operations, Biologics CMC, Brandywine Parkway, West Chester PA 19380, USA
*Email: renata.varga02@tevapharm.com

8.1 Introduction

Peptides are an emerging class of molecules that have gained special interest in the pharmaceutical and biotech industries over the past few decades as therapeutic and diagnostic agents. Their high selectivity, potency, minimal toxicity and indigenous properties allow for applications in complex therapeutic avenues of neurology, endocrinology, cardiovascular disease, infectious diseases and oncology.[1,2] The solid-phase synthesis approach has revolutionized peptide drug development, enabling large-scale manufacture of complex peptides and peptidomimetics with greater purity.[3] As of 2017, more than 60 peptide therapeutics have been approved in the US, EU and Japanese markets and the number of peptides per year entering clinical studies has improved significantly from one to 20 in a span of 40 years.[4,5]

Based on their size and structure, peptides lie in the middle of the traditional small molecules and large biomolecules. Similar to small chemical

Drug Discovery Series No. 72
Peptide Therapeutics: Strategy and Tactics for Chemistry, Manufacturing and Controls
Edited by Ved Srivastava

Published by the Royal Society of Chemistry, www.rsc.org

drugs, peptides can be manufactured using chemical synthesis, thus lowering the overall production costs and complications compared with protein-based biopharmaceuticals. However, they differ from small molecules in terms of the characterization methods used to assess product quality, which often consist of an array of advanced chemical, biophysical and bioanalytical assays. This difference in analytical approach is mainly due to the higher-order spatial arrangements exhibited by peptides in formulated drug products.[6]

The three-dimensional structure of peptides is similar to that of proteins and can be described with four levels of structural elements (Figure 8.1). *Primary structure* is the amino acid sequence, comparable to the molecular formula of a small chemical entity. *Secondary structure* refers to the spatial arrangement of amino acid residues that are nearby in the sequence. Some of these arrangements are regular, giving rise to a periodic structure. α-Helix, β-strand or β-sheet, turns and random coils are the most prominent elements of secondary structure. *Tertiary structure* refers to the spatial arrangement of amino acid residues that are far apart in the sequence and to the pattern of disulfide bonds. However, not all peptides display a real tertiary structure, and especially smaller sequences rarely show higher than secondary structural arrangement. Larger proteins can exhibit a fourth structural level, the *quaternary structure*, which refers to the spatial arrangement of polypeptide subunits. These subunits can be identical or different polypeptide chains. Peptides in the focus of this chapter usually do not exhibit

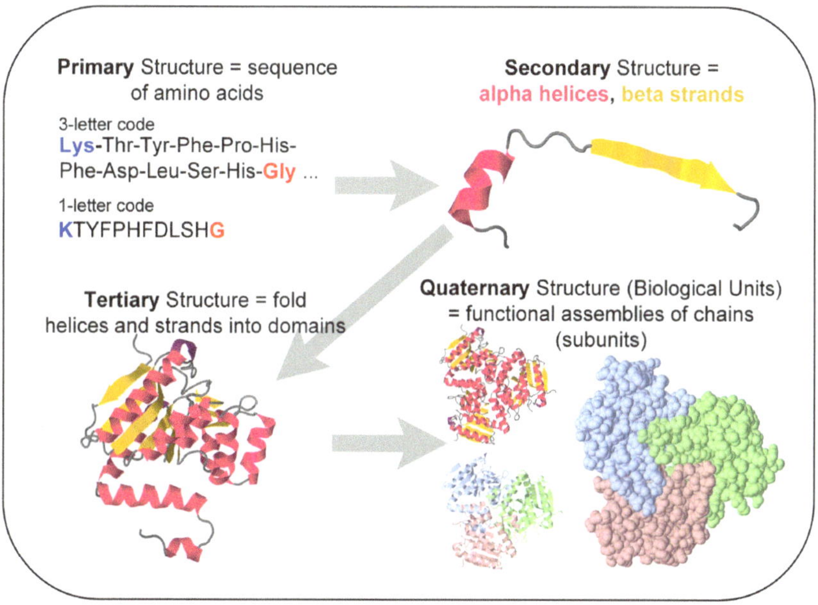

Figure 8.1 Four levels of protein and peptide three-dimensional structure. Reproduced from ref. 88 [http://proteopedia.org/wiki/index.php/Image:Protein-structure-4-levels-III-flat.png] under the terms of a CC BY-SA 3.0 license [https://creativecommons.org/licenses/by-sa/3.0/].

such subunits; however, they can self-associate into higher "mers," such as dimers, trimers or oligomers. These higher-level species can also be called as higher-order aggregates.

When compiling the CMC sections in a regulatory submission of peptide drug products, the structural description always needs to include the characterization and understanding of the higher-order conformation of the particular peptide under development. A simple molecular formula, solubility information and other physicochemical properties are not sufficient to describe the structure of peptides completely. Regulatory agencies encourage applicants to provide information on peptide characterization across the different levels of structures using a combination of analytical methods (often orthogonal methodologies). Also, it becomes even more important in the case of a generic development where a high comparability of formulated generic peptides to the reference listed drug (RLD) needs to be established not just at the amino acid level but also at higher-order structure levels. The following sections describe each higher-order structure level in more detail, providing analytical methodologies to choose from when designing the characterization and comparability studies.

Note that in this chapter peptides are considered, by definition, to contain fewer than 100 amino acid residues and are well distinguished from proteins.

8.2 Primary Structure

All peptides are composed of amino acids. These amino acids are either from the 20 naturally occurring amino acids or chemically modified unnatural amino acids. The uniqueness of the linear sequential arrangement of these amino acids, also known as residues, enables a peptide to function properly and is called as the *primary structure* of the peptide. The amino acids are linked together through a chemical bond called the *peptide bond*, also referred to as the *amide link*. The peptide bond forms a fundamental backbone or chain, supporting the peptide conformation. A distinct feature of the peptide bond is the rigid planar structure formed by delocalization of electrons from the nitrogen into the carbonyl group, resulting in double bond character between the carbonyl carbon and the adjacent nitrogen. This planar construct along with the arrangement of amino acid side-chain groups contributes to the overall peptide conformation. The angular restrictions around the C^{α}–N and C–C^{α} linkages of the peptide backbone play a key role in the formation of higher-order structures of peptides.[7]

Once the peptide has been synthesized or while it is being synthesized, additional primary structural changes can occur *in vivo* or *in vitro* by enzymatic or chemical modifications to susceptible amino acids, in addition to changes involving bond cleavage or crosslinking reactions. Such alterations can lead to changes in the local or overall higher-order structure of the peptide. These modifications are referred to as post-translational modifications, typically including glycosylation, oxidation, deamidation, glycation, N- or C-terminal clipping, acetylation, isomerization, *etc.*

Analytical techniques applied for the characterization of the primary sequence and the above-mentioned post-translational modifications are outlined in the following subsections.

8.2.1 Molecular Mass Determination

Peptide molecular mass can be determined by injecting it in its native form without any pretreatment to remove other formulation compounds on a reversed-phase (RP) column using a liquid chromatographic (LC) system connected to mass spectrometric (MS) detector. Depending on the size and polarity of the peptide, different soft ionization methods can be used, but most often electrospray ionization (ESI) is applied. This generates multiply charged peptide mass-to-charge (m/z) peaks, which are then deconvoluted into the monoisotopic or average molecular mass of the peptide. The highest resolution and accuracy can be achieved by using a Fourier transform ion cyclotron resonance (FT-ICR) analyzer, which provides mass error measurements of <2 ppm for peptides up to 25 kDa. This mass accuracy is often required for unambiguous characterization, especially when the mass shifts are associated with deamidation ($\Delta = 1$ Da) or the difference between acetylation and trimethylation ($\Delta = 0.0363$ Da) or reduced/oxidized disulfide bonds ($\Delta = 2$ Da).

Recently, linear trapping quadrupole (LTQ) orbitrap analyzers have also been used for high-resolution molecular mass determination. Other mass analyzers such as time-of-flight (TOF) detectors, in spite of their lower resolution capacity, have been used for peptide characterization and precise mass determination. This technique was especially employed in combination with matrix-assisted laser desorption/ionization (MALDI), producing far less charged ions than ESI. For better identification of post-translational modifications, peptides can be fragmented through different methods based on activation using thermal heating (collision with gas) or electron capture/transfer. Electron-capture dissociation (ECD) fragments peptides by ultra-fast dissociation of the amino acid backbone, thus generating several product ions along the backbone of the peptide. In comparison with ECD, thermal activation with collision-induced dissociation (CID) produces fewer fragments with increased amino acid specificity.[8]

8.2.2 Peptide Mapping

Peptide mapping can be used in structural characterization to confirm primary structure, identify post-translational modifications, determine product-related impurities, characterize variants observed in other methods [ion exchange (IEX), size-exclusion chromatography (SEC)] and confirm product integrity in lot-to-lot analysis. In peptide mapping, the peptide is treated with chemicals or enzymes to produce peptide fragments, then these fragments are separated and analyzed by LC-MS. This approach provides very detailed information on low-level post-translational modifications.

However, there are multiple pitfalls in peptide mapping that can affect robustness, reproducibility and accuracy. Method inconsistencies can arise due to incomplete digestions, nonreproducible digestion conditions, nonspecific cleavages (over-digestion), enzyme lot-to-lot variability and nonreproducible chromatographic parameters.[9]

Owing to the complexity and inherent variability of the method, peptide mapping is generally a comparative procedure in which the peptide map of the test sample is compared with that of a reference substance prepared in a side-by-side experiment.

8.2.3 Peptide Sequencing

Peptide sequencing is the practical process of determining the amino acid sequence of all or part of a peptide. This may serve to identify the peptide or characterize its post-translational modifications. Mass spectrometric methods are now the most widely used approach for peptide sequencing and identification but the classic Edman degradation method, developed in the 1960s, still remains a valuable tool for characterizing a peptide's N-terminus.[10] Peptide sequencing by Edman chemistry employs a series of chemical reactions to remove and identify the amino acid residue present at the N-terminus of the polypeptide chain, *i.e.* the residue with a free α-amino group. As the reaction progresses, the next residue in the sequence is made available and subjected to the same round of chemical reactions. Reiteration of this process reveals the sequence of the entire polypeptide.[11]

The number of methods available for peptide C-terminal amino acid analysis is much smaller than that for N-terminal analysis. The most common method is to add carboxypeptidases to a solution of the peptide, take samples at regular intervals and determine the terminal amino acid by analyzing a plot of amino acid concentrations against time. However, these methods lack sensitivity, unlike the N-terminal Edman degradation.[12]

8.2.4 Amino Acid Analysis

The traditional amino acid analysis (AAA) consists of a hydrolysis reaction as the first step, where the peptide sample is heated in 6 M hydrochloric acid at 100–110 °C for 24 h or longer. Peptides with many bulky hydrophobic groups may require longer heating periods. However, these conditions are so vigorous that some amino acids, such as serine, threonine, tyrosine, tryptophan, glutamine and cysteine, are degraded. There are several methods to prevent or reduce these losses, such as different heating times and addition of thiol reagents or phenol to the hydrolysis. After hydrolysis, the amino acids can be separated by ion-exchange chromatography, then derivatized to facilitate their detection (post-column derivatization), or more commonly the amino acids are first derivatized and then resolved by reversed-phase high-performance liquid chromatography (RP-HPLC) (with precolumn derivatization). Typical reagents for derivatization are ninhydrin, *o*-phthalaldehyde (OPA) and fluorescamine.

By this traditional amino acid analysis we can quantify the approximate level of each amino acid, rather than the ratio of each amino acid, but we cannot obtain sequence information.

8.2.5 Determination of D-Amino Acids

D-Amino acids do not occur naturally and therefore may pose a safety risk when found in peptide therapeutics. D-Enantiomers can be incorporated during solid-phase peptide synthesis or can be generated by racemization during synthesis or storage.

Various strategies have been developed for the differentiation of enantiomers by chromatography, such as (i) indirect methods involving derivatization of a mixture of target enantiomers by a chiral derivatizing agent, which generates diastereomeric compounds that are physicochemically different and therefore can be resolved chromatographically on achiral chromatographic columns; (ii) direct methods in which enantiomers are separated on achiral stationary phases with chiral mobile phase additives; and (iii) direct methods in which the chiral separations are achieved by using a chiral stationary phase. The basis of the direct methods is the intermediate formation of diastereomeric associates between the chiral selector moiety and the two enantiomers of the chiral analytes.[13]

Over the years, HPLC and supercritical fluid chromatography (SFC) with diverse types of chiral stationary phases have become the preferred methodologies for resolving enantiomers and racemic mixtures. Where applicable, gas chromatography (GC) remains a powerful analytical tool for chiral separation owing to its exceptionally high efficiency. In chiral GC-MS applications, the peptides are hydrolyzed in a deuterated environment (DCl/D_2O) to be able to correct for racemization during hydrolysis and then hydroxyl and amino functional groups are derivatized with methanol and trifluoroacetic anhydride, respectively, prior to the GC-MS separation on a chiral GC column.

8.3 Secondary Structure

Proteins and most peptides exhibit secondary structure due to interaction between the atoms of the backbone *via* hydrogen bonding and cooperative behavior. These structures were first predicted by Pauling and Corey and later observed in atomic detail by X-ray crystallography.[14] Secondary structure can be determined by backbone torsion angles φ (phi) for N–C^{α} and ψ (psi) for C^{α}–C bonds, defining the spatial orientation of peptide backbone in the form of Ramachandran plots. The third angle ω (omega) is generally fixed owing to the planarity of the peptide bond. In the course of recent advances in bioinformatics and with increasing knowledge of polypeptide structures, advanced high-fidelity Ramachandran plots with 3D depiction are being evaluated.[15]

Secondary structures such as helices and strands are formed when backbone phi and psi angles have repeating values. The *α-helix* is most abundantly found in globular polypeptides and exhibits dihedral geometry with $\varphi \approx -60^\circ$, $\psi \approx -40^\circ$. It forms intra-chain hydrogen bonds resulting in a compact rod-like structure with 3.6 amino acids per turn. The side chains extend outwards from the helical axis and are involved in interacting with other peptide domains or small molecules. Some amino acids such as Ala, Gln, Met, Lys and Leu present a high propensity for α-helix formation, whereas Gly, Pro, Tyr and Ser show poor helix-forming tendencies.[16,17] Shorter helical segments of 3.10 helix and π-helix are less commonly seen in peptides.[18]

β-Sheets are another major class of secondary structures that are formed by a number of β-strands stabilized by inter-chain hydrogen bonds. Most β-strands exhibit $\varphi \approx -120^\circ$, $\psi \approx \sim 120^\circ$ spatial orientation, wherein the peptide bonds and side chains of adjacent residues point in opposite directions and away from the backbone resulting in a pleated structure. β-Sheets can be composed of parallel, antiparallel or mixtures of parallel and antiparallel adjacent polypeptide segments. β-Sheet structures find significance in molecular stability, formation of quaternary structure, protein–protein interactions and peptide and protein aggregation.[19,20]

Turns are those regions of the polypeptide chain that allow the protein to change direction, enabling it to fold onto itself. They are composed of three or four residues and form a U-shaped structure linked together by an intra-turn backbone hydrogen bond.[21] Pro and Gly are often found in the turns which redirect the polypeptide chain towards the central core to form a compact structure.[22] *Loops* contain more than five residues and also contribute to changing the overall backbone direction. However, unlike turns they exhibit irregular intra-loop hydrogen bonding. Turns and loops are important in protein folding and their presence on the surfaces is believed to aid ligand binding, molecular recognition, protein–peptide interactions and interactions with external solutes and even present sites for post-translational modifications.[23]

Secondary structural features are often studied using spectroscopic techniques such as far-UV circular dichroism (CD), Fourier transform infrared (FTIR), Raman and nuclear magnetic resonance (NMR) spectroscopy. These analytical methods can provide detailed information about the peptide and protein structures and also facilitate qualitative comparison between the spectra under different study conditions. Here we highlight some of the key features of each analytical technique, including sample preparation, advantages and limitations and basic data interpretation for peptide characterization.

8.3.1 Far-UV CD Spectroscopy

Circular dichroism is observed in chiral molecules due to differential absorption of left and right circularly polarized light. Amide bonds exhibit a dominant CD effect in the far-UV region (260–180 nm) and undergo $n \rightarrow \pi^*$ and $\pi \rightarrow \pi^*$ transitions upon absorbing the circularly polarized light. Based on the geometry of the peptide backbone, these transitions can have varying intensities and energies at a particular wavelength. This results in a signature CD spectrum for α-helix, β-sheets, β-turns and poly-L-proline II structures.[24] The α-helix structure exhibits intense CD bands, with two deep minima around 222 nm ($n\rightarrow\pi^*$) and 208 nm ($\pi\rightarrow\pi^*$) and a maximum at 191–195 nm ($\pi\rightarrow\pi^*$).[25] The band observed at 222 nm is attributed to the strong hydrogen bonding interactions of the helical conformation. A representative CD spectrum of an α-helical dominant peptide is given in Figure 8.2A. The CD spectra of β-sheets display weaker signals than that of a helical structure, with a shallow minimum near 215 nm ($n\rightarrow\pi^*$) and a deep

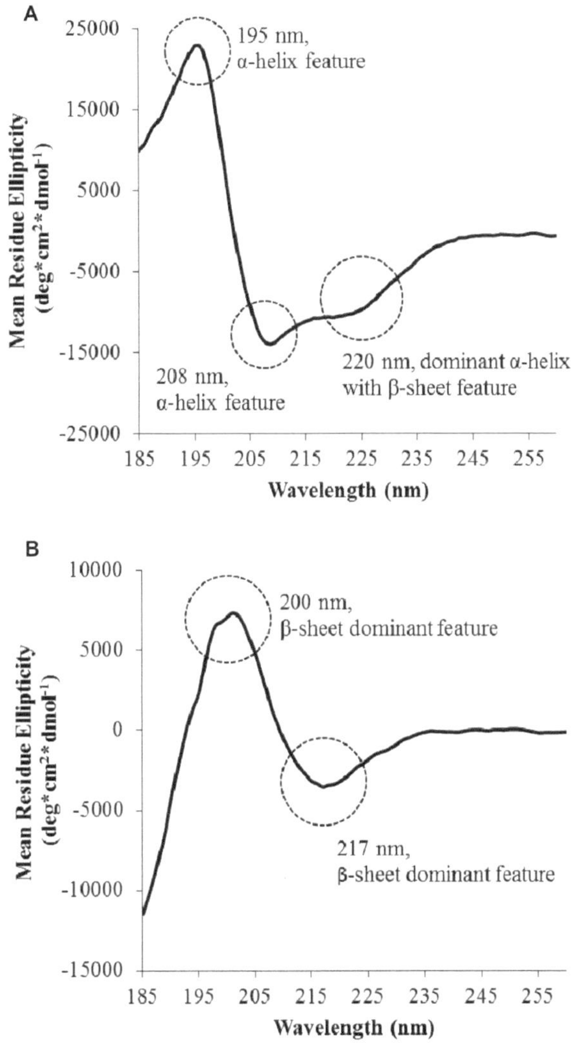

Figure 8.2 (A) Representative far-UV CD spectrum highlighting the characteristic CD regions of a hormonal peptide with a dominant α-helix structure. (B) Representative far-UV CD spectrum of a β-sheet anti-parallel polypeptide showing the classic β-sheet structural features.

maximum between 195 and 200 nm ($\pi \rightarrow \pi^*$), as can be seen in Figure 8.2B. Irregular peptides and denatured polypeptides show a strong minimum around 195–200 nm and a plateau-like maximum or minimum between 215 and 230 nm.

One may not always find these characteristic CD features when characterizing peptide samples owing to their short structures. The CD spectrum is highly influenced by the dominance of a particular type of secondary structure in the peptide, the impact of other contributing domains in the molecule and interference of external environment such as solvents and excipients. There are cases where both α-helix and β-sheet structures are present and the CD spectrum is dominated by α-helical components, thus showing bands at 222, 208 and 190–195 nm.[26] At times, only a single broad minimum around the 210–220 nm region may be observed owing to their overlapping effects. Aromatic side chains, S–S disulfide bonds and Trp stacking have significant contributions in the far-UV region that can offset the peptide-based signals, resulting in a low-intensity profile. Solvent and buffer transparency is critical since the background can cause very large high-tension (HT) voltage when measuring below 200 nm, thus resulting in noisy data with poor accuracy. Buffers and excipients that absorb at lower wavelengths or have chiral centers are undesirable and need to be removed by buffer exchange or diluted to lower concentrations prior to analysis. Water is an ideal diluent for far-UV CD analysis. Preservatives such as *m*-cresol and phenol are commonly added to injectable peptide drug products and can cause significant interference in the far-UV region. They need to be removed by dialysis to obtain a good-quality CD spectrum. High amounts of chloride salts should be avoided and instead can be replaced by fluoride, sulfate or phosphate salts. Many papers provide a list of commonly used buffers and reagents suitable for CD analysis.[24,27]

Background effects can also be minimized by using narrow-pathlength quartz cuvettes. Cells with pathlengths <1 mm are often used for far-UV CD measurements and require very small sample volumes of <0.35 mL. Baseline scans must be performed using only the buffer or solvent and subtracted from sample spectra to account for any absorbance. It is therefore imperative that the cells are cleaned thoroughly after every run to avoid contamination and residual carryover. It is highly recommended for cells with pathlengths of 1 mm or less to be routinely calibrated using a solution of an accurately known absorbance (*e.g.* potassium chromate). This can help to decrease artifacts in the spectral data.

A wide range of concentrations can be used in CD analysis. Concentrations can range from 0.01 to 5 mg mL^{-1} but caution must be exercised not to saturate the detector at higher concentrations. It is typical for the absorbance plot to increase at lower wavelengths during a scan provided that it does not go over the value of 2 and the HT voltage is <800 V (the voltage value can differ for different instrument makes). At high peptide concentrations, aggregation and light-scattering effects may occur, which can negatively affect the CD readings. To obtain a good far-UV CD profile of peptides with >10 amino acids, it is suggested to use a 1 mm pathlength cell with a concentration of 0.2–0.25 mg mL^{-1}. Another aspect is the accuracy of the peptide concentration, which is very important when estimating the secondary structural content in a sample or for comparing samples of different concentrations. Several methods for estimating accurate protein concentrations can be found in the literature.[26,28]

Most CD measurements are performed under ambient conditions or at different temperatures (*e.g.* thermal study) using an external circulating water bath or Peltier probe inserted inside the cell. The instrument must be flushed with nitrogen prior to and during the run to avoid oxidation and damage to the lamp source. Measurement parameters need to be optimized using a series of variables such as scan speed, time constant (integration time), bandwidth and number of scans for an overlay to obtain a good CD signal with minimum noise. A scanning speed in the range 10–50 nm min^{-1} and a digital integration time (DIT) of 1–3 s together determine the average interval of the CD signal. It is often recommended to keep the value of the multiplication of these two factors at <200. The bandwidth should be ≤1 nm, keeping in mind that a lower bandwidth often results in a well resolved spectra. At times for smaller peptides the bandwidth can be increased to 1.5–2 nm in order to obtain better intensity profiles. The data pitch (in the range 0.1–0.5 nm) determines number of data points acquired during the scan and affects the sensitivity and overall run time. Obtaining multiple accumulations, which are later averaged together by the software, helps to improve the signal-to-noise ratio of the profile. As a general rule, around 4–10 accumulations can provide a good CD spectrum of small peptides. Higher accumulations need longer analysis times and often a compromise needs to be made to obtain decent profiles within a given time of an analyst. Newer technologies in this field have allowed the use of autosamplers and automation tools for performing time-consuming, high-throughput CD analysis with minimum human supervision.

The first step in the CD data processing is the baseline/buffer subtraction, after which the CD signal is converted to mean residue molar ellipticity $[\theta]$ in $^{\circ}\,cm^2\,dmol^{-1}$ to normalize for the molar concentrations of amino acid residues and the cell pathlength in centimeters. This allows a comparison of profiles with minor differences in concentrations and also different peptides with varying numbers of amino acids. It is essential that the CD profile should begin at zero. In order to achieve this, the baseline in the beginning portion of the graph (from 260 to 240 nm) is aligned to zero using the instrument software tools. As far as possible, one should refrain from data smoothing and over-manipulation of the spectra. Several algorithms and curve-fitting applications have been constructed to estimate quantitatively the secondary structural composition. CONTIN, SELCON, VARSLC, CLUSTER, CDNN and SVD are some traditionally reliable methods and newly emerging programs such as BeStSel are also being investigated for secondary structure estimations.[24,29–31] Caution must be exercised when reporting the values generated by these programs as they are relative numbers generated using databases of solved X-ray structures of known proteins and often require spectral acquisition as far as 170 nm for better accuracy. Qualitative overlay is a widely accepted approach for CD data presentation, where the CD spectra are normalized and overlaid for comparability purposes. For unbiased spectral comparisons, one could compute the mean spectrum using intermediate precision measurements and set upper and lower boundaries based on two or three standard deviations at each point.[32] If the sample spectrum lies within these limits, it can be deemed comparable. Additional data analysis such as extracting a set of key numerical values from the spectra (such as the ratio of $[\theta]_1$ and $[\theta]_2$ for the two minima in α-helix-dominant peptides), *g*-factor assessment and statistical quantification of spectral difference may also be performed.[33,34]

Synchrotron CD is an extension of the conventional CD that has similar working principles to those of X-ray synchrotron technology. Its main advantage lies in recording CD data below 190 nm in the presence of buffers, detergents and lipid components with an improved signal-to-noise ratio.[27] Other less popular spectroscopic techniques that show potential applications in monitoring secondary structural changes in peptides include vibrational circular dichroism (VCD) and Raman optical activity (ROA). VCD involves differential absorption of left and right circularly polarized IR radiation, unlike electronic radiation that is used in the conventional (electronic) circular dichroism (ECD). This results in amide band shapes that are less influenced by solvents, aromatic residues and polysaccharides.[35] VCD can be used for high-concentration protein and peptide samples in solution and also in forms of soluble and insoluble aggregates and product-related particles.[36] ROA measures the minute differences in vibrational Raman scattering from left and right circularly polarized light from chiral molecules and has recently been gaining potential for secondary structure characterization.

8.3.1.1 Applications of Far-UV CD Spectroscopy

Far-UV CD spectroscopy is a widely used tool in the pharmaceutical industry and academic research laboratories for the characterization of the secondary structure of peptides, proteins and monoclonal antibodies that often guide their biological function. Specifically in regulatory filing of peptide therapeutics, far-UV CD profile comparability can be shown between drug substance and drug product batches to demonstrate robust manufacturing controls, trend stability samples and also to exhibit sameness of structure between generic and reference listed drug (RLD) molecules. Figure 8.3 shows the CD comparability of three lots of a commercially available peptide pharmaceutical product from three different countries, demonstrating high secondary structure comparability of the peptide in the three products.

Formulation composition, excipient–peptide or device–peptide interactions and manufacturing process steps, for example membrane filtration and lyophilization, can have an influence on the higher-order structure of

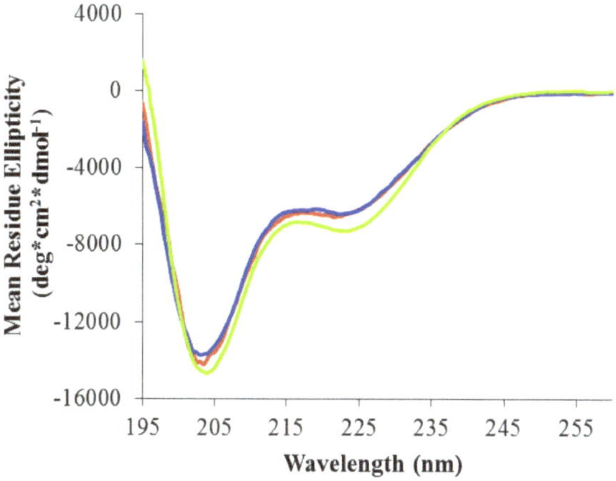

Figure 8.3 Far-UV CD comparability of three different lots of a therapeutic peptide from the USA (blue trace), EU (green trace) and Canada (red trace). Small spectral differences indicate minimal lot-to-lot variability in the overall secondary structure of the peptide drug product.

peptides. Certain lyophilized peptide formulations tend to exhibit poor physical stability, resulting in changes in higher-order structure compared with solution formulations. Lyophilized teriparatide, upon reconstitution at higher concentrations, revealed significant precipitation and transformation of α-helical structure into β-sheet folds compared with the bulk peptide solution analyzed before lyophilization.[37] The fibrillation phenomenon has been observed in insulin formulations subjected to external stress conditions such as agitation and microenvironment alterations. Stepwise changes in the secondary structure of insulin during aggregation and fibril formation were detected by CD analysis.[38]

Another application of CD is to study the effects of chemical modification of peptides, such as lipidation and PEGylation, on the overall structure. CD analysis has shown helix stabilization of glucagon-based peptides modified by acylation with a long-chain saturated lipid.[39] PEGylated antimicrobial peptides characterized by CD showed no significant disruption of secondary structure domains involved in binding to antimicrobial membranes.[40,41] Forced degradation studies are essential for studying the effects of extreme conditions such as light, heat, humidity and agitation to establish the degradation mechanism of therapeutic peptides and polypeptides. Heat treatment at 40 °C for 8 weeks and one ICH cycle of UV and white light treatment on a peptide showed changes in secondary structure when analyzed by far-UV CD (Figure 8.4) and FTIR (see Figure 8.6A and B in Section 8.3.2) spectroscopy.

Although CD spectroscopy is considered a low-resolution technique with poor accuracy and reliability, a high-sensitivity advanced CD instrument does find tremendous utility in rapid structure analysis and spectra comparability of peptide drug products.

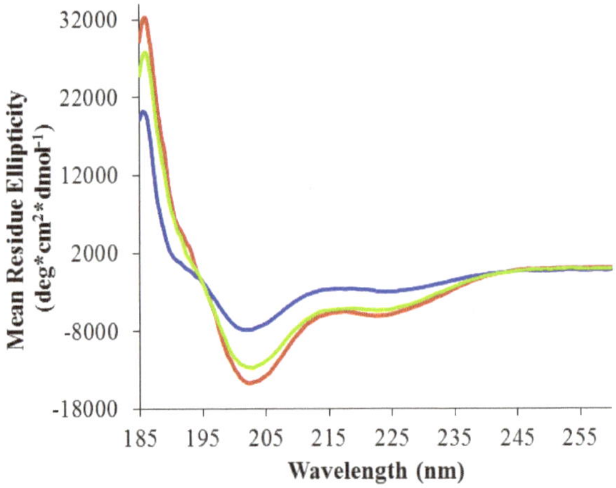

Figure 8.4 Far-UV CD stability study showing the effect of physical stress conditions on the secondary structure of a therapeutic peptide. Unstressed peptide sample is presented in red trace showing key responses at 222, 202 and 186 nm. Light stressed peptide (green trace) showed slight differences in CD intensities. The heat-stressed sample (blue trace) showed significant differences at these characteristic wavelengths compared with the unstressed sample. The results indicate adequate specificity and sensitivity of the method to detect differences in peptide conformations subjected to stress conditions.

8.3.2 FTIR Spectroscopy

FTIR spectroscopy is another effective tool for monitoring secondary structure features and differences in peptides and it is often used in combination with CD spectroscopy. FTIR analysis can be performed on small sample sizes in different states, including solid, crystal, semisolids and clear or turbid liquids. Unlike CD, buffers and excipients have a minimal effect in obtaining a high signal-to-noise absorption spectrum using FTIR spectroscopy. Peptide spectra can be obtained in a wide range of environments ranging from aqueous solutions to organic solvents or from encapsulating polymers to a lipid bilayer system.[42] Its measurement time is very fast compared with CD or NMR methods and requires very little sample preparation.

IR radiation induces vibrational transitions of the repeating units such as stretching, bending and torsion of bonds and a majority of them occur in the mid-IR region between 4000 and 400 cm^{-1}. The wavenumber and intensity of IR radiation absorbed by the peptide are recorded as the IR absorption spectrum. Nine characteristic IR bands have been reported in polypeptides, of which amide I and II are the dominant transitions.[43] Table 8.1 summarizes the band assignments of the main amide vibrations. Amide I is a strong band seen between 1700 and 1600 cm^{-1} by virtue of C=O stretching along with smaller contributions from in-phase bending of N–H bonds and stretching of C–N bonds of the peptide backbone. The amide II band is

Table 8.1 Characteristic amide vibration bands observed from polypeptide repeat units for samples prepared in H_2O by FTIR and Raman spectroscopy.

Characteristic band	Wavenumber/cm^{-1}	Associated with
Amide A	3500	N–H stretching
Amide B	3100	N–H stretching
Amide I	1600–1700	Stretching vibration of C=O
Amide II	1480–1580	Coupled C–N stretching and N–H bending vibrations
Amide III	1230–1300	Coupled C–N stretching and N–H bending vibrations
Amide IV	625–770	O–C–N bending
Amide V	640–800	Out-of-plane N–H bending
Amide VI	540–600	Out-of-plane C=O bending
Amide VII	200	Skeletal mode

slightly more complex than the amide I band and is primarily a result of N–H bond bending vibrations. Since both C=O and N–H bonds are involved in inter- and intramolecular H-bonding, both the amide I and II bands are sensitive to secondary structure changes. Amide I bands in the range 1620–1640 cm^{-1} have been found to contribute to β-sheet structure. α-Helical conformations show an amide I band in the region 1650–1658 cm^{-1}; however, some α-helical peptides have been reported to exhibit the amide I band at lower than 1650 cm^{-1}. A band position around 1640–1650 cm^{-1} signified random coils or unordered structures.[44]

The FTIR spectrum (Figure 8.5A) of a model monoclonal antibody (red trace) depicts distinct amide I, II and III bands at the characteristic wavenumbers 1650, 1550 and 1450–1400 cm^{-1}, respectively. Some of these features may not be observed in smaller proteins and especially peptides. A hormonal peptide (blue) does exhibit strong amide I and II bands but much weaker amide III bands, while another peptide (green) shows only an amide I stretching band. Figure 8.5B shows an overlay of the second-derivative FTIR spectrum of three polypeptides where the IR band features appear sharper and well resolved.

FTIR analysis can be performed using either of two platforms: transmission mode or attenuated total reflectance (ATR) mode. In most transmission mode instruments, the peptide sample is placed between two IR-transparent discs at concentrations >3 mg mL^{-1} in aqueous buffer solutions. Calcium fluoride discs are extensively used for this purpose, where the blank or sample is sandwiched between two discs and mounted onto a cell holder. It is critical to avoid air bubbles and to maintain the same distance between the discs for blank and sample measurements to obtain reproducible spectra.[45] There are other transmission mode configurations available from different vendors that employ a high-throughput format with either a flow-through cell or a standardized IR-transparent multi-well microplate design. In the ATR mode, the sample is placed directly in contact with a zinc selenide, germanium or diamond crystal. This method can be used for dilute

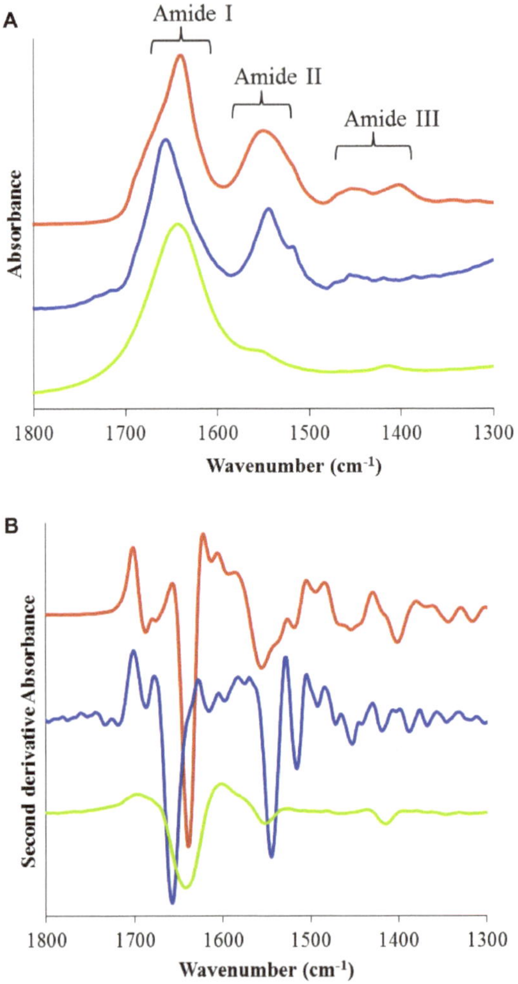

Figure 8.5 (A) Representative FTIR spectra showing the IR band features of amide I, amide II and amide III stretching clearly seen in a monoclonal antibody (red trace) sample. These bands weaken as the polypeptide structure becomes smaller in size. In a larger peptide hormone (blue trace), amide I and II bands are observed whereas amide III is very subtle. A mid-sized therapeutic peptide exhibited only a strong amide I band (green trace). (B) Second-derivative FTIR spectra of the same proteins and peptides with improved band resolution.

solutions and also for lyophilized forms of peptides or peptides bound to excipients (*e.g.* lipids). The sample size requirement for FTIR measurements is very small, often less than 100 μL.

When performing FTIR measurements, the instrument needs to be stabilized with continuous purging of dry air (free of carbon dioxide) to minimize water vapor interference. A high-sensitivity mercury cadmium telluride

(MCT) detector requires cooling with liquid nitrogen and it may need to be replenished if the instrument is used for longer time periods. It is important to check the alignment of the sample cell stage to allow maximum energy throughput across an empty cell. The transmission cell or ATR crystal surface should be clean of any residual peptides or particles and can be cleaned using water or ethanol. An absorption IR spectrum is collected from 400 to 4000 cm^{-1}. To obtain a good-quality signal, multiple scans can be performed for each measurement and the final spectrum is generated as their average. A resolution of 2–4 cm^{-1} should provide sufficient data points per run with an acceptable signal-to-noise ratio.

Data processing involves subtraction of the background and buffer from the sample spectrum. For accurate subtraction, it is important to have exactly the same buffer composition as for the sample, especially for peptide solutions in water in order to subtract the significant water contribution. The cell pathlength, temperature, number of scans, resolution, *etc.*, need to be kept as identical as possible. The sample spectrum can also be corrected for atmospheric water vapor bands using the water vapor absorption spectrum. Caution must be exercised when recording the FTIR spectrum of synthetic peptides owing to interference from trifluoroacetic acid (TFA). TFA is commonly used during solid-phase peptide synthesis and during purification, as a result of which traces of TFA can be present in the peptide drug substance. The strong vibration of the COO^- group in TFA shows an IR band around 1673 cm^{-1}, which can significantly mask the amide I peptide band.[46] Many different methods for the removal of TFA have been suggested, such as washing with a dialysis membrane, lyophilization in the presence of HCl, chromatography or using systematic spectral subtraction.[47,48] It is possible to obtain a higher signal-to-noise ratio for the FTIR signals, especially when analyzing low-concentration peptides (<3 mg mL^{-1}), by exchanging the peptide hydrogens with deuterium. This approach can minimize the effects of water interference in the amide stretching region and can also provide information on peptide dynamics by monitoring the hydrogen–deuterium exchange rates.[49]

The resultant peptide spectrum should ideally show the following features for any further data analysis: a strong amide I band (1650 cm^{-1}), maybe presence of amide II bands (1330–1230 cm^{-1}) and C–H stretching bands (3000–2850 cm^{-1}), and should have a flat baseline between 2200 and 1800 cm^{-1} with a gradual increase below 1800 cm^{-1}. Data are usually truncated to 1800–1400 cm^{-1} where the major amide peaks are sensitive to the peptide secondary structure. Second-derivative spectra are often calculated using a Savitzky–Golay polynomial to analyze minute differences. Spectral deconvolution using Fourier self-deconvolution (FSD) may also be performed to improve the resolution between peaks and assess the contributions of different overlapping amide bands. Deconvolution applied with an appropriate data-fitting function can be used to estimate the secondary structure composition of peptides and proteins.[44] Quantitative analysis of the FTIR amide I, II and III band shapes and positions is possible using newer multivariate tools such principal component analysis (PCA), partial least-squares (PLS) analysis and artificial neural networks (ANN) to elucidate the peptide conformation.[50]

Homo-polypeptides have been used to study the relationship between the IR amide band position and peptide secondary structure. These peptides fold into distinct homogeneous secondary structures depending on the solvent conditions. However, it is difficult to interpret qualitatively an IR spectrum for the structural components of α- and β-turns, *etc.*, since they show overlapping bands. Extensive mathematical deconvolution, derivative spectra and difference spectra are useful tools for improving resolution and assigning structural components to each band.

8.3.2.1 Applications of FTIR Spectroscopy

Peptide–lipid membrane interactions can be studied using FTIR spectroscopy. Magainins and their analogs are 23-amino acid peptides with potential antimicrobial and anticancer properties. FTIR spectroscopy showed that in neutral pH aqueous solution, magainin peptides exist in random coils, indicated by the amide I band at 1642 cm^{-1}. It adopted an ordered structure with dominant α-helices (band at 1647 cm^{-1}) with minor β-sheets (band at 1632 cm^{-1}) in the presence of negatively charged lipid vesicles. This aided a better understanding of its antimicrobial mechanism of action.[51,52] FTIR studies on amyloid-β (Aβ) peptide, consisting of around 40 amino acids, showed aggregation behavior resulting from changes in its secondary conformation. In dilute solutions, monomers of Aβ existed as a random coil at physiological pH and temperature. In contrast, in a membrane-mimicking environment such as sodium dodecyl sulfate (SDS) or lithium dodecyl sulfate (LiDS) micelles or in the presence of phospholipids, the FTIR spectra obtained using deuterated reagents showed more ordered structures of the Aβ peptide with α-helical and β-sheet features. This study also supported a strong relationship between peptide concentration and secondary structure.[53]

FTIR spectroscopy can be used during formulation development to understand the effects of excipients and the manufacturing process on the peptide structure. Exenatide, a 39-amino acid glucagon-like peptide analog, was studied by FTIR spectroscopy in its freeze-dried form and upon encapsulation within polymer microspheres.[54] Microencapsulation involves a series of steps in which the peptide is subjected to harsh processing such as exposure to organic solvents and high temperatures, which can affect its structure. Second-derivative spectra were obtained and compared qualitatively to establish that the encapsulation process had no detrimental effect on the exenatide structure. The stability of a peptide can be examined by an FTIR analysis in which the peptide was exposed to stressed conditions of heat and light (for details, see the far-UV CD Section 8.3.1) and a side-by-side analysis was performed with an unstressed native sample (Figure 8.6A). The second-derivative spectra showed better resolution between the peaks and

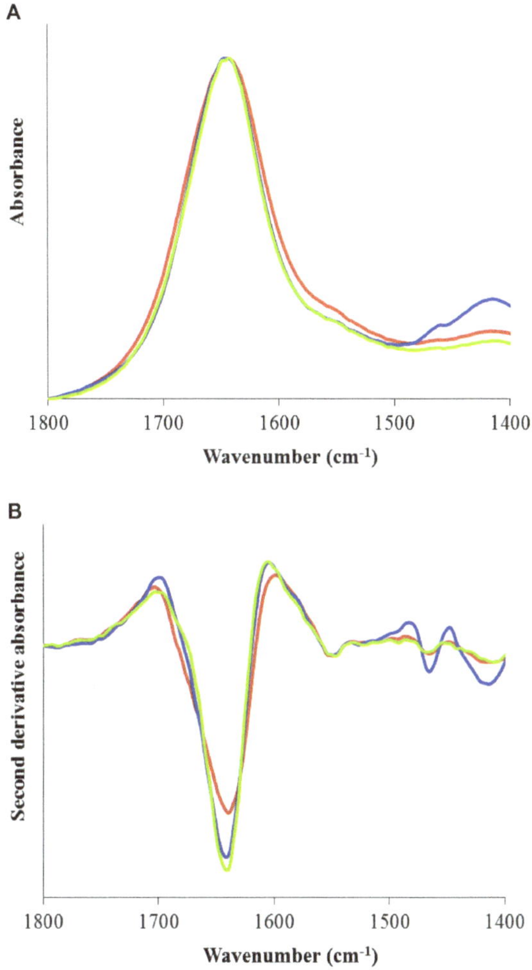

Figure 8.6 (A) FTIR spectra showing the results of a stability study of a therapeutic peptide upon exposure to stress conditions. The unstressed native sample is presented as the red trace, a light stressed sample as the green trace and a heat-stressed sample as the blue trace. The stressed samples exhibited different band intensities and positions compared with the unstressed sample whereas the heat-stressed sample showed major differences in the amide III peak region. (B) Second-derivative FTIR spectra of the same stressed and unstressed peptide samples. The differences between the samples in terms of band intensity and position are more distinct in this data representation.

also highlighted the differences in band intensities and wavenumber shifts (Figure 8.6B). Heat stress (blue trace) showed a significant impact, especially in the amide I and amide III band stretching regions. The light-degraded sample showed differences in the amide I band similar to those for the heat-stressed sample, but the differences were minor.

8.3.3 Raman Spectroscopy

Raman spectroscopy is very similar to IR spectroscopy, being a vibrational technique; however, it differs from the IR technique in certain key fundamental ways, thus serving as an excellent complementary technique. Raman spectroscopy depends on a change in polarizability of the vibrating molecule, whereas IR spectroscopy depends on a change in the dipole moment as a result of molecular vibrations. Raman spectroscopy measures the relative frequencies at which a sample scatters radiation, whereas IR spectroscopy measures absolute frequencies at which a sample absorbs radiation. FTIR spectroscopy is sensitive to heteronuclear functional group vibrations and polar bonds, especially –OH stretching in water (which can interfere with the amide I band signals). Raman spectroscopy, on the other hand, is sensitive to homonuclear molecular bonds, *i.e.* molecules that have a center of symmetry, and for example it can distinguish between C–C, C=C and C≡C bonds. Water is a much weaker Raman scatterer and a possible fluorescence quencher, and thereby creates minimal interference with the strong amide I signal of peptide molecules.

The term Raman effect was first coined by C. V. Raman when he discovered the inelastic light-scattering phenomena of molecules in a deceptively simple but important experiment. It is observed when the incident monochromatic light interacts with matter and scatters light at a different frequency. Physicists have used this spectroscopic technique to correlate the changes in scattered light frequencies due to the vibrations of molecules and rotations of their structure. The Raman spectral pattern is known as the fingerprint of the molecule.

The principle of Raman scattering is briefly illustrated in the electronic–vibronic energy diagram in Figure 8.7. Raman scattering is a two-photon process involving light–molecule interactions. When a monochromatic laser beam is illuminated, the molecule absorbs a photon and transitions into a virtual excited state. A photon is emitted after it has exchanged energy with the atoms of the molecule. Depending on the frequency, the phenomenon can be classified as *Rayleigh scattering* when the emitted energy is the same as the incident radiation or *Stokes Raman scattering* when the emitted photon energy is weaker than the incoming photon energy, or much weaker *anti-Stokes Raman scattering*, observed when the incoming photon gains energy from the molecule and is emitted at a higher energy level. The Stokes Raman light is useful and measured in a conventional Raman experiment.[55]

Several vibrational modes can be used for the analysis of peptide structure. Most characteristic bands are associated with the amide group and are summarized in Table 8.1. Similarly to IR band stretching, the three signals amide I, amide II and amide III are of main interest for the identification of different peptide backbone conformations. The amide III band, which is weak in IR spectra, appears to be very strong in Raman spectra. Secondary structure can be elucidated by monitoring the amide I and III band positions and intensities at 1662–1655 and 1272–1264 cm^{-1} for α-helix and 1674–1672

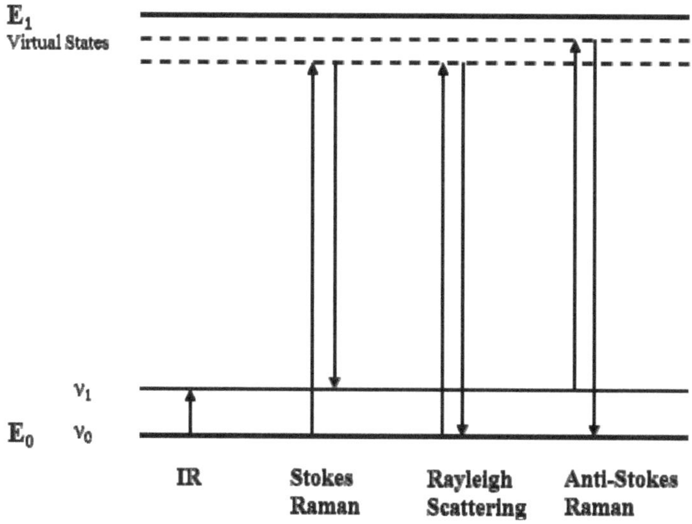

Figure 8.7 Energy diagram of vibrational transitions between the vibrational energy levels corresponding to the processes of IR absorption, Rayleigh scattering and Stokes and anti-Stokes Raman scattering. E_0 and E_1 are the electronic ground and excited states, respectively, and ν_0 and ν_1 are the vibrational ground and excited states, respectively.

and 1242–1227 cm^{-1} for β-sheet polypeptides. Raman spectroscopic characterization can be applied to peptides in various states: aqueous (dilute) solutions, precipitates, fibrils, amorphous aggregates, solids and crystals. Raman spectroscopy offers many advantages in polypeptide structure analysis, as it minimizes the contributions from water bending vibration bands, which can obscure the amide I stretching in IR spectroscopy and thereby affect the structure prediction accuracy.[56] In addition to the characterization of the amide bands, there are additional features in the Raman spectra that permit the description of the environment of numerous amino acid chains. These include the amino acid side-chain modes (*e.g.* tryptophan doublet at 1360 and 1340 cm^{-1}, tyrosine doublet at 860 and 833 cm^{-1}) and sulfur-containing residues in different physical states [C–S stretching with H in the *trans* position of the S atom at 640–680 cm^{-1} and C–S stretching with C in the *trans* position of the S atom at 740–760 cm^{-1} and S–S disulfide bond stretching at 508–515 cm^{-1} (GGG), 523–528 cm^{-1} (TGT) and 540–545 cm^{-1} (TGT)]. With Raman optical activity (ROA), the motif and fold patterns of larger peptides and proteins in solution can also be investigated.

Since Raman spectroscopy is based on the Stokes-scattering photons, which are very weak signals, it can pose certain challenges, especially when measuring low-concentration polypeptides or smaller-sized peptides. However, this issue can be resolved by modern Raman instruments such as FT-Raman, UV resonance Raman, visible dispersive Raman, pulsed Raman, two-dimensional stimulated resonance Raman, *etc.*, which have revolutionized protein and peptide analysis, offering rapid, highly sensitive, robust solutions for analyte concentrations of <1 mg mL^{-1}.[57] Fluorescence is a major problem and can obscure the Raman spectral quality. To minimize fluorescence, longer wavelength laser excitation is generally used. However, this can prolong the data acquisition time, resulting in overheating of the thermosensitive peptide samples.[58] Even though Raman spectroscopy could serve as a tremendous complementary technique to far-UV CD and FTIR spectroscopy to study secondary structure of peptides, it has not yet been widely applied.

8.3.4 NMR Spectroscopy

NMR spectroscopy – a nondestructive chemical characterization technique – utilizes electromagnetic radiation in the radiowave region. While most forms of spectroscopic techniques rely on the interaction of radiation with electrons, traditional NMR spectroscopy studies how the protons in the atomic nuclei of a molecule absorb radiofrequency electromagnetic radiation in a magnetic field. However, only those nuclei that have appropriate spin properties will be active, including ^{1}H, ^{13}C and ^{15}N, these being the species of greatest interest in the case of peptide and protein characterization. Since each nucleus in a peptide drug will be in a different spatially unique position, its ability to interact with the probing electromagnetic radiation will vary and be dependent on its local physicochemical environment. Hence the resonant frequency of each nucleus will be different and will result in an observable unique shift (chemical shift) in the NMR spectrum, which can eventually be linked to the relative position of a nucleus within the molecular structure of the peptide. Generation of the NMR spectrum can result in valuable characteristic information that can be used as a complementary method for the assessment of peptide conformation. Even though short linear peptides often adopt random coil conformations in solution, NMR has provided evidence for the presence of stable secondary structure, including β-turns and regions of α-helix in peptides in aqueous solution, even though these flexible peptides undergo rapid interconversion among many low-energy conformations.

During a basic one-dimensional (1D) pulsed NMR experiment, the sample is placed in a strong magnetic field, pulsed with radiofrequency radiation and the response of the system is recorded. The resonant frequency of a given nucleus is determined by the strength of the magnetic field and the magnetogyric ratio of the nucleus. Greater spectral dispersion and increased sensitivity can be achieved by increasing the magnetic field strength. The measured parameter is the chemical shift, *i.e.* the positions of resonances in the NMR spectrum, reflecting the relevant nuclei's sensitivity to the chemical environment. Differences in the chemical shifts of nuclei are caused by differences in their shielding from the external magnetic field by the electronic environment, therefore resulting in a slightly different resonant frequency for the particular nucleus.[59]

During sample preparation, paramagnetic impurities (*e.g.* dissolved oxygen) must be avoided and it is desirable to avoid or minimize contact of the sample with metal surfaces (*e.g.* spatulas, HPLC tubing). The sample concentration should be in the order of 1–5 mM at least for two-dimensional (2D) NMR studies. Poor solubility and aggregation tendencies in certain NMR solvents may cause challenges during sample preparation of peptides. Aggregation can be mitigated by adjusting the pH, ionic strength and solute concentration or by inclusion of detergents. In protic solvents such as water or methanol, the peptide backbone and side-chain amide protons exhibit proton exchange leading to broadening of their resonances. The exchange process is lowest around pH 3 for backbone amide protons and around pH 6 for hydroxyl protons. The actual pH used for NMR studies will also depend on the isoelectric point of the peptide. High pH values should be avoided in order to minimize disulfide bond exchange and degradation. For field-frequency locking, mixtures of protic and deuterated solvents (90 : 10) are often used. Dimethyl sulfoxide (DMSO) is also a popular choice for preparing peptide samples since it does not show chemical exchange reactions. The choice of temperature will be influenced by the thermal stability of the peptide. Linewidths are narrower at higher temperatures because of a decrease in the overall rotational correlation time of the molecule. As a result, it is suggested to record spectra at the highest temperatures based on sample stability. However, caution needs to be exercised when using protic and deuterated solvents where the exchange reaction increases with increase in temperature, causing broadening or even loss of signals. Spectrometers operating below 500 MHz are generally suitable only for the investigation of smaller molecules, such as peptides up to about 10–20 amino acid residues in length.[59] Peptides with >20 amino acid residues can be analyzed using higher field instruments that have a better dispersion mechanism.

A wealth of information can be obtained from the interactions of peptide side groups with the local environment and the overall peptide conformation using pH titration and solvent titration curves.[60,61]

Trifluoroethanol (TFE) has often been used to induce secondary structure (usually helix) in peptides that otherwise would not have any stable structure in aqueous solution. TFE is not simply a helix-inducing solvent, but rather should be considered as a helix-enhancing co-solvent, and stabilizes native peptides that have a helix-forming propensity.[62] Hydrogen exchange rate measurements give information on the solvent accessibility of exchangeable protons, which can be important in studying processes such as peptide folding and dynamics. The formation of hydrogen bonds in ordered structures, such as α-helices and β-sheets dramatically reduces the hydrogen–deuterium exchange rates of the amide protons. Usually a 2D NMR experiment is necessary to confirm the assignment of the slowly exchanging protons.

2D NMR is an extension of the 1D method into a second frequency axis, thus spreading the resonances in two dimensions and facilitating spectral analysis. The principal 2D ^{1}H-NMR experiments used for peptides are double-quantum filtered correlation spectroscopy (DQF-COSY), total correlation spectroscopy (TOCSY), homonuclear Hartmann–Hahn spectroscopy

(HOHAHA), nuclear Overhauser effect spectroscopy (NOESY) and rotating frame Overhauser effect spectroscopy (ROESY). The first two methods rely on through-bond correlations *via J*-coupling and the last two on through-space correlations.[59]

One of the advantages of NMR over crystallographic methods of structure determination is that it is possible to obtain information on the dynamic behavior of molecules. However, NMR spectroscopy is typically a very time-consuming experiment, especially when conformational information is required and a 2D experiment needs to be run. There are certain limitations of NMR spectroscopy in studying formulated peptides, such as the minimum concentration requirement of 1–5 mM and interference from excess of NMR-active excipients. Even though NMR spectroscopy is a useful high-resolution technique for higher-order structure studies, its application can be challenging for peptide drug product characterization and nearly impossible for comparability studies of generic peptide products at very low concentrations.

8.4 Tertiary Structure

The next level in higher-order structures of peptides and proteins is the overall 3D arrangement arising from side-chain interactions of the amino acid groups allowing higher levels of folding of the secondary structural elements. Tertiary folded structures are more compact but still flexible with hydrophobic groups oriented towards the core and hydrophilic groups on the surface. The three-dimensional structure is stabilized by different inter- and intramolecular interactions including disulfide bridges, hydrogen bonds, ionic interactions and even weak van der Waals forces. It is indeed the tertiary conformation that gives the polypeptide its unique structure, physicochemical properties and biological activity.[7]

Owing to the small size, not all peptides exhibit tertiary structure. There are certain peptides that do exhibit special tertiary structural elements, *e.g.* exendin-4 with its so-called Trp-cage structure,[63] but in most cases only solvent exposure and microenvironment of certain spectroscopically active amino acids (Trp, Tyr and Phe) can be studied and compared, if needed.

8.4.1 Near-UV CD Spectroscopy

Near-UV CD between 360 and 260 nm is used to study the effects of peptide side-chain changes on their local environments. The basics of CD spectroscopy are described in Section 8.3.1. Aromatic amino acid side chains such as Tyr, Trp, Phe and Cys are major contributors to the near-UV CD spectrum of peptides. Trp shows a peak above 280 nm and a shoulder feature between 290 and 305 nm. Tyr shows a distinct maximum around 275–280 nm and often its fine structures are overlapped by contributions from Trp in the long-wavelength tail. Phe bands are weaker but show a sharp fine structure between 255 and 270 nm. Cys can be observed at around 320 nm as a weak band in addition to contributions from disulfide bonds as a broad band near 260 nm.

Owing to the weak absorbance observed in the near-UV range, more material is required to produce a sufficient CD signal. The sample may be diluted in the buffer of choice at 1 mg mL^{-1} and analyzed using a 10 mm cuvette. Most physiological buffers and excipients pose no interference in the near-UV CD measurements. Figure 8.8 compares near-UV CD features found in a monoclonal antibody with two peptide molecules. The monoclonal antibody (red) showed a distinct maximum at 294 nm for Trp with a series of minima and maxima for CD activity by Tyr (270–290 nm) and Phe (250–270 nm) residues. A mid-sized peptide (blue) showed a wide valley around 270 nm that could be from surface exposed Tyr and Phe residues. A smaller peptide (green) exhibited signature peaks attributed to the two Phe residues in its structure.

As already mentioned, most peptides do not possess a prominent tertiary structure. Weak transitions are observed in therapeutic peptides corresponding to the individual aromatic groups which are accessible and absorb strongly. Near-UV CD spectra show a dependence on the choice of solvent and can affect the intensity of the signal. For instance, the near-UV CD profile of a hormonal peptide changes with differences in buffer composition and pH (Figure 8.9).

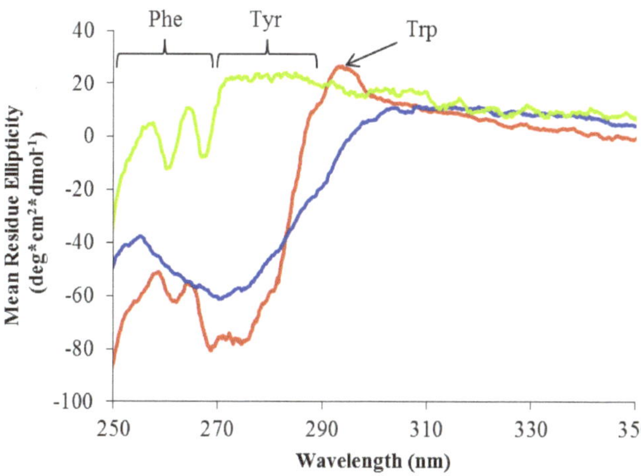

Figure 8.8 Near-UV CD spectral overlay of a monoclonal antibody (red trace) showing characteristic peaks for each of the representative amino acid regions. These features can be very subtle in smaller proteins and peptides. A mid-sized peptide (blue trace) and a smaller peptide (green trace) showed only those amino acid features which were surface exposed and accessible for CD analysis.

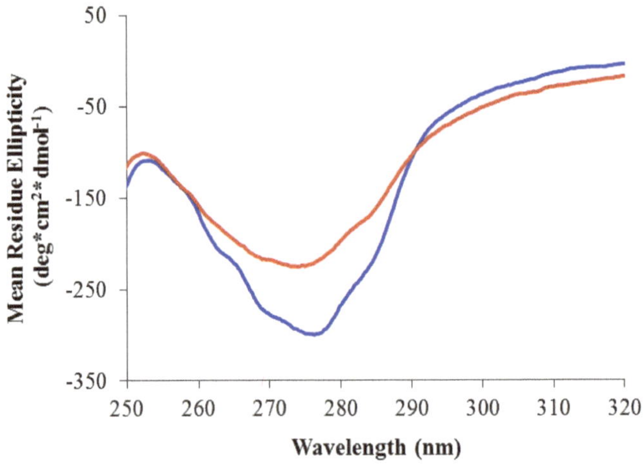

Figure 8.9 Near-UV CD spectra of a hormonal peptide showing the effects of buffer composition and pH on the intensity of the CD signal. The red trace represents 1 mg mL^{-1} peptide dissolved in dilute hydrochloric acid solution (pH 3) and the blue trace represents 1 mg mL^{-1} dissolved in phosphate-buffered saline (pH 7).

8.4.2 Intrinsic Fluorescence Spectroscopy

Even though most peptides do not possess a prominent tertiary structure, local environmental characteristics or differences can be studied by intrinsic fluorescence spectroscopy. In fluorescence emission spectroscopy, light of a certain wavelength is absorbed, causing an electron to jump from a ground state to an excited state. As the electron relaxes back to its ground state, energy is lost and the system emits light with a different (longer) wavelength and lower energy. Fluorescence is exhibited by molecules that have highly conjugated double bonds. In proteins and peptides, the aromatic amino acids tryptophan, phenylalanine and tyrosine exhibit UV light-induced intrinsic fluorescence. Trp-containing peptides have higher molar absorptivities compared with peptides containing Phe or Tyr and thus exhibit intense fluorescence signals at longer wavelengths. The physicochemical behavior of the fluorophore is extremely sensitive to its physicochemical environment. Changes in solvent polarity, for example exposure to or exclusion from water, can affect the fluorescent signal and provide information on the location of these particular amino acids if they are predominantly on the surface or in the internal folded region of the peptide. Thus, intrinsic fluorescence spectroscopy can be combined with other structure-elucidating techniques to investigate the overall conformation of therapeutic peptides.

In peptides, steady-state fluorescence can be used to monitor the intrinsic fluorescence spectra of Trp, Tyr and Phe residues and also the extrinsic fluorescence of associated dyes, at a given wavelength. Time-resolved fluorescence spectroscopy provides a measure of the time dependence of the fluorescence intensity after a picosecond excitation pulse, which can also be made as a function of emission wavelength. The time dependence of fluorescence emission is much more sensitive to changes in the environment around these intrinsic fluorophores than steady-state measurements of peak emission wavelength or intensity. In addition to this increased sensitivity to fast polypeptide conformational changes or interactions, the technique can also be performed using polarized light to provide time-resolved fluorescence anisotropy measurements.

The indole ring of tryptophan (Trp) is highly sensitive to its microenvironment, including properties such as solvent polarity, viscosity, pH and excipients. This is due to the large redistribution of electrons in the asymmetric indole ring upon absorption of photons. The intrinsic fluorescence intensity and peak position of Trp-containing peptides can be used to study overall conformational changes upon exposure to different solvent conditions. When completely exposed to an aqueous environment, Trp residues typically fluoresce in the 348–355 nm range. Any shift in the fluorescence wavelength maximum (λ_{max}) and also changes in intensity and fluorescence lifetime position will indicate changes in peptide conformation. In general, blue shift (lower wavelength) indicates the movement of aromatic side chains to a less polar (more hydrophobic) environment, whereas red shift (to higher wavelength) indicates movement to more polar environment. Specific interactions with other side chains and solutes may further modify the position of this emission maximum.

Even though Tyr and Phe exhibit intrinsic fluorescence, they show significantly lower quantum yields than Trp and are insensitive to changes in surroundings. As a result, they do not play a key role in tertiary structure studies. However, these groups can be monitored in peptides that do not contain Trp residues. In such cases, their fluorescence is not suppressed by the dominant Trp emission and their reduced environmental sensitivity may still be of some use. Tyr fluorescence in peptides can be of special interest because of its strong pH dependence.

The appropriate wavelengths for the exclusive excitation of each of the amino acid fluorophores are provided in Table 8.2 along with the corresponding emission wavelength window. Usually 1 nm spectral resolution and a 1 s integration time are chosen for initial runs, but can be further optimized, if needed. For example, the analysis of Phe fluorescence may require much wider slit width settings owing to its weak emission.[64] A typical fluorescence emission spectrum of tryptophan is given in Figure 8.10 and the weak emission spectrum of a phenylalanine-containing peptide is given in Figure 8.11.

Table 8.2 Excitation and emission wavelengths to study exclusive excitation of Trp-, Tyr- or Phe-containing peptides.

Amino acid fluorescence	Excitation wavelength/nm	Emission wavelength/nm
Tryptophan	295	300–400
Tyrosine	280	290–390
Phenylalanine	250–260	265–365

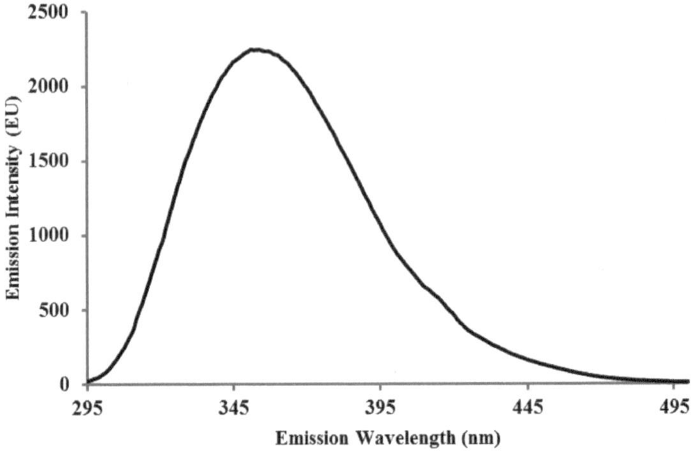

Figure 8.10 Representative intrinsic fluorescence spectrum of a tryptophan-containing peptide. Tryptophan-containing peptides always possess an emission spectrum dominated by Trp emission at around 348–355 nm.

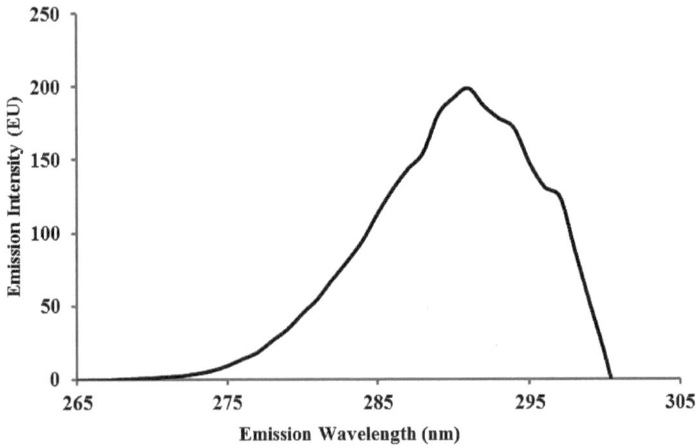

Figure 8.11 Representative intrinsic fluorescence spectrum of a peptide containing only phenylalanine. Peptides without a tryptophan or tyrosine can be studied for the intrinsic fluorescence of Phe, resulting in emission in the range 280–295 nm.

This technique would not give a great deal of characterization information, because if the peptide contains Trp, its emission will suppress any other signals. However, for comparability studies intrinsic fluorescence spectroscopy can be a valuable tool to demonstrate comparable/similar local environments of Trp (and Tyr or Phe) residues in the peptides under investigation. For instance, the intrinsic fluorescence of Trp was used in teriparatide lyophilized formulations to study the drug stability. When the reconstituted solution was analyzed during storage conditions, the λ_{max} showed a blue shift where the wavelength shifted from 354 to 350 nm, suggesting a tendency for solvent protection, and the substantial increase in fluorescence intensity observed was attributed to the formation of larger aggregates.[37] Hence for Trp-containing peptides intrinsic fluorescence can be a valuable technique to monitor the effects of solvents or excipients (injectables) and process conditions (lyophilization, spray drying) on the overall structure and peptide stability.

8.4.3 Differential Scanning Calorimetry

Differential scanning calorimetry (DSC) measures the heat capacity of a protein or peptide solution as a function of temperature at constant pressure, which is used to determine the enthalpy function of the system. Thermodynamic changes occur when a polypeptide undergoes folding or unfolding, just as other chemical reactions. These thermal transitions can be measured directly using liquid DSC. More specifically, one can determine the enthalpy of the changing system by measuring the molar heat capacity at constant pressure, C_p, over a range of temperatures.

DSC operates in differential mode, which means that the heat capacity of the peptide in aqueous solution is measured relative to the heat capacity of the corresponding buffer. As the peptide sample is heated, it unfolds, which is represented by the heat capacity curves. The area under the curve represents the enthalpy of unfolding. The temperature at the maximum of the curve is the transition temperature (also known as melting point, T_m) and the difference in the heat capacities of the native and unfolded states defines the temperature dependence of the enthalpy and entropy functions and, thus, the thermal dependence of peptide stability.[65] DSC also provides information about the modes of peptide unfolding, *i.e.* whether the unfolding process is two-state or multi-state. Two-state unfolding is the simplest model in which no significant intermediate states are observed. It can be defined by van't Hoff's equation and is an indication of the sharpness of the profile. The ratio of van't Hoff's enthalpy (ΔH_{VH}) and experimental calorimetric enthalpy (ΔH) can provide information regarding the mode of unfolding and the proportion of polypeptide structure that melts as a cooperative unit. If the value of the ratio is one, the observed transition follows a two-state model and if the ratio is less than one then it is a multi-state model with complex intermediate transitions corresponding to the unfolding of different domains.[66]

DSC can be used for peptide characterization by measuring the heat capacity profile of a 1 mg mL^{-1} peptide solution with reference to its buffer upon heating at 1 °C min^{-1} over the temperature range 20–100 °C. This technique finds potential applications in understanding the mechanisms of peptide–excipient interactions. It may also be used for comparability studies to demonstrate similarities or differences in unfolding features and melting temperatures of the peptides under investigation. However, it is essential to keep in mind that this technique may not be sensitive to detect the weak thermal transitions of small peptides. For such molecules, measuring far-UV CD spectra as a function of temperature can be used to study thermal denaturation.[67]

8.4.4 Disulfide Bonds

Covalent disulfide linkages play an important role in maintaining the overall structure–function and stability of proteins and certain peptides. These bonds can occur within a polypeptide chain (intra-chain bonding) and between two different chains in the same peptide (inter-chain bonding). However, disulfide bonds can also go astray, resulting in altered higher-order structures or aggregates *via* disulfide scrambling or exchange between other disulfide bonds or free cysteine residues in the same peptide or different peptides.

The location of disulfide bonds in peptides and proteins can be determined by the comparison of reduced and non-reduced peptide mapping results using mass spectrometry. Based on the amino acid sequence and the position of cysteine residues, the missing peptides in non-reduced mapping can lead to the position of the bond.

8.5 Quaternary Structure – Self-association

Self-association of peptides – including dimerization, oligomerization and formation of higher-order aggregates or irreversible fibrils – is frequently observed in pharmaceutical peptide formulations and can be accompanied by conformational distortion (unfolding or misfolding). Self-association is more frequently caused by non-covalent interactions such as hydrophobic or ionic interactions and hydrogen bonding; however, it can also occur due to non-specific covalent interactions.

The early stages of self-association generally result in reversible aggregates that are partly unfolded monomers or reversible oligomers. These reversible aggregates exist in equilibrium and will eventually dissociate in monomers *in vivo* due to dilution, circulation motion and moderate heating under the physiological conditions. For example, insulin can coexist in solution as monomers, dimers and hexamers. The equilibrium kinetics between the oligomer and monomer forms can be utilized to adjust the pharmacokinetics and mechanism of peptide drug delivery. The formation of reversible aggregates can sometimes improve the stability of the peptide formulation against chemical degradation and the formation of irreversible aggregates (*e.g.* fibrils). These soluble aggregates can be characterized for their size, structure, molecular weight and morphology. There is a preference to use *in situ* analytical methods to minimize the effects of sample handling and instrumentation on the delicate aggregate structures. Reversible aggregation can further result in the formation of irreversible aggregates under stress conditions.

Irreversible aggregation is a serious problem and results in the formation of unordered high molecular weight aggregates and precipitates. These aggregates are significant risk factors for immunogenicity responses, loss of efficacy, increased turbidity and viscosity of formulations and can result in the formation of particles, thus compromising the drug product quality. Irreversible aggregates are monitored and controlled during peptide manufacture, where the peptide is susceptible to physical stress conditions such as temperature changes, light, shaking and surface contact. Such aggregates also need to be studied during long-term storage and during forced degradation studies to understand the mechanism and kinetics of aggregation.

Therapeutic peptides show a tendency to form fibril-like irreversible aggregates, most likely due to their high structural flexibility. These structures have been studied using orthogonal techniques such as spectroscopic and imaging methods and reveal the presence of a β-sheet conformation together with fiber morphology. CD finds only limited use in aggregate analysis owing to the insoluble nature of the samples; however, it can be used in the early growth phases of the fibril aggregates to provide information on the secondary and tertiary structures. FTIR spectroscopy can help to confirm the presence of β-sheet fibrils and also to detect the presence of polymorphism. However, FTIR spectroscopy is challenged by its low resolution and high-concentration requirements. The size of the fibrils can also exceed the motion limit for solution NMR; however, recent progress in the use of solid-state NMR has permitted the high-resolution structure characterization of amyloid fibrils. High-resolution imaging tools such as atomic force microscopy and transmission electron microscopy have also been used to study oligomeric and fibrillar assemblies.[6]

The following subsections provide a brief overview of available analytical methodologies for studying peptide self-association. There are always advantages and disadvantages of each technique, hence the selection of a technique needs to be based on the purpose of the study, whether it is for product quality control (*i.e.* included in the specification) or whether it is for in-depth product understanding *via* thorough characterization. For the last purpose, the application of more than one, preferably orthogonal, technique is recommended.

8.5.1 Size-exclusion Chromatography with Traditional UV Detection (SEC-UV)

Size-exclusion chromatography (SEC), also called gel filtration chromatography, has been used widely in the biopharmaceutical industry for the purity analysis of peptides and proteins. This is the preferred assay to monitor and control self-association and is typically included in product specifications. Soluble higher-order aggregates, oligomers, monomers and fragments can be separated on a size-exclusion column where separation occurs on the basis of molecular size and hydrodynamic volume.[68] Most SEC stationary phases are typically made of homogeneous porous particles that cannot physically retain high molecular weight species such as aggregates and are therefore eluted first, followed by mid-sized species such as monomers. Fragments and buffer solutes are highly retained in the pores of the stationary phase and are eluted later in the chromatographic run. The column separation efficiency is monitored as the resolution between different molecular weight peaks and depends greatly on pore size, particle size, column dimensions and mobile phase composition. Columns with smaller pore sizes in the range 80–120 Å are useful for the size characterization of peptides. SEC employed with UV detection is a simple, inexpensive and commonly used technique where data are collected either between 210 and 220 nm (specific for peptide linkages) or at 280 nm (specific for aromatic side chains). Molecular weight can be indirectly determined by using commercially available gel filtration peptide–protein mixed standards and by plotting a calibration curve of molecular weight *versus* retention volume of their peak elution. Gel filtration standard mixtures also help determine the column efficiency and are often included as part of the system suitability. One drawback of SEC analysis is that peptides and their molecular weight variants can interact non-specifically with the column resin, resulting in underestimation of higher-order aggregates.[69] Modifications in column chemistry to make the resin biologically inert, use of higher salt concentrations or inclusion of L-arginine, or organic solvents in the mobile phase buffer have been studied to minimize this issue.[70] It is therefore important during in-depth characterization to use this method in conjugation with other aggregate-determining techniques such as analytical ultracentrifugation (AUC), asymmetric flow field-flow fractionation (AF4), hydrodynamic chromatography or native mass spectrometry to confirm the results obtained from SEC.[68,71]

8.5.2 Size-exclusion Chromatography Coupled with Multi-angle Light-scattering Detection (SEC-MALS)

Static light scattering is an excellent tool to determine molar mass (M), size (r_g) and the second virial coefficient (A_2) but gives no information about distribution, unless combined with a separation technique. SEC-UV is a relative method to determine molecular weight, relying on calibration standards. However, it assumes the molecular conformation and density to be same as those of the standards and can inaccurately estimate the molecular weight since the retention time is dependent on the hydrodynamic volume and not on molecular mass. SEC-MALS is a more powerful technique that can measure the absolute molecular mass and molecular size in solution, determine the oligomeric states, detect, characterize and quantify peptide aggregates, assess heterogeneity of the SEC peaks, characterize peptide conjugates and also determine protein/peptide extinction coefficients.

In a typical MALS experiment, solutions are illuminated by a laser light inducing an oscillating dipole moment in macromolecules that re-radiates the light, due to Rayleigh scattering. The more polarizable the macromolecule, the larger is the induced dipole and hence the greater the intensity of the scattered light. In order to analyze the scattering from a solution of macromolecules, it is necessary to know their polarizability relative to the surrounding medium, which is measured by a differential refractometer, typically coupled with the MALS detector. When there are many macromolecules in a solution, each scatters light, whereas the intensity of the scattered light is proportional to the concentration of the macromolecules and also to the molar mass. If peptide or protein monomers form a dimer, the originally independently moving two species, scattering incoherent light, will scatter coherent light, which is twice as intense as the total light scattered by the two independent monomers.[72]

Batch mode or in-line MALS has been widely adopted to assess the aggregation status of proteins and peptides. When coupling an analytical SEC column with in-line MALS, a UV detector or a refractive index detector is used as a widespread approach to determine the size distribution of oligomeric states and for the direct molecular mass and conformation analysis of aggregates. Recently, ion-exchange chromatography connected with MALS (IEC-MALS) has been used for the characterization of aggregates in concentrated peptide solutions without column overloading.[73] MALS also finds application in understanding the kinetics of irreversible and reversible self-association processes in insulin formulations.[74,75]

8.5.3 Sedimentation Velocity (SV)/Sedimentation Equilibrium (SE) – Analytical Ultracentrifugation (AUC)

Analytical ultracentrifugation (AUC) is one of the orthogonal methods to SEC, which determines the average molecular weight and distribution of aggregate species without dilution using a different separation mechanism. Even though AUC has a much lower throughput than other techniques, it is very sensitive to assess the stability of molecular aggregates at very low peptide concentrations.

In the AUC method, a large centrifugal force is applied to the sample, which results in separation of different molecular weight species based on density. Larger peptide species with higher densities tend to sediment at the bottom of the cell. The sedimentation pattern can be detected in real time using absorbance, refractive index, interference and fluorescence detectors.[76] The sedimentation coefficient is measured using the *sedimentation velocity* and the shape of the moving boundary and modeled to predict the hydrodynamic shape of the peptide and its oligomeric states. *Sedimentation equilibrium* measurements are used for the determination of molar mass and to study self-associations and heterogeneous interactions.[77]

Even though sedimentation analysis is a standard technique for the aggregation analysis of therapeutic proteins, it is not that widely used for peptides. Owing to their small mass, it can be challenging to determine the aggregation profile of peptides that distribute into monomers and small oligomers. However, it was possible to characterize the self-association of humanin, a 24-amino acid peptide, and other shorter neuroprotective peptides in different buffer conditions using AUC.[78]

8.5.4 Sub-visible Particles by Light Obscuration

AUC and SEC can be used to analyze peptide aggregates that are smaller than 0.1 µm. The light obscuration method is the gold standard technique employed in the biopharmaceutical industry for quantitating the number of sub-visible particles between 2 and 100 µm. This method can measure the size and number of particles, but lacks a differentiating ability to tell the difference between the nature of particles, *e.g.* proteinaceous particles, foreign material, microbubbles or silicone oil droplets, that can be found in protein and peptide solutions.[79]

In the light obscuration method, particles are analyzed based on the amount of light blocked from reaching the detector. The liquid sample flows through a narrow capillary tube and is subjected to a beam of light (laser). As particles pass through the laser beam, a certain amount of light is obscured and casts a shadow across the photodetector. This provides a rapid estimate of the concentration and size of micron-sized particles.

Light obscuration is the primary method described in USP ⟨788⟩ to determine particulate matter in injections and is typically part of product specifications to monitor and control particles in parenterals. However, this technique is not frequently used for in-depth characterization studies, owing its limitations and high sample volume requirement.

8.5.5 Sub-visible Particles by Micro Flow Imaging (MFI)

Flow imaging systems where a digital image is taken of a particle flowing through the optical system is relatively new technology compared with light obscuration. These images visualize the morphology, optical properties and size of the particles that flow through the system. Using this type of data, silicone oil droplets, microbubbles and some foreign material may be differentiated from the protein or peptide particles. However, in this technique the translucency of the particles can affect the accuracy of particle sizing.[80]

The flow imaging systems are best suited as a characterization tool, whereas the light obscuration method can be used for routine screening and lot release. The combination of the two techniques with certain advanced instrument settings has enabled a clearer picture to be gained of the distribution of the size and particle type of sub-visible particles present in peptide and protein samples.

8.5.6 Dynamic Light Scattering (DLS)

DLS is another useful tool for characterizing the hydrodynamic radius of peptides and is also available in a high-throughput setup. It can measure colloidal particles in the range from 1–2 nm to 3–5 µm. However, its application to smaller sized peptides may be limited owing to its poor sensitivity and accuracy at low concentrations.

DLS characterizes molecules and particles based on their Brownian motion in solution. It measures the amount of light scattered by the moving molecules and results in time-dependent fluctuations in scattered light intensity detected by a fast photon counter. The fluctuations show a direct correlation to the rate of diffusion of the molecule through the solvent, which is related in turn to the particle hydrodynamic radius.[81]

DLS has been used to evaluate the compatibility of preservatives for multi-use formulations and time-resolved DLS measurements have also been used to establish the association kinetics and reversibility of peptide self-association. Wang *et al.* characterized the reversible transformations in the oligomeric state of liraglutide, a 37-residue glucagon-like peptide, by measuring changes in the hydrodynamic diameter and molecular weight by DLS.[82]

8.5.7 Asymmetric Flow Field-flow Fractionation (AF4)

Asymmetric flow field-flow fractionation is an emerging one-phase chromatographic technique to characterize self-association. High-resolution separation is achieved in a very thin flow channel against which a perpendicular force field is applied. The flow and sample are confined within a channel between two plates that are separated by a spacer foil. The spacer foil has a typical thickness of 100–500 μm. The upper channel plate is impermeable; however, the bottom channel plate is permeable and made of a porous frit material. An ultrafiltration membrane with a typical size barrier of 10 kDa covers the bottom plate to prevent the sample from penetrating the channel. This size barrier can be replaced with a 1 or 5 kDa membrane to accommodate smaller peptides, but there is still a huge loss of material through the membrane.

Within the flow channel, the laminar flow of the liquid creates a parabolic flow profile – the stream moves more slowly closer to the boundary edges than it does at the center of the channel flow. When the perpendicular force field is applied to the laminar stream, the proteins or peptides are driven towards the boundary layer. Diffusion caused by Brownian motion, in turn, creates a counteracting motion. Smaller particles, which have relatively large diffusion coefficients, tend to reach an equilibrium position higher up in the channel, where the longitudinal flow is faster. Thus, the velocity gradient flowing inside the channel separates different sizes of particles. Small proteins elute from the channel before large proteins – the opposite of SEC separation in which the large molecules elute first.

In AF4, there is no stationary phase to interact with the samples and therefore separation occurs in a gentle manner without involving much shearing forces. Sample preparation is simple and large insoluble protein and peptide aggregates can be detected with good resolution. However, it is a less mature technique than chromatography and involves optimization of multiple method parameters (channel dimensions, types of membranes and flow rate) to be used reliably for sample analysis. It can be coupled with multiple detection systems such as UV-visible, MALS or MS. Even though there are very limited applications for peptide characterization, a few groups have studied the aggregation of glucagon, a 29-residue peptide hormone, using AF4-MALS, static light scattering (SLS) and DLS techniques.[83,84]

8.5.8 Compositional Gradient Multi-angle Light Scattering (CG-MALS)

CG-MALS is gaining increasing importance as a protein and peptide characterization tool that operates on the basis of static light scattering (SLS). The intensity of light scattered from macromolecules in solution is proportional to the product of the concentration and the weight-averaged molar mass (M_w). As peptide complexes are formed, M_w increases. In a reversible association, the ratio of peptide complexes to peptide monomers reaches an equilibrium value that depends on the initial concentration of each species, and also on the buffer conditions. Under changing compositions and concentrations, different ratios will lead to differing values of M_w. By analyzing a series of SLS measurements made under various compositions and concentrations, it is possible to determine which complexes are being formed and their respective binding affinities in the form of the association constant, K_a, or the dissociation constant, $K_d = 1/K_a$.[85]

Attri *et al.* characterized the oligomerization of insulin in the presence and absence of Zn by using CG-MALS and CG-DLS. In the absence of Zn, insulin appears to follow an isodesmic association model with a minimum K_D of 40 μM at pH 7.2. In the presence of Zn and at neutral pH, insulin is known to form hexamers, with hexamer association complete beyond ~0.3 mg mL^{-1}. The present study showed that insulin hexamers further self-associate isodesmically with a K_D of just under 100 μM.[86,87]

8.6 Conclusion

This chapter provides a brief overview of the different analytical methodologies currently available for studying the higher-order structure of peptides. It also provides some pointers on how to perform certain studies or what precautions need to be taken while performing these analyses to obtain reliable and valuable results.

A few key information points are summarized here as a concluding remark for this chapter. In designing higher-order structure studies, one should always consider the advantages and limitations of each technique beyond its primary applicability, *i.e.* best suited for quality controlling (release and stability), characterizing or comparing peptides. The basic recommendation is to always apply more than one technique to study a certain peptide higher-order structure level and preferably to use orthogonal methods.

Table 8.3 summarizes the key recommendations for scientists designing these studies.

Table 8.3 Summary of higher-order structure techniques along with general recommendations.

Structural level	Analytical technique	Quality control	Characterization	Comparability
Primary structure	Molecular mass[a]	Yes	Yes	Yes
	Peptide mapping[a]	No	Yes	Yes
	Peptide sequencing	No	Yes	Yes
	Amino acid analysis	Yes	Yes	Yes
	Enantiomeric purity (D-amino acids in synthetic peptides)	Yes	Yes	Yes
Secondary structure	Far-UV CD spectroscopy[a]	No	Yes	Yes
	FTIR spectroscopy[a]	No	Yes	Yes
	Raman spectroscopy	No	Yes	Yes
	NMR spectroscopy	No	Yes	Yes
Tertiary structure	Near-UV CD spectroscopy	No	Yes	Yes
	Intrinsic fluorescence spectroscopy[a]	No	No	Yes
	DSC	No	Yes	Yes
	Disulfide bonds	No	Yes	Yes
Quaternary structure – self association	SEC-UV[a]	Yes	Yes	Yes
	SEC-MALS[a]	No	Yes	Yes
	SV/SE-AUC	No	Yes	Yes
	Light obscuration	Yes	No	No
	MFI	No	Yes	Yes
	DLS	No	Yes	Yes
	AF4	No	Yes	Yes
	CG-MALS	No	Yes	Yes

[a]Analytical techniques recommended as primary methods of choice in the design of structural characterization and comparability studies applicable to all sizes of therapeutic peptide products. Other techniques can provide supportive information, if needed.

References

1. K. Fosgerau and T. Hoffmann, *Drug Discovery Today*, 2015, **20**, 122–128.
2. V. J. Hruby and T. O. Matsunaga, *Synthetic peptides: a user's guide*, 2002, **2**, 292–376.
3. T. Uhlig, T. Kyprianou, F. G. Martinelli, C. A. Oppici, D. Heiligers, D. Hills, X. R. Calvo and P. Verhaert, *EuPa Open Proteomics*, 2014, **4**, 58–69.
4. J. L. Lau and M. K. Dunn, *Bioorg. Med. Chem.*, 2018, **26**, 2700–2707.
5. A. A. Kaspar and J. M. Reichert, *Drug Discovery Today*, 2013, **18**, 807–817.
6. S. M. D'Addio, J. R. Bothe, C. Neri, P. L. Walsh, J. Zhang, E. Pierson, Y. Mao, M. Gindy, A. Leone and A. C. Templeton, *J. Pharm. Sci.*, 2016, **105**, 2989–3006.
7. S. A. Berkowitz and D. J. Houde, in *Biophysical Characterization of Proteins in Developing Biopharmaceuticals*, Elsevier, 2015, pp. 1–21.
8. J. D. Tipton, J. C. Tran, A. D. Catherman, D. R. Ahlf, K. R. Durbin and N. L. Kelleher, *J. Biol. Chem.*, 2011, DOI: 10.1074/jbc.R111.239442.

9. S. M. Koza, Peptide Mapping 101: Essential Tools for Effective Development and Characterization, Part1: Introduction to Peptide Mapping, http://www.waters.com/webassets/cms/events/docs/PeptideMappingWebinarFinal.pdf, (accessed July 2018).
10. M. Mann, *Clin. Chem.*, 2016, **62**, 293–294.
11. J. B. Smith, *Peptide Sequencing by Edman Degradation*, 2001.
12. J. P. K. Weder and H. D. Belitz, in *Encyclopedia of Food Sciences and Nutrition*, ed. B. Caballero, Academic Press, 2nd edn., 2003, pp. 4805–4818.
13. I. Ilisz, A. Péter and W. Lindner, *TrAC, Trends Anal. Chem.*, 2016, **81**, 11–22.
14. D. Eisenberg, *Proc. Natl. Acad. Sci. U. S. A.*, 2003, **100**, 11207–11210.
15. S. A. Hollingsworth and P. A. Karplus, *Biomol. Concepts*, 2010, **1**, 271–283.
16. L. Regan, *Proc. Natl. Acad. Sci. U. S. A.*, 1997, **94**, 2796–2797.
17. J.-P. Schermann, in *Spectroscopy and Modeling of Biomolecular Building Blocks*, ed. J.-P. Schermann, Elsevier, Amsterdam, 2008, pp. 251–296.
18. C. Tonlolo and E. Benedetti, *Trends Biochem. Sci.*, 1991, **16**, 350–353.
19. Y. Dou, P.-F. Baisnée, G. Pollastri, Y. Pécout, J. Nowick and P. Baldi, *Bioinformatics*, 2004, **20**, 2767–2777.
20. J. S. Nowick, *Acc. Chem. Res.*, 2008, **41**, 1319–1330.
21. C. Venkatachalam, *Biopolymers*, 1968, **6**, 1425–1436.
22. H. Lodish, A. Berk, S. L. Zipursky, P. Matsudaira, D. Baltimore and J. Darnell, *Molecular Cell Biology*, W. H. Freeman, New York, 4th edn, 2000, Section 3.1.
23. A. M. C. Marcelino and L. M. Gierasch, *Biopolymers*, 2008, **89**, 380–391.
24. N. J. Greenfield, *Anal. Biochem.*, 1996, **235**, 1–10.
25. Y. Wei, A. A. Thyparambil and R. A. Latour, *Biochim. Biophys. Acta Proteins Proteom.*, 2014, **1844**, 2331–2337.
26. S. R. Martin and P. M. Bayley, in *Calcium-Binding Protein Protocols: Volume 2: Methods and Techniques*, Springer, 2002, pp. 43–55.
27. S. M. Kelly, T. J. Jess and N. C. Price, *Biochim. Biophy. Acta Proteins Proteom.*, 2005, **1751**, 119–139.
28. S. M. Kelly and N. C. Price, *Curr. Protein Pept. Sci.*, 2000, **1**, 349–384.
29. P. Manavalan and W. C. Johnson Jr, *Anal. Biochem.*, 1987, **167**, 76–85.
30. G. Böhm, R. Muhr and R. Jaenicke, *Protein Eng. Des. Sel.*, 1992, **5**, 191–195.
31. A. Micsonai, F. Wien, É. Bulyáki, J. Kun, É. Moussong, Y.-H. Lee, Y. Goto, M. Réfrégiers and J. Kardos, *Nucleic Acids Res.*, 2018, **46**, W315–W322.
32. H. Bierau, G. E. Tranter, D. D. LePevelen, C. E. Giartosio and C. S. Jone, *BioProcess Int.*, 2008, 52–59.
33. B. R. Baker and R. L. Garrell, *Faraday Discuss.*, 2004, **126**, 209–222.
34. N. N. Dinh, B. C. Winn, K. K. Arthur and J. P. Gabrielson, *Anal. Biochem.*, 2014, **464**, 60–62.
35. G. Shanmugam and P. L. Polavarapu, *Proteins: Struct., Funct., Bioinf.*, 2006, **63**, 768–776.
36. R. P. Nagarkar, B. M. Murphy, X. Yu, M. Cornell Manning and W. A. Al-Azzam, *Curr. Pharm. Biotechnol.*, 2013, **14**, 199–208.
37. G. Merutka, B. M. Murphy, R. W. Payne, G. A. Wilson, J. E. Matsuura, C. S. Henry and M. C. Manning, *Eur. J. Pharm. Biopharm.*, 2016, **99**, 84–93.

38. Q.-X. Hua and M. A. Weiss, *J. Biol. Chem.*, 2004, **279**, 21449–21460.
39. B. P. Ward, N. L. Ottaway, D. Perez-Tilve, D. Ma, V. M. Gelfanov, M. H. Tschop and R. D. Dimarchi, *Mol. Metab.*, 2013, **2**, 468–479.
40. Y. Imura, M. Nishida and K. Matsuzaki, *Biochim. Biophys. Acta*, 2007, **1768**, 2578–2585.
41. C. J. Morris, K. Beck, M. A. Fox, D. Ulaeto, G. C. Clark and M. Gumbleton, *Antimicrob. Agents Chemother.*, 2012, **56**, 3298–3308.
42. P. I. Haris and D. Chapman, *Biopolymers*, 1995, **37**, 251–263.
43. J. Kong and S. Yu, *Acta Biochim. Biophys. Sin.*, 2007, **39**, 549–559.
44. P. Garidel and H. Schott, *BioProcess Int.*, 2006, **4**, 48–55.
45. H. Yang, S. Yang, J. Kong, A. Dong and S. Yu, *Nat. Protoc.*, 2015, **10**, 382.
46. W. K. Surewicz, H. H. Mantsch and D. Chapman, *Biochemistry*, 1993, **32**, 389–394.
47. V. V. Andrushchenko, H. J. Vogel and E. J. Prenner, *J. Pept. Sci.*, 2007, **13**, 37–43.
48. L. E. Valenti, M. B. Paci, C. P. De Pauli and C. E. Giacomelli, *Anal. Biochem.*, 2011, **410**, 118–123.
49. P. I. Haris, *Pharm. Pharmacol. Commun.*, 1999, **5**, 15–25.
50. K. E. Wilcox, E. W. Blanch and A. J. Doig, *Biochemistry*, 2016, **55**, 3794–3802.
51. M. Jackson, H. H. Mantsch and J. H. Spencer, *Biochemistry*, 1992, **31**, 7289–7293.
52. D. J. Hirsh, J. Hammer, W. L. Maloy, J. Blazyk and J. Schaefer, *Biochemistry*, 1996, **35**, 12733–12741.
53. A. Wahlstrom, L. Hugonin, A. Peralvarez-Marin, J. Jarvet and A. Graslund, *FEBS J.*, 2008, **275**, 5117–5128.
54. R. Liang, X. Li, Y. Shi, A. Wang, K. Sun, W. Liu and Y. Li, *Int. J. Pharm.*, 2013, **454**, 344–353.
55. E. Smith and G. Dent, *Modern Raman Spectroscopy: A Practical Approach*, John Wiley & Sons, 2013.
56. N. C. Maiti, M. M. Apetri, M. G. Zagorski, P. R. Carey and V. E. Anderson, *J. Am. Chem. Soc.*, 2004, **126**, 2399–2408.
57. Z. Q. Wen, *J. Pharm. Sci.*, 2007, **96**, 2861–2878.
58. A. Rygula, K. Majzner, K. M. Marzec, A. Kaczor, M. Pilarczyk and M. Baranska, *J. Raman Spectrosc.*, 2013, **44**, 1061–1076.
59. M. G. Hinds and R. S. Norton, *Mol. Biotechnol.*, 1997, **7**, 315–331.
60. M. P. Williamson, in *Spectroscopic Methods and Analyses*, Springer, 1993, pp. 69–85.
61. J. E. Nielsen, in *Annual Reports in Computational Chemistry*, Elsevier, 2008, vol. 4, pp. 89–106.
62. F. Sonnichsen, J. Van Eyk, R. Hodges and B. Sykes, *Biochemistry*, 1992, **31**, 8790–8798.
63. J. W. Neidigh, R. M. Fesinmeyer, K. S. Prickett and N. H. Andersen, *Biochemistry*, 2001, **40**, 13188–13200.
64. M. R. Liyanage, K. Bakshi, D. B. Volkin and C. R. Middaugh, in *Therapeutic Peptides*, Springer, 2014, pp. 237–246.

65. W. W. Streicher and G. I. Makhatadze, in *Protein Folding Protocols*, Springer, 2007, pp. 105–113.
66. I. B. Durowoju, K. S. Bhandal, J. Hu, B. Carpick and M. Kirkitadze, *J. Visualized Exp.*, 2017, DOI: 10.3791/55262.
67. N. J. Greenfield, *Nat. Protoc.*, 2006, **1**, 2527.
68. S. Fekete, A. Beck, J.-L. Veuthey and D. Guillarme, *J. Pharm. Biomed. Anal.*, 2014, **101**, 161–173.
69. K. Štulík, V. Pacáková and M. Tichá, *J. Biochem. Biophys. Methods*, 2003, **56**, 1–13.
70. T. Arakawa, D. Ejima, T. Li and J. S. Philo, *J. Pharm. Sci.*, 2010, **99**, 1674–1692.
71. A. Revillon, *J. Liq. Chromatogr. Relat. Technol.*, 1994, **17**, 2991–3023.
72. M. Chen, Characterizing Solution Properties of Biopharmaceuticals Using Light Scattering Combined with Orthogonal Analytical Approaches, https://www.wyatt.com/library/webinars/characterizing-solution-properties-of-biopharmaceuticals-using-light-scattering-combined-with-orthogonal-analytical-approaches.html, (accessed July, 2018).
73. H. Amartely, O. Avraham, A. Friedler, O. Livnah and M. Lebendiker, *Sci. Rep.*, 2018, **8**, 6907.
74. A. K. Attri, C. Fernández and A. P. Minton, *Biophys. Chem.*, 2010, **148**, 28–33.
75. M. H. Jensen, P.-O. Wahlund, J. K. Jacobsen, B. Vestergaard, M. van de Weert and S. Havelund, *J. Chromatogr. B*, 2011, **879**, 2945–2951.
76. J. L. Cole, J. W. Lary, T. P. Moody and T. M. Laue, *Methods Cell Biol.*, 2008, **84**, 143–179.
77. J. Lebowitz, M. S. Lewis and P. Schuck, *Protein Sci.*, 2002, **11**, 2067–2079.
78. T. Arakawa, T. Niikura, Y. Kita and F. Arisaka, *Drug Discov. Ther.*, 2009, 208.
79. S. K. Singh, N. Afonina, M. Awwad, K. Bechtold-Peters, J. T. Blue, D. Chou, M. Cromwell, H. J. Krause, H. C. Mahler, B. K. Meyer, L. Narhi, D. P. Nesta and T. Spitznagel, *J. Pharm. Sci.*, 2010, **99**, 3302–3321.
80. A. Ríos Quiroz, G. Québatte, F. Stump, C. Finkler, J. Huwyler, R. Schmidt, H.-C. Mahler, A. V. Koulov and M. Adler, *Anal. Chem.*, 2015, **87**, 6119–6124.
81. P. J. Wyatt, *Anal. Chim. Acta*, 1993, **272**, 1–40.
82. Y. Wang, A. Lomakin, S. Kanai, R. Alex and G. B. Benedek, *Mol. Pharmaceutics*, 2015, **12**, 411–419.
83. G. Yohannes, M. Jussila, K. Hartonen and M.-L. Riekkola, *J. Chromatogr. A*, 2011, **1218**, 4104–4116.
84. C. C. Hoppe, L. T. Nguyen, L. E. Kirsch and J. M. Wiencek, *J. Biol. Eng.*, 2008, **2**, 10.
85. D. Some, E. Hitchner and J. Ferullo, *American Biotechnology Laboratory®*, 2009, **27**, 16.
86. D. Some, *Biophys. Rev.*, 2013, **5**, 147–158.
87. A. K. Attri, C. Fernandez and A. P. Minton, *Biophys. Chem.*, 2010, **148**, 23–27.
88. E. Martz, Protein-structure-4-levels-III-flat, http://proteopedia.org/wiki/index.php/Image:Protein-structure-4-levels-III-flat.png, (accessed August, 2018).

CHAPTER 9

Impurity Characterization and Quantification by Liquid Chromatography–High-resolution Mass Spectrometry

KUI ZENG,*[a] MICHAEL T. BOYNE,[b] TIM K. TOBY[b] AND CONNIE RUZICKA[a]

[a] Division of Pharmaceutical Analysis, US Food and Drug Administration, 645 South Newstead Avenue, St Louis MO 63110, USA;
[b] COUR Pharmaceuticals Development Company, 2215 Sanders Road, Northbrook IL 60062, USA
*Email: kui.zeng@fda.hhs.gov

9.1 Introduction

Therapeutic peptides have become an important and fast-growing segment in the pharmaceutical industry. Consequently, the number of peptide-related Abbreviated New Drug Applications (ANDAs) is dramatically increasing at the Office of Generic Drugs at the US Food and Drug Administration (FDA).[1,2] However, these molecules are more complex in terms of chemistry, manufacturing and controls (CMC) than most small molecules, leading to challenges in the drug development process. For example, peptide-related therapeutics tend to have additional process impurities such as amino acid (AA) deletions or insertions and they are also susceptible to a variety of degradation pathways.[3,4] Understanding the impurity profile of

Drug Discovery Series No. 72
Peptide Therapeutics: Strategy and Tactics for Chemistry, Manufacturing and Controls
Edited by Ved Srivastava

Published by the Royal Society of Chemistry, www.rsc.org

peptide drug products is important to ensure product quality and safety and avoid potential adverse events associated with impurity-induced immunogenicity. Standard peptide impurity profiling methods using liquid chromatography (LC) online with ultraviolet (UV) detection (LC–UV) demonstrate significant limitations for this application. For example, UV detection does not demonstrate the specificity required to differentiate some peptide-related impurities and LC–UV cannot unequivocally identify impurities by retention time and UV absorbance alone or quantify impurities that coelute. Hence the more sensitive and specific strategy of coupling high-resolution mass spectrometry (HRMS) to LC (LC–HRMS) is better suited for characterizing and quantifying peptide-related impurities.

The challenges associated with the manufacture and approval for human use of generic peptide products percolate beyond drug development and CMC into the sphere of regulatory review. The difficulties are partly due to the complexity of the drug substances, but can also be attributed to a lack of general guidelines for establishing method validation and performing comparative analytical analysis for peptide drugs. The complexity of peptide products is simply a result of the biology and chemistry involved in their manufacture. For example, side reactions, incomplete coupling and truncations may occur during the commonly used solid-phase peptide synthesis (SPPS) method, creating a number of different but highly related impurities. Numerous degradation pathways (*e.g.* oxidation, deamidation, isomerization and/or truncation), possible secondary structure formation and aggregation can also occur during the synthesis of peptide products. Therefore, the AA composition of therapeutic peptides and their impurities may contain partially identical sequences, which complicates their detection and differentiation. Importantly, because these peptide-related impurities may affect the quality, safety and efficacy of peptide drugs, sensitive and specific detection is necessary. However, the current International Council on Harmonisation of Technical Requirements for Registration of Pharmaceuticals for Human Use (ICH) guidelines regarding the testing and quality of pharmaceutical products do not specifically include peptides. Additionally, the method validation guidance that is already established may not translate directly to peptide drug products owing to the lack of suitable internal standards and/or reference materials. Typically, generic drug applications of peptides rely on comparisons with approved drug products or Reference Listed Drugs (RLDs) to demonstrate therapeutic equivalency and establish quality control assessments.

Recently, the FDA published a *Draft Guidance for Industry: ANDAs for Certain Highly Purified Synthetic Peptide Drug Products That Refer to Listed Drugs of rDNA Origin*;[5] the rules that govern the regulatory acceptability of peptide-related impurities are as described in the following:

> "In general, for the peptides covered by this guidance, applicants should identify each peptide-related impurity that is 0.10 percent of the drug substance or greater. [1.] For each peptide-related impurity common to the proposed generic synthetic peptide and its RLD, the level of the

peptide-related impurity in the proposed generic synthetic peptide is not more than that found in the RLD; [2.] The new specified peptide-related impurity level for the proposed generic synthetic peptide is no more than 0.5 percent of the drug substance; and [3.] For each new specified peptide-related impurity that is not more than 0.5 percent of the drug substance, the ANDA applicant should characterize the impurity. Further, the ANDA applicant should provide justification for why such impurity does not affect the safety of the proposed generic synthetic peptide (including with respect to immunogenicity) and why it does not affect its effectiveness."

Furthermore, the FDA recommends that applicants apply sensitive and high-resolution analytical procedures (*e.g.* LC–HRMS)[6] to detect and characterize peptide-related impurities in a proposed generic synthetic peptide in comparison with the RLD.

9.1.1 Impurities of Peptides

Peptide drug product impurities can be divided into two categories: either process or product related (Table 9.1). The process-related impurities present in a peptide product differ based on the specific manufacturing process used to create the product. There are two general routes by which most therapeutic peptides are manufactured: (1) peptides are recombinantly expressed in host cells and then subsequently purified; and (2) peptides are chemically manufactured *via* synthetic organic chemistry. Process-related impurities in recombinant peptides resulting from the former manufacturing route mostly stem from the biological context of the specific expression system involved and include host cell proteins and DNA, media residues, inducers, antibiotics, fetal bovine serum, antifoam, growth factors, trypsin and insulin. Other impurities originate from post-expression processing and purification reagents such as column leachable.

Process-related impurities in synthetic peptides manufactured by chemical reaction schemes include deletion and insertion of amino acids related to incomplete fluorenylmethyloxycarbonyl (Fmoc) deprotection and excess use of amino acid reagents, incomplete removal of protecting groups, oxidation/reduction and side- and end-chain impurities. Generally, the impurities resulting from the primary peptide drug product manufacturing strategy can be minimized using an appropriate and well-controlled manufacturing process.

There are currently no ICH guidelines regarding the testing and quality of peptide products, although the FDA has recently published a draft Guidance for Industry.[5] However, ICH Guidances Q3A, Q3B and Q6B are still applied to the development of peptide drugs. According to this guidance, the developer should summarize the impurities observed during the manufacture and/or stability studies of the new drug substances and product.[7–9] Three thresholds for impurities were defined in this guidance: the Reporting Threshold (RT), Identification Threshold (IT) and Qualification Threshold (QT).

Table 9.1 Typical peptide process-related impurities/contaminants during manufacture and peptide product-related impurities/degradants mainly during storage.

Impurity	Mechanism of formation and additional information
Amino acid deletion	Incomplete coupling due to incomplete removal of the protecting group of the last coupled amino acid or insufficient activation of the incoming amino acid
Amino acid insertion	Insertion of an additional amino acid into the peptide sequence due to excess of Fmoc-protected amino acids and improper washing
Truncation	Missing one or more amino acid residues at either the N- or C-terminal end due to precipitation of resin beads during synthesis
Incomplete protecting group removal	Results in protecting groups covalently attached to peptide sequences
Oxidation/reduction	Met, Cys, Tyr, Trp oxidation induced by temperature, pH, metal ions and buffers during synthesis and/or storage
Diastereoisomerization/ racemization	Asp, Asn at pH ≥8, more commonly occurs during synthesis. Alteration of L-amino acids to D,L mixtures. Degradation occurs through succinimide formation
Side- and end-chain impurities	Amination in the side chain of Arg with a mass difference of +15 Da. Other examples include acetylation, histamine-substituted guanidine with a mass difference of +126 Da
Dimers	Self-association of hydrophobic amino acid residues in peptides to form aggregates
Peptide counter ions	Trifluoroacetic acid (TFA) in the SPPS process, for example
Structurally unrelated contamination	The contaminant can have a completely different sequence than the desired peptide product
β-Elimination	Susceptible residues under alkaline conditions (pH ≥8) are Cys, Lys, Phe, Ser, Thr
Diketopiperazine, pyroglutamate	Diketopiperazine formation occurs if Pro or Gly residues are at position 1 or 2. Pyroglutamic acid forms if Gln is the N-terminal amino acid residue
Deamidation	Asn and Gln direct hydrolysis at pH <3. Intramolecular cyclization can occur at pH >6
Isomerization	Rapid isomerization of Asp at pH <5
Proteolysis	Peptide backbone proteolysis near N- and C-termini upon exposure to harsh conditions such as low pH, high temperature or enzymes
Aggregation and precipitation	Association of hydrophobic amino acid residues to form aggregates and precipitation occurs if on a macroscopic scale
Disulfide exchange	Change in conformation due to reactions between disulfides within a peptide chain
Denaturation	Modification of the peptide structure that changes its physical, chemical and biological properties, which occurs in the presence of denaturing agents

RT refers to the limit above which an impurity should be reported. IT, on the other hand, is the limit above which a degradation product or impurity should be structurally characterized and confidently identified. The QT is the limit above which the impurity should be qualified, meaning that data must be collected and evaluated to establish the biological safety of the impurity or the given impurity profile at the level(s) specified. For example, peptide-related substances in chemically synthesized peptide active pharmaceutical ingredients (APIs) are expected to adhere to an RT of 0.1%, IT of 0.5% and QT of 1.0%.

9.1.2 LC–HRMS

Over the last 20 years, LC–MS has become the gold standard for peptide and protein characterization studies. The basic functional modules of a typical LC–MS system that can be found in proteomics laboratories across industries and academic disciplines are illustrated in Figure 9.1. Note the inclusion of LC on the front end as a means of analyte separation in the time domain. Various chromatographic media can be incorporated in LC columns to separate analytes in a sample based on their physical and chemical properties (*e.g.* molecular size in size-exclusion chromatography or hydrophobicity in reversed-phase chromatography), which permits the sensitive detection of isolated analytes during downstream MS and more facile spectral analysis after measurement. The mass spectrometer itself is a complex system of several components that facilitate ionization at the source, ion funneling and transit *via* ion optics, ion mass-to-charge ratio (*m*/*z*) determination in various common mass analyzers and systems for ion detection.[10] Since the routine use of MS began to take hold in the 1950s, decades of technical development and instrumentation improvement have made LC–MS a ubiquitous state-of-the-art technique for qualitative and quantitative analytical applications. One of the primary advantages of MS-based methods is that no *a priori* knowledge of the composition of the analyte is technically required, unlike in UV analysis; hence LC–MS is ideally suited for discovery and characterization studies for a wide range of molecules.[11] The principles and applications of LC–MS have been thoroughly explored in many books and articles.[11–13] Additionally, a number of best practice guides have been developed for daily use, accurate mass measurement and reliable quantification.[14–16]

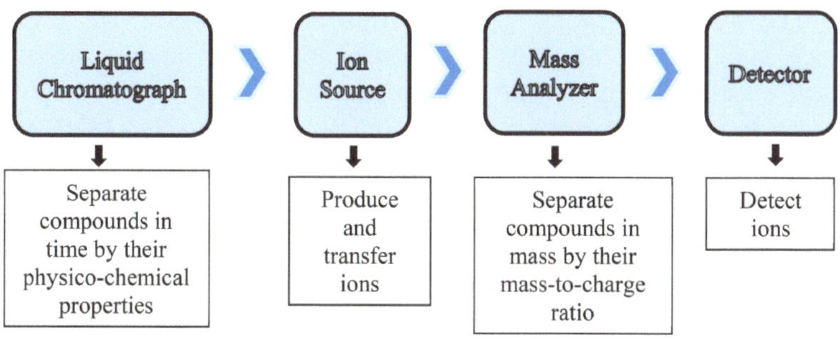

Figure 9.1 Basic functional modules of an LC–MS system.

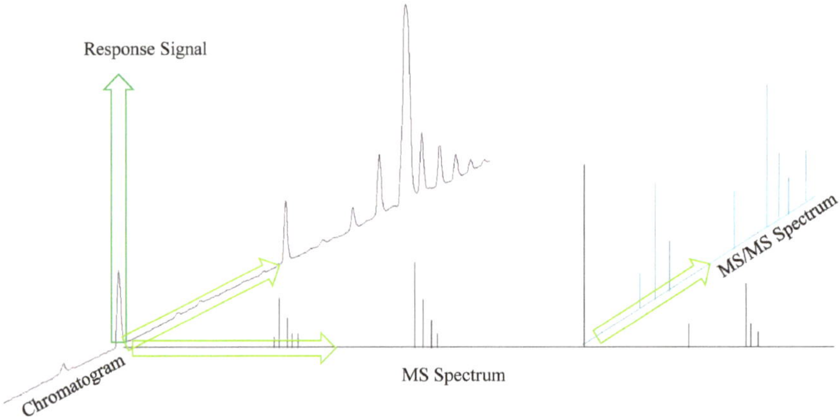

Figure 9.2 The three-dimensional separation power of the LC–HRMS technique to detect and identify each component in a complex mixture.

LC–HRMS combines two highly selective but functionally different techniques to isolate and measure the analyte(s) of interest in highly complex mixtures: LC differentiates compounds by their physical and/or chemical properties, whereas MS differentiates compounds by their unique *m*/*z* ratios. Furthermore, most commercial mass spectrometers are configured to perform tandem mass spectrometry (MS/MS or MS^2), a technique that facilitates protein and peptide identification by submitting detected gas-phase ions to fragmentation followed by another round of mass analysis. In this manner, the identities of polypeptide chains can be ascertained from the known, unique masses of their AA constituents. By combining MS with a chromatographic technique for preanalytical separations, sample components can be effectively separated both in time (LC) and by mass (MS) to isolate, identify and quantify virtually any component in any mixture. With LC–HRMS, finding that proverbial "needle in a haystack" can literally be possible – that is, if a high enough degree of sample prefractionation and online separation has been implemented in coordination with a mass spectrometer with the appropriate duty cycle speed, dynamic range and resolution required for the analyte of interest.[11,17,18] Figure 9.2 illustrates the multidimensional separation and detection capabilities of this technique that have made LC–HRMS one of the most important and highly utilized analytical techniques in the biopharmaceutical industry.[19] The high sensitivity, superb resolving power, mass accuracy and specificity of LC–HRMS are well suited for the comprehensive characterization and quantification of peptides and peptide-related impurities.

9.1.3 Characterization

Characterization is a widely used term in drug discovery and development, especially in the context of complex API drugs such as peptides.[20,21] According to ICH Guidances Q6A and Q6B, characterization of a drug product (which includes the determination of physicochemical properties, biological activity, safety, purity and impurities) by appropriate techniques is applied to establish relevant specifications.[9,22] The adequate characterization of physicochemical properties includes the determination of the composition, physical properties and primary structure of the desired product. In the Common Technical Document (CTD) for CMC drug application submission, a characterization section is required for drug substances where the API and impurity structures must be elucidated. In this chapter, peptide impurity characterization by LC–HRMS is focused on peptide primary structure sequence identification and confirmation. MS on the intact peptide and subsequent top-down analysis (*i.e.* MS/MS) will be performed for peptide characterization.

HRMS on an intact peptide is integral for the determination of the monoisotopic mass and subsequent identification and confirmation of the molecular formula. Peptide primary sequence identification and characterization by MS have two general approaches: (1) bottom-up analysis (aka peptide mapping) and (2) mass measurement of the intact peptide and subsequent MS/MS analysis.[23] Although the bottom-up peptide mapping approach is an essential method in proteomics research, the use of top-down MS/MS using fragmentation techniques such as electron-transfer dissociation (ETD) for peptide sequencing has more recently gained popularity.[4,23,24] Overall, most therapeutic peptides are of an appropriate size for facile application of top-down approaches for primary sequence characterization and identification.[4,25]

Importantly, intact and top-down MS/MS do not require *a priori* knowledge of the peptide sequence. Peptide impurities or degradation products are separated in *m*/*z* space, allowing for their identification and further peptide structure sequencing using top-down MS/MS. The theoretical fragment ion nomenclature in MS/MS is shown in Figure 9.3. At the minimum, the MS/MS measurements involve peptide ion fragmentation along the peptide amide backbone to produce b- and y-ion series by collision-induced dissociation (CID) and or higher energy collisional dissociation (HCD). If the sequence coverage is low, alternative fragmentation techniques such as ETD may be used, albeit at the expense of fragment ion abundance.

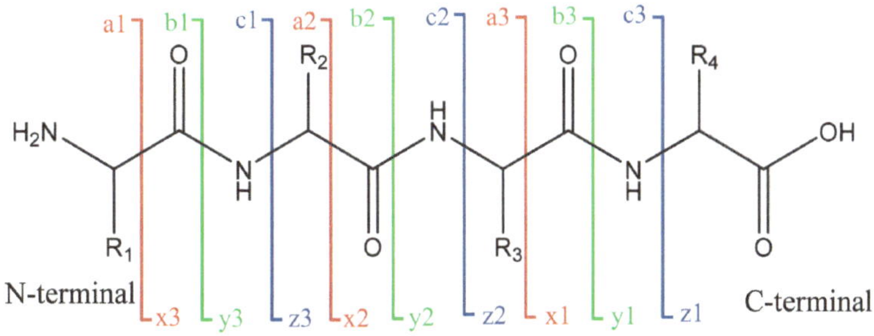

Figure 9.3 Peptide MS/MS ion fragmentation nomenclature.

ETD is particularly useful for peptides that contain a significant number of basic residues and are highly positively charged (not less than three positive charges). In addition, ETD can be used for selective cleavage of disulfide bonds.[25] Generally, ETD cleaves the amide backbone to produce c- and z-ion series as shown in Figure 9.3. In some cases, full coverage cannot be obtained owing to preferred fragmentation at particular sites or high chemical noise from the background or other interferences. In such cases, the combination of bottom-up approaches such as peptide mapping in conjunction with MS/MS can provide more complete sequence information.[21] Furthermore, MS detection features high sensitivity that allows for relatively low levels of impurity or degradation products to be identified and characterized.

9.1.4 Quantification

Quantification and control of process impurities or degradation products in a drug substance or a drug product should be performed after structure identification and confirmation. Each New Drug Application (NDA) and ANDA must include criteria and quantification analytical procedures in the control section of the CMC CTD.[9,22] The analytical procedures are necessary to ensure the identity, strength, quality, purity and potency of the drug substance and drug product. Parameters that may be evaluated during method development are specificity, linearity, limit of detection (LOD), limit of quantitation (LOQ), range, accuracy and precision.[26] The LC–HRMS method must be fully evaluated to meet this quantification purpose.[6] Furthermore, LC–HRMS intact peptide methods have an advantage for peptide quantification: the peptide species tend to be multiply charged so the summing of the signals of multiple ions can be used to improve sensitivity.[27]

9.1.5 Case Studies

In this chapter, three peptide drugs, salmon calcitonin (sCT), bivalirudin (BV) and exenatide (ET), were used as model systems to evaluate the suitability and validate the use of LC–HRMS for peptide impurity characterization and quantification. The case studies included identification, characterization, separation and quantification of the peptides and any peptide-related impurities present. The LC–HRMS method was validated for selectivity, sensitivity, accuracy, precision and linearity according to ICH Q2 analytical guidelines. The case studies demonstrated the proof of concept of the suitability of LC–HRMS method for the analysis of peptide impurities and ensuring the quality of peptide drug products.

sCT is a peptide with the following 32-AA sequence: Cys-Ser-Asn-Leu-Ser-Thr-Cys-Val-Leu-Gly-Lys-Leu-Ser-Gln-Glu-Leu-His-Lys-Leu-Gln-Thr-Tyr-Pro-Arg-Thr-Asn-Thr-Gly-Ser-Gly-Thr-Pro-NH_2. Two formulations, a nasal spray and an injectable, are available from both the originator and follow-on manufacturers and the drug products are used treat metabolic bone diseases such as osteoporosis, Paget's disease and osteoarthritis.[28] The USP monograph lists the known related peptides of the sCT drug substance as Salmon Calcitonin Related Compound A (RSA, *N*-acetyl-salmon calcitonin) and Salmon Calcitonin Related Compound B (RSB, salmon calcitonin-glycine).[29] The assay acceptance criterion for the drug substance is between 90 and 105%, with individual impurities no more than (NMT) 3.0% and total impurities NMT 5.0% of the total area of all (HPLC–UV) chromatographic peaks. Although the acceptance criterion of the salmon calcitonin drug product assay for the nasal solution is between 80 and 115%, there is no specific requirement for impurities.

BV is a peptide with 20 AA and its sequence is Phe-Pro-Arg-Pro-Gly-Gly-Gly-Gly-Asn-Gly-Asp-Phe-Glu-Glu-Ile-Pro-Glu-Glu-Tyr-Leu. It is a thrombin-specific anticoagulant,[30] intended for administration by injection only. Potential process impurities of BV include the deletion or insertion of glycine between positions 4 and 9 [des-glycine (DG) or plus-glycine (PG)], deamidation of the asparagine at position 9 to form aspartic acid/isoaspartic acid [β-aspartic analog (BA)] and a truncated form, bivalirudin [12–20].

ET is a synthetic peptide amide with 39 AA in the following sequence: His-Gly-Glu-Gly-Thr-Phe-Thr-Ser-Asp-Leu-Ser-Lys-Gln-Met-Glu-Glu-Glu-Ala-Val-Arg-Leu-Phe-Ile-Glu-Trp-Leu-Lys-Asn-Gly-Gly-Pro-Ser-Ser-Gly-Ala-Pro-Pro-Pro-Ser-NH_2. This drug product regulates glucose metabolism and insulin secretion and is approved for the treatment of diabetes mellitus type 2.[31] ET has two different formulated products on the market, Byetta and Bydureon. Potential process impurities of ET include the loss of glycine at position 2 and double-coupled alanine at position 18.

9.2 Experimental

9.2.1 Materials

Peptide drug products were either purchased from a pharmacy or requested from the manufacturer unless noted otherwise. Salmon calcitonin and its related impurity standards were purchased from American Peptide Company (Sunnyvale, CA, USA). The exenatide peptide standard was purchased from GenScript (Piscataway, NJ, USA). All reference standards were used as received after verifying the mass and sequence by using intact MS and MS/MS. Chemical reagents including acetonitrile (ACN), methanol, water, acetone and formic acid were purchased from Fisher Scientific (Fair Lawn, NJ, USA). Dichloromethane (DCM) and trifluoroacetic acid (TFA) were purchased from Sigma-Aldrich (St Louis, MO, USA). All mobile phase solvents were of Optima LC–MS grade and other solvents/chemicals were of ACS grade and used as received.

9.2.2 Sample Preparation

Stock standard solutions of the peptides and their related impurities were prepared at 1 $mg\,mL^{-1}$ in diluent (water–acetonitrile, 9 : 1). Stock standard solutions were aliquoted and stored at −80 °C until use. Individual intermediate stock standard solutions were prepared as needed by dilution of the stock standard solutions. The highest working solution of the peptide was prepared at 10 μM by mixing the individual intermediate stock standard solutions with the appropriate amount of diluent. Working standard solutions were then serially diluted to generate standard solutions at 0.1, 0.25, 0.5, 1.0, 2.5, 5.0 and 10.0 μM. Stock and immediate solutions were prepared in 2 mL microcentrifuge tubes (low retention and siliconized). Samples for injection were kept in 2 mL HPLC glass vials or with a 250 μL polypropylene insert.

Salmon calcitonin nasal solution samples (formulation strength, 2200 IU mL^{-1}, 0.37 mg mL^{-1}, 107 μM) were used as received. Bivalirudin for injection lyophilized solid was prepared at a concentration of 0.1 mg mL^{-1} (45.9 μM) in the diluent. Exenatide drug products are formulated with a unique continuous-release microsphere delivery system. A solution of exenatide was prepared at 2 mg mL^{-1} (478 μM) in diluent.[32] Briefly, one vial of exenatide (2 mg dose) (about 45 mg) was dissolved in 0.3 mL of DCM in a 1.5 mL Eppendorf vial, then 0.7 mL of acetone was added to the mixture and the vial was vortexed. After centrifugation at 11 752 *g* for 5 min (Eppendorf Model 5424 centrifuge), the supernatant was discarded. The residue was washed three times with 0.9 mL of mixed solvent (3:1 acetone–DCM) by repeating the above process. The residue was dissolved in diluent after 2 h sitting at room temperature. The sample working solution (40 μM) was prepared by mixing 84 μL of the sample stock solution with 914 μL of diluent.

9.2.3 Instrumentation

Two LC–HRMS instruments were used to analyze the peptide impurities. *Instrument 1:* HPLC separation was achieved on an Agilent 1290 HPLC system consisting of a Model 1290 binary pump, thermostat and autosampler (Agilent, Santa Clara, CA, USA), coupled to an Agilent 6520B Q-TOF mass spectrometer that was used for HRMS analysis. Data were acquired with Mass Hunter Data Acquisition (version B.04) and analyzed by Mass Hunter Qualitative Analysis (version B.04) software (Agilent). *Instrument 2:* A Thermo-Accela-HPLC system coupled to a Thermo Q-Exactive mass spectrometer (Thermo Fisher Scientific, Waltham, MA, USA) was used. LC–HRMS data were acquired and analyzed using Thermo Xcalibur (version 3.1) software.

9.2.4 LC–HRMS Analysis Conditions

Peptides were separated using a reversed-phase HPLC column with a low concentration of organic acid present in the mobile phase. Peptide samples were analyzed and detected by MS using an electrospray ionization source in the positive ion mode. Adsorption/desorption interactions of peptides with reversed-phase columns were sensitive to the concentration of organic solvent present in the mobile phase and isocratic elution was not efficient for the HPLC separation of the three peptides tested here. Hence, consistent with the observations of others, gradient elution was preferred, even if the gradient was very shallow (<1% min^{-1} of organic phase).[33,34]

HPLC separation was performed using a Waters Acquity UPLC BEH C_{18} column (100×2.1 mm i.d., 130 Å, 1.7 µm film thickness) in the three case studies presented. HPLC mobile phase A consisted of 0.1% formic acid in water and mobile phase B was 0.1% formic acid in acetonitrile. Gradient elution at a flow rate of 0.3 mL min^{-1} was performed with 5% B for 2.0 min, increased to 40% B over 37 min, then increased to 90% B at 40 min and held there for 2 min before decreasing to 5% B at 43 min. The column was re-equilibrated at 5% B for an additional 4 min. The total run time was 47 min. The column temperature was maintained at 30 °C and the autosampler was chilled to 4 °C. A dual electrospray ionization source was operated in the positive ion mode to acquire full-scan mass spectra from *m*/*z* 50 to 2000 with a scan rate of 4.0 spectra per second. The source gas temperature was set at 325 °C, gas flow rate at 12 L min^{-1}, nebulizer at 20 psi and skimmer 1 at 65.0 V. Additional source parameters were set with V_{cap} at 4000 V, fragmenter at 175 V and octopole RF peak at 750 V. The sample injection volume was 10 µL.

9.2.5 Data Analysis

Data analysis was performed using Mass Hunter Qualitative Analysis software for instrument 1 and Thermo Xcalibur software for data collected with instrument 2. Peptides and their relative impurities were analyzed and confirmed by their individual *m*/*z* extracted chromatogram, full MS spectra, isotopic patterns, charges and tandem mass spectra. Assignments of MS/MS fragments were assisted using ProSight PTM 3.0 (Northwestern University, Evanston, IL, USA). Linear regression was used to generate a calibration curve(s) from standards and calculate the concentrations of quality control and unknown samples. The fit equation of the calibration curve was in the $y = mx + b$ format, where y is the peak area of the analyte, x is the concentration of the analyte, m is the slope and b is the intercept of the regression line. Peak areas were obtained from the extracted ion chromatograms (EIC) by integrating peaks of the three highest charge states and three isotope ions at a mass extraction window (MEW) of 50 ppm. The EIC peak area was used for linear regression. Peak areas obtained from either total ion chromatograms (TICs) or EIC were used to study peptide drug impurities.

9.2.6 Method Validation

Method validation followed the FDA Guidance for Industry *Analytical Procedures and Methods Validation for Drugs and Biologics, 2015.*[26] Linearity was established by generating a calibration curve with seven non-zero calibrators covering the expected concentration range. Each calibrator had five determinations. Each tested concentration in the linear range featured precision within ±15%. The limit of detection (LOD) was calculated using the equation LOD = 3.3 × SD of noise/slope and the limit of quantification (LOQ) was calculated using the equation LOQ = 10 × SD of noise/slope. The accuracy of the method for each peptide was determined by spiked recovery experiments at three different concentration spikes (low, middle and high relative to the linear curve: 0.3, 0.9, 3.0 μM). Peak areas obtained from the EIC were used for method validation.

9.2.7 Selection of Optimized HPLC Conditions

HPLC separation of peptides was achieved using reversed-phase C_{18} columns with 0.1% formic acid in water (A) and 0.1% formic acid in acetonitrile (B). LC separation in an LC–HRMS method was further optimized by modifying the column temperature and the gradient slope. LC–HRMS TIC analysis of exenatide at different column temperatures with a gradient of 5–40% B from 2 to 37 min was investigated. Column temperatures were tested at 35, 45, 50, 55 or 60 °C. Increasing the column temperature improved both the peak shape and separation. A lower column temperature of 55 °C was selected without sacrificing resolution. For the three case studies, the column temperature was kept at 55 °C and the slope of the elution gradient was varied as follows: a, 5–40% B from 2 to 37 min; b, 10–40% B from 2 to 37 min; c, 5–30% B from 2 to 20 min, then 30–40% B from 20 to 40 min; d, 5–25% B from 2 to 15 min, then 25–40% B from 15 to 40 min. Gradient conditions were optimized to yield a symmetric peak shape and baseline resolution where possible.

9.2.8 Optimization of MS Parameters

MS systems should be calibrated and optimized for mass accuracy, resolution, sensitivity and reproducibility. No LC–HRMS system should be used without passing a system suitability procedure. System suitability procedures are essential for reliable MS quantification measurements. The *Guide to Achieving Reliable Quantitative LC–MS Measurements* is a publication dedicated to optimizing the LC–HRMS interface and the mass spectrometer.[16] This guide recommends optimizing several key MS parameters associated with the nebulization and desolvation processes in addition to mass analyzer parameters.

MS parameters were optimized for each specific peptide to ensure consistent MS detection. For example, potential interfering adduct ions of specific peptides can often be broken up by using a higher declustering potential (DP) voltage. Another example is that the higher molecular weight for exenatide compared with the other peptides studied necessitated a higher gas temperature for suitable mass spectra. In addition, for exenatide, the nebulizer pressure and fragmenter voltage were adjusted to achieve an appropriate response. The following settings were adjusted from the method previously described in the analysis section for the detection of bivalirudin and calcitonin: source parameters: gas temperature, 325–350 °C; gas flow rate, 12.0–13.0 L min^{-1}; nebulizer, 20–40 psi; scan source parameters: V_{cap}, 4000 V; fragmenter, 175–210 V; skimmer 1, 65.0–80.0 V; octopole RF peak, 750 V.

9.2.9 Characterization

The intact peptide monoisotopic mass was determined by HRMS MS1 full scan through direct infusion or with LC elution. To interpret the ions in the mass spectra better, the sample concentration and solvent should be specified if sample introduction into the instrument is conducted *via* direct infusion. The peptide primary structure sequence identification and confirmation in our case studies were performed by top-down analysis (*i.e.* MS/MS of the intact peptide).

9.3 Results and Discussion

9.3.1 Characterization – Intact Peptide MS for Accurate Monoisotopic Mass, MS/MS for Peptide Sequence

The intact peptide monoisotopic mass was obtained through HRMS MS1 full scan. The positive peptide charge was primarily hydrogen protonated as $M + nH^+$, where n is the number of charges. Figure 9.4 shows the HRMS isotopic distribution mass spectra of salmon calcitonin, bivalirudin and their related impurities. The spectra on the left correspond to the +4 charge states of salmon calcitonin and related impurities from the four synthetic standards: salmon calcitonin (sCT), *N*-acetylcalcitonin (AC), Glu14-calcitonin (Glu14) and Glu20-calcitonin (Glu20). Glu14 and Glu20 are sCT deamidation forms at positions 14 and 20; these two species are isobaric in mass. The monoisotopic mass peak is the first peak of the isotopic distribution peaks. The charge of isotopic distribution peaks can be calculated by the ratio of 1 to the mass difference of two neighboring peaks. The experimental monoisotopic mass equals the charge value times the value of the monoisotopic mass peak minus the monoisotopic mass of H^+. Below is an example of the calculation of the experimental monoisotopic mass through isotopic distribution mass spectra of salmon calcitonin:

The charge of the sCT peak is 4.

The mass difference of two neighboring peaks

$$= 858.6795 - 858.4283 = 0.2512$$

The charge $n = 1/0.2512 = 4.0$

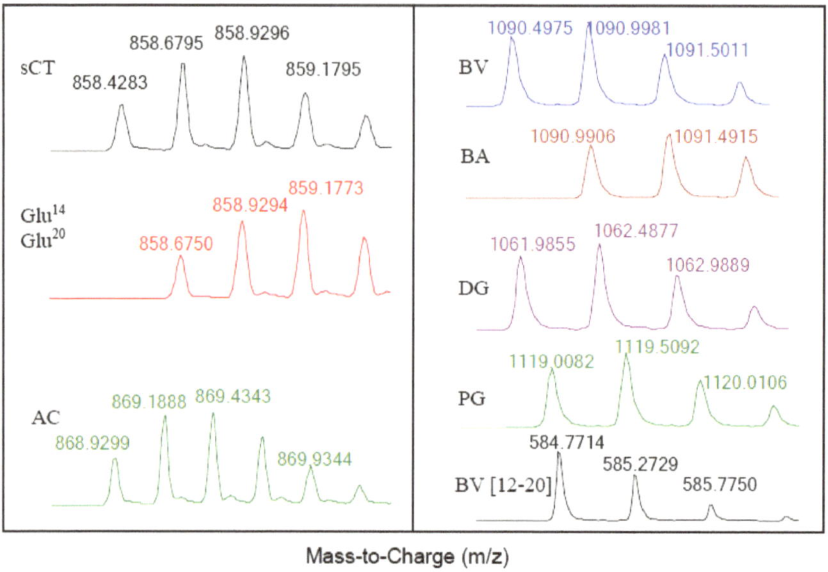

Figure 9.4 High-resolution mass spectra of salmon calcitonin, bivalirudin and their related impurities. Left: the +4 charge states of sCT and its related impurities are shown. Right: the +2 charge states of BV and its related impurities are shown.
Reproduced from ref. 6 with permission from Springer Nature, Copyright 2015.

The monoisotopic mass peak value of sCT is 858.4283 at charge 4 H^+. The experimental monoisotopic mass value of sCT is 3429.6841.

$$\text{The monoisotopic mass of } H^+ = 1.00728$$

$$\text{The experimental monoisotopic mass of sCT} = (858.4283 - 1.00728) \times 4 = 3429.6841 \text{ Da}$$

Table 9.2 gives the experimental monoisotopic mass of each tested peptide. The experimental monoisotopic mass values of sCT, AC and Glu14/20 were 3429.6841, 3471.6905 and 3430.6709 Da, respectively. The experimental values were consistent with heoretical values within 10 ppm error. In Figure 9.4, the deamidated species (Glu14/20) and their corresponding +0.984 Da mass shifts relative to sCT are clearly distinguished by a positive shift in *m*/*z* from 858.4283 to 858.6750 for the +4 charge state.

The spectra on the right in Figure 9.4 correspond to the +2 charge states of BV and its related impurities: BV, BA, DG, PG and BV [12–20]. The experimental monoisotopic mass values observed were 2178.9804, 2179.9666,

Table 9.2 Monoisotopic mass of each tested peptide from an LC–HRMS experiment. Reproduced from ref. 6 with permission from Springer Nature, Copyright 2015.

Compound	Monoisotopic mass, theoretical/Da	Monoisotopic mass, experimental/Da	Delta mass from theoretical/Da	Mass error/ppm
sCT	3429.7133	3429.6841	−0.0292	−8.5
Glu14/20	3430.6973	3430.6709	−0.0264	−7.7
AC	3471.7239	3471.6905	−0.0334	−9.6
BV	2178.9858	2178.9804	−0.0054	−2.5
DG	2121.9644	2121.9564	−0.0079	−3.8
PG	2236.0073	2236.0018	−0.0054	−2.5
BV [12–20]	1167.5336	1167.5282	−0.0053	−4.6
BA	2179.9698	2179.9666	−0.0032	−1.5
ET	4184.0273	4184.0517	+0.0244	+5.8

2121.9564, 2236.0018 and 1167.5282 Da, respectively, which were calculated from deconvolution of the mass spectra (see Table 9.2). The +0.984 Da mass shift associated with deamidation (BV to BA) can clearly be distinguished from the BV species at *m*/*z* (+2 charge) 1090.4975 to 1090.9906 (see Figure 9.4). The experimental monoisotopic mass for exenatide was 4184.0517 Da, which is consistent with the theoretical mass within 6 ppm error (see Table 9.2). In all three case studies, the experimental values were consistent with the theoretical values within 10 ppm error.

Further characterization of peptide structure sequence can be accomplished by using top-down MS/MS. The high-resolution tandem mass spectrum of salmon calcitonin is shown in Figure 9.5 using CID as the ion fragmentation method. In Figure 9.5, spectrum (1) is the raw tandem mass spectrum of the salmon calcitonin parent ion *m*/*z* 858.4356 at +4 charge, in which complex ions include multiple charges and isotopic ions were observed. To assign the MS/MS fragments to the peptide structure manually or assisted by software programs, it is necessary to simplify the raw tandem mass spectrum to a single mass and charge by deconvolution and deisotope. Spectrum (2) is the deconvoluted and deisotoped tandem mass spectrum with b ions labeled in red. Assignment of the MS/MS fragments was conducted using ProSightPTM 2.0 (Northwestern University) and confirmed manually.

Accurate monoisotopic mass determination in combination with MS/MS helps to determine the AA composition and sequence confirmation, which can help to identify any variation from the predicted peptide sequence. Figure 9.6 shows the amino acid sequence confirmation for salmon calcitonin and its related impurities by LC–MS/MS. The red lines identify the MS/MS fragments generated by CID fragmentation of the +4 charge states, blue lines are the MS/MS fragments generated by fragmentation of the b_{31} ion and the green lines are the MS/MS fragments generated by ETD fragmentation of the +4 charge states. The MS/MS fragmentation maps for each of the calcitonin related compounds showed the sequence coverage with observed fragmentation labeled as colored lines and retained disulfide

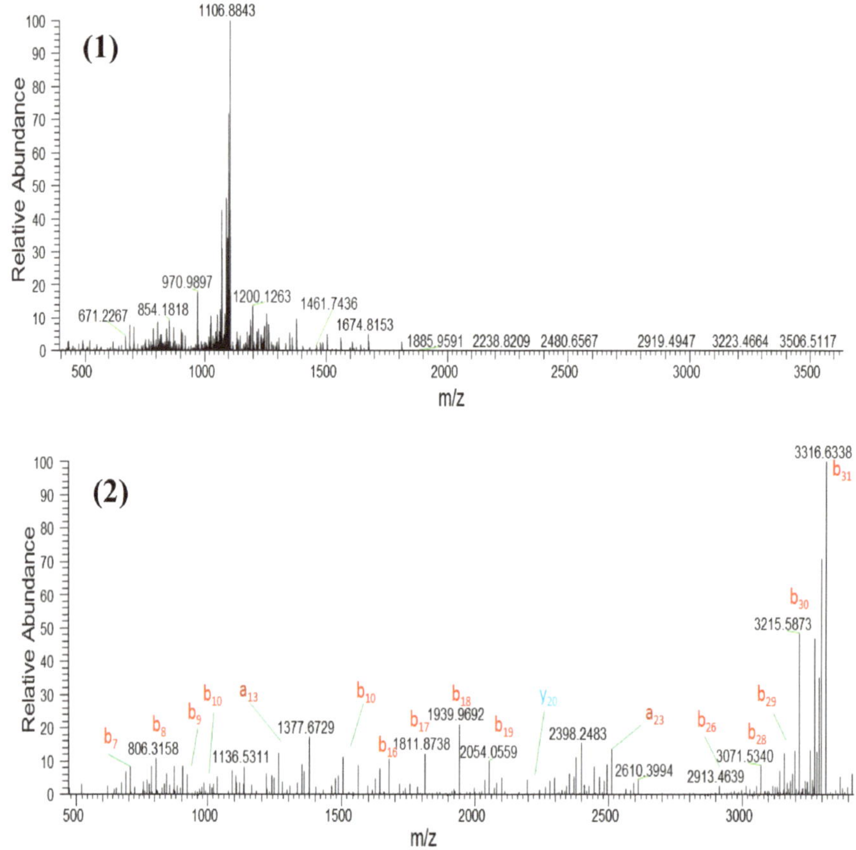

Figure 9.5 Typical high-resolution tandem mass spectra of salmon calcitonin. (1) Raw tandem mass spectrum of salmon calcitonin parent ion at *m*/*z* 858.4356 at +4 charge. (2) Deconvoluted and deisotoped tandem mass spectrum; a, b ions are labeled in red.

bonds shown in black. Full sequence coverage was observed by using CID and ETD ion dissociation techniques. For therapeutic peptides, full sequence coverage can be achieved by using primarily the CID technique. If any sequence issues, such as a missing sulfide bond, deamidation or change in the site of modification, were to occur, this approach would be able to identify them readily. Table 9.3 identifies typical peptide structure modifications and their associated mass change in MS.

Accurate monoisotopic mass determination in combination with MS/MS was also used to determine the AA composition and sequence confirmation for peptide reference standards. Figure 9.7 shows the identifying sites of unexpected modifications of the reference standards. For example, the monoisotopic mass for the USP Salmon Calcitonin Related Compound B reference standard (RSB, salmon calcitonin glycine amide) was shifted 1 Da. This peptide was sequenced by MS/MS and the mass shift was localized to

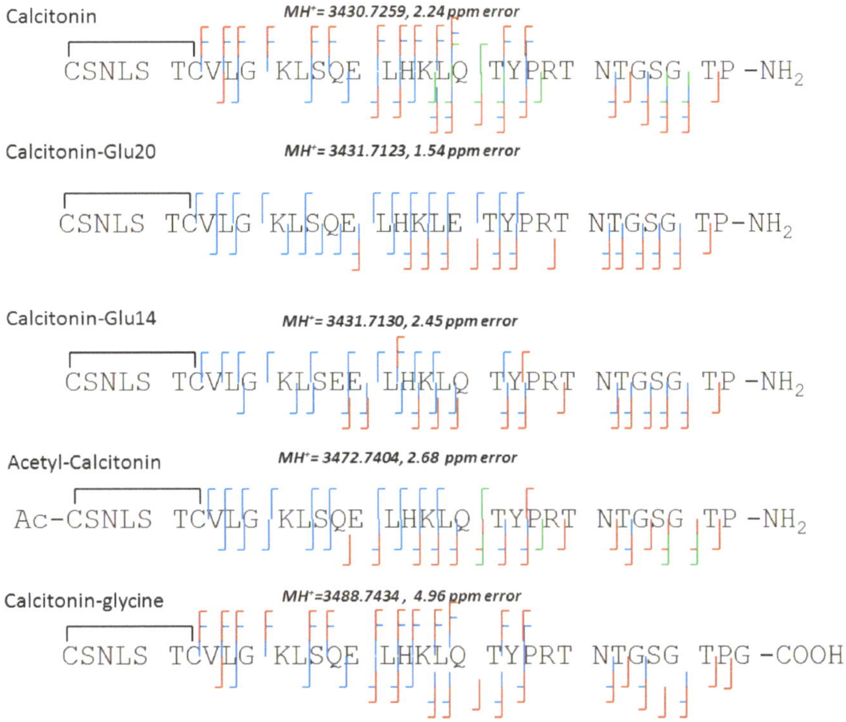

Figure 9.6 Amino acid sequence confirmation for salmon calcitonin and its related impurities by LC-MS/MS. Black lines: intact disulfide bond. Red lines: MS/MS fragments generated by CID fragmentation of the +4 charge states. Blue lines: MS/MS fragments generated by fragmentation of the b_{31} ion. Green lines: MS/MS fragments generated by ETD fragmentation of the +4 charge states.

Table 9.3 Typical peptide modification and its mass change in MS.

Modification	Delta mass	AA residue	Delta formula
Acetylation	+42.0106	N-terminus	$+C_2H_2O$
Deamidation	+0.9840	Asn, Gln	−NH +O
Oxidation	+15.9949	Met, Cys, His, Trp	+O
Sulfation	+79.9568	Ser, Thr, Tyr	$+SO_3$
Disulfide bond formation	−2.0157	Cys–Cys	$-H_2$
Methylation	+14.0157	Tyr, Lys	$+CH_2$
Phosphorylation	+79.9663	Ser, Thr, Tyr	$+HPO_3$
Lysine truncation	−128.09496	C-terminus	$-C_6H_{12}N_2O$
Carbamylation	+43.0058	N-terminus, Lys	+CHNO
Pyroglutamate	−43.0058	Glu, Gln	$-H_2O$, $-NH_3$
Racemization	0	Asp, Asn	0

the C-terminus, indicating a terminal carbocyclic acid instead of the expected amide (NH_2 was replaced with OH) in the USP standard. In addition, the calculated monoisotopic mass of the synthesized salmon calcitonin USP

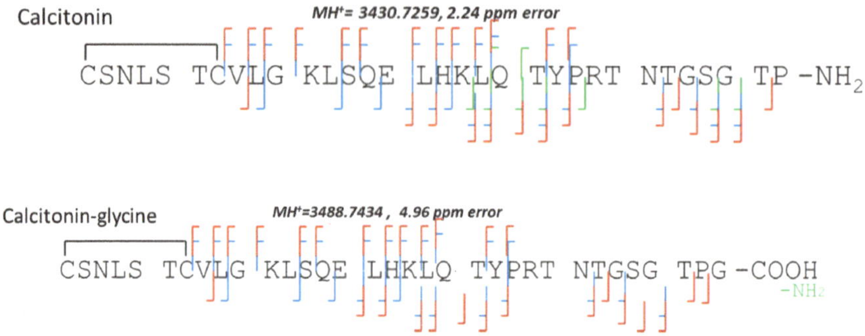

Figure 9.7 LC–HRMS identification of sites of unexpected modifications. Salmon calcitonin: disulfide form. Actual: mass shifted +2 Da; there was no disulfide bridge. Salmon calcitonin–glycine amide. Actual: mass shifted +1 Da; the C-terminus was carboxylic acid instead of the expected amide.

standard was 2 Da heavier than expected, indicating that the expected disulfide bond was not formed. The calcitonin sequence shown in Figure 9.4 was the USP standard replacement batch, where MS and MS/MS data were used to confirm and match the sequence. The composite fragmentation map verified the connectivity of almost all the peptide bonds for all the synthesized standards, indicating that the peptide standards lacked a disulfide bond and were deamidated. Red and blue flags indicate b/y ions generated from the CID experiments and the green flags indicate c/z ions generated from the ETD experiments.

9.3.2 Separation and Detection of Peptide Impurities by Both LC and HRMS

A peptide and its related impurities can be effectively separated by HPLC according to retention time (RT) and isolated by MS using specific *m*/*z* ratios. Figure 9.8 is the LC–HRMS TIC chromatogram of the salmon calcitonin drug product and its related impurities. The main peak at RT 24.927 min is calcitonin and the other peaks at different RTs were calcitonin-related impurities.

Additional peptide separation by *m*/*z* by HRMS was obtained by exploring the mass spectra of each LC–HRMS TIC peak. Figure 9.9 shows the LC–HRMS mass spectrum of the calcitonin TIC peak at RT 24.9 min. Calcitonin MS ion peaks were observed for four different charge states. By zooming into the peaks of one of the four charges, for example +4, six ion peaks with different *m*/*z* were observed, as shown in Figure 9.9b. All peaks showed typical peptide isotope *m*/*z* distributions. Hence it is important to identify each ion peak for LC–HRMS peptide qualification and quantification.

In-source ion fragmentation, adduction and reaction should be taken into consideration for qualification and quantitation. In-source reactions include oxidation, sodium adduction and dehydration products. Table 9.4 shows typical common adducts or cluster ions formed in LC–MS. Other

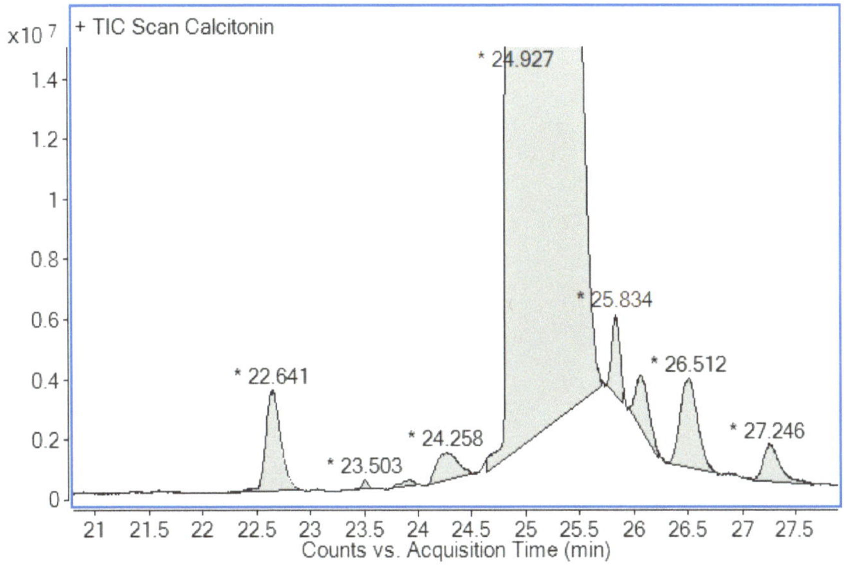

Figure 9.8 LC–HRMS TIC chromatogram of salmon calcitonin drug product and its related impurities.

interference or contaminant ions can be observed in the LC–MS system. A good reference for non-target ion assignment is the paper "Interferences and Contaminants Encountered in Modern Mass Spectrometry."[35] After an in-depth analysis, all six ion peaks observed in Figure 9.9b were assigned. The ion of salmon calcitonin was confirmed at *m*/*z* 858.8651 (+4). The remaining five ions in the zoomed-in region of Figure 9.9b were not impurity ions but were calcitonin adducts/clusters. The following five ions were identified: (1) the ion at *m*/*z* 864.3621 (+4) was a sodium adduct with a delta mass of 21.9880, (2) the ion at *m*/*z* 868.1038 (+4) was a potassium adduct with a delta mass of 37.9570, (3) the ion at *m*/*z* 869.8575 (+4) was a two-sodium adduct with a delta mass of 43.9700, (4) the ion at *m*/*z* 872.0943 (+4) was an NH_4Cl salt cluster with a delta mass of 52.9172 and (5) the ion at *m*/*z* 875.3529 (+4) was a three-sodium adduct with a delta mass of 65.9516.

The combination of LC separation and selective HRMS detection was able to achieve baseline resolution of reference standard BV and its related impurities reference standards (Figure 9.10). From top to bottom, the TIC chromatograms of BV and its related impurities are EIC of BV [12–20], EIC of DG, EIC of BV, EIC of BA and EIC of PG. The EIC chromatogram of BA has two peaks, BA and BV, which were separated by time. BA and BV cannot be separated by MS because they share one isotopic ion peak at *m*/*z* 1090.99 in charge 2 mode.

A robust LC–HRMS quantification method can be achieved through baseline resolution provided by peak retention time separation on a LC column, analyte *m*/*z* separation and detection with a MS analyzer. Caution

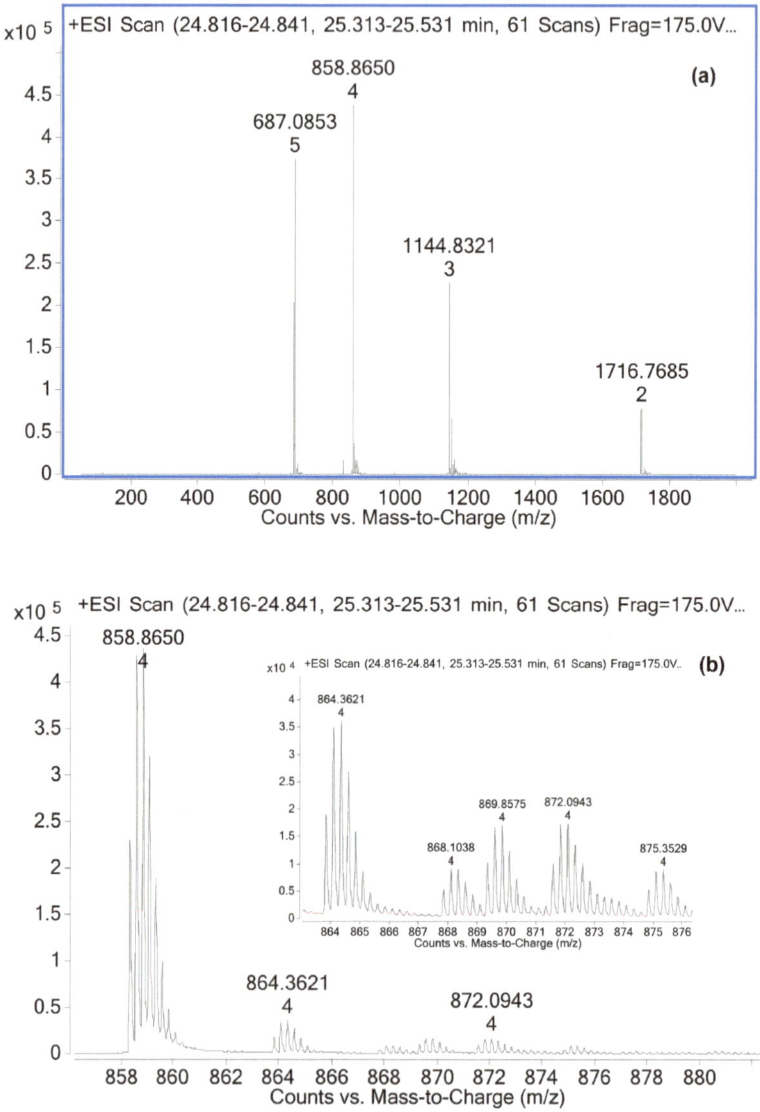

Figure 9.9 LC–HRMS mass spectra of calcitonin TIC peak at RT 24.9 min. (a) Mass spectrum of salmon calcitonin showing ion peaks in a different charge. (b) Mass spectrum of salmon calcitonin by zooming in charge +4 peaks. Six ion peaks were observed. All peaks showed a typical peptide isotope *m*/*z* distribution. The ion of salmon calcitonin was confirmed at *m*/*z* 858.8651(+4). The other five ions zoomed in are not impurity ions but calcitonin adducts/clusters. The ion at *m*/*z* 864.3621 (+4) is a one-sodium adduct with a delta mass of 21.9880. The ion at *m*/*z* 868.1038 (+4) is a one-potassium adduct with a delta mass of 37.9570. The ion at *m*/*z* 869.8575 (+4) is two-sodium adduct with a delta mass of 43.9700. The ion at *m*/*z* 872.0943 (+4) is an NH_4Cl salt cluster with a delta mass of 52.9172. The ion at *m*/*z* 875.3529 (+4) is a three-sodium adduct with a delta mass of 65.9516.

Table 9.4 Typical common adducts and clusters ions formed in LC–MS.

$[M+H]^+$ alternative ions formed	Delta mass	Cause	Adduct[a]/ionic
Sodium $[Na^+ \leftrightarrow H^+]$	+21.9819	Sodium salts	Ionic
Potassium $[K^+ \leftrightarrow H^+]$	+37.9559	Potassium salts	Ionic
Ammonia $[NH_4{}^+ \leftrightarrow H^+]$	+17.0265	Ammonia/NH_4OH	Adduct
Lithium $[Li^+ \leftrightarrow H^+]$	+6.0082	Lithium salts	Ionic
Water $[H_3O^+ \leftrightarrow H^+]$	+18.0106	Water/acids	Adduct
Methanol $[CH_3OH_2{}^+ \leftrightarrow H^+]$	+32.0262	Methanol/solvent	Adduct
Acetonitrile $[CH_3CNH^+ \leftrightarrow H^+]$	+41.0265	Acetonitrile/solvent	Adduct
Cluster $[M+NH_4Cl+H^+]$	+53.0032	NH_4Cl salt adducts	Adduct

[a]Adducts can often be broken up by using a higher DP voltage.

must be exercised when selecting the MS ion peaks for both qualitative and quantitative purposes because some ions might originate from sources other than the sample itself, such as in-source fragmentation and adduction. Furthermore, the ionization efficiency can differ between ions. For peptide impurity identification through MS detection, ions produced by in-source fragmentation and adduction should be excluded. For peptide impurity quantification using an MS EIC method, the ionization efficiency and system suitability should be evaluated first. Ions selected for the EIC chromatogram should be adjusted by the observed *m*/*z* and RT window and ions that overlap/interfere with other species should be excluded for better selectivity. Ionization efficiency is related to many factors including, but not limited to, instrument hardware, instrument operating parameters, sample concentration and mobile phase composition. Peptide separation and ionization are extremely sensitive to the concentration of the organic solvent in the mobile phase and shallow gradient elution is preferable. Precise and accurate LC–HRMS results depend on better separated impurities and consistent ionization efficiency.

9.3.3 LC–HRMS Method Validation

In the presented case studies, method validation followed the FDA Guidance.[26]

9.3.3.1 Linearity

Quantitation of peptide drugs and their related impurities was achieved using the MS response peak areas obtained from the EIC chromatograms. Two EIC mass extraction windows (MEWs) (20 and 100 ppm) were used to assess the impact of this parameter on the analysis. Table 9.5 shows the linearity (in terms of linear regression correlation coefficient R^2) and sensitivity (linear curve slope) of salmon calcitonin, bivalirudin, exenatide and their related impurities. All linearity R^2 values were >0.99 for all tested peptides in both the 20 and 100 ppm MEWs. Hence these results showed that the LC–HRMS method was suitable for peptide quantification in terms of linearity. Higher

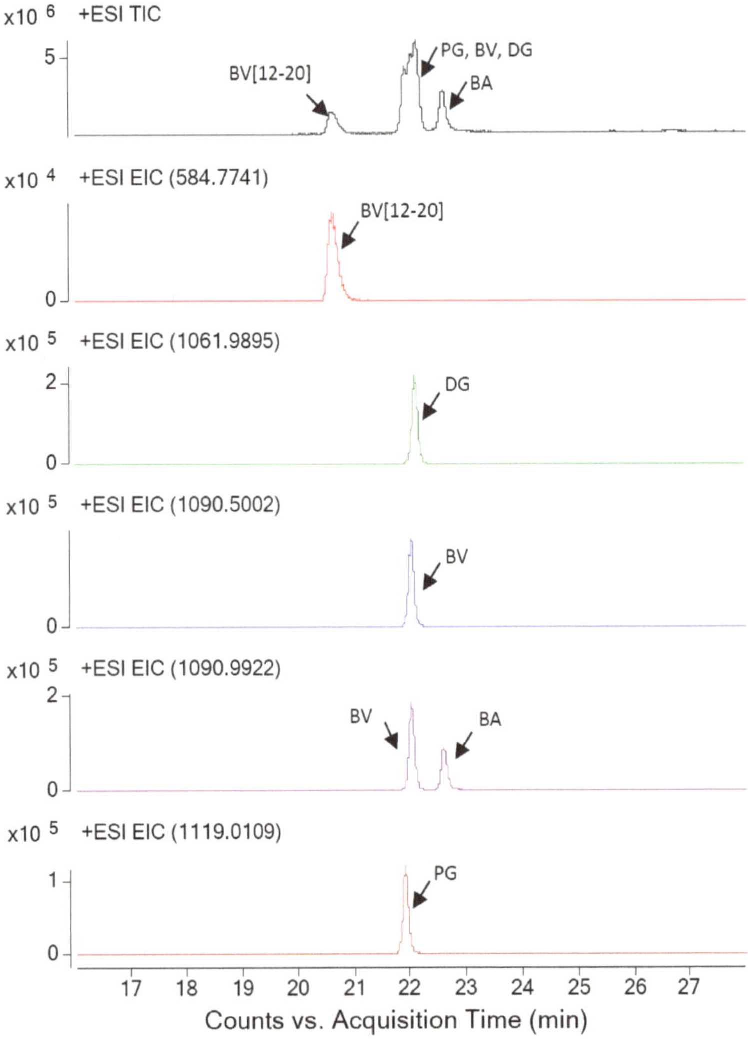

Figure 9.10 LC–HRMS TIC and EIC chromatograms of reference standard BV and its related impurities reference standards. From top to bottom: TIC of BV and its related impurities, EIC of BV [12–20], EIC of DG, EIC of BV, EIC of BA and EIC of PG. The EIC chromatogram of BA has two peaks: BA and BV.
Reproduced from ref. 6 with permission from Springer Nature, Copyright 2015.

sensitivity was obtained for the MEW at 100 pm because more ions were included in the peak area measurements. However, in some cases a narrower MEW(s) may be required to increase the selectivity or specificity.

Table 9.5 contains the linearity and sensitivity data for salmon calcitonin, bivalirudin, exenatide and their related peptides. The selected *m*/*z* value

Table 9.5 Linearity (R^2) and sensitivity (linear curve slope) of salmon calcitonin, bivalirudin, exenatide and their related impurities. Reproduced from ref. 6 with permission from Springer Nature, Copyright 2015.

	MEW[a] 20 ppm		MEW 100 ppm		
Compound	Slope ($S1 \times 10^6$)	R^2	Slope ($S2 \times 10^6$)	R^2	Sensitivity ratio $S2/S1$
sCT	1.905	0.995	5.320	0.996	2.79
Glu14	1.649	0.996	4.284	0.997	2.60
AC	2.116	0.993	5.572	0.994	2.63
BV	1.234	1.000	3.107	0.999	2.52
DG	1.962	1.000	8.268	0.999	4.21
PG	1.032	0.999	4.868	0.999	4.72
BV [12–20]	8.424	0.999	2.586	0.999	3.07
BA	1.073	0.995	2.708	0.994	2.52
ET	1.234	0.999	3.116	0.999	2.53

[a]MEW, mass extraction window.

around which the MEW was set for each salmon calcitonin related compound was determined based on selectivity and where the highest response was obtained, which were as follows: sCT at *m*/*z* 858.8681, Glu14 at *m*/*z* 859.1135 and AC at *m*/*z* 869.3684. Salmon calcitonin and its related impurities (Glu14 and AC) featured linear responses using either MEW over the concentration range 0.1–10 μM. The sensitivity of each compound with a 100 ppm MEW was several times higher ($S2/S1$ range from 2.6 to 2.79) than with a 20 ppm MEW. For bivalirudin, the extracted ion *m*/*z* for each related compound was also determined based on selectivity and highest response: BV at *m*/*z* 1090.5002, DG at *m*/*z* 1061.9895, PG at *m*/*z* 119.0109, BV [12–20] at *m*/*z* 584.7741 and BA at *m*/*z* 1090.9922. A linear response was observed for bivalirudin and its related impurities (DG, PG, BV [12–20] and BA) using either MEW (Table 9.5) in the concentration range 0.1–10 μM. The sensitivity of each compound with a 100 ppm MEW was several times higher ($S2/S1$ range from 2.5 to 4.7) than with a 20 ppm MEW. Finally, the selected ion *m*/*z* at 1047.5150 was chosen for exenatide based on selectivity and the highest response observed. Over the concentration range 0.25–10 μM, exenatide featured a linear response using either MEW. The sensitivity with a 100 ppm MEW was 2.5 times higher ($S2/S1 = 2.53$) than with a 20 ppm MEW. Because the wider MEW did not compromise selectivity, precision and accuracy, an EIC at a 50 ppm MEW was used for subsequent sample analysis.

9.3.3.2 *Precision*

The precision of the LC–HRMS response was determined using multiple injections of a sample and evaluated in terms of relative standard deviation (RSD). Table 9.6 summarizes the precision results of five injections from the quantitation of bivalirudin and its related impurities. RSD values of <3.0% were obtained for all test concentrations, except for 0.1 μM, using data from either the 20 or 100 ppm MEW. Furthermore, the RSD values were no more than (NMT) 2.0% at test concentrations from 1 to 10 μM. This NMT 2.0%

Table 9.6 Precision (RSD/%) of bivalirudin and its related impurities. Reproduced from ref. 6 with permission from Springer Nature, Copyright 2015.

	MEW[a] 20 ppm					MEW 100 ppm				
Concentration/μM	BV	DG	PG	12–20	BA	BV	DG	PG	12–20	BA
0.1	3.49	3.27	1.55	2.89	4.51	4.57	1.21	1.23	2.72	5.1
0.25	2.41	1.09	1.74	0.73	1.58	1.21	1.2	2.52	0.76	1.42
0.5	0.88	0.27	0.42	1.26	2.1	1.09	0.88	1.39	0.51	1.9
1	0.86	1.6	0.98	0.52	1.26	0.99	0.95	0.96	0.88	0.99
2.5	0.75	0.84	0.72	0.74	0.78	0.67	0.28	0.32	1.42	2.01
5	0.65	0.52	0.41	0.68	0.55	0.52	0.59	0.94	0.94	1.6
10	0.40	0.24	0.21	0.45	0.86	0.51	0.56	0.52	0.40	0.66

[a]MEW, mass extraction window.

RSD meets the system suitability criteria of most drug HPLC assay methods. Similar results were obtained with the other two peptide drugs and their related impurities (data not shown).

9.3.3.3 *Accuracy*

The accuracy of related impurities measurements for the LC–HRMS method was determined based on spiked recoveries using three different concentration spikes within the experimental linear range. Related impurities of salmon calcitonin and bivalirudin were spiked into their respective API solutions at 0.1, 0.3 and 1% of the API experimental concentration, which corresponded to spikes of 0.3, 0.9 and 3.0 μM, respectively. Table 9.7 shows that the spiked recoveries of sCT-related impurities Glu14 and AC were in the ranges 85–114 and 98–100% at spiked concentrations 0.3 and 3.0 μM, respectively. The spiked recoveries of bivalirudin-related impurities DG, PG, BV [12–20] and BA were in the ranges 102–110 and 100–105% at spiked concentrations 0.3 and 3.0 μM, respectively. The spiked recovery range at the 0.1% API level was within ±15% and at the 1% API level was within ±5%. The recovery results suggested that the LC–HRMS method gave an accurate measurement of the concentrations of the peptide impurities over the concentration ranges tested.

9.3.3.4 *Limit of Detection and Limit of Quantification*

The limits of detection (LODs) and quantitation (LOQs) for the related impurities of salmon calcitonin and bivalirudin using the LC–HRMS method are summarized in Table 9.7. The LODs for bivalirudin's related impurities DG, PG, BV [12–20] and BA were 0.01, 0.01, 0.01 and 0.04 μM, respectively. The LOQs of bivalirudin-related impurities DG, PG, BV [12–20] and BA were 0.04, 0.04, 0.03 and 0.11 μM respectively. The LODs for Glu14-calcitonin and acetyl-calcitonin were 0.03 and 0.04 μM, equivalent to 0.03 or 0.04% of label claim, respectively. The LOQs of Glu14-calcitonin and acetyl-calcitonin were 0.08 and 0.11 μM, corresponding to 0.07 and 0.1% of the label claim,

Table 9.7 Accuracy (spike recovery/%) and sensitivity (LOD, LOQ) of the related impurities of salmon calcitonin and bivalirudin. Reproduced from ref. 6 with permission from Springer Nature, Copyright 2015.

	Spike recovery/%			Sensitivity/μM	
Compound	0.3 μM	0.9 μM	3.0 μM	LOD[a]	LOQ[a]
Glu14	85.1	88.6	97.9	0.03	0.08
AC	113.9	95.1	100.5	0.04	0.11
DG	102.4	98.7	100.1	0.01	0.04
PG	103.4	98.2	100.1	0.01	0.04
BV [12–20]	96.1	103.2	99.8	0.01	0.03
BA	110.3	98.2	105	0.04	0.11

[a]LOD, limit of detection; LOQ, limit of quantification.

respectively. The LOD and LOQ values obtained show that peptide-related impurities present at less than 0.1% of the peptide drug API concentration can be detected by the LC–HRMS method.

9.3.4 Peptide Drug Product Impurities Determination

The LC–HRMS method was used to evaluate the impurities of peptide drugs commercially available in the US market. These three drugs cover a wide AA sequence, varying from 20 to 39, with salmon calcitonin having a disulfide bond.

Table 9.8 summarizes the impurities detected in a salmon calcitonin sample obtained in the marketplace and evaluated using the LC–HRMS method. The impurity results were calculated as a relative percentage of the peak area from the EIC. The total amount of related impurities for the salmon calcitonin nasal solution used was determined as 2.64%. In addition to the USP known related impurities (Glu and AC) that were identified and quantitated, 10 additional related impurities were also detected with the LC–HRMS method. The content of each of these 10 related impurities was NMT 0.5%, which is less than the identification threshold established in the ICH guidance. Additional related impurities were detected in the salmon calcitonin nasal solution when using the LC–HRMS method compared with the USP HPLC–UV method.[29]

The peptide impurity results obtained by the LC–HRMS method for bivalirudin and exenatide products are summarized in Table 9.9. The concentration of total related impurities detected in the bivalirudin injection product was 1.5%. The known impurities (DG, PG, BV [12–20] and BA) were identified and quantitated, but also three additional unknown related impurities were detected. The concentration of total related impurities detected for the exenatide extended release microsphere product was 9.2%. Although the experimental data are not shown here, the quantity of impurities detected in the exenatide extended release microsphere product is likely associated with the extraction procedure used to prepare the drug product samples. The increased quantity of impurities detected in the microsphere

Table 9.8 Impurity results (% area) of LC–HRMS of US marketplace salmon calcitonin. Reproduced from ref. 6 with permission from Springer Nature, Copyright 2015.

Peak	RRT[a]	% area
Unknown 1	0.91	0.39
Unknown 2	0.97	0.18
sCT	1	97.36
Glu	1.06	0.41
AC	1.09	0.73
Unknown 3	1.13	0.15
Unknown 4	1.17	0.12
Unknown 5	1.32	0.04
Unknown 6	1.38	0.11
Unknown 7	1.45	0.02
Unknown 8	1.46	0.27
Unknown 9	1.48	0.05
Unknown 10	1.5	0.15
Total impurities		*2.64*

[a]RRT, relative retention time.

Table 9.9 Impurity results (% area) of LC–HRMS of US marketplace bivalirudin and exenatide. Reproduced from ref. 6 with permission from Springer Nature, Copyright 2015.

Drug	Peak	RRT[a]	% area
Bivalirudin	Unknown 1	20.3	0.2
	BV [12–20]	20.64	0.1
	PG	21.85	0.7
	BV	21.92	98.5
	DG	22.1	0.2
	Unknown 2	22.4	0.1
	BA	22.47	0.1
	Unknown 3	23.4	0.1
	Total impurities		*1.5*
Exenatide	Exenatide	25.328	90.8
	Unknown 1	26.794	3.8
	Unknown 2	27.854	5.4
	Total impurities		*9.2*

[a]RRT, relative retention time.

product is likely an artifact associated with degradation during the extraction procedure used to isolate the peptides from the drug product microsphere formulation.

9.3.5 Total Ion Chromatogram (TIC) or Extracted Ion Chromatogram (EIC)

The percentage of drug substance and potential impurities can be calculated using peak areas from both the LC–HRMS TICs and EICs if the API and its related impurities can be baseline separated chromatographically. In this case, HRMS serves as a robust and sensitive detector. For example, using TIC area normalization, the impurity content in exenatide was ~9.2%, whereas the EIC area normalization method determined an impurity content of ~8.8% (data not shown here). TIC and EIC calculations were consistent and within 1% of each other. The choice of whether to use either TIC and EIC for quantitative studies should be made on a case-by-case basis and determined in a quality-controlled environment with subsequent method validation studies.

If baseline separation of the API and its related impurities cannot be resolved *via* HPLC, as was observed in the case of bivalirudin and its related impurities, the EIC method should be used for qualitative and quantitative purposes to circumvent the issue of coeluting peptides. In cases where peptides coelute, HRMS serves as a highly selective analyzer and sensitive detector. Peak areas from EIC chromatograms were obtained by integrating peaks of the selected charge states and isotope ions at selected mass resolution. It has been reported that the method linearity R^2 was >0.99 when choosing one or multiple charge states and/or isotope ions.[36] Furthermore, the linearity result was not affected by selecting mass resolution provided that ion peaks separated by *m*/*z*. For best practice, selection of the three highest charge states and three isotope ions at a resolution of 50 ppm MEW is recommended.

9.3.6 EIC Mass Extraction Window

EIC chromatograms are obtained by extracting the selected isotope ions at different charge states within the selected mass resolution range. This mass resolution range is also called the mass extraction window (MEW) in EIC chromatograms. The EIC MEW is an important parameter for MS separation in order to quantitate a drug product and any impurities present. Two EIC MEWs were examined: 20 and 100 ppm. Salmon calcitonin, bivalirudin, exenatide and their related impurities had linear and precise responses with both MEWs over the concentration range 0.1–10 μM (see Tables 9.5 and 9.6, respectively). The sensitivity with a 100 ppm MEW was 2.5–4.7 ($S2/S1$) times higher than with a 20 ppm MEW. A wider MEW results in a higher MS response due to the inclusion of more ions. The MEW should be optimized for specific peptides with consideration of selectivity, sensitivity, linearity, precision, accuracy and robustness.

9.3.7 LC–HRMS Compared With LC–UV

The LC–HRMS method has several inherent advantages over traditional HPLC–UV approaches, including the ability to provide both qualitative and quantitative data within a single experiment.[37,38] Particularly for peptide drugs, LC–HRMS can easily detect peptides with high sensitivity over a wide linear dynamic range even in the absence of a UV chromophore.

The information-rich HRMS data can be used for accurate molecular weight determinations and, in combination with online MS/MS, to determine AA composition and confirm the AA sequence. In addition, LC–HRMS can be used to detect and characterize related impurities, both known and unknown, in a sample. As demonstrated in the bivalirudin case study, chromatographic separation alone could not baseline resolve the bivalirudin and its related impurities, hence the LC–HRMS method was used to detect coeluting peaks selectively so that that accurate quantitation could be performed. In addition, more peptide-related impurities were detected in the salmon calcitonin nasal solution using the LC–HRMS method than with the USP HPLC–UV-based method.

9.4 Conclusion

The case studies presented using salmon calcitonin, bivalirudin and exenatide demonstrated the suitability of an LC–HRMS method for monitoring peptide drug product quality. Calcitonin and its related impurities displayed linear responses over the range 0.1–10 μM. Linear regression correlation coefficient R^2 values for salmon calcitonin, Glu14-calcitonin and acetyl-calcitonin were 0.995, 0.996 and 0.993, respectively. Intra-assay precision in term of RSD was NMT 2.0% at test concentrations from 1 to 10 μM. The accuracy of the method was 98–100% as measured by spiking 1% of Glu14-calcitonin and acetyl-calcitonin into a stock calcitonin solution. Limits of detection for calcitonin impurities Glu14-calcitonin and acetyl-calcitonin were 0.03 and 0.04% of the label claim, respectively, indicating that an impurity present at <0.1% of the drug product API concentration could be detected. Method validation studies analyzing bivalirudin and exenatide drug products exhibited similar results to those for salmon calcitonin with respect to high selectivity, sensitivity, precision and linearity.

Overall, LC–HRMS methods can be used to obtain quantitative results, with high selectivity, sensitivity, precision and suitable linearity, and also provide qualitative information, including characterization of peptide structure and sequence, within a single experiment, as demonstrated in the presented case studies for the analysis of salmon calcitonin, bivalirudin and exenatide. Furthermore, the LC–HRMS method is advantageous because it can separate, detect, identify and quantitate more impurities than traditional HPLC–UV methods. In summary, LC–HRMS is a fit-for-purpose approach that can be used as part of the analytical framework to ensure proper quality control of peptide drug products.

Disclaimer

This chapter reflects the views of the authors and should not be construed to represent FDA's views or policies.

Acknowledgements

Internal funding for this work was provided by the CDER Critical Path Program, Office of Pharmaceutical Quality and the Office of Generic Drugs.

References

1. J. Ermer and M. Vogel, *Biomed. Chromatogr.*, 2000, **14**, 373–383.
2. A. A. Kaspar and J. M. Reichert, *Drug Discovery Today*, 2013, **18**, 807–817.
3. V. S. Dorpe, M. Verbeken, E. Wynendaele and B. De Spiegeleer, *J. Bioanal. Biomed.*, 2011, **S6**, 1–15.
4. A. C. Gucinski and M. T. Boyne, 2nd, *Rapid Commun. Mass Spectrom.*, 2014, **28**, 1757–1763.
5. FDA, *ANDAs for Certain Highly Purified Synthetic Peptide Drug Products That Refer to Listed Drugs of rDNA Origin, Guidance for Industry*, 2017.
6. K. Zeng, I. Geerlof-Vidavisky, A. Gucinski, X. Jiang and M. T. Boyne, *AAPS J.*, 2015, **17**, 643–651.
7. ICH, *ICH Q3A(R2) Impurities in New Drug Substances* 2006, 15.
8. ICH, *ICH Q3B(R2) Impurities in New Drug Products*, 2006, 16.
9. ICH, *ICH Topic Q6B Specifications: Test Procedures and Acceptance Criteria for Biotechnological/Biological Products*, 1999, 20.
10. J. P. Savaryn, T. K. Toby and N. L. Kelleher, *Proteomics*, 2016, **16**(18), 2435–2443.
11. R. Y. Willoughb, E. Sheehan and S. Mitrovich, *A Global View of LC/MS: How to Solve Your Most Challenging Analytical Problems*, Chem-Space Associates, Inc., 2002.
12. J. J. Pitt, *Clin. Biochem. Rev.*, 2009, **30**, 19–34.
13. G. Siuzdak, *The Expanding Role of Mass Spectrometry in Biotechnology*, MCC Press, San Diego, 2003.
14. V. Barwick, J. Langley, T. Mallet, B. Stein and K. Webb, *Best Practice Guide for Generating Mass Spectra*, 2006, 27.
15. K. Webb, T. Bristow, M. Sargent and B. Stein, *Methodology for Accurate Mass Measurement of Small Molecules Best Practice Guide*, 2004, 26.
16. *Guide to Achieving Reliable Quantitative LC-MS Measurements*, ed. M. Sargent, RSC Analytical Methods Committee, 2013, p. 61.
17. D. F. Hunt, R. A. Henderson, J. Shabanowitz, K. Sakaguchi, H. Michel, N. Sevilir, A. L. Cox, E. Appella and V. H. Engelhard, *Science*, 1992.
18. R. A. Henderson, H. Michel, K. Sakaguchi, J. Sakaguchi, E. Appella, D. F. Hunt and V. H. Engelhard, *Science*, 1992.
19. H. Wei, A. A. Tymiak and G. Chen, *Bioanalysis*, 2013, **5**, 1299–1313.
20. L. C. Wu, F. Chen, S. L. Lee, A. Raw and L. X. Yu, *Int. J. Pharm.*, 2017.
21. S. M. D'Addio, J. R. Bothe, C. Neri, P. L. Walsh, J. Zhang, E. Pierson, Y. Mao, M. Gindy, A. Leone and A. C. Templeton, *J. Pharm. Sci.*, 2016, **105**(10), 2989–3006.

22. EMEA, *ICH Topic Q6A Specifications: Test Procedures and Acceptance Criteria for New Drug Substances and New Drug Products: Chemical Substances*, 2000, p. 32.
23. W. Cui, H. F. Rohrs and M. L. Gross, *Analyst*, 2011, **136**(19), 3854–3864.
24. J. R. Yates, C. F. Ruse and A. Nakorchevsky, *Annu. Rev. Biomed. Eng.*, 2009.
25. A. Szajek, B. Gregg, M. Verlander, H. Rode, I. Eggen and A. Swietlow, *Pharmaceutical Technology*, 2014, 38.
26. FDA, *Guidance for Industry: Analytical Procedures and Methods Validation for Drugs and Biologics*, 2015, 15.
27. L. Dillen and F. Cuyckens, *Bioanalysis*, 2013, **5**, 1145–1148.
28. C. H. Chesnut, M. Azria, 3rd, S. Silverman, M. Engelhardt, M. Olson and L. Mindeholm, *Osteoporos Int*, 2008, **19**, 479–491.
29. USP 36-NF 31 (United States Pharmacopeia Convention), 2013, **36**, 2736–2741.
30. M. D. Reed and D. Bell, *Pharmacotherapy*, 2002, **22**, 105S–111S.
31. R. S. Cvetkovic and G. L. Plosker, *Drugs*, 2007, **67**, 935–954.
32. B. Liu, Q. Dong, L. Shi, M. Wang, C. Li, Y. Wu, X. Yu, Y. Shan, Y. Chen, W. Kong and J. Shen, *Chem. Res. Chinese Universities*, 2010, **26**, 33–37.
33. M. I. Aguilar and M. T. Hearn, *Methods, Enzymol.*, 1996, **270**, 3–26.
34. S. Fekete, A. Beck, J. L. Veuthey and D. Guillarme, *J. Pharm. Biomed. Anal.*, 2012.
35. B. O. Keller, J. Sui, A. B. Young and R. M. Whittal, *Anal. Chim. Acta*, 2008, **627**, 71–81.
36. A. C. Gucinski and M. T. Boyne, 2nd, *Anal. Chem.*, 2012, **84**, 8045–8051.
37. E. N. Fung, M. Jemal and A. F. Aubry, *Bioanalysis*, 2013, **5**, 1277–1284.
38. R. Ramanathan and W. Korfmacher, *Bioanalysis*, 2012, **4**, 467–469.

CHAPTER 10

Mass Spectrometric Analysis for the Quality Control of Peptide Conjugate Therapeutics

OSAMA CHAHROUR,*[a] ANNA MELJON[b] AND DIEGO F. COBICE[c]

[a] Spectroscopy Group, Almac House, 20 Seagoe Industrial Estate, Craigavon BT63 5QD, Northern Ireland, UK; [b] School of Biological Sciences, Institute for Global Food Security, Queen's University, University Road, Belfast BT7 1NN, Northern Ireland, UK; [c] Biomedical Sciences Research Institute (BMSRI), Ulster University, Cromore Road, Coleraine BT52 1SA, Northern Ireland, UK

*Email: osama.chahrour@gmail.com

10.1 Analysis of PEGylated Peptides by Mass Spectrometry

10.1.1 Introduction

Naturally occurring peptides and proteins play an important role in inter- and intracellular signal transduction, physiological homeostasis and immune response, serving as acute modulators of a wide range of body functions. Many of these physiologically important compounds, such as hormones, growth factors, cytokines, enzymes and antibodies, have been sequenced and hence can be synthetically manufactured. As the biological activity of these molecules is regulated by a precise "key and lock"

Drug Discovery Series No. 72
Peptide Therapeutics: Strategy and Tactics for Chemistry, Manufacturing and Controls
Edited by Ved Srivastava

Published by the Royal Society of Chemistry, www.rsc.org

mechanism (receptor activation or inhibition, proteolytic release of signalling compounds), the peptide-derived drugs can very precisely target the disease. The specificity reduces the risk of side effects and provides excellent selectivity; however, other features characteristic of bioactive peptides, such as a short half-life, low stability and high immunogenicity, complicate their use as therapeutics. The half-life of peptide drugs can be as short as a few minutes,[1] which can be a major challenge for maintaining effective levels in the patient's blood. The loss of the active pharmaceutical ingredients (APIs) is mostly related to indigenous enzymatic activity (digestion, hydroxylation) or exogenous renal ultrafiltration. This can be addressed by increasing the concentration of the pharmaceuticals or more frequent administration of the API. The maximum level of the API dose is tightly limited by its toxicity and possible side effects, while frequent administration may not be practical, especially in the treatment of chronic disorders.

Various strategies are implemented to reduce constraints related to peptide drug properties. The most common include amino acid manipulation to decrease proteolytic degradation and immunogenicity, encapsulating peptides in vesicles to improve drug delivery (liposomes) and binding peptides to a stable ligand to improve the half-life. The ligand-binding methodology involves covalent or noncovalent attachment of the peptide to the ligand moiety such as a fatty acid, albumin-binding molecule, carbohydrate or polymer such as poly(ethylene glycol) (PEG) chains. PEGylation is an excellent example of a delivery system that has scientific and multi-billion dollar commercial importance owing to the remarkable improvement in the circulatory half-lives of therapeutic proteins and peptides.[2–4]

PEGylation is the attachment of a PEG chain to a compound of interest such as an API. The concept of PEGylation was introduced in the 1970s by Frank Davies and colleagues at Rutgers University.[5] In this initial work, coupling a PEG moiety with bovine liver catalase significantly reduced the immunogenicity and susceptibility for proteolytic digestion of the enzyme. Simultaneously, the biological function was largely maintained – the loss of the activity was reduced by 10% compared with the original enzyme. Consecutive studies demonstrated further advantages of PEGylation such as reduced hepatic clearance. Since then, PEGylation has become an established method for improving the pharmacokinetic profile of peptide drugs. Examples of PEGylated drugs approved by the US Food and Drug Administration (FDA) are listed in Table 10.1.

10.1.2 Structure and Physicochemical Properties of Poly(ethylene Glycol)

PEGs are a family of polyethers consisting of poly(ethylene glycol) monomers linked in a straight or branched chain containing a variable number of

Table 10.1 Examples of PEGylated drugs approved by the FDA.

Year of approval	API (PEG conjugate)	Indication	Trade name	Company	PEG species (type of chain, MW)
1990	PEGadenosine deaminase[6]	Severe combined immunodeficiency (SCID)	Adagen	Enzon	Linear, multiple chains, ~5 kDa
1994	PEGasparaginase[7,8]	Acute lymphoblastic leukaemia	Oncaspar	Enzon	Linear, multiple chains, ~5 kDa
2002	PEGinterferon-α-2a[8,9]	Hepatitis C	Pegasys	Schering-Plough	Branched, 40 kDa
2002	PEGfilgrastim (GCSF)[10]	Neutropenia	Neulasta	Amgen	Linear, 20 kDa
2003	PEGvisomant (PEGgrowth hormone receptor antagonist)[11]	Acromegaly	Somavert	Pharmacia & Upjohn	Several linear chains, 5 kDa
2004	PEGaptanib[12]	Wet age-related macular degeneration (AMD)	Macugen	Eyetech, Pfizer	Branched, 40 kDa
2007	mPEG-epoetin-β[13]	Anaemia	Mircera	Hoffman La Roche	Linear, 30 kDa
2008	Certolizumab PEGol[14,15]	Crohn's disease, Arthritis	Cimzia	UCB	Branched, 40 kDa
2010	PEGuricase (PEGloticase)[16,17]	Gout	Puricase, Krystexxa	Savient	Linear, multiple chains, 10 kDa Random lysine conjugation
2014	Naloxegol (PEGylated naloxone)[18,19]	Opioid-induced constipation (OIC)	Movantik	AstraZeneca	Linear, 1 kDa
2014	PEGinterferon β-1a[20,21]	Relapsing forms of multiple sclerosis	Plegridy	Biogen	Linear, 20 kDa
2015	PEGylated antihaemophilic factor[22]	Haemophilia A	Adynovate	Baxalta	Branched, 20 kDa

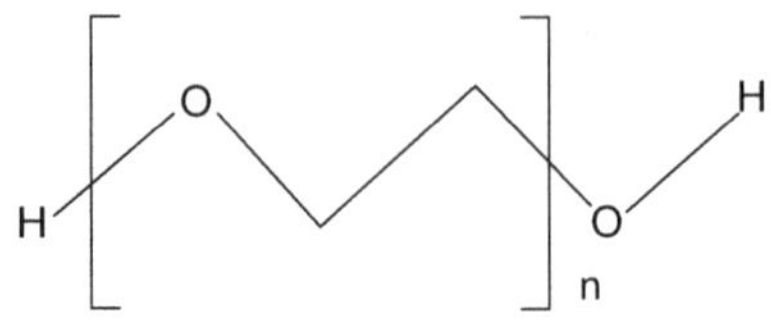

Figure 10.1 Structure of poly(ethylene glycol) (PEG).

monomers (Figure 10.1). The resulting molecule is then covalently attached to the compound of interest such as a peptide in a process called PEGylation. Commercially available PEGs are polydispersed mixtures, containing polymeric species of different chain lengths. The average molecular weight of the mixture is denoted in the PEG nomenclature as a number. For example, PEG 600 contains a mixture of polymers with an average molecular weight of 600 g mol^{-1}.

As PEG mixture can vary by the average molecular weight (M_w) and the degree of distribution of the chain lengths, further parameters were introduced to provide a more accurate characteristic. A PEG is therefore defined by the average molecular weight (M_w) and the number-average molecular weight (M_n). M_n is calculated by summing the masses of each of the individual polymer species and dividing by the number of species. Furthermore, the ratio M_w/M_n determines the polydispersity distribution and is called the polydispersity index (PDI).

PEG is highly soluble in water and the interaction of the polymer with water is aided by two lone electron pairs present on oxygen atom in each ethylene glycol monomer. Water molecules form bonds with the lone electrons present on the ether oxygen; research has shown that the ratio of water coordination is approximately equal to 2–3 water molecules per one monomer.[23] The high hydrophilicity of PEG increases the solubility of PEG-conjugated compounds, and a better solubility of the API improves the delivery and distribution, enhancing the bioavailability of therapeutics. Furthermore, the water surrounding a PEG molecule forms a hydration shell, which expands the volume of the polymer; this process is also referred to as an increased hydrodynamic radius. Hydrated PEG attached to the peptide effectively creates steric hindrance, which provides protection from microfiltration, degrading enzymes and immune recognition (Figure 10.2). Even though these effects are expected to increase with increase in size of the PEG, there is no defined value of molecular weight above which this outcome can be clearly observed. This is related to the property of PEG chains, which, even when hydrated, retain some degree of flexibility. This feature is most likely responsible for the phenomena of "reptation" (Latin *reptile*, creeping), allowing relatively large PEG molecules to pass through small structures such as glomerular pores.[24] Effective inhibition of microfiltration is observed only with PEGs with molecular weight exceeding 30 kDa.[5] To summarize, a carefully selected size of PEG conjugated to the peptide can significantly improve the pharmacokinetic properties of decelerating renal clearance, proteolytic digestion and immunogenicity.

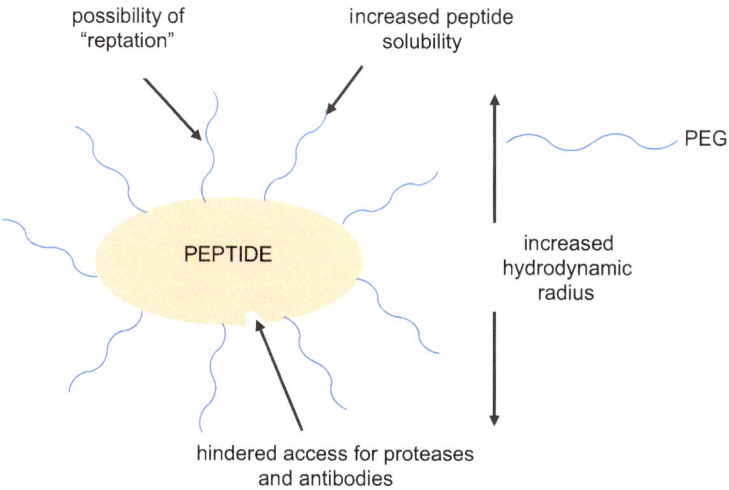

Figure 10.2 Schematic view of a PEGylated protein. The polymer increases the hydrodynamic radius by creating steric hindrance. The increased size of the conjugate reduces accessibility for proteolytic enzymes and renal clearance.

10.1.3 Toxicity of PEG and PEGylated Molecules

Since the 1950s, PEG has been shown to have a very low toxicity *in vivo* and *in vitro*. The most toxic compound is the one with the lowest molecular weight, *i.e.* poly(ethylene glycol) itself. Even for this compound the LD_{50} value in guinea pigs was established as 6.6 g kg^{-1}. The toxicity was related to oxalic acid, the main metabolite of ethylene glycol.[25] For a compound with a higher molecular weight, PEG 4, the LD_{50} value was even higher and was determined as 15–22 g kg^{-1} in rodents. As the molecular weight of the PEG increased further, to 10 kDa, the LD_{50} values exceeded 50 g kg^{-1} in rats.[26] Chronic exposure to the PEG also indicates the relative safety of PEG administration – rodents fed with PEG 1400 and PEG 400 for 2 years (0.006 and 0.002 g kg^{-1} per day, respectively) experienced no adverse effects. Only intravenous application of a high concentration of low molecular weight PEG caused renal lesions. This was due to the deposition of oxalate crystals in the kidneys.[27] Although based on toxicological studies PEG has been approved by the FDA as Generally Recognized As Safe (GRAS), recent studies showed an accumulation of PEG and PEG-peptide molecules in the cell vacuolar system.[28,29] This finding, along with the fact that PEG is not metabolized in the body, highlights the need for a more detailed study of the metabolic impact of PEGs.

10.1.4 PEG Conjugation Strategies

Various coupling methods can be employed to conjugate PEGs and peptides and a number of factors must be considered when choosing the PEGylation strategy. The length and structure (linear, branched, cyclic) of the PEG moiety and the number of PEG molecules used (single or multiple PEGylations) will have an impact on the characteristics of the resulting polypeptide. Linear PEGs are easier to control in terms of the chain length and dispersity; however, branched species will have the advantage of improving the peptide's stability and solubility and decreasing the viscosity and could further increase the conjugate's half-life. The downside is an elaborate synthesis procedure, which significantly increases the cost of the polymer.

Also, the location of the conjugation on the peptide is important for its physiological activity and interactions with other biological agents. In the first-generation PEGylation, the polymeric molecule or molecules were randomly grafted to the therapeutic. This strategy is useful if the PEGylation sites do not interfere with the activity of the API and serve only as a route to increasing the half-life *via* increased molecular size. The downside of this approach is that the resulting material is heterogeneous and proteins with different numbers and sizes of attached PEGs may have different chemical and physical properties. Also, the lack of uniformity creates a challenge for the reliable quality control required for the release of the API to patients. Even with these limitations, this technology was successfully used in the development of the first PEGylated drugs, namely Adagen and Oncaspar, manufactured by Enzon.

Further advances in conjugation technology allowed the PEG to be introduced in a specific location of the peptide; this approach is known as "second-generation PEGylation." Although accurate positioning of the polymer is not always essential and random PEGylation is still widely used, second-generation PEGylation provides better control of the polypeptide's properties. This may be particularly important when the modification is in the proximity of the active site on the protein. Further development of the PEGylation strategy is aimed at enhancing a drug's bioactivity by the introduction of branched or combed PEG chains.[6]

10.1.5 PEGylated Pharmaceuticals – Analytical Strategies

As opposed to the majority of APIs where the molecule of interest has a single, defined molecular weight, PEG consists of a wide range of scattered masses corresponding to the polymer's chains. This polydispersed nature of the PEG creates a major challenge when it comes to developing analytical methodology to test PEGylated biopharmaceuticals. Mass spectrometry (MS), *via* accurate mass measurement, permits differentiation and defines PEG chains, thus providing a useful tool for the characterization of PEGylated APIs.

The analysis of PEGylated biopharmaceuticals is performed using two main techniques: MALDI-TOF (matrix-assisted laser desorption/ionization time-of-flight), which generates an MS fingerprint of the analyte, and separation-based LC-MS (liquid chromatography-mass spectrometry), which allows the characterization of separate compounds in the mixtures.

In MALDI, the analyte is combined with a suitable matrix, which contains an energy-absorbing agent. The resulting solution is applied onto a metal plate to dry, then the dried residue is illuminated with a laser. The laser beam transfers the energy *via* the matrix onto the sample, which in turn is transformed into a gaseous state. During vaporization, the energy from the laser also ionizes the analyte, which subsequently can be subjected to MS analysis. This technique is considered a "soft" ionization technique, in which the analyte retains low levels of residual energy, reducing the fragmentation of the molecule.

As the entire sample mixture is analysed in MALDI, the result shows a comprehensive profile of the analysed sample. This approach is suitable for qualitative characterization of the analyte; however, for a more detailed analysis, LC-MS or high-performance liquid chromatography (HPLC)-MS is used. During LC-MS, the MS analysis is preceded by a separation step carried out using LC. To introduce the analyte into the system, the compound is dissolved in a solution that is injected into the LC column. The compounds present in the sample mixture are separated as they pass through the stationary phase, propelled by the liquid mobile phase. The mobile phase consists of solvents that maintain the analyte in a soluble form, enabling interaction with the stationary phase. The stationary phase is a solid material usually packed inside a metal column that has the ability to retain the analytes with different strengths. The retention depends on the affinity (usually hydrophobicity, ionic charge, size of molecules) of the stationary phase to the analyte molecules. As PEG is highly soluble in water, the analysis is usually performed using reversed-phase chromatography with an apolar mobile phase and a nonpolar stationary phase.[30] Separation of PEGylated compounds can also be achieved by exploiting the differences between the PEG chains; such an approach is used in size-exclusion chromatography (SEC).[31]

Following the LC separation, the compounds are passed to a mass spectrometer. First, the analytes are charged in an ionization source. In the analysis of PEGylated peptides, the most commonly employed ionization technique is electrospray ionization (ESI). In this method, a differential electric potential is applied to a metal capillary tube. The solution passing through the capillary tube gains electric charge and is released from the capillary as a charged aerosol. As the droplets of electrospray fly in the direction of the detector, the solvents evaporate from the droplets and the charge is passed onto the molecules of the analyte. In ESI, electrically neutral compounds gain charge by forming an adduct with charged molecules, effectively forming quasi-molecular ions. In the positive ion mode it includes protons (H^+) (protonation), metal cations (sodium, potassium), salt moieties (ammonium) or charged solvent molecules (acetonitrile, methanol, water). In the negative ion mode the charge can be obtained by the loss of a proton from the analyte molecule or by forming adducts with anions such as chloride, formate or acetate. To improve the ionization, weak acids or bases can be added to the mobile phase to increase the concentration of available ions in the solution. Depending on the molecular structure, an analyte can form adducts with one or multiple ions of the same type or with several ions of different species. The formation of multi-species adducts reduces the spectral peak intensity as the signal of the ionized compound is divided between several adducts.

PEG and PEGylated analyte molecules, due to their size, are prone to the formation of multiple adducts, adding to an already complex spectrum of the poly(ethylene glycol) (Figure 10.3). To reduce the number of available ion species, metal salts such as sodium citrate can be added to the mobile phase. These provide an abundance of single-species ions that force the formation of single-type adducts; for example, in the positive ionization mode sodium citrate would increase the generation of sodium cation adducts.

As only one species of ions is available, PEGylated molecules form multiply charged uniform adducts. These charged analytes have lower *m*/*z* ratios than the singly charged counterparts, which allows the detection of high-mass molecules within the acquisition window of a regular MS instrument. However, compounds with a very large number of charges have a correspondingly lower isotopic spacing, which can affect the accuracy of *m*/*z* measurements. To address this problem, charge-stripping agents such as amines (*e.g.* triethylamine) can be added to the LC flow. The concentration of charge-stripping agent can be adjusted to obtain the most suitable degree

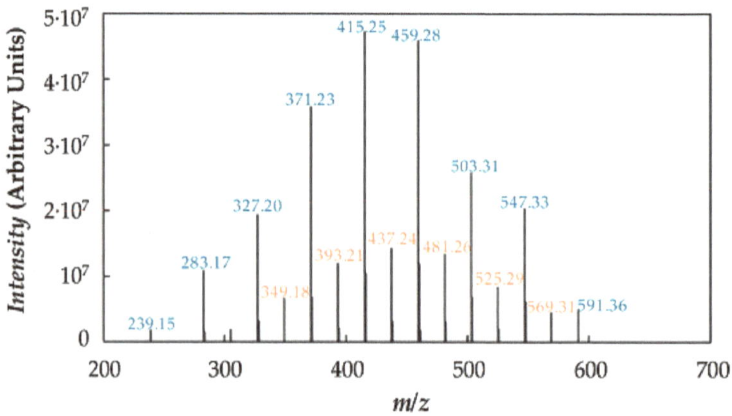

Figure 10.3 Example of LC-MS spectrum of PEG 400. The characteristic 44 amu peak spacing corresponds to an ethylene monomer difference between PEG chains. In blue are annotated protonated $[M+H]^+$ PEG peaks; in orange are annotated sodium adducts $[M+Na]^+$. The characteristic mass difference of 22 amu between consecutive $[M+H]^+$ ions and $[M+Na]^+$ corresponds to the difference between a hydrogen cation (1 amu) and a sodium cation (23 amu). Note: presented is the spectrum of unconjugated PEG; however, the same characteristic features are observed in the spectra of PEGylated compounds.[35]
Reproduced from ref. 32 (https://doi.org/10.3390/nano8010016) under the terms of a CC BY 4.0 license (https://creativecommons.org/licenses/by/4.0/).

of ionization. The drawback of this approach is that the charge-stripping agents can impede the chromatographic separation; *e.g.* the high pH of amines is not compatible with the majority of silica columns used in the LC. To overcome this constraint, the post-column infusion (PCI) technique is used.[32] A solution containing the charge-reducing agent is added at a constant flow to the post-column eluent *via* a T-piece connector. This does not affect the chromatographic separation while maintaining a desired intensity of ionization.

Figure 10.4 shows the mass spectrum of PEGylated protein acquired with different concentrations of triethylamine (TEA) delivered by post-column infusion. When no TEA was added, the most prominent feature was charge envelopes of highly charged ions (low-m/z ions were the most abundant). An increased concentration of TEA led to a decrease in ion charge and an increase in the m/z values of detected ions. The MS software can convert the m/z values for multiply charged analytes into the analyte mass values; Figure 10.5 shows the deconvoluted spectrum of a PEGylated protein with a characteristic 44 amu spacing reflecting PEG monomers.

The most frequently used analysers are ion trap and quadrupole/time-of-flight (QTOF) instruments. These detectors provide high-resolution (HR) data acquisition which permits the detection of m/z values to several decimal places. Accurate mass measurement is particularly useful to discriminate between the isotopes of the same compounds or different compounds with very similar molecular masses. The mass spectral data obtained with HR instruments can be extracted and processed to determine the polydispersity values of selected PEGylated peptides. Using relevant MS software, the m/z values for each mass spectral peak representing one PEG species (chain length) can be selected and an extracted ion chromatogram (EIC) can be generated. The area of each resulting EIC peak represents the intensity (population) of the corresponding polymeric species and its numerical value, together with the peak mass, can be further entered into calculations of the polydispersity index (PDI), the number-average weight (M_n) and the weight-average molecular weight (M_w). The relevant equations are

$$M_n = \Sigma M_i N_i / \Sigma N_i \tag{10.1}$$

$$M_w = \Sigma M_i^2 N_i / \Sigma M_i N_i \tag{10.2}$$

$$\text{PDI} = M_w / M_n \tag{10.3}$$

where M_i is the mass of each PEG polymer species (i) and N_i is the peak intensity of each species (the peak area). M_n describes the most abundant polymer species or the apex of the polymeric distribution. M_w is another way of describing the molar mass of a polymer where a heavy molecule (longer chain) will have a larger contribution than a smaller molecule. PDI refers to the degree of the scatter of PEG species (Figure 10.6).

The proximity of PDI to 1 reflects the closeness of the PEG polymeric scatter to a uniform distribution. Higher polydispersity values indicate a

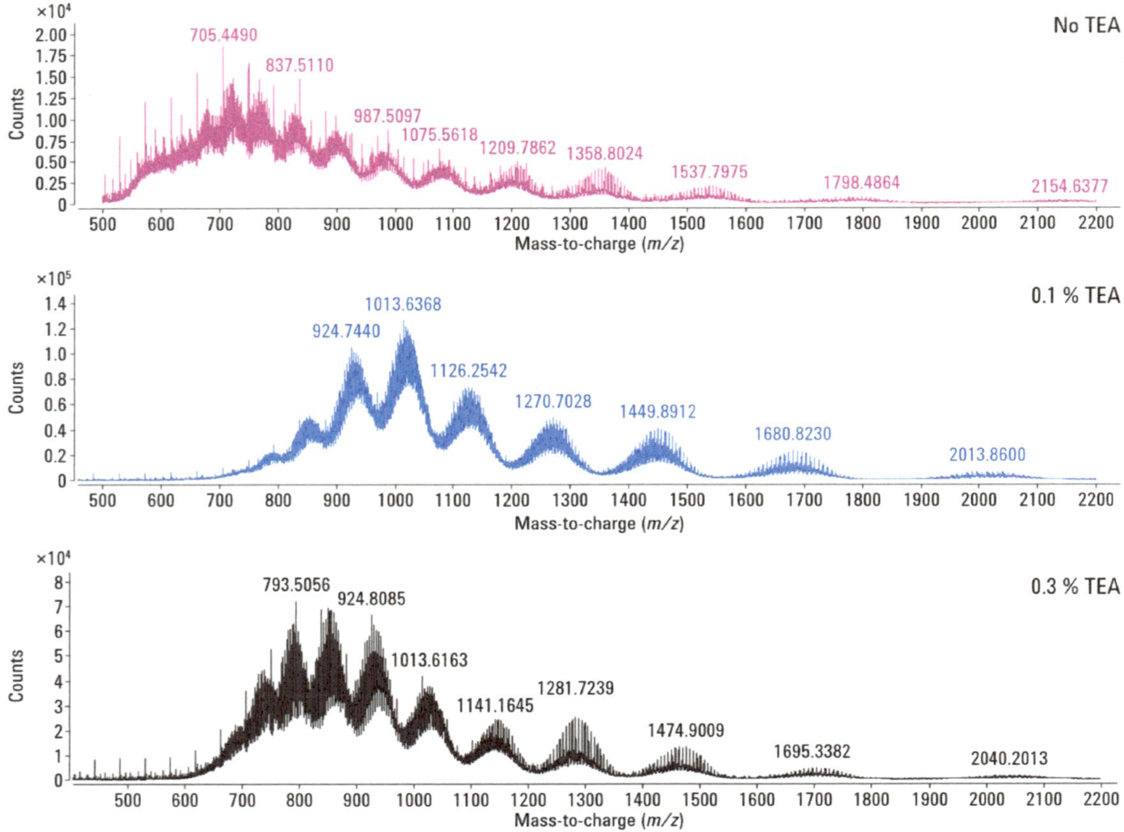

Figure 10.4 Mass spectrum of 10 kDa mPEG-*p*-nitrophenylcarbonate (mPEG-PNP). The data were acquired at different concentrations of TEA using Agilent Q-TOF.
Reproduced from ref. 153 with permission from Agilent Technologies, Inc., Copyright © Agilent Technologies, Inc. 2012.

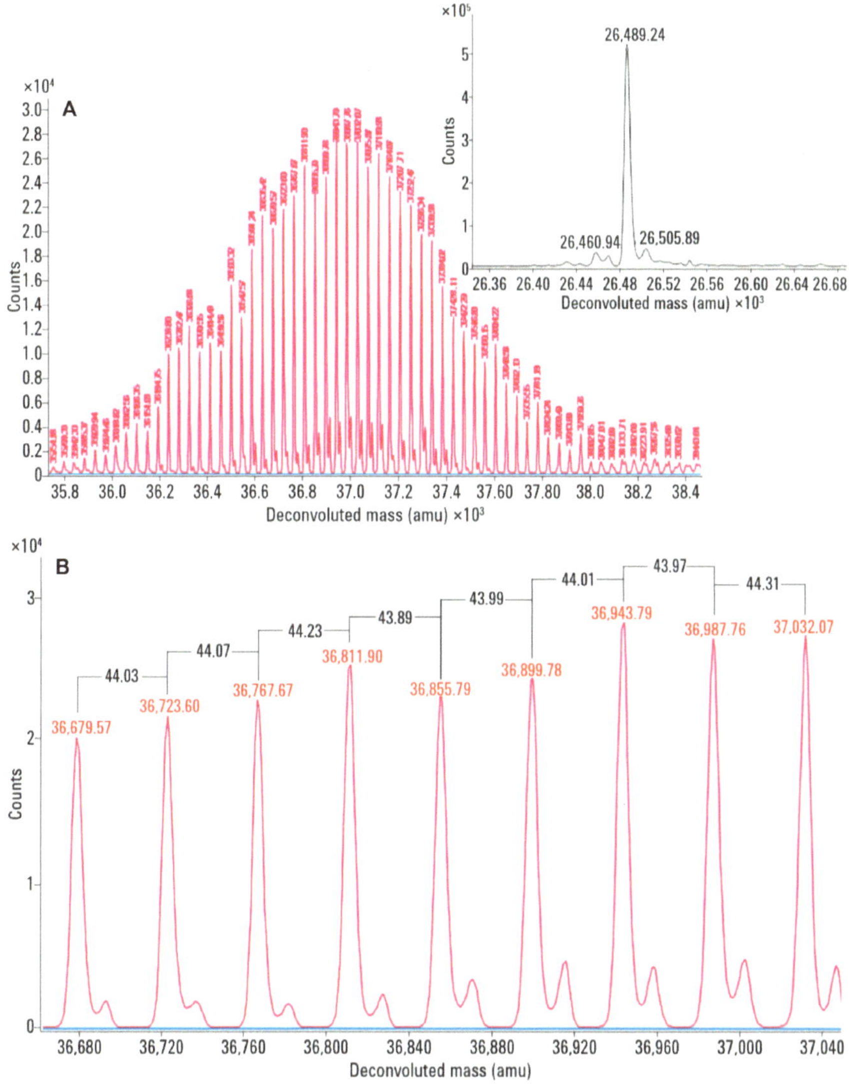

Figure 10.5 Deconvoluted mass spectra of mPEG-PNP. Data acquired under 0.3% TEA (full spectrum shown in Figure 10.4). (A) Full deconvoluted spectrum; (B) zoomed spectrum with accurate mass difference between PEG chains. The data were acquired using Agilent Q-TOF.
Reproduced from ref. 153 with permission from of Agilent Technologies, Inc., Copyright © Agilent Technologies, Inc. 2012.

wider spread of PEG chains in the analysed material. This may be not desirable as the different lengths of the PEG moieties can have a potential impact on the biochemical and pharmacological characteristics of the drug. The evaluation of polydispersity is therefore crucial during the synthesis of

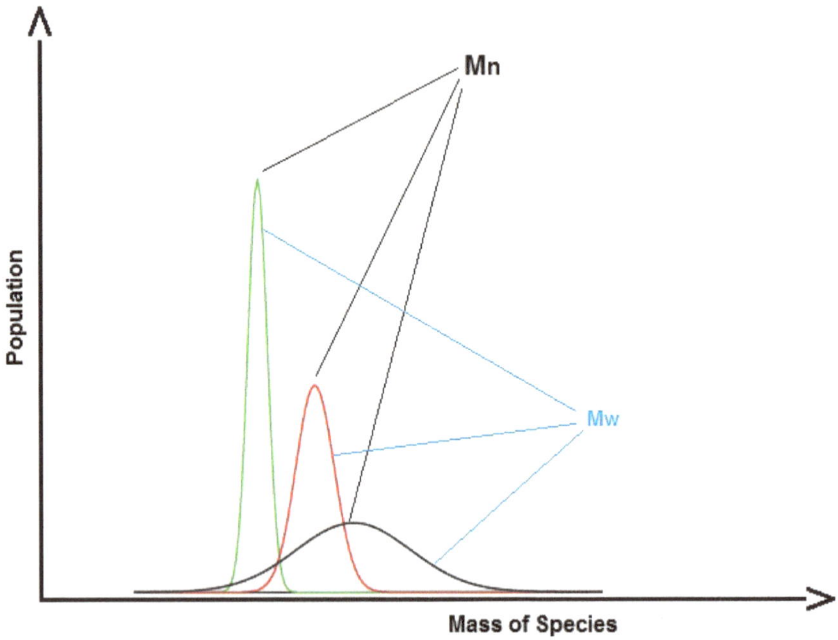

Figure 10.6 Mass distribution of three polymers. The green polymer has the lowest PDI due to the narrow distribution, hence the difference between M_n and M_w is minimal. The black polymer has a high PDI, hence the difference between M_n and M_w is larger.

the API to ensure that the released drug has the required properties. The heterogenicity of PEG is also monitored in the drug stability studies to evaluate potential degradation of the polyethylene chains.

Currently, there are no established limits for the polydispersity of PEG in modified therapeutics. Defining the PDI value for the entire class of therapeutics will be challenging as there are wide varieties between the PEG species used for the APIs. Evaluation of the impact of the chain length on the activity of the drug and the polydispersity would have to be assessed separately for each therapeutic.

MS analysis provides characteristics of the entire PEGylated molecule; defining the actual site of PEGylation requires a slightly different approach. In tandem mass spectrometry (MS/MS or MS^2) analysis, the molecule of interest passes through the first MS analyser, where the *m*/*z* values of the ions of interest are selected. These are called parent ions and are passed to the MS collision chamber, where they are subjected to fragmentation, the resulting fragment ions (referred to also as daughter ions) being transferred to a second MS analyser. The fragmentation can be achieved in several ways; for structural elucidation of molecules, the most common technique is collision-induced dissociation (CID). In this approach, the analyte is exposed to the molecules of an inert gas such as nitrogen, argon or helium.

The molecules of the inert gas collide with the analyte molecules, producing MS/MS fragment ions that can subsequently be detected by the second MS analyser.

Peptide chains, when subjected to CID, fragment in a predictable manner across the backbone bounds. The resulting fragment ions are denoted according to Niemann terminology, where a letter indicates the broken bond and terminus contained in the fragment. Fragment ions *y* and *b* are produced by breaking the peptide bonds, whereas *x*, *z*, *a* and *c* ions are generated by dissociation of bonds in proximity of the α-carbon. Each ion also has a number that indicates the number of α-carbons in the chain. The fragmentation can be further complicated by the loss of water and ammonia from the MS/MS fragments. In CID, the least stable bonds are peptide bonds, hence the tandem mass spectrum of the peptide is dominated by *y* and *b* ions, with *a*, *c*, *x* and *z* ions present at lower levels. The principles of the MS/MS analysis of peptides also apply to the analysis of PEGylated peptide therapeutics. An example of a PEGylated peptide with possible dissociation sites is shown in Figure 10.7.

The resulting tandem mass spectrum is extremely complex, hence the interpretation of the daughter ions and translation into the peptide sequence are performed using relevant MS sequencing software. The algorithm determines the distances between the ions and matches the results with amino acids and *m*/*z* values of their potential modifications. The peak intensities of the analysed sequence are compared with the peak intensities of the theoretical sequence and a probability score is calculated. The theoretical sequence may be that of the unmodified peptide; the fragment ion with a mass difference between the theoretical and empirical values equal to the PEG moiety indicates the PEGylation site.

An example of the use of PEGylated peptide spectra to confirm the PEGylation site was discussed by Lu *et al.*[33] The parent ion chosen was the

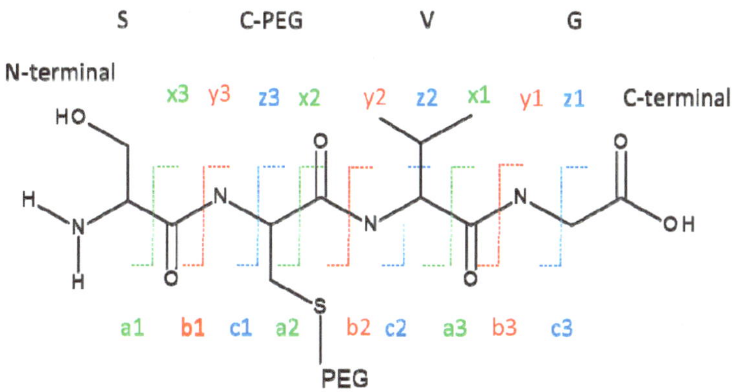

Figure 10.7 Schematic view of PEGylated tetrapeptide: serine–cysteine–valine–glycine, abbreviated as S–C–V–G, with PEGylated cysteine (C-PEG). The theoretical MS/MS fragmentation is indicated.

doubly charged ion of *m/z* 510.7 and subjected to fragmentation. The resulting daughter ions with suggested peak assignments confirmed that the PEG moiety was grafted to cysteine *via* a 208 Da linker. Moreover, mass elevation was not observed for *a* or *b* ions, which further corroborates the site of the modification.

To summarize, LC-MS/MS sequencing is a reliable tool for a confirmatory analysis regarding the position of the PEG moiety in the peptide chain. Moreover, the application of MS/MS reaches beyond the assessment of the accuracy of PEG grafting. MS/MS data can supply information regarding other factors potentially detrimental to the drug efficacy, such as alterations of the peptide chain, *e.g.* oxidation, amino acid deletion, insertion or truncation of the peptide chain. The sequencing data, along with calculation of the polydispersity, provide a comprehensive evaluation of the quality of PEGylation regarding the synthesis and the stability of drug products. While mass spectral analysis can generate all the required spectral information, the calculation of the polydispersity is labour intensive as processing of the data is usually performed with the aid of external spreadsheets. Today, the major challenge for large-scale analysis lies in developing software capable of automated data processing.

10.2 Analysis of Antibody–Peptide Conjugates by Mass Spectrometry

10.2.1 Introduction

The chemical addition of peptides to proteins and other biopolymers is one of the most essential methods for the synthesis of bioconjugates.[34,35] The addition creates agonist or antagonist conjugates that specifically target the intended receptor. For example, voltage-gated sodium channel Na $_V$1.7 can be inhibited using peptide GpTx-1 extracted from tarantula venom and then conjugated using PEG linkers to a suitable channel-specific antibody. Such inhibition helps in pain management and the development of a new generation of analgesics.[36] In addition to pain management, conjugation of monoclonal antibodies (mAbs) to linear synthetic peptide substrates can be used for multiple therapeutic applications such as immunogenic response modulation or peptide drug delivery.[37,38] Antibody–macrocyclic peptide conjugates have also been used as antimicrobial agents, to direct the bioactive-cyclic peptide to the targeted Gram-negative bacteria. The conjugation is achieved enzymatically using sortase A that embeds these bioactive peptides onto an *Escherichia coli*-targeted monoclonal antibody.[39] The attachment to mAbs minimizes peptide limitations, *e.g.* low stability, low specificity and off-target toxicity, allowing their use as therapeutic agents *via* exploitation of the unique selectivity and safety profile of mAbs. More than 40 IgGs and IgG derivatives have been approved to treat several diseases such as cancer and inflammatory diseases, and near 30 more are currently being investigated in late clinical trials.[40] From a structural point of view,

IgGs (chimeric, humanized or human) are tetrameric glycoproteins with molecular weights around 150 kDa. In terms of their molecular structure, they consist of two heavy chains (HCs, *ca.* 50 kDa) and two light chains (LCs, *ca.* 25 kDa), as illustrated, for example, in Figure 10.8.[41]

The conjugation of peptides to proteins can be achieved using numerous chemistries. Conjugation usually commences with the functionalization of the mAb by either attachment of a bifunctional linker, reduction of inter-chain disulfides or oxidation (for carbohydrate conjugation), followed by reaction with the peptide or with a linker species.[42] A well-established involves activation of the protein by glutaraldehyde, followed by the addition of a peptide with at least one free amino group, including the N-terminus. Borohydride reagents are then used to reduce the imines formed, stabilizing them and preventing any reverse reactions. This method, which can be performed in one-pot or two-step protocols, is relatively reliable and hence

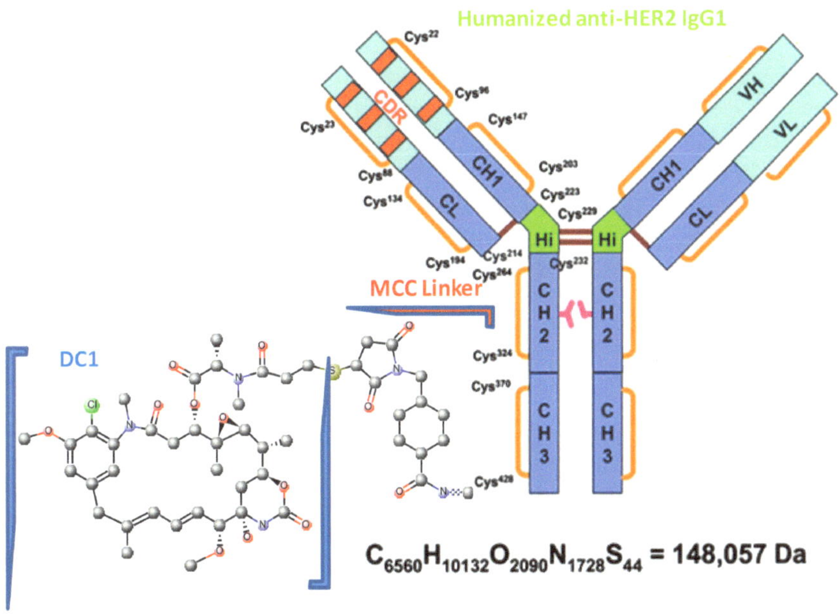

Figure 10.8 Ado-trastuzumab emtansine is an HER2-targeted ADC that contains the humanized anti-HER2 IgG1, trastuzumab, covalently linked to the microtubule inhibitory drug DM1 (a maytansine derivative) *via* the stable thioether linker MCC [4-(*N*-maleimidomethyl)cyclohexane-1-carboxylate]. Emtansine refers to the MCC–DM1 complex. The antibody trastuzumab is a well-characterized recombinant monoclonal antibody product produced by mammalian (Chinese hamster ovary) cells and the small-molecule components (DM1 and MCC) are produced by chemical synthesis. Ado-trastuzumab emtansine contains an average of 3.5 DM1 molecules per antibody.
Reproduced from ref. 41 with permission from American Chemical Society, Copyright 2012.

commonly used. However, it yields heterogeneous products. To obtain more defined conjugates, heterobifunctional linkers with variable lengths and structures were introduced. Most heterobifunctional linkers contain an *N*-hydroxysuccinimide ester (NHS) function and a maleimide functional group. The NHS group mediates the attachment of the linker to lysine moieties at the carrier mAb.[43] On the other hand, the free maleimide group can be attacked by a thiol functional group, mostly of a cysteine amino acid. This amino acid is usually introduced to the chosen peptide sequence, merely to allow the conjugation as the peptide used may not contain cysteine suitable for use. The thiol can also be obtained by the reduction of a disulfide bond in the peptide/protein. Cleavage of the disulfide bond, however, may lead to unwanted modification of the higher-order structures such as secondary and tertiary structures. Unfortunately, maleimide-mediated attachments have significant disadvantages. They are inclined to cause side reactions, such as hydrolysis, which decrease the conjugation yield and may lead to unstable conjugates.[44] Furthermore, solvent-accessible maleimide–thiol adducts show a rapid loss of the linker *via* maleimide exchange with reactive thiols in blood plasma.[45,46] Lately, some unconventional maleimide linkers have been described that displayed improved stability of the respective conjugates.[47,48] However, these linkers seem not to be available commercially. Approaches to achieve site-specific conjugation have been developed; however, either a recombinant protein with site-specific mutations is necessary or other complex variants of "click chemistry" must be applied.[49]

Since the yield of the conjugation reactions is heterogeneous with respect to both the distribution and loading of the drug species on the mAb, strict quality control of such products is necessary. The analytical testing of such variable conjugates is difficult. Matrix-assisted laser desorption/ionization time-of-flight mass spectrometry (MALDI-TOF-MS) is one of the preferred method for determining the conjugation density of peptide carriers.[50,51] Another powerful method is high-resolution MS with ESI. Although the resolution of different conjugation degrees is possible, the method is infrequently applied owing to expensive instrumentation and complex data processing.

The heterogeneity of the products is a major issue because even if the number of peptides per carrier protein were the same, the sites where the peptides would be attached may be variable. This heterogeneity is challenging from both manufacturing process control and analytical development perspectives.[52] Recent efforts to tackle this issue have included both process development strategies and the use of protein engineering. In a designed approach, inter-chain cysteines have been selectively exchanged with serine residues and cysteines have been rationally introduced to optimize drug conjugation with well-defined stoichiometry and minimize the disruption to the mAb structure and epitope binding.[53]

Key information that is critical in assessing the clinical efficacy of an antibody–peptide conjugate are the target site specificity and the binding properties of the antibody, stability of the linker and the conjugated peptide (drug) species, drug potency and both the distribution and average number of peptide species on the antibody.[54] These requirements underline the importance of understanding the physicochemical properties of antibody–peptide conjugates and choosing appropriate analytical platforms to evaluate and monitor them during development, manufacture and subsequent storage.[55]

10.2.2 Drug (Peptide)-to-antibody Ratio (DAR)

One of the most important quality attributes of antibody–peptide conjugates is the average number of peptide drugs that are conjugated, because this determines the amount of "payload" that can be delivered to the targeted cell and can directly affect both safety and efficacy.[56] A plethora of methods have been proposed for measuring this feature, depending on the properties of the peptide drug and how the drug is linked to the protein (*i.e.* site of conjugation and structure of the linker).[57] After the conjugation process, antibody–peptide conjugates consist of a mixture of mAbs linked to different numbers of drugs molecules. The average number of drugs loaded onto the mAbs, expressed as the drug-to-antibody ratio (DAR), represents one of the most important quality control features.[58] The DAR delineates the strength of the antibody–peptide conjugates and is the weighted summation of the different drug-loaded antibody species. DAR is directly involved in the *in vivo* drug delivery; consequently, DAR is a key feature in determining efficacy and patient safety. Several analytical methodologies have been developed and implemented for DAR measurements, based on the properties of the mAbs and payload.[59] For example, different extinction coefficients of the mAb and the payload allow the determination of DAR by UV–Vis spectroscopy as changes in the hydrophobicity of mAbs upon payload conjugation make the separation of species with a different number of payloads achievable by hydrophobic interaction chromatography (HIC).[60] The use of HIC is mostly limited to cysteine-conjugated species. DAR estimation can also be achieved by the use of ion-exchange chromatography (IEC). Nevertheless, intact LC-MS is the preferred method based on its simple strategy for determining DAR according to the increment in mass variation as a result of payload conjugation.[61] Mass spectrometry-based platforms provide deep molecular-level information for DAR and payload distribution profiles compared with DAR characterizations by LC-UV.[62,63] In intact LC-MS analysis, sample pre-treatment (*e.g.* deglycosylation) is often required in order to reduce the heterogeneity of the conjugates before MS analysis. To improve the MS resolution of the different conjugates, fragments can be generated either chemically or enzymatically, improving accuracy in DAR measurements and providing information on payload distribution.

For example, Terral *et al.*[62] used IdeS digestion and dithiothreitol (DTT) disulfide bond reduction to generate the Fd and Fc fragments of the heavy chain and light chain of the antibody before LC-MS analysis and obtained information on payload distribution. Another advantage of LC-MS analysis of the fragmented conjugates is that the DARs of intact and fragments can be cross-verified.[62,63] Measurements of DAR by intact LC-MS are based on the assumption that the payload conjugation and mAb fragmentation have an insignificant impact on the overall ionization efficiency of the analytes. DAR characterization by orthogonal approaches is important to guarantee accurate DAR calculation. Regarding LC-MS methodology design for intact conjugates analysis, the front-end LC separation and MS detection can be conducted in an online manner using reversed-phase (RP) LC or SEC under MS-compatible conditions, and also in an offline approach using HIC separation.[64] Even though RP-LC-MS is widely used in proteomics, it is particularly challenging for the characterization of cysteine-conjugated species because both light and heavy chains of cysteine-conjugated antibody are associated through noncovalent interactions and RP-LC can disrupt noncovalent complexes into antibody fragments.[65] For the characterization of cysteine-conjugated species, the current approaches for DAR analysis include HIC, SEC-MS and two-dimensional LC-MS.[65] Bagal *et al.*[66] utilized native protein MS to characterize cysteine-conjugated species without the disruption of antibody complex structures, thus obtaining accurate DAR measurements. Recent platforms using ion mobility (IM)-MS offer improved analysis. The ion drift tube in IM-MS permits the separation of ions with different sizes and shapes in the gas phase, providing valuable information at a molecular level of the protein complexes.[66] Huang *et al.* further extended the application of IM-MS in DAR characterization using conventional RP-LC in the platform workflow.[67] The value of RP-LC-IM-MS considerably improved the protein MS signals *via* ion "clean-up" in the gas phase. Additionally, drift-time measurements provided unique fingerprints for lot-to-lot DAR reproducibility.[68]

10.2.3 Drug (Peptide) Distribution

In addition to the valuable information of the average DAR, multiple platforms have been utilized to assess the distribution of drug-linked forms (*e.g.* fraction of antibodies containing zero, one, two, . . ., *n* peptides).[69] This is a key feature as different forms would have different pharmacokinetic (PK) and toxicological properties. Conjugates formed through lysine residue linkages are not easily separated by LC because of their high degree of heterogeneity.[70] Several reports have been published on the use of MS to characterize drug distribution, with the first report in 1991 using UV MALDI-TOF-MS.[71] MALDI-TOF-MS possesses poor mass accuracy for large molecules and its limited resolution does not provide the desired resolution for species with different drug loads.[71] More recent publications, such as those describing analyses of T-DM1 (trastuzumab-MCC-DM1) and thio-trastuzumab-DM1,31 huN901-SPP-DM1,32 and C242-DM4,33 have used LC-MS with ESI coupled to TOF or triple-quadrupole mass detectors.[72–74] These platforms benefit from higher mass accuracy and resolution than can be achieved using MALDI-TOF-MS.

10.2.4 Analysis of Unconjugated Peptide

Another key quality aspect of bioconjugates is the quantity of free (unconjugated) peptide, which poses concerns regarding potential efficacy and toxicological issues. Residual traces of unconjugated peptide or peptide-related impurities may remain in the final product as a result of an incomplete removal in purification steps downstream of the conjugation reaction.[75] Related forms of the unconjugated peptide, such as linker–peptide species or other degradation products, may also be released during storage of the conjugate. For that reason, it is crucial to develop suitable platforms to characterize and quantify free peptide species.[76]

RP-HPLC-MS platforms are suitable for the identification and quantitation of various drugs (peptides) released from the conjugate. This technique is the gold standard choice for force degradation studies. Several published studies have suggested that MS analysis helped to identify several unconjugated drugs as the free drug or drug linker. Furthermore, RP-HPLC-MS can be applied to evaluate the *in vitro* stability of the conjugate and monitor the free drug species released in plasma[77] or in cultured tumour cell lines.[78]

10.2.5 Peptide Mapping

Peptide mapping is a powerful tool for the assessment of protein modifications such as amino acid substitution, pyroglutamic acid formation, oxidation, deamidation, isomerization, glycation, disulfide shuffling, peptide bond cleavage and crosslinking. Enzyme-induced cleavage can be employed to detect drug-containing peptides.[79] Labelled peptides with a hydrophobic drug tag would have different chemical interactions than their unmodified forms during RP-HPLC analysis, leading to different retention times.[64] Protease activity could also be affected by conjugation, *e.g.* trypsin will not cleave at drug-induced modified lysines, leading to larger and more hydrophobic peptides. Exclusive chromatographic profiles of the enzymatic digests of a bioconjugate can be used to confirm product identity, characterize sites of peptide binding and assist with product consistency.[80]

RP-HPLC-MS of antibody–drug conjugate (ADC) peptides provides key structural information on conjugation sites. This information could be important in cases where modification of residues in the complementarity-determining regions (CDRs) might correlate with an effect on mAb binding to the target antigen.[53]

As most antibodies have numerous surface-accessible residues that can potentially react, conjugation through lysine side-chain amines typically results in a complex mixture of conjugate species that differ in the number of drugs attached and also in the sites of drug linkage.

To summarize, recent advances in MS, including improvements to TOF analysers, Orbitraps™ and Fourier transform ion cyclotron resonance

(FT-ICR) capabilities, have allowed the more accurate characterization of DAR and conjugation sites. The latest developments in HPLC column chemistries now permit enhanced separations of the complex peptide mixtures derived from mAbs during development and manufacture.

10.3 Quantitative Determination of Peptides and Their Conjugates by Mass Spectrometry

In addition to the qualitative aspects of MS-based peptide analysis such as primary sequence identification and determination of post-translational modifications (PTMs), the absolute and relative quantification of target peptides is an important complementary application. Quantitation can be used to identify specific changes between control samples and multiple experimental conditions during a stability study or to compare a reference product with a new generic formula.[81–84] The quantitative determination of stable isotopes utilizes the increase in mass due to mass tags containing stable isotopes. The common workflow is to tag peptide-containing samples with equivalent reagents, one of which includes a heavy mass tag and the other a light mass tag or no tag. The labelled samples are analysed by MS after being mixed and fractionated. The peaks in the mass spectra reveal the ratio of the two different isotopic or mass tag variants. The ratio is then used to identify the protein or peptide relative abundances.[85–87]

10.3.1 Quantitative Analysis of Thiol Oxidation

Sulfhydryl groups are useful targets for peptide conjugation and labelling since they are present in most proteins. Creating conjugates *via* crosslinking of sulfhydryl groups is more selective than with primary amines since they are less abundant in peptide sequences. The number of available (*i.e.* free) sulfhydryl groups can be easily controlled or modified as they can be generated by reduction of native disulfide bonds or they can be introduced into molecules through reaction with primary amines using sulfhydryl-addition reagents, such as 2-iminothiolane (Traut's reagent), SATA, SATP or SAT(PEG)4. The labelling and conjugation of peptides can be achieved *via* sulfhydryl-reactive chemical groups including haloacetyls, maleimides, aziridines, acryloyls, arylating agents, vinylsulfones, pyridyl disulfides, TNB-thiols (derivatives of 5-thio-2-nitrobenzoic acid) and disulfide reducing agents. Most of these groups conjugate to sulfhydryls by either alkylation (usually the formation of a thioether bond) or disulfide exchange (formation of a disulfide bond).

A characteristic of cysteine- and methionine-containing peptides is the tendency of these residues to oxidize during manufacture and storage. Susceptibility to oxidation is sequence dependent and sometimes even minimal exposure to air can lead to oxidation. The use of degassed or deoxygenated solvents and maintaining peptide solutions at acidic pH is recommended for minimizing oxidation since the rate of oxidation increases with increase in pH, *i.e.* under neutral or basic conditions. Reversible thiol oxidation of cysteine residues to disulfide, sulfenic or *S*-nitroso forms is considered in many cases as a post-translational modification that regulates protein functions in many intracellular catalytic and signalling processes. Even with confirmed examples of the significance of cysteine oxidation as a protein role controller, many instances of cysteine's thiol redox are repeatedly considered as an exotic phenomenon or instability and debated as a protein function controller. A systematic and comprehensive determination of all oxidized cysteine residues in a given sequence under different storage/biological conditions is known as redox proteomics or analysis. Redox analysis is technically challenging because of the labile nature of the SH group's oxidation reactions and the lack of tools to detect the oxidized residues directly and distinguish them from reduced forms.[88–90]

MS-based redox analysis is achieved by treating peptide samples with a light isotope-coded affinity tag (ICAT), followed by a thiol reducing agent and then a heavy ICAT variant (Figure 10.9). ICATs are biotinylated iodoacetamide derivatives (or acrylamide derivatives) that react with the sulfhydryl groups of denatured peptides' cysteine side chains (as this group of reagents can only be applied to proteins with a cysteine moiety).[91] The intermediate bridge of these reagents, which contains the isotopic carbon or hydrogen atoms used for differential labelling, connects the iodoacetamide moiety to a biotin group for affinity capture of the derivatized peptides onto a streptavidin-based column (Figure 10.10). The biotin–streptavidin affinity capture allows cysteine-containing peptides to be isolated from the complex sample mixture, thereby considerably reducing the number of different peptides/molecules introduced into the mass spectrometer ion source.

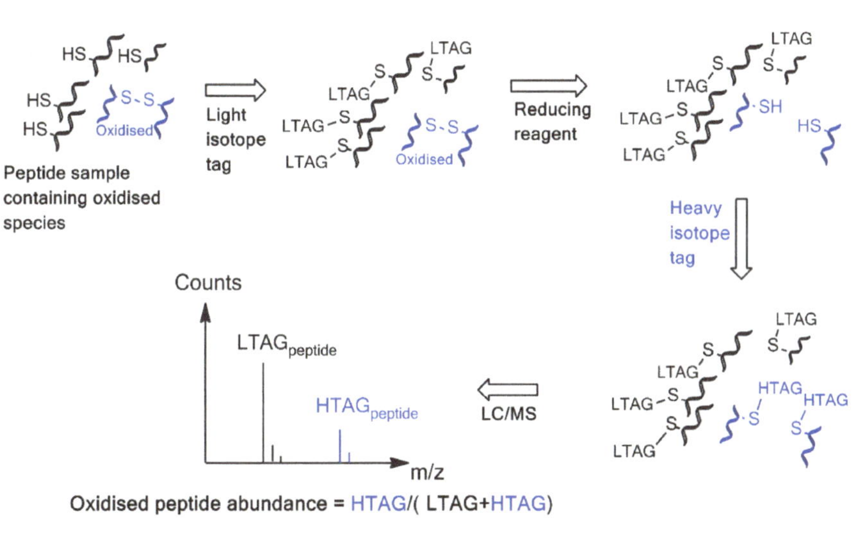

Figure 10.9 Determination of thiol oxidation state using ICAT (OxICAT).

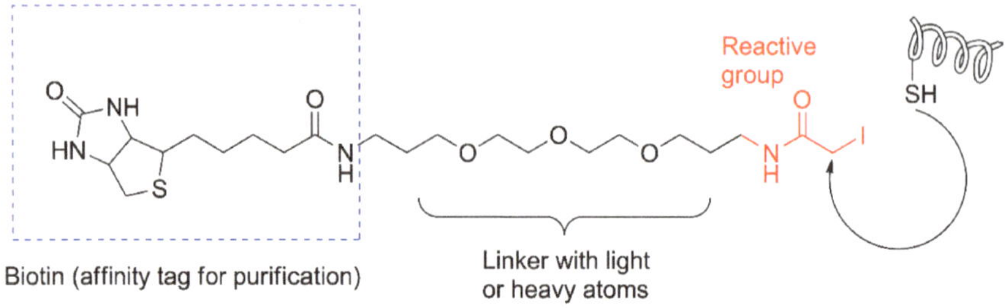

Figure 10.10 Structure and reactivity of ICAT reagents.

The linker bridge is available in two forms, one normal/light version and one heavy form, in which hydrogen or carbon atoms are replaced with deuterium/^{13}C.

In the first step of the oxidation state ICAT (OxICAT) workflow, the peptide is denatured to gain access to all reduced cysteines and label them irreversibly with light ICAT. Next, all reversibly oxidized thiols are reduced using the strong thiol reductant tris(2-carboxyethyl)phosphine (TCEP) and then are tagged with heavy ICAT. This procedure generates two forms of chemically identical proteins that differ only in the mass of their ICAT labels (9 Da more for each heavy ICAT label) depending on their initial redox state. Trypsin digestion and affinity purification will simplify the analysed mixture of the ICAT-labelled peptides and the subsequent MS analysis for quantification and MS/MS for peptide identification. The OxICAT method detects all reversibly oxidized cysteines, but higher oxidation states such as sulfinic and sulfonic acids will not be detected and compared. However, replacement of the nonspecific thiol reducing agent TCEP with more specific reagents such as glutaredoxin or ascorbate will allow the use of the OxICAT workflow to detect modifications such as glutathionylations or nitrosylations, respectively.[92,93]

ICAT reagents can also be used to compare two different samples derived from two different sources such as a reference peptide drug product *versus* a generic or a reference peptide API material against a certain batch. The compared samples are treated with light or heavy ICAT tags and then quantitatively combined (and proteolysed in the case of proteins). Peptides bound to ICAT tags are selectively isolated on streptavidin columns and analysed by MS. The relative abundance of the peptide is determined by the ratio of the signal intensities of the heavy and light tagged peptide pairs.[85,94]

Peptides labelled using detuerated ICAT reagents, however, are liable to chromatographic peak separation, resulting in an inconsistent ratio across the peak. Zhang and co-workers revealed that the substitution of ^{1}H for ^{2}D changes the retention time of the labelled species in the RP-HPLC separation. They demonstrated that even a baseline separation of the light and heavy labelled counterparts can be obtained.[95,96] This problem has been overcome by using ^{13}C labelling of the heavy ICAT reagent. Another issue is that the presence of the biotin group mass in the spectrum may interfere with a database search and its hydrophobicity causes peptides to elute in a relatively narrow retention time window. A cleavable ICAT (cICAT) reagent has therefore been introduced, wherein the biotin can be removed by acid treatment before MS analysis.[97,98] Multiple applications of the cICAT have been reported, indicating a major improvement in this labelling approach in comparison with the original reagent.[99–103] The mass difference between light and heavy tags (9 Da) prevents confusion between oxidized methionine and two cysteines labelled with ICAT reagents in the workflows. The new generations of ICAT have less tag fragmentation, thereby improving MS/MS data quality and significantly increasing the number of identified proteins with higher confidence per experiment. However, the high cost of the aforementioned reagents could limit the more widespread application of the ICAT approach. The major disadvantage of ICAT reagents is that some proteins do not have a cysteine residue and therefore are excluded from this type of analysis. Another practical drawback is that the release of the biotinylated peptides from the streptavidin column is not quantitative for low-abundance peptides.[85,104,105]

10.3.2 Absolute Quantitation by Mass Spectrometry (AQUA Strategy)

Small organic molecules can be quantified by isotope dilution MS using stable isotope (^{2}H, ^{13}C, ^{15}N, ^{18}O)-labelled analogues as internal standards. The AQUA (absolute quantification) workflow was introduced in 2003 and can be described as a modification of isotope dilution MS that includes the use of synthetic labelled peptide as internal standards.[106] An AQUA peptide is merely a synthetic peptide matching the amino acid sequence of the peptide of interest and includes at least one isotope-labelled amino acid, creating an increase in mass. When the sample is spiked with an AQUA peptide, the original native peptide and the synthetic peptide elute together chromatographically and ionize with the same efficiency. The AQUA peptide and the native peptide are easily resolved in the mass spectrometer due to the mass shift.[85,107–109] The type of heavy amino acid used determines the mass increase of the synthetic peptide: +4 Da (L-alanine-$^{13}C_3$,^{15}N), +6 Da (L-proline-$^{13}C_5$,^{15}N, L-valine-$^{13}C_5$,^{15}N), +7 Da (L-isoleucine-$^{13}C_6$,^{15}N, L-leucine-$^{13}C_6$,^{15}N), +8 Da (L-lysine-$^{13}C_6$,$^{15}N_2$) or +10 Da (L-arginine-$^{13}C_6$,$^{15}N_4$, L-phenylanaline-$^{13}C_9$,^{15}N). Because the quantity of the AQUA peptide is known, the absolute quantity of the targeted peptide can be determined using the MS peak areas of the AQUA and the native peptide.

Determination of the absolute quantities of proteins requires complete cleavage into the component peptides by chemical or enzymatic digestion. From the digest, a monitor peptide sequence is designated to be a quantitative marker for the protein. The quantitation is accomplished by spiking with a labelled AQUA version of the same sequence as the internal standard. Single ion monitoring (SIM) chromatograms or extracted ion

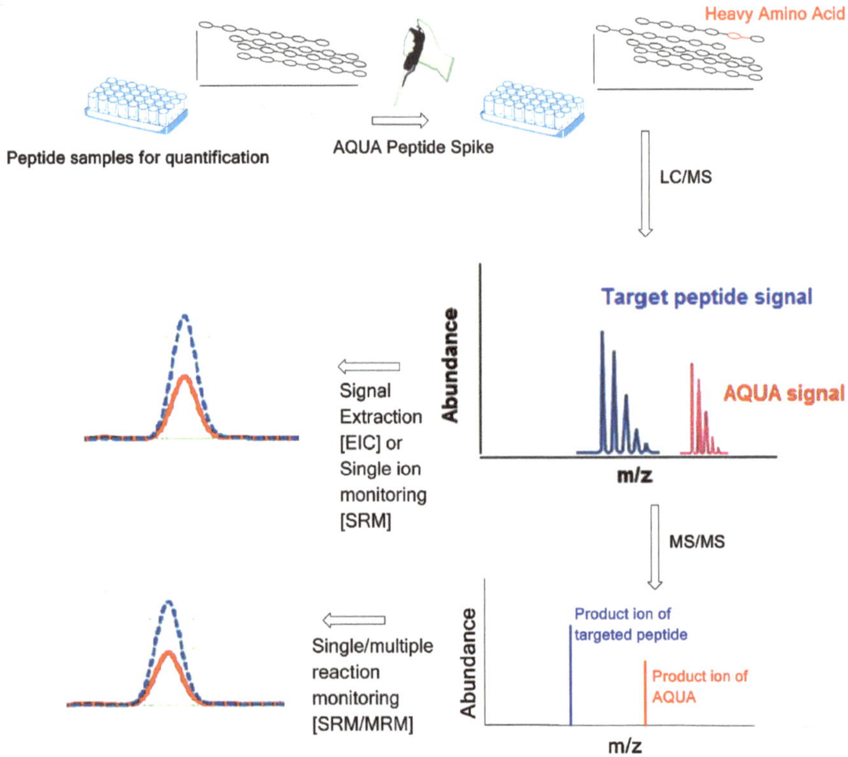

Figure 10.11 AQUA peptide workflow for absolute quantitation.

chromatograms (EICs) are generated for the native peptide and the synthetic AQUA internal standard. Peak ratios are used to calculate the quantity of the original peptide (Figure 10.11). In lieu of SIM and EIC, the internal standard is analysed by MS/MS to determine the fragmentation patterns that will aid in designing selected reaction monitoring (SRM), in which a transition from precursor ion to product ion is measured. Sample analysis using LC-SRM experiments reveals the abundance of the specific transitions for both the native peptide and its labelled analogue. This subcategory of isotope dilution for absolute quantitation (AQUA) is known as stable isotope dilution–multiple reaction monitoring mass spectrometry (SID-MRM-MS).[110–113]

One of the most important advantages of the AQUA strategy is its ability to compensate for loss during sample preparation and introduction into the LC-MS instrument because the internal standard is spiked at the beginning of the workflow, hence errors do not affect the ratio of the native and the AQUA peptide. Nevertheless, the high cost of internal standard peptides has restricted the application and development of SID-MRM-MS methods.[114] A more economic strategy for absolute quantitation has recently been developed using element-tag labelling and inductively coupled plasma (ICP)-MS.[115,116]

10.3.3 Quantitative Analysis by Inductively Coupled Plasma Mass Spectrometry (ICP-MS)

ICP-MS does not directly facilitate peptide quantitation because of the decomposition of the organic structure in the 7000 K plasma ion source. Nevertheless, elements present in peptides such as sulfur and phosphorus can be quantified by ICP-MS and the results utilized to calculate concentrations of the parent peptides.[116] Alternatively, ICP-MS-detectable elements are incorporated into the targeted peptides by forming a covalent bond with bifunctional chemical labelling reagents (containing metals). The goal of the different approaches is to detect peptides by ICP-MS and consequently utilize the instrument's high sensitivity and matrix robustness to quantify them. The stoichiometries of the chemical labels must be previously established to allow quantitative determination. ICP-MS is a promising complementary technique to ESI- and MALDI-MS analysis because the plasma atomizes compounds and reduces the sample complexity, offering high sensitivity for the quantification of the targeted elements, a wide linear dynamic detection range (up to nine orders of magnitude) and minimal matrix effects, especially when low-flow sample introduction techniques such as capillary HPLC or nano-HPLC are used.[117–119]

10.3.3.1 Hetero-natural Element Tags

Many peptides contain functional groups with trace elements such phosphorus, sulfur, selenium and iodine that can be exploited as tags for quantification. Sulfur is a constituent of cysteine and methionine, hence it is present in most peptides. Quantification of peptides with a known number of sulfur-containing amino acids and an established amino acid sequence can be performed *via* sulfur determination. Selenium, present in selenocysteine and selenomethionine of the selenoproteins, and phosphorus, occurring as post-translational modifications in proteins, can also be used to quantify the corresponding proteins.[120–122]

10.3.3.1.1 Phosphorus. Phosphorus detection by ICP-MS is challenging owing to the low ionization efficiency in an argon-based plasma and various polyatomic interferences such as $^{14}N^{16}O^{1}H$ that negatively affect the limit of quantitation. Phosphorus is monoisotopic (main isotope ^{31}P), hence it is not possible to detect a different isotope with less interference. Nevertheless, the removal of interference is possible *via* the use of collision/reaction cell instruments or high-resolution magnetic sector detectors that allow the utilization of phosphorus for the quantification of peptides. Hyphenating ICP-MS with separation techniques such as capillary LC and SEC is crucial for peptide quantification in complex mixtures *via* phosphorus detection (Figure 10.12).[123,124]

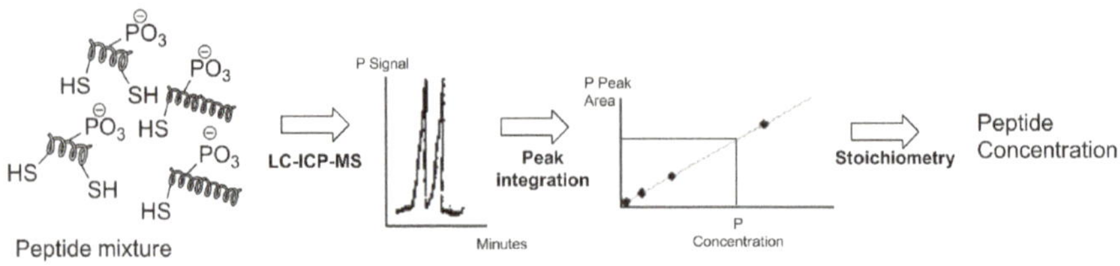

Figure 10.12 Quantitative analysis of peptides using ICP-MS and phosphorus as a hetero-natural element tag.

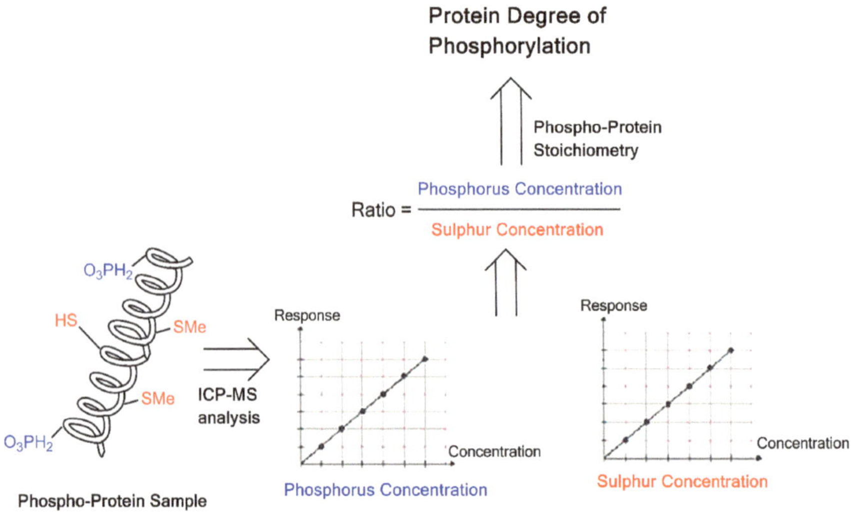

Figure 10.13 Determination of the degree of phosphorylation of a peptide using ICP-MS.

In addition to protein quantification, the degree of peptide phosphorylation can be determined through the measurement of phosphorus. Bandura *et al.* described the application of a dynamic reaction cell ICP-MS instrument to eliminate phosphorus polyatomic interferences using oxygen as reaction gas.[125] Phosphorus is measured as the resultant oxide ($^{31}P^{16}O^{+}$) and sulfur, from cysteine and methionine, can also be measured in parallel as $^{32}S^{16}O^{+}$, and the ratio of the two measurements allows the estimation of the degree of phosphorylation of the peptide (Figure 10.13). The method was used to determine the degree of phosphorylation for regular α-casein (8.8 phosphorus moieties per molecule) and β-casein (4.9 phosphorus moieties per molecule) and the results obtained were concordant with the structure of the proteins.[125,126] In the same way, Becker *et al.* described the analysis of small sample amounts using flow injection hyphenated with high-resolution ICP-MS to determine the total phosphorus content in proteins. The results obtained were concordant with the values specified for the certified reference material used for method validation.[127]

10.3.3.1.2 Sulfur. Sulfur is a promising natural tag for peptide quantification by ICP-MS as it is present in most peptides due to the sulfur-containing amino acids. This approach allows the absolute quantification of proteins or peptides once their sulfur stoichiometry has been clarified by the complementary techniques of ESI- or MALDI-based MS analysis.[128,129] Although >90% of human proteins can in theory be determined on the basis of their sulfur content, sulfur detection by ICP-MS is hampered by the low ionization in the ICP source due to the high first ionization potential (10.36 eV). Another challenge in sulfur measurement is the polyatomic interferences on sulfur's most abundant isotopes at *m*/*z* 32 and 34. The main interferences originate from O_2^+ species: $^{16}O^{16}O^+$ and $^{16}O^{18}O^+$. Less abundant ion interferences such as $^{31}PH^+$, $^{14}N^{18}O^+$ and $^{15}N^{16}OH^+$ on $^{32}S^+$ also have to be considered for specific and accurate sulfur detection. The use of high-resolution detectors such as magnetic sector instruments easily resolves different interferences. However, most commercially available ICP-MS instruments are based on a collision cell design rather than magnetic sector, and the removal of interferences becomes more challenging. To address this challenge, oxygen is infused to the collision cell to induce the ion–molecule (S^+ and O_2) reaction in order to obtain the SO^+ species and shift the monitored mass to an interference-free region. The separation of peptides by reversed-phase capillary LC coupled to ICP-MS as sulfur detector for quantitative measurements was demonstrated.[130–132] Wind *et al.* described the quantification of insulin, as a model protein, using sulfur as a natural elemental tag and thiamine as internal standard.[129]

10.3.3.1.3 Selenium. Selenium is present in the amino acids selenomethionine and selenocysteine, which are often incorporated into proteins under sulfur-deficient conditions.[133] *In silico* studies suggest that the human proteome may contains at least 25 selenoproteins.[134–136] Hence selenium is not suitable for large-scale protein quantification; however, it is a selective tag for only a limited, well-defined number of proteins. Schaumlöffel *et al.* introduced the application of special nebulizer for the coupling of capillary LC and ICP-MS to quantify selenoproteins.[137] This approach achieved 100–150 times lower detection limits for selenium in comparison with results obtained with conventional HPLC setups and ICP-MS detection.

10.3.3.2 Chemical Labelling Using Bifunctional Chelating Agents

Chemical reactions can be used to incorporate elemental labels, making the targeted peptide detectable by ICP-MS. Controlled chemical labelling allows better detection limits compared with natural tags since elements that exhibit high sensitivity and are not prone to interferences can be chosen for tagging. Heavy metal nanoparticles or lanthanides containing chelating agents[138,139] are used as bifunctional chelating tags to allows ICP-MS-based quantitation of most peptides even if they do not have a natural elemental tag (Figure 10.14). Alternatively, an indirect labelling strategy *via* labelled antibodies that bind to the targeted peptides (antigens) can be used to obtain a complex with an ICP-MS-detectable label.[140–143] The immunochemistry-based indirect approach is ideal for laser ablation ICP-MS detection and has been implemented for imaging applications in cancer research and biomarker identification.[144,145]

Zhang *et al.* described the first method for the quantitation of thyroid-stimulating hormone (TSH) in serum utilizing a chemical labelling workflow.[146] Eu^{3+}-labelled TSH antibody was used for the formation of a protein complex (antibody–antigen) with the TSH. Consequently, the complex was determined in serum samples using ICP-MS. The TSH antibody was labelled using N^1-(*p*-isothiocyanatobenzyl)diethylenetriamine-N^1,N^2,N^3,N^4-tetraacetate–Eu^{3+} as a chelating bifunctional agent. The comparison between the previously established radioimmunoassay and the ICP-MS assay showed a good correlation ($r^2 = 0.9639$). The concept was further developed to allow the quantitation of multiple peptides simultaneously. Immobilized antibodies embedded on a special plate or holder are used to capture the targeted peptides and isolate them from the sample mixture. The captured

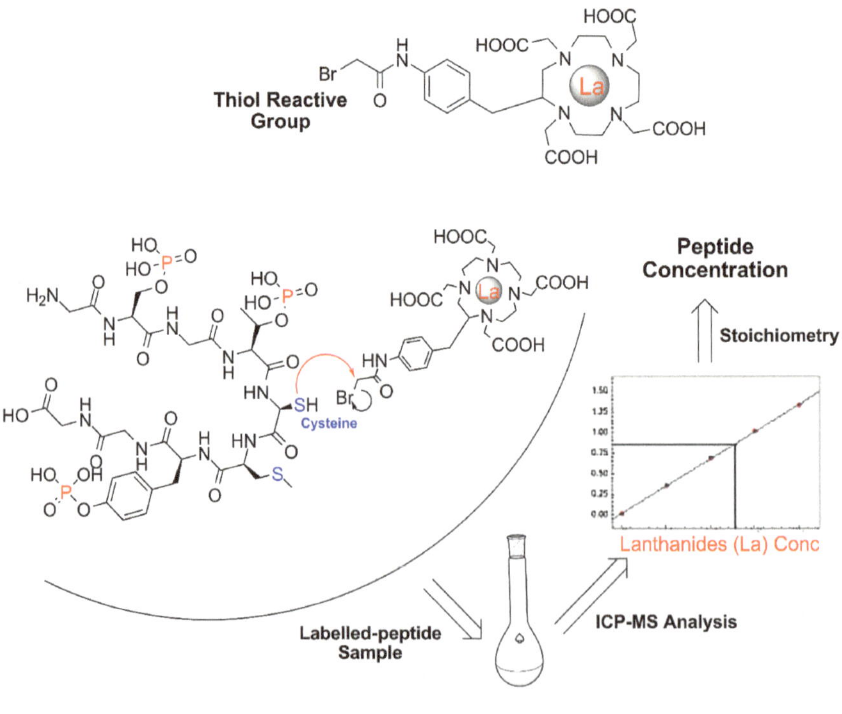

Figure 10.14 Absolute quantification of a peptide using bifunctional chelating reagents.

peptides are quantified by treating them with labelled antibody followed by ICP-MS detection. Eu-labelled antibodies can be prepared using Wallace AutoDELFIA reagents, originally designed for fluorescence-based detection. Antibodies with ICP-MS-detectable elements can also be customized using NANOGOLD reagents. Using this methodology, different peptides and also peptide–peptide interactions can be quantitatively measured by ICP-MS-based elemental analysis.[147] The simultaneous determination of four oncogenic proteins, *i.e.* human stem cell factor receptor c-Kit, kinase BCR/Abl, myeloid cell surface antigen CD33 and integrin receptor VLA-4, in a human leukaemia cell line using ICP-MS was described. Secondary antibodies were conjugated to Au, Sm, Eu and Tb using the NANOGOLD and Wallace AutoDELFIA reagents. The method demonstrated that the immunophenotyping by ICP-MS is comparable to that of flow cytometric analysis (FACS).[148]

Element-coded affinity tag (ECAT) reagents react with the SH group of the amino acid cysteine in different peptides allowing direct labelling without the need to implement antibodies in the workflow. In addition to the bromoacetamido group that has high thiol reactivity, the ECAT molecule also contains a bifunctional chelating group, 1,4,7,10-tetra-azacyclododecane-*N*,*N*,*N*,*N*-tetraacetic acid (DOTA), that holds ICP-MS-detectable elements. The ECAT reagents were further developed into a metal-coded affinity tag (MeCAT) that has a malemide functional group for the thiol connection and a structural position that can be substituted with chemical groups for affinity purification (Figure 10.15). A special subclass of MeCAT was patented and it features a further variable position that allows the introduction of different functional groups. Comparable to the ICAT reagent, the MeCAT reagent can include a biotin residue at a variable position, which allows the exploitation of biotin–avidin affinity chromatography. Hence crudely labelled protein in complex mixtures can be purified and distinctively enriched.[149–152]

To summarize, the transition in MS peptide analysis from qualitative to quantitative allows the determination of relative abundances between different states of samples (or absolute quantities in the case of the AQUA experiment). Consequently, comparative analysis of protein content between two products will help in revealing bio-similarity and verifying label claims. The possibility of multiplexing and comparing up to 10 different states within one experiment adds flexibility to the experimental design, without

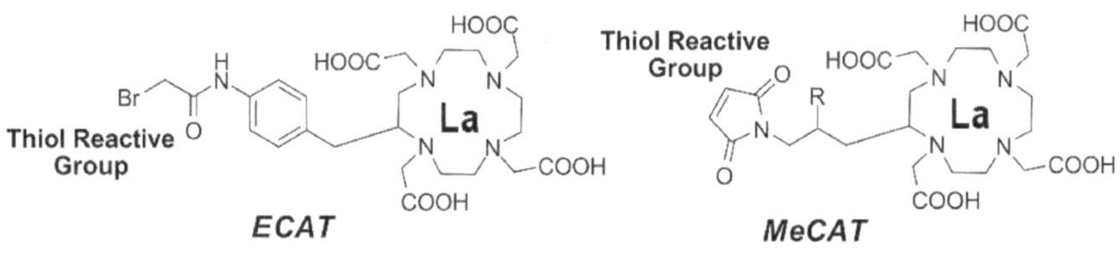

Figure 10.15 Structures of element-coded affinity tag (ECAT) and metal-coded affinity tag (MeCAT).

complicating MS data analysis. ICP-MS has now been established as a powerful complementary tool to be used in parallel with ESI or MALDI proteomics. It gives quantitative elemental data while molecule-specific MS (ESI and MALDI) aids compound characterization and identification. The scientific community has to establish how this powerful technique can help in solving challenges that are difficult to address with ESI or MALDI instruments, which are still the established standards within the field of peptide quantitation. The increasing number of ICP-MS-based proteomic publications indicates that a good number of groups have recognized the importance of the technique and are working hard to optimize it as a useful complementary tool for peptide research.

10.4 Conclusion

The technological improvements in analytical chemistry, especially high-resolution mass spectrometry (HRMS), have allowed the development of methods to elucidate the efficacy, safety and stability of peptide conjugates during GMP testing. For example, the quality of PEGylated peptides can be evaluated *via* qualitative and quantitative MS analysis to ensure that the PEGylated therapeutics have the desired properties such as molecular mass and polydispersity. This is critical to support recent advances in PEGylation manufacturing technology, which has evolved from random conjugation of short-chain polymers to controlled, site-directed insertion of high molecular weight, more homogeneous PEG moieties. Antibody–peptide conjugate products have also benefited from the improved HRMS analytical data. In comparison with the parent "unconjugated" antibody, the antibody–peptide conjugate has an increased level of complexity that arises from the heterogeneity of the conjugation. Improvements in technology have offered the tools to characterize these molecules better, supporting process and formulation development by developing reliable methods needed for QC lot-release and stability testing. The average DAR, drug loading distribution and unconjugated drug and impurity characterization *via* peptide mapping can be determined using HRMS and then used for product and process development, lot release and stability testing.

The Quality-by-Design (QbD) joint initiative between the FDA and the pharmaceutical industry concerning therapeutic biotechnology products recommended that manufacturers should have a QbD platform for the therapeutic drug during the development phases. They recommended focusing on the critical quality attributes (CQAs), *e.g.* physicochemical properties, that influence safety and efficacy. For peptide conjugates, it is becoming clear that there will be a distinct set of CQAs such as size variants, process impurities such as those related to the drug and the conjugation and product isoforms. These CQAs can be used to demonstrate reproducibility and set the product specifications during development. CQAs will help in understanding how process conditions affect these critical attributes and then suitable limits on critical process parameters can be established. With the combination of the supporting analytical methodologies, knowledge of the manufacturing process and a robust quality system, safe and efficient peptide conjugates can be manufactured for clinical and commercial use.

In addition, relative or absolute MS-based quantitation of peptides and peptide conjugates is very useful for assessing stability, label claims and batch-to-batch variability. Different strategies exploiting stable isotope labelling and isotope dilution are possible, each having specific strengths and weaknesses. ICP-MS has recently emerged as a complementary technique to the previous methods, allowing absolute protein quantification using suitable elemental-based tags.

References

1. W. C. Duckworth, R. G. Bennett and F. G. Hamel, *Endocr. Rev.*, 1998, **19**, 608–624.
2. S. B. van Witteloostuijn, S. L. Pedersen and K. J. Jensen, *ChemMedChem*, 2016, **11**, 2474–2495.
3. A. Zorzi, S. J. Middendorp, J. Wilbs, K. Deyle and C. Heinis, *Nat. Commun.*, 2017, **8**, 16092.
4. S. N. S. Alconcel, A. S. Baas and H. D. Maynard, *Polym. Chem.*, 2011, **2**, 1442.
5. T. Yamaoka, Y. Tabata and Y. Ikada, *J. Pharm. Sci.*, 1994, **83**, 601–606.
6. *J. Am. Coll. Toxicol.*, 1983, **2**, 17–34.
7. *Int. J. Toxicol.*, 2006, **25**, 121–138.
8. H. F. Smyth, C. P. Carpenter and C. S. Weil, *J. Am. Pharm. Assoc., Sci. Ed.*, 1955, **44**, 27–30.
9. D. G. Rudmann, J. T. Alston, J. C. Hanson and S. Heidel, *Toxicol. Pathol.*, 2013, **41**, 970–983.
10. A. Bendele, J. Seely, C. Richey, G. Sennello and G. Shopp, *Toxicol. Sci*, 1998, **42**, 152–157.
11. M. S. Hershfield, R. H. Buckley, M. L. Greenberg, A. L. Melton, R. Schiff, C. Hatem, J. Kurtzberg, M. L. Markert, R. H. Kobayashi, A. L. Kobayashi and A. Abuchowski, *N. Engl. J. Med.*, 1987, **316**, 589–596.
12. S. S. Patel and P. Benfield, *Clin. Immunother*, 1996, **5**, 492–496.
13. D. L. Barnard, *Curr. Opin. Investig. Drugs*, 2001, **2**, 1530–1538.
14. P. Bailon, A. Palleroni, C. A. Schaffer, C. L. Spence, W. J. Fung, J. E. Porter, G. K. Ehrlich, W. Pan, Z. X. Xu, M. W. Modi, A. Farid, W. Berthold and M. Graves, *Bioconjug. Chem.*, 2001, **12**, 195–202.
15. C. L. Vogel, M. Z. Wojtukiewicz, R. R. Carroll, S. A. Tjulandin, L. J. Barajas-Figueroa, B. L. Wiens, T. A. Neumann and L. S. Schwartzberg, *J. Clin. Oncol.*, 2005, **23**, 1178–1184.
16. A. Guttmann, S. Krasnokutsky, M. H. Pillinger and A. Berhanu, *Ther. Adv. Drug Saf.*, 2017, **8**, 379–388.
17. P. J. Trainer, W. M. Drake, L. Katznelson, P. U. Freda, V. Herman-Bonert, A. J. van der Lely, E. V. Dimaraki, P. M. Stewart, K. E. Friend, M. L. Vance, G. M. Besser, M. O. Thorner, C. Parkinson,

A. Klibanski, J. S. Powell, A. L. Barkan, M. C. Sheppard, M. Maldonado, D. R. Rose, D. R. Clemmons, G. Johannsson, B.-Å. Bengt-Åke Bengtsson, S. Stavrou, D. L. Kleinberg, D. M. Cook, L. S. Phillips, M. Bidlingmaier, C. J. Strasburger, S. Hackett, K. Zib, W. F. Bennett, R. J. Davis and J. A. Scarlett, *N. Engl. J. Med.*, 2000, **342**, 1171–1177.
18. I. Melnikova, *Nat. Rev. Drug Discovery*, 2005, **4**, 711–712.
19. E. Floettmann, K. Bui, M. Sostek, K. Payza and M. Eldon, *J. Pharmacol. Exp. Ther.*, 2017, **361**, 280–291.
20. P. K. Coyle, S. Shang, Z. Xiao, Q. Dong and C. Castrillo-Viguera, *Mult. Scler. Relat. Disord.*, 2018, **22**, 134–138.
21. C. Madsen, *Brain Behav.*, 2017, **7**, e00696.
22. A. L. Dunn, S. P. Ahuja and E. S. Mullins, *Haemophilia*, 2018, **24**, e84–e92.
23. F. M. Veronese and G. Pasut, *Drug Discovery Today*, 2005, **10**, 1451–1458.
24. FDA and CdER, Movantik (naloxegol) Capsules Label., 2014, 1–19, https://www.accessdata.fda.gov/drugsatfda_docs/label/2014/204760s000lbl.pdf, (accessed 5 July 2018).
25. FDA, Summary Basis for Regulatory Action- Adynovate., 2015, 1–30, https://www.fda.gov/ucm/groups/fdagov-public/@fdagov-bio-gen/documents/document/ucm474903.pdf, (accessed 5 July 2018).
26. R. Webster, V. Elliott, B. K. Park, D. Walker, M. Hankin and P. Taupin, in *PEGylated Protein Drugs: Basic Science and Clinical Applications*, ed. F. M. Veronese, Birkhäuser Basel, Basel, 2009, pp. 127–146.
27. A. Abuchowski, J. R. McCoy, N. C. Palczuk, T. van Es and F. F. Davis, *J. Biol. Chem.*, 1977, **252**, 3582–3586.
28. V. Crupi, M. Jannelli, S. Magazu', G. Maisano, D. Majolino, P. Migliardo and R. Ponterio, *J. Mol. Struct.*, 1996, **381**, 207–212.
29. P. Caliceti and F. M. Veronese, *Adv. Drug Delivery Rev.*, 2003, **55**, 1261–1277.
30. C. Crafts, I. Acworth and D. Thomas, *LCGC North Am.*, 2012, (Suppl), 40.
31. W. D. Chey, L. Webster, M. Sostek, J. Lappalainen, P. N. Barker and J. Tack, *N. Engl. J. Med.*, 2014, **370**, 2387–2396.
32. M. A. Marcos, D. Cabaleiro, M. J. G. Guimarey, M. J. P. Comuñas, L. Fedele, J. Fernández and L. Lugo, *Nanomater*, 2017, **8**, 18.
33. X. Lu, P. C. Gough, M. R. DeFelippis and L. Huang, *J. Am. Soc. Mass Spectrom.*, 2010, **21**, 810–818.
34. B. Hughes, *Nat. Rev. Drug Discovery*, 2010, **9**, 665–667.
35. G. Houen, *Peptide Antibodies: Methods and Protocols*, 2015, pp. 1–6.
36. K. Biswas, T. E. Nixey, J. K. Murray, J. R. Falsey, L. Yin, H. Liu, J. Gingras, B. E. Hall, B. Herberich, J. R. Holder, H. Li, J. Ligutti, M.-H. J. Lin, D. Liu, B. D. Soriano, M. Soto, L. Tran, C. M. Tegley, A. Zou, K. Gunasekaran, B. D. Moyer, L. Doherty and L. P. Miranda, *ACS Chem. Biol.*, 2017, **12**, 2427–2435.
37. B. M. Foxwell, H. A. Band, J. Long, W. A. Jeffery, D. Snook, P. E. Thorpe, G. Watson, P. J. Parker, A. A. Epenetos and A. M. Creighton, *Br. J. Cancer*, 1988, **57**, 489–493.

38. G. Y. Berguig, A. J. Convertine, S. Frayo, H. B. Kern, E. Procko, D. Roy, S. Srinivasan, D. H. Margineantu, G. Booth, M. C. Palanca-Wessels, D. Baker, D. Hockenbery, O. W. Press and P. S. Stayton, *Mol. Ther.*, 2015, **23**, 907–917.
39. F. Touti, G. Lautrette, K. Johnson, J. Delaney, A. Wollacott, H. Tissire, K. Viswanathan, Z. Shriver, S. Mong, A. Mijalis, O. Plante and B. Pentelute, *ChemBioChem*, 2018, DOI: 10.1002/cbic.201800295.
40. F. Erni and R. W. Frei, *J. Chromatogr. A*, 1978, **149**, 561–569.
41. A. Beck, S. Sanglier-Cianférani and A. Van Dorsselaer, *Anal. Chem.*, 2012, **84**, 4637–4646.
42. J. F. DiJoseph, M. M. Dougher, D. C. Armellino, D. Y. Evans and N. K. Damle, *Leukemia*, 2007, **21**, 2240–2245.
43. A. C. J. Lee, J. E. Powell, G. W. Tregear, H. D. Niall and V. C. Stevens, *Mol. Immunol.*, 1980, **17**, 749–756.
44. R. P. Lyon, J. R. Setter, T. D. Bovee, S. O. Doronina, J. H. Hunter, M. E. Anderson, C. L. Balasubramanian, S. M. Duniho, C. I. Leiske, F. Li and P. D. Senter, *Nat. Biotechnol.*, 2014, **32**, 1059–1062.
45. L. N. Tumey, M. Charati, T. He, E. Sousa, D. Ma, X. Han, T. Clark, J. Casavant, F. Loganzo, F. Barletta, J. Lucas and E. I. Graziani, *Bioconjug. Chem.*, 2014, **25**, 1871–1880.
46. B.-Q. Shen, K. Xu, L. Liu, H. Raab, S. Bhakta, M. Kenrick, K. L. Parsons-Reponte, J. Tien, S.-F. Yu, E. Mai, D. Li, J. Tibbitts, J. Baudys, O. M. Saad, S. J. Scales, P. J. McDonald, P. E. Hass, C. Eigenbrot, T. Nguyen, W. A. Solis, R. N. Fuji, K. M. Flagella, D. Patel, S. D. Spencer, L. A. Khawli, A. Ebens, W. L. Wong, R. Vandlen, S. Kaur, M. X. Sliwkowski, R. H. Scheller, P. Polakis and J. R. Junutula, *Nat. Biotechnol.*, 2012, **30**, 184–189.
47. S. D. Fontaine, R. Reid, L. Robinson, G. W. Ashley and D. V. Santi, *Bioconjug. Chem.*, 2015, **26**, 145–152.
48. J. P. M. Nunes, M. Morais, V. Vassileva, E. Robinson, V. S. Rajkumar, M. E. B. Smith, R. B. Pedley, S. Caddick, J. R. Baker and V. Chudasama, *Chem. Commun.*, 2015, **51**, 10624–10627.
49. I. Abbas, T. Schwaar, F. Bienwald, M. Weller, I. M. Abbas, T. Schwaar, F. Bienwald and M. G. Weller, *Methods Protoc.*, 2017, **1**, 2.
50. S. S. Lateef, S. Gupta, L. P. Jayathilaka, S. Krishnanchettiar, J.-S. Huang and B.-S. Lee, *J. Biomol. Tech.*, 2007, **18**, 173–176.
51. O. B. Torres, R. Jalah, K. C. Rice, F. Li, J. F. G. Antoline, M. R. Iyer, A. E. Jacobson, M. N. Boutaghou, C. R. Alving and G. R. Matyas, *Anal. Bioanal. Chem.*, 2014, **406**, 5927–5937.
52. H. Liu, G. Gaza-Bulseco, D. Faldu, C. Chumsae and J. Sun, *J. Pharm. Sci.*, 2008, **97**, 2426–2447.
53. A. Beck, E. Wagner-Rousset, D. Ayoub, A. Van Dorsselaer and S. Sanglier-Cianférani, *Anal. Chem.*, 2013, **85**, 715–736.
54. J. Lambert, *Curr. Opin. Pharmacol.*, 2005, **5**, 543–549.
55. A. M. Wu and P. D. Senter, *Nat. Biotechnol.*, 2005, **23**, 1137–1146.

56. F. Hudecz, M. C. Garnett, T. Khan and R. W. Baldwin, *Biomed. Chromatogr.*, 1992, **6**, 128–132.
57. R. V. Chari, K. A. Jackel, L. A. Bourret, S. M. Derr, B. M. Tadayoni, K. M. Mattocks, S. A. Shah, C. Liu, W. A. Blättler and V. S. Goldmacher, *Cancer Res.*, 1995, **55**, 4079–4084.
58. D. Willner, P. A. Trail, S. J. Hofstead, H. D. King, S. J. Lasch, G. R. Braslawsky, R. S. Greenfield, T. Kaneko and R. A. Firestone, *Bioconjug. Chem.*, 1993, **4**, 521–527.
59. L. M. Hinman, P. R. Hamann, R. Wallace, A. T. Menendez, F. E. Durr and J. Upeslacis, *Cancer Res.*, 1993, **53**, 3336–3342.
60. K. J. Hamblett, P. D. Senter, D. F. Chace, M. M. C. Sun, J. Lenox, C. G. Cerveny, K. M. Kissler, S. X. Bernhardt, A. K. Kopcha, R. F. Zabinski, D. L. Meyer and J. A. Francisco, *Clin. Cancer Res.*, 2004, **10**, 7063–7070.
61. P. R. Hamann, L. M. Hinman, C. F. Beyer, D. Lindh, J. Upeslacis, D. A. Flowers and I. Bernstein, *Bioconjug. Chem.*, 2002, **13**, 40–46.
62. G. Terral, A. Beck and S. Cianférani, *J. Chromatogr. B*, 2016, **1032**, 79–90.
63. R. Y.-C. Huang and G. Chen, *Drug Discovery Today*, 2016, **21**, 850–855.
64. A. Safavy, J. A. Bonner, H. W. Waksal, D. J. Buchsbaum, G. Y. Gillespie, R. Arani, D.-T. Chen, M. Carpenter, K. P. Raisch and K. P. Raisch, *Bioconjug. Chem.*, 2003, **14**, 302–310.
65. R. J. Sanderson, M. A. Hering, S. F. James, M. M. C. Sun, S. O. Doronina, A. W. Siadak, P. D. Senter and A. F. Wahl, *Clin. Cancer Res.*, 2005, **11**, 843–852.
66. D. Bagal, J. F. Valliere-Douglass, A. Balland and P. D. Schnier, *Anal. Chem.*, 2010, **82**, 6751–6755.
67. R. Y.-C. Huang, E. G. Deyanova, D. Passmore, V. Rangan, S. Deshpande, A. A. Tymiak and G. Chen, *J. Am. Soc. Mass Spectrom.*, 2015, **26**, 1791–1794.
68. L. Wang, G. Amphlett, W. A. Blättler, J. M. Lambert and W. Zhang, *Protein Sci.*, 2005, **14**, 2436–2446.
69. A. C. Lazar, L. Wang, W. A. Blättler, G. Amphlett, J. M. Lambert and W. Zhang, *Rapid Commun. Mass Spectrom.*, 2005, **19**, 1806–1814.
70. J.-P. Stephan, P. Chan, C. Lee, C. Nelson, J. M. Elliott, C. Bechtel, H. Raab, D. Xie, J. Akutagawa, J. Baudys, O. Saad, S. Prabhu, W. L. T. Wong, R. Vandlen, F. Jacobson and A. Ebens, *Bioconjug. Chem.*, 2008, **19**, 1673–1683.
71. M. M. Siegel, I. J. Hollander, P. R. Hamann, J. P. James, L. Hinman, B. J. Smith, A. P. Farnsworth, A. Phipps, D. J. King and M. Karas, *Anal. Chem.*, 1991, **63**, 2470–2481.
72. S. K. Jung, K. H. Lee, J. W. Jeon, J. W. Lee, B. O. Kwon, Y. J. Kim, J. S. Bae, D.-I. Kim, S. Y. Lee and S. J. Chang, *mAbs*, 2014, **6**, 1163–1177.
73. A. Wakankar, Y. Chen, Y. Gokarn and F. S. Jacobson, *mAbs*, 2011, **3**, 161–172.
74. I. Hollander, A. Kunz and P. R. Hamann, *Bioconjug. Chem.*, 2008, **19**, 358–361.

75. Z. Sosic, D. Houde, A. Blum, T. Carlage and Y. Lyubarskaya, *Electrophoresis*, 2008, **29**, 4368–4376.
76. Y. V. Kovtun, C. A. Audette, Y. Ye, H. Xie, M. F. Ruberti, S. J. Phinney, B. A. Leece, T. Chittenden, W. A. Blättler and V. S. Goldmacher, *Cancer Res.*, 2006, **66**, 3214–3221.
77. J. A. Francisco, C. G. Cerveny, D. L. Meyer, B. J. Mixan, K. Klussman, D. F. Chace, S. X. Rejniak, K. A. Gordon, R. DeBlanc, B. E. Toki, C.-L. Law, S. O. Doronina, C. B. Siegall, P. D. Senter and A. F. Wahl, *Blood*, 2003, **102**, 1458–1465.
78. S. O. Doronina, B. A. Mendelsohn, T. D. Bovee, C. G. Cerveny, S. C. Alley, D. L. Meyer, E. Oflazoglu, B. E. Toki, R. J. Sanderson, R. F. Zabinski, A. F. Wahl and P. D. Senter, *Bioconjug. Chem.*, 2006, **17**, 114–124.
79. R. S. Greenfield, T. Kaneko, A. Daues, M. A. Edson, K. A. Fitzgerald, L. J. Olech, J. A. Grattan, G. L. Spitalny and G. R. Braslawsky, *Cancer Res.*, 1990, **50**, 6600–6607.
80. E. Maeda, K. Urakami, K. Shimura, M. Kinoshita and K. Kakehi, *J. Chromatogr. A*, 2010, **1217**, 7164–7171.
81. S.-E. Ong and M. Mann, *Nat. Chem. Biol.*, 2005, **1**, 252–262.
82. K. A. Neilson, N. A. Ali, S. Muralidharan, M. Mirzaei, M. Mariani, G. Assadourian, A. Lee, S. C. van Sluyter and P. A. Haynes, *Proteomics*, 2011, **11**, 535–553.
83. J. Huang, F. Wang, M. Ye and H. Zou, *J. Chromatogr. A*, 2014, **1372C**, 1–17.
84. A. M. N. Silva, R. Vitorino, M. R. M. Domingues, C. M. Spickett and P. Domingues, *Free Radical Biol. Med.*, 2013, **65**, 925–941.
85. T. Nakamura and Y. Oda, *Biotechnol. Genet. Eng. Rev.*, 2007, **24**, 147–164.
86. E. Ciccimaro and I. A. Blair, *Bioanalysis*, 2010, **2**, 311–341.
87. M. von Bergen, N. Jehmlich, M. Taubert, C. Vogt, F. Bastida, F.-A. Herbst, F. Schmidt, H.-H. Richnow and J. Seifert, *ISME J.*, 2013, **7**, 1877–1885.
88. G. Chiappetta, S. Ndiaye, A. Igbaria, C. Kumar, J. Vinh and M. B. Toledano, *Methods Enzymol.*, 2010, **473**, 199–216.
89. N. Brandes, D. Reichmann, H. Tienson, L. I. Leichert and U. Jakob, *J. Biol. Chem.*, 2011, **286**, 41893–41903.
90. P. Liu, H. Zhang, H. Wang and Y. Xia, *Proteomics*, 2014, **14**, 750–762.
91. S. P. Gygi, B. Rist, S. A. Gerber, F. Turecek, M. H. Gelb and R. Aebersold, *Nat. Biotechnol.*, 1999, **17**, 994–999.
92. L. I. Leichert, F. Gehrke, H. V. Gudiseva, T. Blackwell, M. Ilbert, A. K. Walker, J. R. Strahler, P. C. Andrews and U. Jakob, *Proc. Natl. Acad. Sci. U. S. A.*, 2008, **105**, 8197–8202.
93. S. García-Santamarina, S. Boronat, A. Domènech, J. Ayté, H. Molina and E. Hidalgo, *Nat. Protoc.*, 2014, **9**, 1131–1145.
94. B. A. Petriz, C. P. Gomes, L. A. O. Rocha, T. M. B. Rezende and O. L. Franco, *J. Cell. Physiol.*, 2012, **227**, 885–898.

95. R. Zhang, C. S. Sioma, R. A. Thompson, L. Xiong and F. E. Regnier, *Anal. Chem.*, 2002, **74**, 3662–3669.
96. R. Zhang, C. S. Sioma, S. Wang and F. E. Regnier, *Anal. Chem.*, 2001, **73**, 5142–5149.
97. J. Li, H. Steen and S. P. Gygi, *Mol. Cell. Proteomics*, 2003, **2**, 1198–1204.
98. J. Wu, Q. Lin, T. K. Lim, T. Liu and C.-L. Hew, *J. Virol.*, 2007, **81**, 11681–11689.
99. K. C. Hansen, G. Schmitt-Ulms, R. J. Chalkley, J. Hirsch, M. A. Baldwin and A. L. Burlingame, *Mol. Cell. Proteomics*, 2003, **2**, 299–314.
100. E. C. Yi, X.-J. Li, K. Cooke, H. Lee, B. Raught, A. Page, V. Aneliunas, P. Hieter, D. R. Goodlett and R. Aebersold, *Proteomics*, 2005, **5**, 380–387.
101. W. W. Wu, G. Wang, S. J. Baek and R.-F. Shen, *J. Proteome Res.*, 2006, **5**, 651–658.
102. J. Qu, W. J. Jusko and R. M. Straubinger, *Anal. Chem.*, 2006, **78**, 4543–4552.
103. J. Qu and R. M. Straubinger, *Rapid Commun. Mass Spectrom.*, 2005, **19**, 2857–2864.
104. A. I. Nesvizhskii, O. Vitek and R. Aebersold, *Nat. Methods*, 2007, **4**, 787–797.
105. A. Fares, C. Nespoulous, M. Rossignol and J.-B. Peltier, *Methods Mol. Biol.*, 2014, **1072**, 609–620.
106. S. A. Gerber, J. Rush, O. Stemman, M. W. Kirschner and S. P. Gygi, *Proc. Natl. Acad. Sci. U. S. A.*, 2003, **100**, 6940–6945.
107. W. X. Schulze and B. Usadel, *Annu. Rev. Plant Biol.*, 2010, **61**, 491–516.
108. C. E. Eyers, C. Lawless, D. C. Wedge, K. W. Lau, S. J. Gaskell and S. J. Hubbard, *Mol. Cell. Proteomics*, 2011, **10**, M110.003384.
109. A. N. Kettenbach, J. Rush and S. A. Gerber, *Nat. Protoc.*, 2011, **6**, 175–186.
110. G. E. Craft, A. Chen and A. C. Nairn, *Methods*, 2013, **61**, 186–218.
111. A. Prakash, D. M. Tomazela, B. Frewen, B. Maclean, G. Merrihew, S. Peterman and M. J. Maccoss, *J. Proteome Res.*, 2009, **8**, 2733–2739.
112. V. Lange, P. Picotti, B. Domon and R. Aebersold, *Mol. Syst. Biol.*, 2008, **4**, 222.
113. N. L. Anderson, N. G. Anderson, T. W. Pearson, C. H. Borchers, A. G. Paulovich, S. D. Patterson, M. Gillette, R. Aebersold and S. A. Carr, *Mol. Cell. Proteomics*, 2009, **8**, 883–886.
114. M. Modzel, H. Płóciennik, M. Kielmas, Z. Szewczuk and P. Stefanowicz, *J. Proteomics*, 2014, **115C**, 1–7.
115. X. Wang, X. Wang, W. Qin, H. Lin, J. Wang, J. Wei, Y. Zhang and X. Qian, *Analyst*, 2013, **138**, 5309–5317.
116. A. Sanz-Medel, M. Montes-Bayón, J. Bettmer, M. Luisa Fernández-Sanchez and J. Ruiz Encinar, *TrAC, Trends Anal. Chem.*, 2012, **40**, 52–63.
117. A. Prange and D. Pröfrock, *J. Anal. At. Spectrom.*, 2008, **23**, 432.
118. M. Wang, W.-Y. Feng, Y.-L. Zhao and Z.-F. Chai, *Mass Spectrom. Rev.*, 2010, **29**, 326–348.
119. C. G. Vogiatzis and G. A. Zachariadis, *Anal. Chim. Acta*, 2014, **819**, 1–14.

120. E. Maes, K. Tirez, G. Baggerman, D. Valkenborg, L. Schoofs, J. R. Encinar and I. Mertens, *Mass Spectrom. Rev.*, 2016, DOI: 10.1002/mas.21440.
121. J. Bianga, Z. Touat-Hamici, K. Bierla, S. Mounicou, J. Szpunar, L. Chavatte and R. Lobinski, *J. Proteomics*, 2014, **108**, 316–324.
122. L. Mueller, H. Traub, N. Jakubowski, D. Drescher, V. I. Baranov and J. Kneipp, *Anal. Bioanal. Chem.*, 2014, **406**, 6963–6977.
123. A. Lothian, D. J. Hare, R. Grimm, T. M. Ryan, C. L. Masters and B. R. Roberts, *Front. Aging Neurosci.*, 2013, **5**, 35.
124. E. Maes, W. Brusten, F. Beutels, G. Baggerman, I. Mertens, D. Valkenborg, B. Landuyt, L. Schoofs and K. Tirez, *Rapid Commun. Mass Spectrom.*, 2015, **29**, 35–44.
125. D. R. Bandura, V. I. Baranov and S. D. Tanner, *Anal. Chem.*, 2002, **74**, 1497–1502.
126. D. R. Bandura, O. I. Ornatsky and L. Liao, *J. Anal. At. Spectrom.*, 2004, **19**, 96.
127. J. S. Becker, S. F. Boulyga, C. Pickhardt, J. Becker, S. Buddrus and M. Przybylski, *Anal. Bioanal. Chem.*, 2003, **375**, 561–566.
128. J. Bettmer, M. Montes Bayón, J. R. Encinar, M. L. Fernández Sánchez, M. del R. Fernández de la Campa and A. Sanz Medel, *J. Proteomics*, 2009, **72**, 989–1005.
129. M. Wind, A. Wegener, A. Eisenmenger, R. Kellner and W. D. Lehmann, *Angew. Chem., Int. Ed. Engl.*, 2003, **42**, 3425–3427.
130. S. El Balkhi, J. Poupon, J.-M. Trocello, F. Massicot, F. Woimant and O. Laprévote, *Anal. Chem.*, 2010, **82**, 6904–6910.
131. K. Bierla, J. Bianga, L. Ouerdane, J. Szpunar, A. Yiannikouris and R. Lobinski, *J. Proteomics*, 2013, **87**, 26–39.
132. M. Wang, W. Feng, W. Lu, B. Li, B. Wang, M. Zhu, Y. Wang, H. Yuan, Y. Zhao and Z. Chai, *Anal. Chem.*, 2007, **79**, 9128–9134.
133. N. Metanis and D. Hilvert, *Curr. Opin. Chem. Biol.*, 2014, **22C**, 27–34.
134. V. M. Labunskyy, D. L. Hatfield and V. N. Gladyshev, *Physiol. Rev.*, 2014, **94**, 739–777.
135. G. V. Kryukov, S. Castellano, S. V. Novoselov, A. V. Lobanov, O. Zehtab, R. Guigó and V. N. Gladyshev, *Science*, 2003, **300**, 1439–1443.
136. S. Castellano, V. N. Gladyshev, R. Guigó and M. J. Berry, *Nucleic Acids Res.*, 2008, **36**, D332–D338.
137. D. Schaumlöffel, J. Ruiz Encinar and R. Łobiński, *Anal. Chem.*, 2003, **75**, 6837–6842.
138. P. A. Whetstone, N. G. Butlin, T. M. Corneillie and C. F. Meares, *Bioconjug. Chem.*, 2004, **15**, 3–6.
139. R. Ahrends, S. Pieper, A. Kühn, H. Weisshoff, M. Hamester, T. Lindemann, C. Scheler, K. Lehmann, K. Taubner and M. W. Linscheid, *Mol. Cell. Proteomics*, 2007, **6**, 1907–1916.
140. S. Zhang, C. Zhang, Z. Xing and X. Zhang, *Clin. Chem.*, 2004, **50**, 1214–1221.

141. M. Careri, L. Elviri, A. Mangia and C. Mucchino, *Anal. Bioanal. Chem.*, 2007, **387**, 1851–1854.
142. L. Waentig, S. Techritz, N. Jakubowski and P. H. Roos, *Analyst*, 2013, **138**, 6309–6315.
143. Z. Liu, B. Chen, M. He, X. Zhang, H. Wang and B. Hu, *Analyst*, 2016, **141**, 926–933.
144. P. H. Roos, A. Venkatachalam, A. Manz, L. Waentig, C. U. Koehler and N. Jakubowski, *Anal. Bioanal. Chem.*, 2008, **392**, 1135–1147.
145. C. Giesen, T. Mairinger, L. Khoury, L. Waentig, N. Jakubowski and U. Panne, *Anal. Chem.*, 2011, **83**, 8177–8183.
146. C. Zhang, F. Wu, Y. Zhang, X. Wang and X. Zhang, *J. Anal. At. Spectrom.*, 2001, **16**, 1393–1396.
147. Z. A. Quinn, V. I. Baranov, S. D. Tanner and J. L. Wrana, *J. Anal. At. Spectrom.*, 2002, **17**, 892–896.
148. O. I. Ornatsky, V. Baranov, D. Bandura, S. Tanner and J. Dick, *FASEB J.*, 2006, **20**, A100.
149. G. Schwarz, S. Beck, M. G. Weller and M. W. Linscheid, *Anal. Bioanal. Chem.*, 2011, **401**, 1203–1209.
150. D. Esteban-Fernández, C. Scheler and M. W. Linscheid, *Anal. Bioanal. Chem.*, 2011, **401**, 657–666.
151. A. H. El-Khatib, D. Esteban-Fernández and M. W. Linscheid, *Anal. Chem.*, 2014, **86**, 1943–1948.
152. U. Bergmann, R. Ahrends, B. Neumann, C. Scheler and M. W. Linscheid, *Anal. Chem.*, 2012, **84**, 5268–5275.
153. Agilent Technologies Application Note: "Analysis of PEG and a PEGylated Therapeutic Protein Using an Agilent HPLC-Chip Coupled to a Q-TOF Mass Spectrometer", 2012, https://www.agilent.com/cs/library/applications/5991-1492EN.pdf, accessed August 2018.

CHAPTER 11

Determination of the Identity, Content and Purity of Therapeutic Peptides by NMR Spectroscopy

EDWIN KELLENBACH*[a] AND TORGNY RUNDLÖF[b]

[a] Aspen Oss BV, Kloosterstraat 6, P.O. Box 98, 5340 AB Oss, The Netherlands; [b] Swedish Medical Products Agency, P.O. Box 26, 751 03 Uppsala, Sweden, Email: torgny.rundlof@lakemedelsverket.se
*Email: ekellenbach@nl.aspenpharma.com

11.1 Introduction

NMR spectroscopy has been applied to peptides for a very long time and some rather old and perhaps slightly outdated, but still very useful, books on the topic are available.[1,2] NMR spectroscopy, together with mass spectrometry, is the most powerful technique for structure elucidation. In contrast to mass spectrometry, NMR spectroscopy is a nondestructive technique that also yields information on the three-dimensional structure of peptides. In recent decades, multidimensional, inverse pulsed field gradient stimulated heteronuclear NMR was developed for proteins to increase both sensitivity and resolution.[3] These techniques are equally applicable to peptides. In this chapter, we limit ourselves to analytical and quality control (QC) applications of 1D and 2D ^{1}H and ^{13}C NMR spectroscopy to small (<100 amino acids) therapeutic peptides. Reviews on NMR spectroscopy using less

Drug Discovery Series No. 72
Peptide Therapeutics: Strategy and Tactics for Chemistry, Manufacturing and Controls
Edited by Ved Srivastava

Published by the Royal Society of Chemistry, www.rsc.org

abundant nuclei such as ^{15}N[4] and ^{17}O[5] have appeared elsewhere, in addition to reviews on the 3D structure determination of proteins.[3] Articles on the ^{31}P NMR spectroscopy of phosphorylated peptides[6] and ^{19}F NMR spectroscopy of fluorinated peptides and proteins have also been published.[7–9] ^{19}F NMR spectroscopy will also sensitively detect trifluoroacetate.

11.2 Peptide Identity

Currently, a number of monographs for synthetic peptides included in the European Pharmacopoeia (Ph. Eur.) require amino acid analysis (AAA) according to Moore and Stein[10] as an identity test. Recently, NMR spectroscopy was introduced as an alternative method in the Ph. Eur.[11] The United States Pharmacopeia (USP) contains a method for oxytocin identification by NMR spectroscopy.[12,13] These NMR methods are meant as an alternative to AAA as an identity test for peptides. Briefly, AAA entails the acid hydrolysis of a peptide to yield the related amino acids followed by chromatographic separation, ninhydrin derivatization and detection.[10] The relative amount (ratio) and (absolute) content of amino acids in a peptide can be determined to be used both as an identity test and to quantify peptides. Some drawbacks of this method are the degradation of amino acids by the harsh hydrolysis conditions (typically overnight at 100 °C) and the laborious and time-consuming sample preparation.

On the other hand, NMR analysis in solution is widely recognized, accepted and established as a powerful tool for the structural elucidation and characterization of peptides and proteins.[1,2,14] ^{1}H and ^{13}C NMR spectra, for instance, readily distinguish all 20 naturally occurring amino acids and also unnatural amino acids present in synthetic peptides.

A ^{1}H NMR spectrum of a peptide depends primarily on the relative amounts and identities of the amino acids present in the sequence, *i.e.* the same information as obtained from an AAA.

^{1}H and ^{13}C NMR spectra readily distinguish all 20 naturally occurring amino acids and unnatural amino acids present in synthetic peptides and the random coil chemical proton chemical shifts in water have been compiled.[1,15,16] Quantification of the proton signals of peptides is relatively straightforward. The intensity of the proton signal is directly proportional to the relative number of protons at a particular position in a peptide.[17,18] As a result, the proton NMR spectrum of a peptide depends primarily on the relative amounts and identities of the amino acids present in the sequence. This information concerning the identity of a peptide is the same as that obtained from an AAA. Therefore, from a scientific point of view, NMR spectroscopy is a suitable alternative to AAA.

^{1}H NMR spectroscopy at 400 MHz was recently introduced as an identity test for eight peptides that are described in Ph. Eur. Monographs (Table 11.1). The identification consists of a straightforward and simple comparison with the spectrum of a (well-characterized) peptide standard and no calculations are performed. Even peptides that differ only very slightly can be readily

Table 11.1 Composition of the peptides from Figure 11.1. Reproduced from ref. 13 with permission from © Pharmeuropa Scientific Notes, 2008-1.

Peptide	No. of AA residues	Sequence
Oxytocin	9	H–Cys-Tyr-Ile-Gln-Asn-Cys-Pro-Leu-Gly–NH_2
Desmopressin acetate	9	Mpa–Tyr-Phe-Gln-Asn-Cys-Pro-D-Arg-Gly–NH_2
Gonadorelin acetate	10	Glp–His-Trp-Ser-Tyr-Gly-Leu-Arg-Pro-Gly–NH_2
Gonadorelin diacetate	10	Glp–His-Trp-Ser-Tyr-Gly-Leu-Arg-Pro-Gly–NH_2
Buserelin acetate	10	Glp–His-Trp-Ser-Tyr-D-Ser(^{t}Bu)-Leu-Arg-Pro-Gly–NHEt
Goserelin	10	Glp–His-Trp-Ser-Tyr-D-Ser(^{t}Bu)-Leu-Arg-Pro-Azagly–NH_2
Protirelin	3	Glp–His-Pro–NH_2
Tetracosactide	24	H–Ser-Tyr-Ser-Met-Glu-His-Phe-Arg-Trp-Gly-Lys-Pro-Val-Gly-Lys-Lys-Arg-Arg-Pro-Val-Lys-Val-Tyr-Pro–OH

Table 11.2 Comparison between NMR spectroscopy and AAA. Reproduced from ref. 11 with permission from © Pharmeuropa Scientific Notes, 2008-1.

Method	Advantages	Drawbacks
AAA	Widely available; classical, accepted technique; also applicable for larger peptides/proteins	Destruction of Asn and Gln; Cys, Met and Trp; hydrolysis conditions differ from one peptide to another; time-consuming and laborious sample preparation
NMR	General, robust technique; no hydrolysis; data acquisition fast and interpretation straightforward; allows identification/quantitation of proton-bearing organic counter ions	Less widely available[a]

[a]NMR analyses can, however, be readily outsourced to laboratories working under GMP for prices in the order of hundreds of euros, *i.e.* far less than the price for outsourcing of a classical AAA.

distinguished. Moreover, since no hydrolysis is performed prior to analysis, a distinction can be made between Glu and Gln, Asp and Asn, something that AAA is not able to do. Also, amino acids prone to hydrolysis and/or oxidation, such as Trp and Cys in AAA, can be detected without any problems by NMR spectroscopy. Table 11.2 summarizes the benefits and drawbacks of NMR spectroscopy and AAA.

If we only consider peptides containing the 20 naturally occurring L-amino acids, there is a staggering number (20^{10}) of potential decapeptides, rising to 40^{10} potential decapeptides if D-amino acids are included. Obviously, a strongly discriminating technique is therefore required to distinguish between these for identification. One-dimensional ^{1}H NMR spectroscopy is extremely sensitive to composition sequence and is *at least in principle* capable of discriminating between different sequences with the same amino acid composition whereas, in contrast, conventional AAA by its very nature is *not*.

Figure 11.1 demonstrates the strength of the ^{1}H NMR method by showing the spectra of eight different peptides all recorded in the same way at 400 MHz, a moderate field strength widely available within the pharmaceutical industry. All peptides can be readily distinguished, even in the case of goserelin and buserelin, two peptides showing only very small differences in their sequence (Table 11.1). The method is robust and reproducible and brought great simplification to the final release testing of peptides.

Using ^{19}F, ^{23}Na or ^{35}Cl NMR spectroscopy, counter ions such as trifluoroacetate (TFA), sodium and chloride may be identified and quantified.[19] This is illustrated for TFA in Figure 11.3. ^{19}F NMR spectroscopy may also be employed for the quantification of TFA.[20]

A further example of the selectivity of NMR is presented in Figure 11.4, in which a simple comparison of ^{1}H NMR spectra allows the differentiation of the peptide goserelin acetate from its serine diastereoisomer named goserelin related compound A that differs from the former by the epimerization of a single serine α-carbon (marked with an arrow in Figure 11.4).[11] Note the signal dispersion and the differences in the spectra of the two compounds (major differences are in the boxed areas). This is not feasible by conventional AAA because no distinction is made between L- and D-amino acids by the liquid chromatographic (LC) method.

A D/L modification of a single amino acid can result in large chemical shift differences. For the pharmaceutical nonapeptide buserelin, distinct differences were observed in the ^{1}H,^{13}C-HSQC spectrum when L-His in buserelin was changed to D-His. Spectra were acquired using ~0.5 mg of material that was dissolved in 0.2 mL of D_2O–acetic acid-d_4 (8 : 2) mixture in a 3 mm NMR tube. For the aromatic amino acids, large proton and carbon shift differences of about 0.2 ppm were observed for the histidine residue (Figure 11.5). Small but distinct differences were observed for some of the tryptophan and tyrosine signals. For the α-protons, shift differences of 0.00–0.07 ppm were observed, the largest for the His and Tyr residues (not shown).

The assignment of ^{1}H and ^{13}C chemical shifts of the buserelin D-His stereoisomer was performed by a combination of standard NMR experiments: one-dimensional ^{1}H, two-dimensional ^{1}H,^{1}H-total correlation spectroscopy (TOCSY) and rotating frame nuclear Overhauser spectroscopy (ROESY) and two-dimensional ^{1}H,^{13}C-heteronuclear single quantum coherence (HSQC) and heteronuclear multiple bond correlation (HMBC). The amino acid composition was determined by evaluation of α-proton cross-peaks in the TOCSY spectrum. For each amino acid, magnetization is transferred from the α-proton to other protons in the amino acid side chain. Two different spin-lock delays of 30 and 60 ms were used. At the shorter delay, TOCSY cross-peaks were detected mainly for the β-proton(s), while additional γ and δ cross-peaks were observed at the longer delay. The sequence of amino acids was determined from inter-residue carbonyl cross-peaks in the HMBC spectrum. Figure 11.6 clearly shows that sequence information was detected between all residues except Arg-Pro.

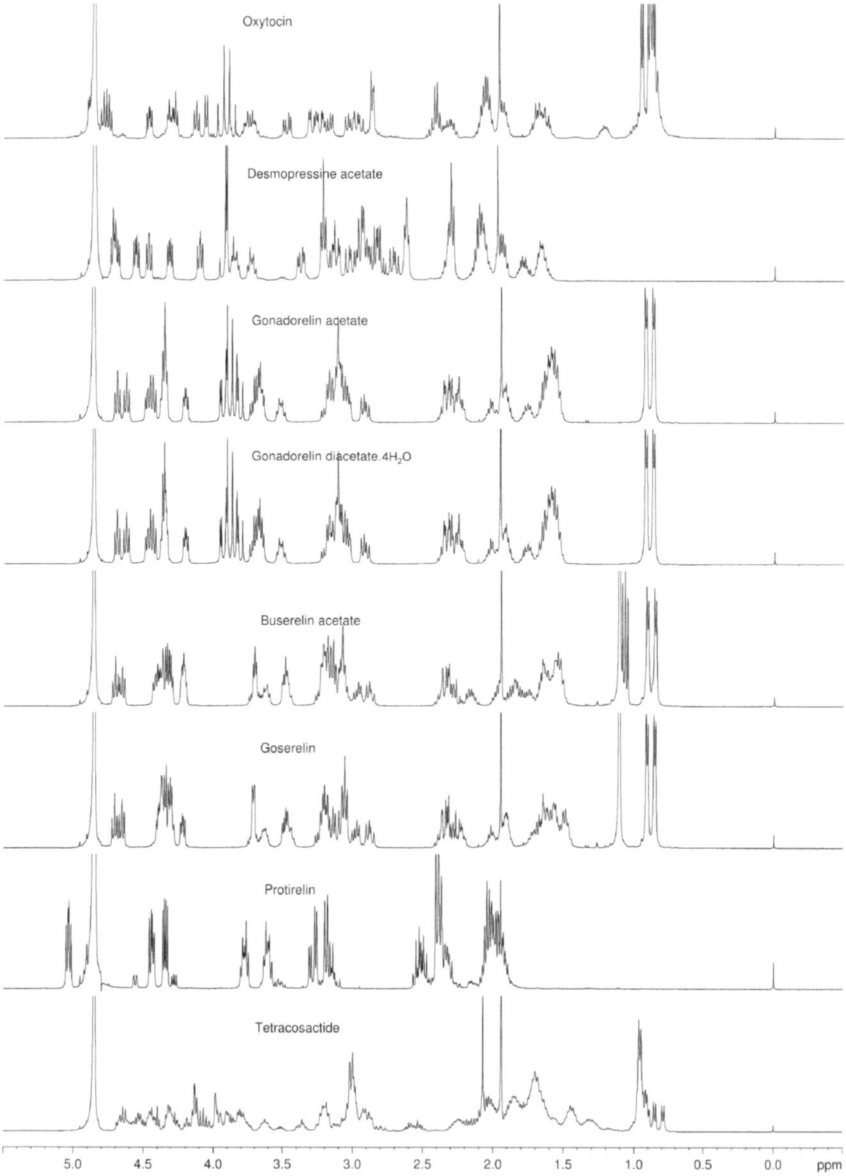

Figure 11.1 Aliphatic region of the 400 MHz ^{1}H NMR spectra of the eight peptides listed in Table 11.1. The spectra were recorded in D_2O in 200 mM phosphate buffer at pH 5. Also the two salt forms of gonadorelin (acetate and diacetate), using the areas of the acetate counter ion (Table 11.2) based on the intensity of the acetate signal. This is illustrated in Figure 11.2. More generally, counter ion identification and stoichiometry can be readily performed when nonexchangeable proton-bearing counter ions are involved.
Reproduced from ref. 11 with permission from © Pharmeuropa Scientific Notes, 2008-1.

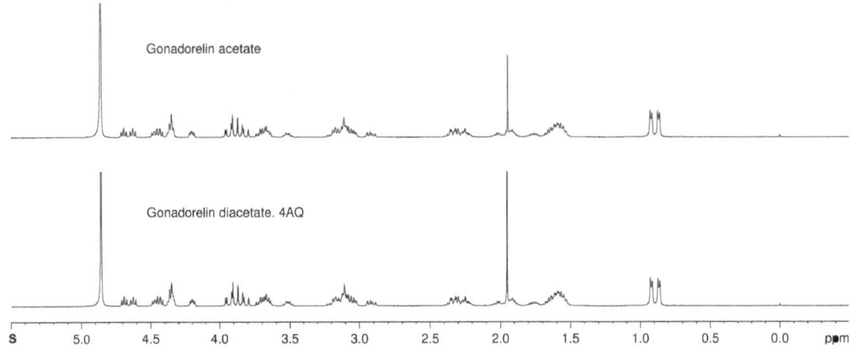

Figure 11.2 ^{1}H NMR spectra of gonadorelin acetate (top) and gonadorelin diacetate (bottom). The intensity of the acetate peak (singlet at about 1.9 ppm) allows distinction between the two peptides.
Reproduced from ref. 11 with permission from © Pharmeuropa Scientific Notes, 2008-1.

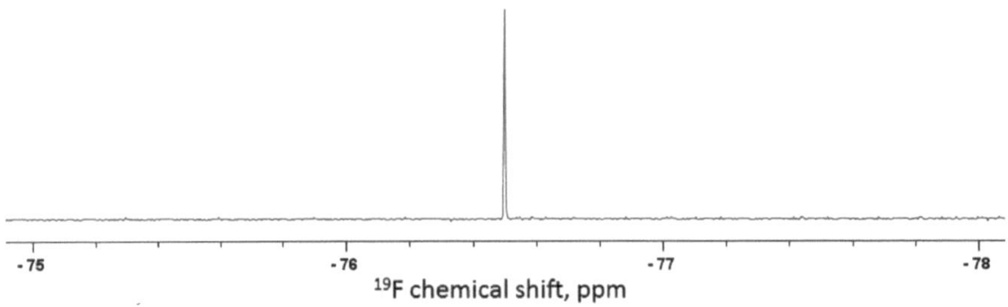

Figure 11.3 ^{19}F NMR spectrum of residual TFA in an aqueous solution of oxytocin. A singlet peak at −76.5 ppm is observed. The NMR sample contained 6 mg of oxytocin material and 1.5 μg of TFA, corresponding to 0.25 mg g^{-1}. The spectrum was acquired on a 600 MHz instrument equipped with a cryoprobe, the total experiment time was 4 min and the signal-to-noise ratio was 200 : 1. Reasonably, 10-fold lower amounts of TFA can be detected under these experimental conditions.

Fortunately, sequence information could be obtained from the ROESY spectrum that showed distinct inter-residue through-space correlation peaks between the Arg α-proton and the Pro β-protons (Figure 11.5). A mixing time of 200 ms was used.

Complete NMR chemical shift assignments are given in Table 11.3. All carbons and nonexchangeable protons were detected and interpreted.

The systematic approach to determining the ^{1}H spin systems for every amino acid, the C–H correlations at each position and inter-residual sequence determination is a powerful way to determine the primary structure of a peptide and also for modifications – and this without the need for reference data!

It was shown in 1988 that human, porcine and bovine insulins (51 amino acids, differing in 1–4 amino acids; Table 11.4) can be readily distinguished both by one-dimensional ^{1}H NMR (Figure 11.7) and by homonuclear

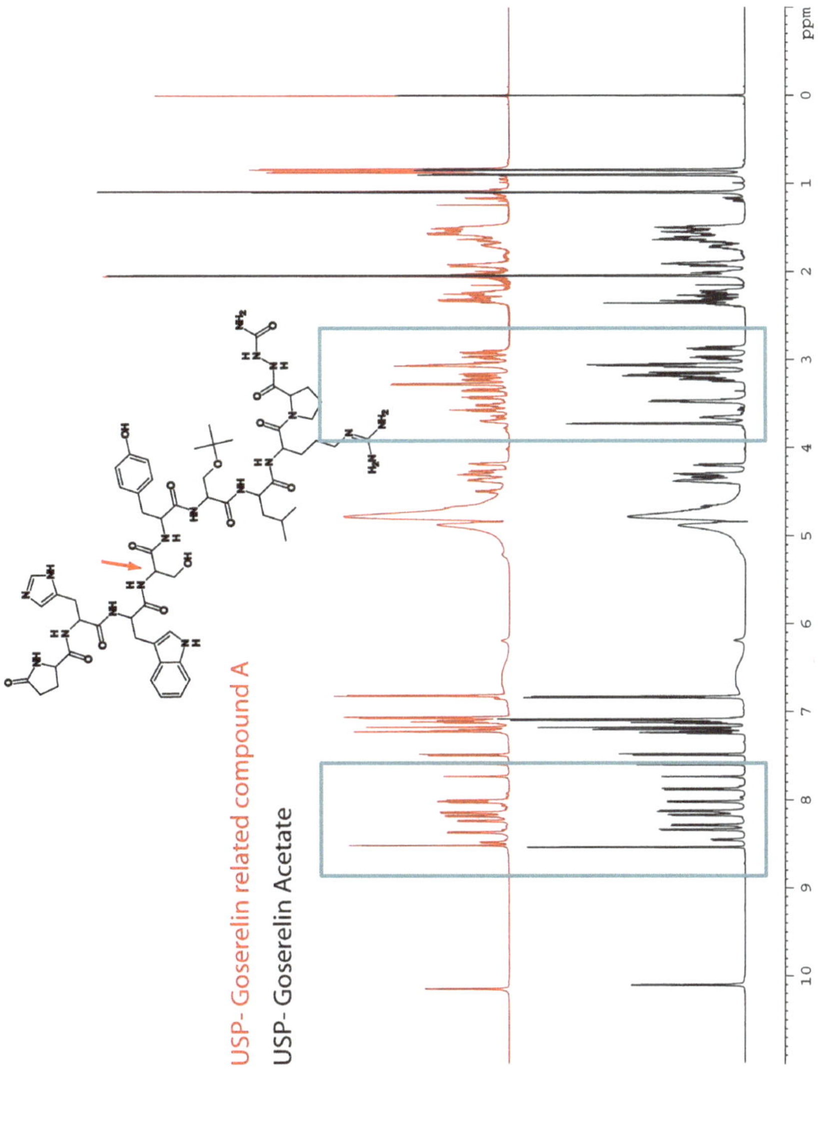

Figure 11.4 ^{1}H NMR (600 MHz, 95% H_2O) spectrum of goserelin acetate and of a serine diastereoisomer (goserelin related compound A). Regions with clear spectral differences are indicated by the boxes.[21]

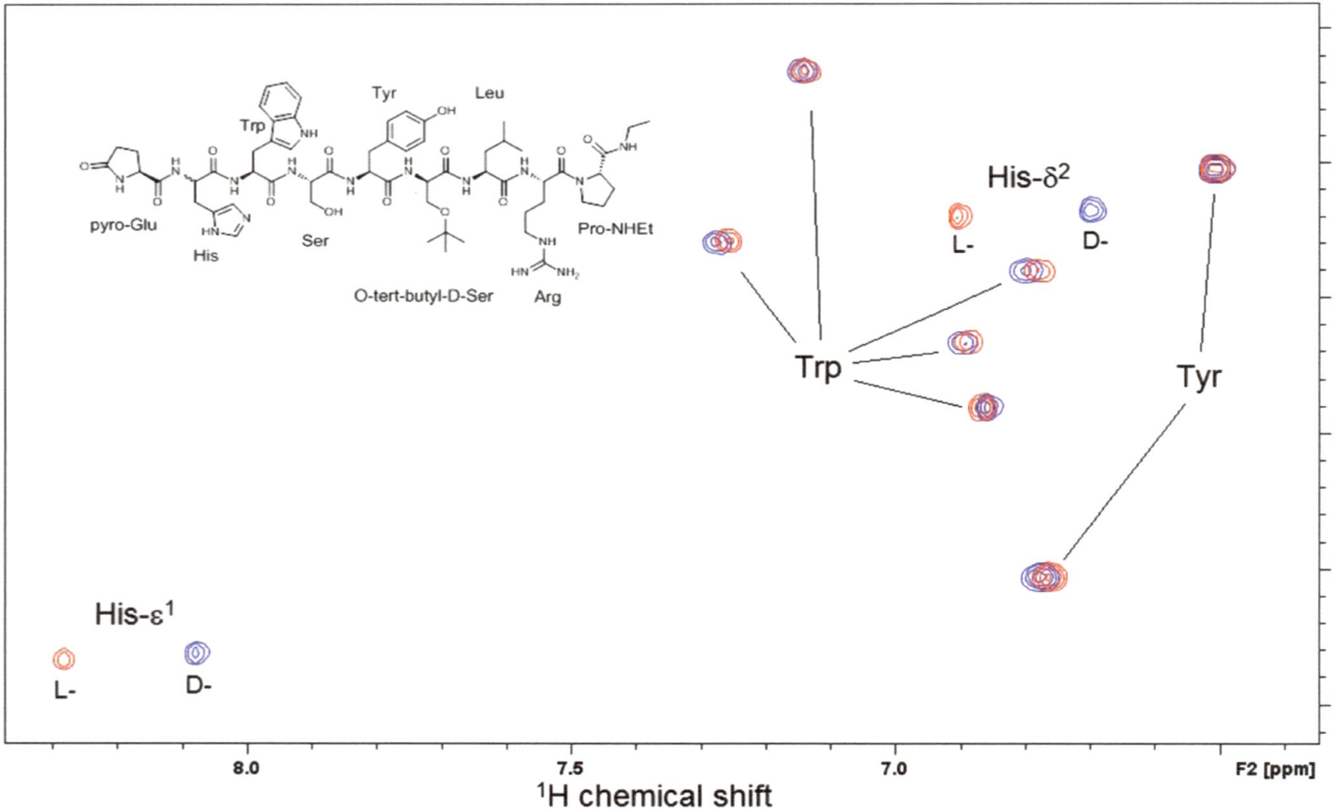

Figure 11.5 Overlay of the aromatic region of $^{1}H,^{13}C$-HSQC spectra for D-His-buserelin (blue) and buserelin (red). The chemical structure of buserelin is shown. Large ^{1}H and ^{13}C chemical shift differences were observed for the D- and L-isomers of the His residue. Smaller differences were observed for some of the signals from the other aromatic amino acid residues.

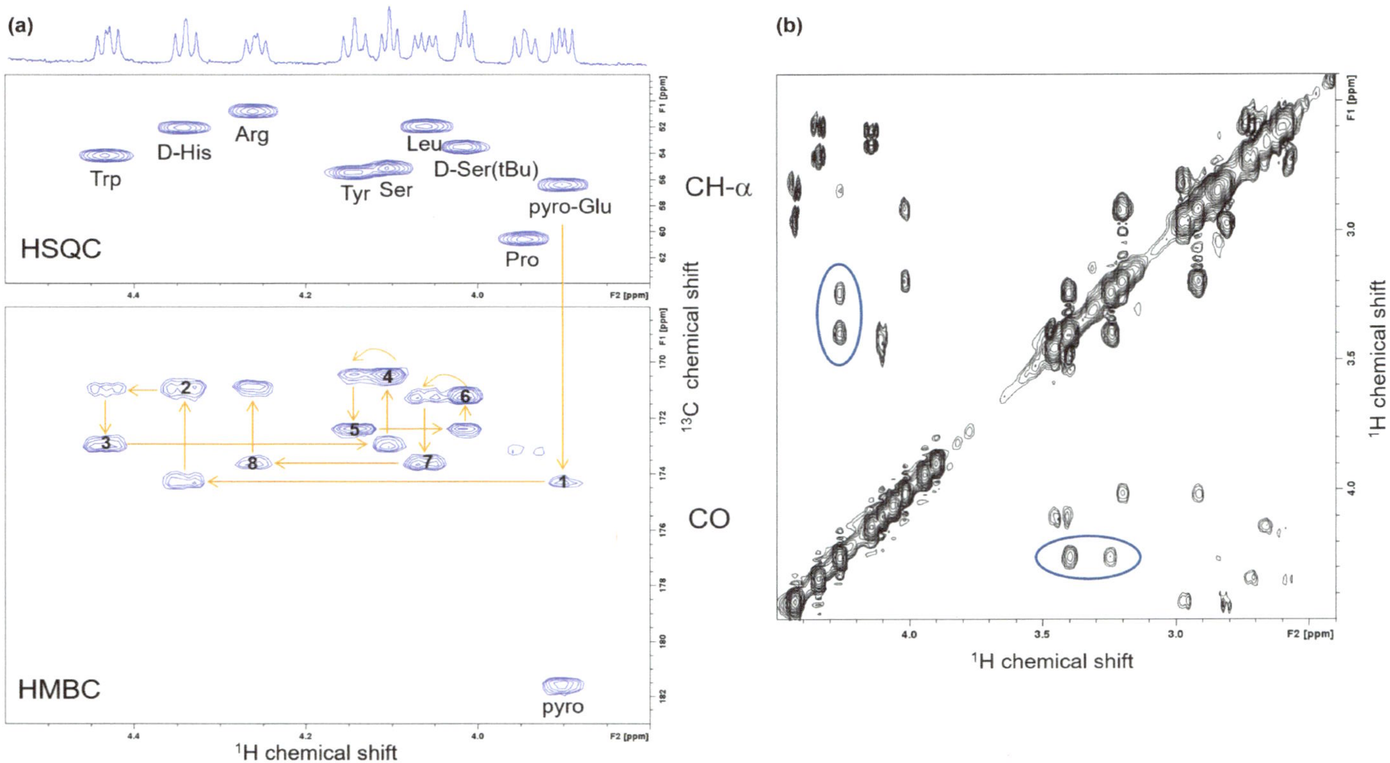

Figure 11.6 Informative two-dimensional NMR spectra for D-His-buserelin. Expansions from (a) the ^{1}H,^{13}C-HSQC (alpha region) and the HMBC (carbonyl region) and (b) the alpha to aliphatic region of the ^{1}H,^{1}H-ROESY spectrum. Sequence information was obtained from cross-peaks between the ^{13}C backbone carbonyl peak and the α-proton of the next residue. The ROESY spectrum showed through-space correlation between Arg and Pro, as indicated by ellipses in (b).

Table 11.3 Interpretation of NMR data for D-His-buserelin in D_2O–acetic acid-d_4 (4 : 1) at 25 °C.

Residue	Shift, δ	α[a]	β	γ	δ	Other	CO	H,H-TOCSY[b] [H,H-ROESY][c]	C,H-HMBC from CO to[d]
pyro-Glu-1	δ_H/ppm	3.90	2.09, 1.42	1.99				β 2.09/1.42; γ 1.99	
	δ_C/ppm	56.4	24.9	28.9			174.3; 181.5 (sc)		**CO: 4.34**/(3.90)/2.09/1.42 CO sc: 3.90/2.09/1.99/(1.42)
D-His-2	δ_H/ppm	4.34	2.72, 2.59			Ar: 8.08, 6.70		β 2.72/2.59	
	δ_C/ppm	52.0	25.9			Ar: 133.1, 116.7, 127.8	170.9		CO: (**4.43**)/4.34/4.26/2.72/2.59/(1.51) Ar: 8.08/2.72/2.59; 8.08/4.34/2.72/2.59
Trp-3	δ_H/ppm	4.43	2.97, 2.82			Ar: 7.28, 7.14, 6.90, 6.86, 6.80		β 2.97/2.82	
	δ_C/ppm	54.2	26.9			Ar: 117.9, 111.5, 121.6, 124.0, 119.0, 136.0, 126.7, 108.5	173.0		CO: 4.43/**4.11**/2.97/2.82 Ar: 7.28; 7.14; 7.28/6.86; 7.28/7.14/6.86/2.97/2.82; 7.28/6.86/4.43/2.97/2.82
Ser-4	δ_H/ppm	4.11	3.45, 3.41					β 3.45/3.41	CO: **4.15**/4.11/3.45/3.41
	δ_C/ppm	55.1	60.9				170.4		
Tyr-5	δ_H/ppm	4.15	2.66, 2.63			Ar: 6.78, 6.51		β 2.66/2.63	CO: 4.15/**4.02**/2.66/2.63 Ar: 2.66/2.63; 6.78/6.51; 6.51/4.15/2.66/2.63
	δ_C/ppm	55.4	36.0			Ar: 130.2, 115.2, 154.5, 127.2	172.3		

O-tert-butyl-D-Ser-6	δ_H/ppm	4.02	3.20, 2.92			*tert*-Butyl: 0.79		β 3.20/2.92	CO: **4.06**/4.02/3.20/2.92 *tert*-Butyl: 3.20/(2.92)/0.79
	δ_C/ppm	53.5	60.7			*tert*-Butyl: 26.1, 74.4	171.2		
Leu-7	δ_H/ppm	4.06	1.32, 1.23	0.59; 0.53				β 1.32/1.23; γ 0.59/0.53	CO: **4.26**/4.06/(1.32)/(1.23)
	δ_C/ppm	51.9	39.6	22.1; 20.3			173.6		
Arg-8	δ_H/ppm	4.26	1.51, 1.39	1.32	2.84			β 1.51/1.39; γ 1.32; δ 2.84 **[3.40/3.24]**[c]	CO: (4.43)/4.34/4.26/2.72/ 2.59/(1.51) Ar: 2.84
	δ_C/ppm	50.8	27.5	39.6	40.4	Ar: 156.3	170.9		
Pro-9	δ_H/ppm	3.95	1.87, 1.54	1.69	3.40, 3.24			β 1.87/1.54; γ 1.69; δ 3.40/ 3.24	CO: (3.95)/**2.85**/1.87/1.54
	δ_C/ppm	60.6	29.1	24.3	47.6		173.2		
N-Ethyl[e]	δ_H/ppm					2.85; 0.76			
	δ_X/ppm					34.2; 13.3			

[a]Atom position.
[b]Correlations from Hα that reports on amino acid identity.
[c]Inter-residual ROE from Hα.
[d]Inter-residual HMBC in bold and weak interactions in parentheses.
[e]C-terminal modification.

Table 11.4 Position and presence of threonine and alanine residues in human, porcine, bovine, glargine, and lispro insulin A and B chains.

Insulin species	A-8	A-10	B-14	B-27	B-30
Human	Thr	Ile	Ala	Thr	Thr
Porcine	Thr	Ile	Ala	Thr	Ala
Bovine	Ala	Val	Ala	Thr	Ala
Glargine	Thr	Ile	Ala	Thr	Thr
Lispro	Thr	Ile	Ala	Thr	Thr

two-dimensional NMR at a relatively moderate field strength of 360 MHz.[22] The sequence features for a selection of insulins are summarized in Table 11.4. As indicated in Table 11.4, there are differences between human, porcine, and bovine insulins. Glargine and lispro are human insulin variants, with identical Ala and Thr distribution. Despite this, the small structural differences between various insulin types are reflected in the Ala and Thr H_α/H_β region of the ^1H, ^1H correlation spectroscopy (COSY) spectra (Figure 11.8).

The insulin materials that were used in Figure 11.7 were prepared in D_2O–acetic acid-d_4 solvent mixture. Similar spectra were also obtained when D_2O–acetonitrile-d_3 was used, as shown in Figure 11.8. The pattern of the 1D ^1H and 2D $^1H,^1H$ correlation spectroscopy (COSY) spectra of human, porcine, bovine, glargine and lispro insulin clearly show differences. Human, glargine, and lispro insulins have very similar peptide sequences, differing only at few positions: A21 (Asn *vs.* Gln) for human *vs.* glargine, and B28-B29 (Pro-Lys *vs.* Lys-Pro) for human *vs.* lispro. Glargine insulin also has two additional Arg residues at positions B31 and B32. Despite the structural similarity, the COSY spectra of these materials show distinct differences for reliable identification of the respective insulin type.

In conclusion, distinction between various insulin types is possible in short time from standard NMR spectra on the material. Comparison to reference spectra is convenient if standardized sample conditions are employed. Human, porcine and bovine insulins can be distinguished by comparison of proton and/or COSY spectra. Notably, the expansions from the proton spectra of human and glargine insulins are very similar, whereas the patterns in the COSY spectra are distinctly distinguishable. The differences between proton as well as COSY spectra of human and lispro insulins are small but fair for both solvent mixtures.

11.2.1 Protamine Sulfate

Protamine sulfate from chum salmon is composed of four major peptides, A, B, C and D (Table 11.5), each containing 30–32 residues of the amino acids arginine, serine, proline, isoleucine, alanine, glycine and valine.[23] NMR spectroscopy can be used to determine the amino acid composition, to obtain sequence information and for quantification of the total amount. Furthermore, there are multiple possible sources of protamine sulfate and the origin of the material can be determined by NMR spectroscopy.

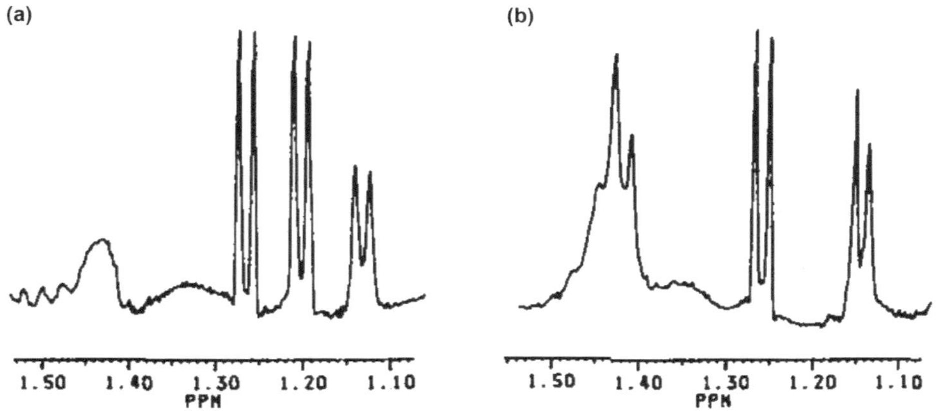

Figure 11.7 High-field region of the 360 MHz 1D ^{1}H NMR spectra of (a) human, (b) porcine and (c) bovine insulin. The species can easily be distinguished and identified from the spectra.
Reproduced from ref. 22 with permission from John Wiley and Sons, Copyright © 1988 Royal Pharmaceutical Society of Great Britain.

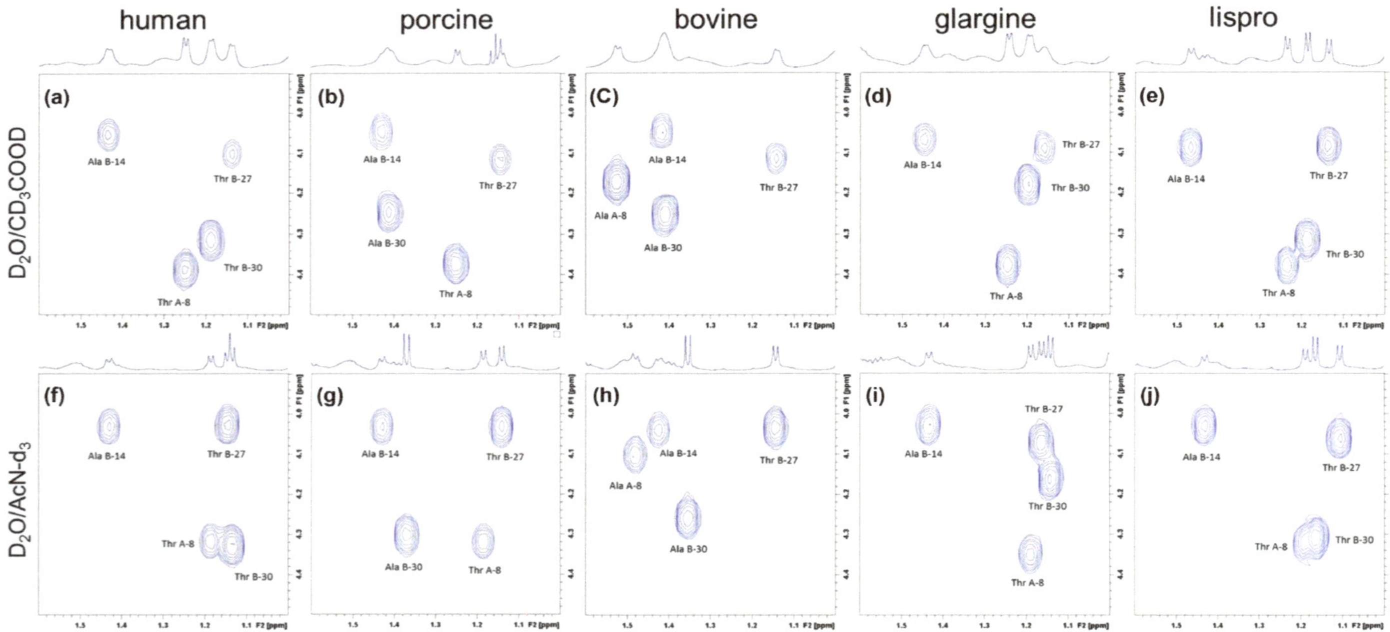

Figure 11.8 Methyl region of proton and COSY spectra for five insulins. The upper row shows spectra in D_2O–acetic acid-d_4 at pH 3 and 37 °C, for (a) human, (b) porcine, (c) bovine, (d) glargine, and (e) lispro insulin. The lower row shows the corresponding spectra (f)–(j) in D_2O–acetonitrile-d_3 2 : 1 mixture at pH 2.5–3 and 25 °C. Samples were prepared using 2–4 mg insulin dissolved in 0.7 ml solvent mixture, and spectra were acquired on a 600 MHz instrument. The chemical shift was set to 0.00 ppm for internal TSP-d_4.

Table 11.5 Single-letter peptide sequences[a] from chum salmon protamine sulfate.

Peptide	Sequence
A	PRRRRRSSSRPIRRRRRPRASRRRRRGGRRRR
B	PRRRRSSSRPVRRRRRPRVSRRRRRRGGRRRR
C	PRRRRSSRRPVRRRRRPRVSRRRRRRGGRRRR
D	PRRRRASRRIRRRRRPRVSRRRRRGGRRRR

[a]P, proline; R, arginine; S, serine; I, isoleucine; A, alanine; G, glycine; V, valine.

A simple test for the determination of the source of the protamine sulfate material was proposed by Gucinski *et al.*[24] An assay was developed based on the ratios of the alanine, glycine and arginine NMR peaks (relative to the arginine C_δ proton signal). The test also confirms that no aromatic amino acids are present, but does not report on the exact molar or weight ratio between peptides A–D. Selected peaks in the proton NMR spectrum were integrated (Figure 11.9) and the experimental integral ratios calculated. The results are reported in Table 11.6, together with the proposed limits. The determined ratios were well within the limits and showed that the source of the material is protamine sulfate from salmon.

As can be seen in Table 11.6, the experimental ratio for arginine $C_\alpha H/C_\delta H$ was lower than the expected ratio of 50%. When the minor arginine $C_\alpha H$ peak at 4.59 ppm (see below) was included in the Arg $C_\alpha H$ area, an experimental ratio of 50.0% was obtained (3.3 + 46.7%).

The amino acid composition and sequence information have been studied by various kinds of homo- and heteronuclear two-dimensional NMR experiments. Two protamine sulfate samples were prepared in D_2O and H_2O, respectively, at a concentration of 2.5 mg mL^{-1}. Spectra were acquired at 25 °C on a 600 MHz NMR spectrometer equipped with a 5 mm H(F)CN cryogenically chilled probe. The amino acids were confirmed by interpretation of side-chain correlation peaks in the two-dimensional $^1H,^1H$-COSY, $^1H,^1H$-TOCSY and $^1H,^{13}C$-HMBC spectra using the D_2O sample. Sequence information was obtained by interpretation of inter-residue correlations in the $^1H,^1H$-NOESY, $^1H,^1H$-ROESY and $^1H,^{13}C$-HMBC spectra using the H_2O sample (allowing detection of exchangeable amide protons). A typical fingerprint of chum salmon protamine sulfate is shown in the HSQC spectrum in Figure 11.10.

The assigned ^1H and ^{13}C chemical shifts were found to correlate well with reference data for the respective amino acids in various proteins.[3] The interpreted NMR data were in accord with the amino acid types and their relative occurrence in the peptide sequences for chum salmon protamine sulfate (Table 11.5): (i) major NMR peaks were from Arg, (ii) larger peaks were observed for Ser, Pro and Gly and (iii) smaller peaks were observed for the less frequent amino acids Val, Ile and Ala (see Figure 11.9, HSQC). No additional signals from other amino acids were observed. Two or more different sets of proton/carbon resonances (spin systems) were observed for Arg, Pro, Ser and Val, reflecting the fact that these amino acids occur at several and structurally different positions in peptides A–D. On the other

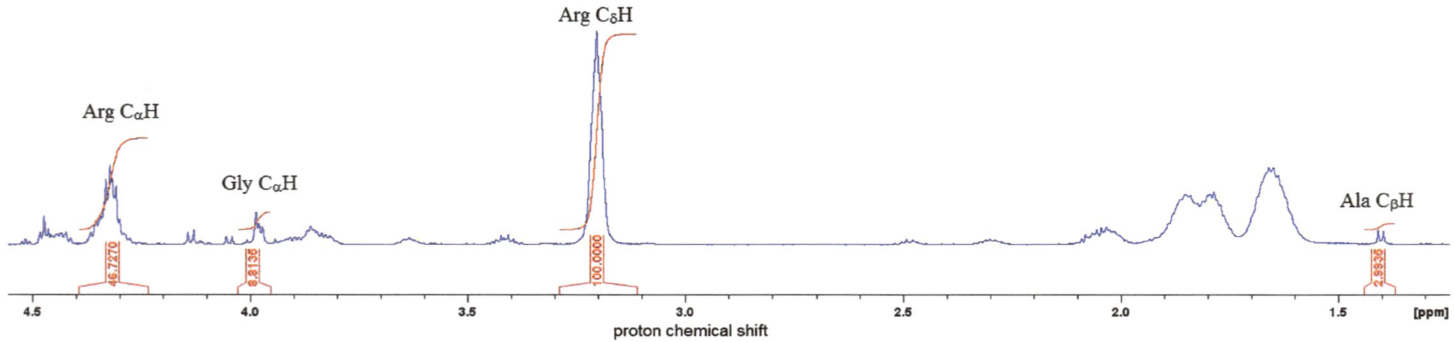

Figure 11.9 Proton NMR spectrum of protamine sulfate from salmon in D_2O solution. The source of the material was confirmed by calculation of selected ratios using peak areas and the number of protons in the respective peak.

Table 11.6 NMR test for salmon source assay.

Amino acids	No. of H ratio	Area ratio/%	Experimental ratio/%	Proposed limits/%[24]
Ala $C_\beta H$/Arg $C_\delta H$	3 : 2	3.0	2.0	1.4–3.4
Gly $C_\alpha H$/Arg $C_\delta H$	2 : 2	8.8	8.8	6.4–12.4
Arg $C_\alpha H$/Arg $C_\delta H$	n.a.[a]	46.7	46.7	45–55

[a]Not applicable.

hand, only one set of resonances was observed for Ile, Ala and Gly (except for the additional carbonyl signal for Gly).

Sequence information was obtained from correlations between adjacent amino acid residues: (a) nuclear Overhauser effect (NOE)/rotating frame Overhauser effect (ROE) correlations from the α-proton (C–H) of one amino acid to an amide proton (N–H) of the next residue and (b) HMBC correlations from a carbonyl (C=O) of one amino acid to the α-proton (C–H) of the next residue. This is illustrated in Figure 11.11.

The observed peptide backbone correlations were -Pro-Arg-, -Arg-Ser-, -Ser-Arg-, -Val-Arg-, -Val-Ser-, -Pro-Val-, -Arg-Val- (from NOE/ROE) and -Gly-Arg-, -Arg-Arg-, -Val-Ser-, -Gly-Gly-, -Ser-Ser- (from HMBC). The NOE/ROE correlations for protamine sulfate are illustrated in the ROESY spectrum shown in Figure 11.12. The arrows indicate inter-residue correlations that were used for confirmation of the peptide sequence.

Correlations between different amino acids are illustrated in Figure 11.13, in agreement with the structures of peptides A–D. No conflicting correlations were observed in the NMR spectra. Note that the HMBC spectrum also allows the detection of correlations between amino acids of the same type, *e.g.* Arg-Arg. Such correlations are present in the HMBC spectrum but are not illustrated in Figure 11.13.

11.3 Peptide Content by qNMR Spectroscopy

In addition to identity determination, quantitative NMR (qNMR) spectroscopy[25] is an excellent technique to determine peptide content.[26] For the content determination of, *e.g.*, reference standards, qNMR spectroscopy is a well-established method described in both the Ph. Eur. and the USP.[27,28] Peptide content can be evaluated relative to an unrelated but well-characterized standard. Related impurities may be distinguished, in contrast to, *e.g.*, Kjeldahl, titration and amino acid analysis; however, for a reference standard, material is usually selected that has undergone an additional purification. Moreover, most methods applied for content declaration are nonspecific: impurities and degradants are not distinguished. Proton-bearing counter ions can be quantified simultaneously. The variation is usually $<1\%$. Limited method development is required for qNMR for this generic method, making it very suitable for peptides applied in personalized peptide-based treatment vaccines.

As an illustrative example, Figure 11.14 shows the determination of gonadorelin acetate using a maleate standard. Five independent weighings of both the peptide and standard were performed.

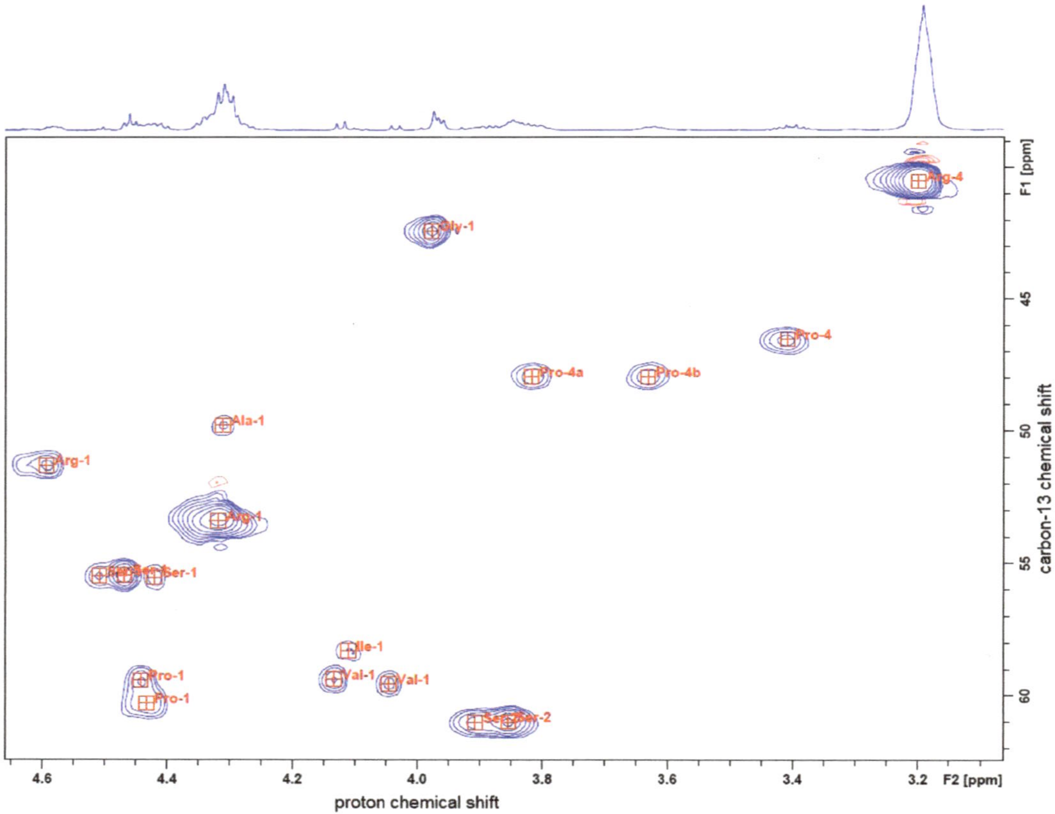

Figure 11.10 Expansion of the two-dimensional $^{1}H,^{13}C$-HSQC spectrum of protamine sulfate in D_2O solution. The spectrum shows the backbone α-H,C peaks for all amino acids (arginine, serine, proline, isoleucine, alanine, glycine and valine) and also some Ser, Pro, Arg side-chain peaks. The corresponding proton spectrum is shown above the HSQC spectrum. The peaks are annotated with the respective amino acid and side-chain positions 1 (α), 2 (β), 3 (γ), 4 (δ), … Note that a certain amino acid type may display multiple peaks, *i.e.* different chemical shifts, as a consequence of different positions in the peptide chains.

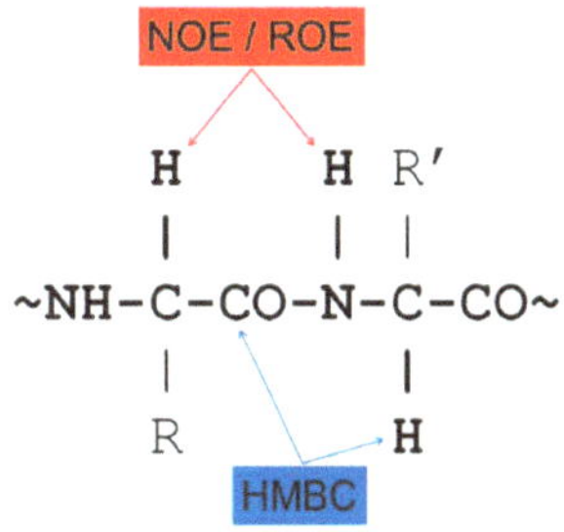

Figure 11.11 NMR inter-residue correlations from NOE, ROE and HMBC experiments provide sequence information.

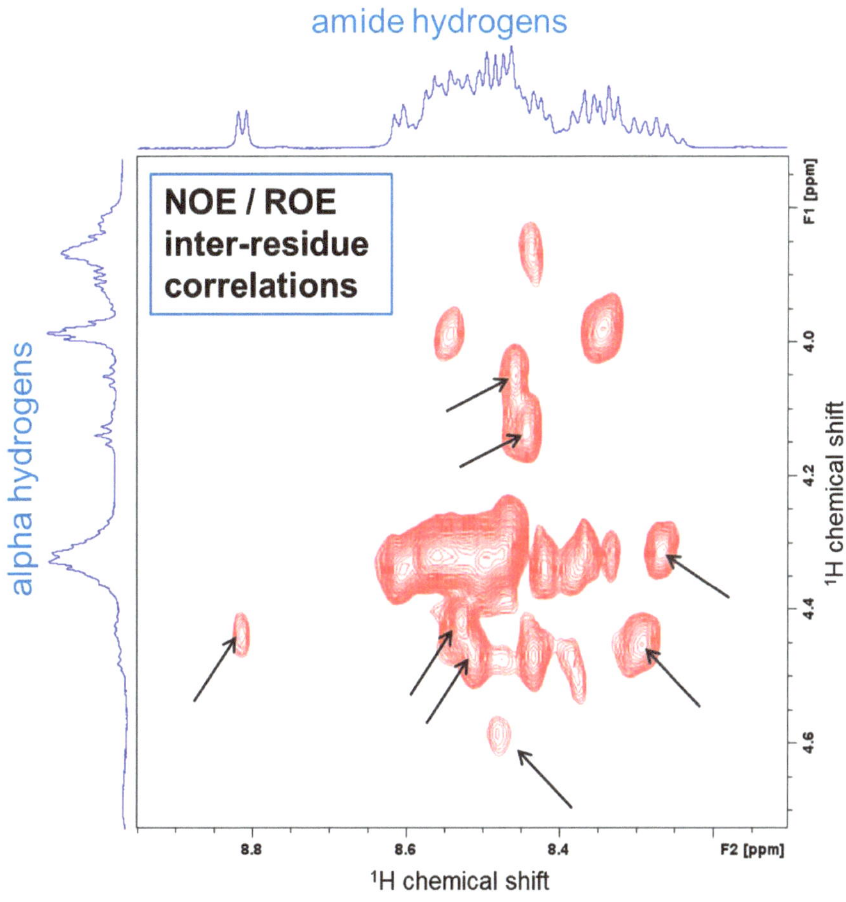

Figure 11.12 Part of the 600 MHz ROESY spectrum for protamine sulfate in H_2O solution at 25 °C. Inter-residue correlations between the alpha and amide hydrogens are indicated by arrows. These cross-peaks provided valuable sequence information.

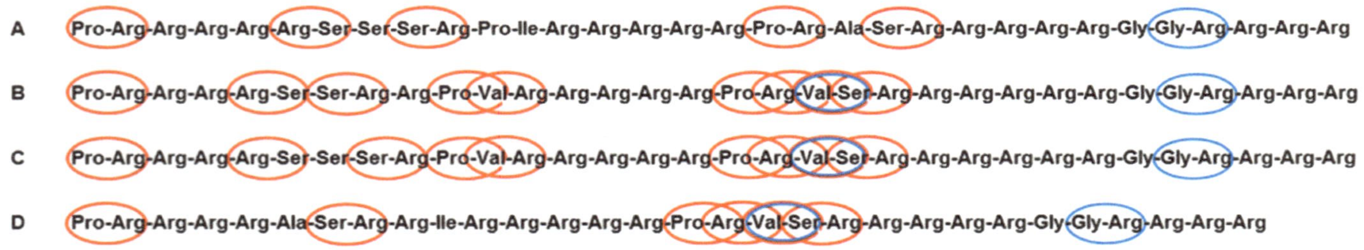

Figure 11.13 Observed inter-residue correlations in NOESY and ROESY spectra (red ellipses) and HMBC spectra (blue ellipses) for protamine sulfate peptides A–D. No other, unexpected, correlations were observed in the NMR spectra.

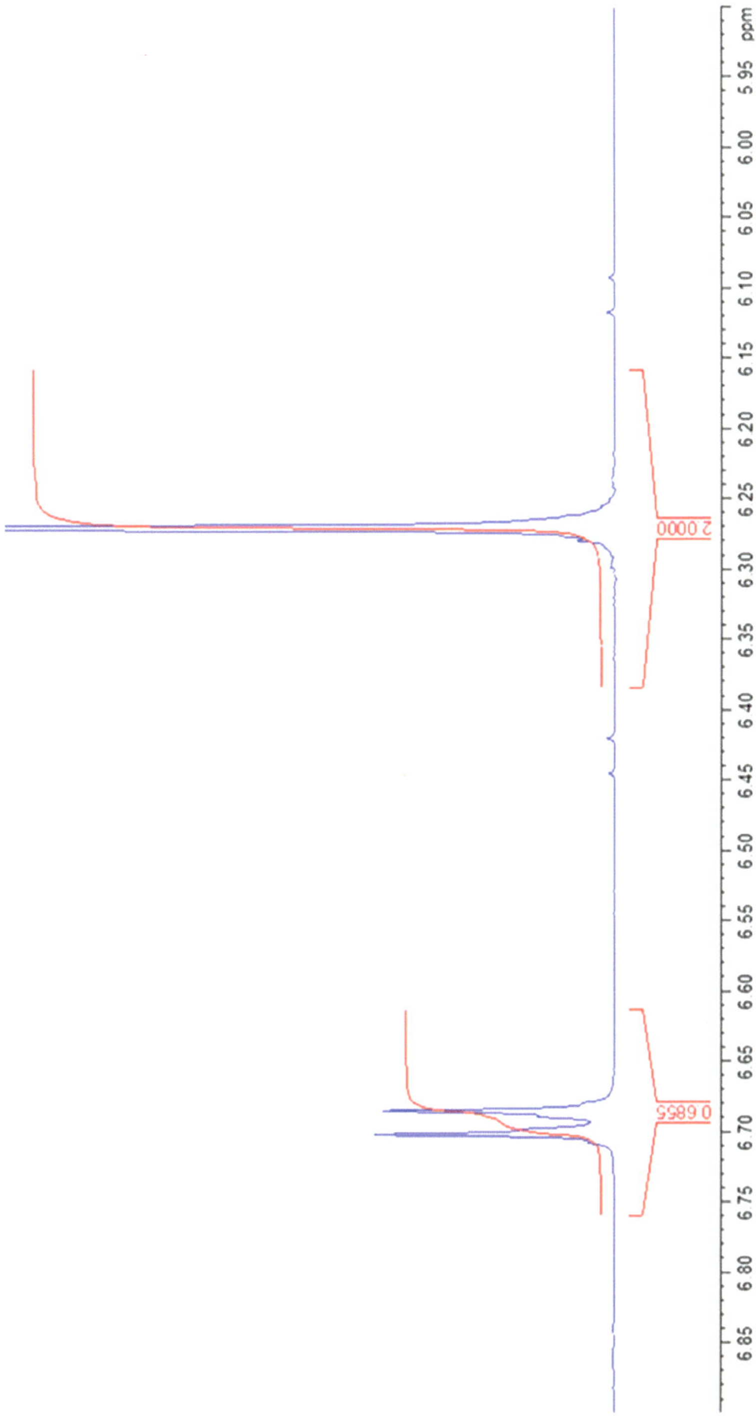

Figure 11.14 Part of the ^{1}H NMR (500 MHz, 95% D_2O) spectrum of gonadorelin acetate and maleate displaying the signals and their integrals used in content determination.

Using the equation

$$P(\text{sample}) = P(\text{std}) \times \frac{\text{MW(sample)}}{\text{MW(std)}} \times \frac{n\text{H(std)}}{n\text{H(sample)}} \times \frac{m\text{(std)}}{m\text{(sample)}} \times \frac{A\text{(sample)}}{A\text{(std)}}$$

where P(sample) = searched purity (wt%), P(std) = purity of standard (wt%), MW = molecular weight (g mol^{-1}), nH = number of equivalent protons in the NMR peak (1, 2, 3, ...), m = weight of material (mg) and A = area of NMR peak (arbitrary units), a content of 89.6% (m/m) with an SD of 0.3% could be calculated. The total experimental time was several hours.

This method can also be applied to assess the concentration of solutions as shown in the example in Figure 11.15. Lanreotide was studied in solutions with direct addition of an exact volume of an internal standard solution of 3-sulfolene. The content was determined as 1.02 mg per vial with an RSD of 0.13% (n = 3). The labeled content was 1.016 mg.

The method is also applicable to larger peptides such as insulin, but some challenges may be encountered. Insulin is a polypeptide with a well-defined structure composed of two peptide chains named A and B. NMR sample preparation is problematic, since insulin is prone to aggregation, which results in poor NMR spectra with very broad signals. Several useful approaches have been reported, *e.g.* using mixtures of water and an organic solvent such as acetic acid[22] or acetonitrile[29,30] at pH 3. Solutions of

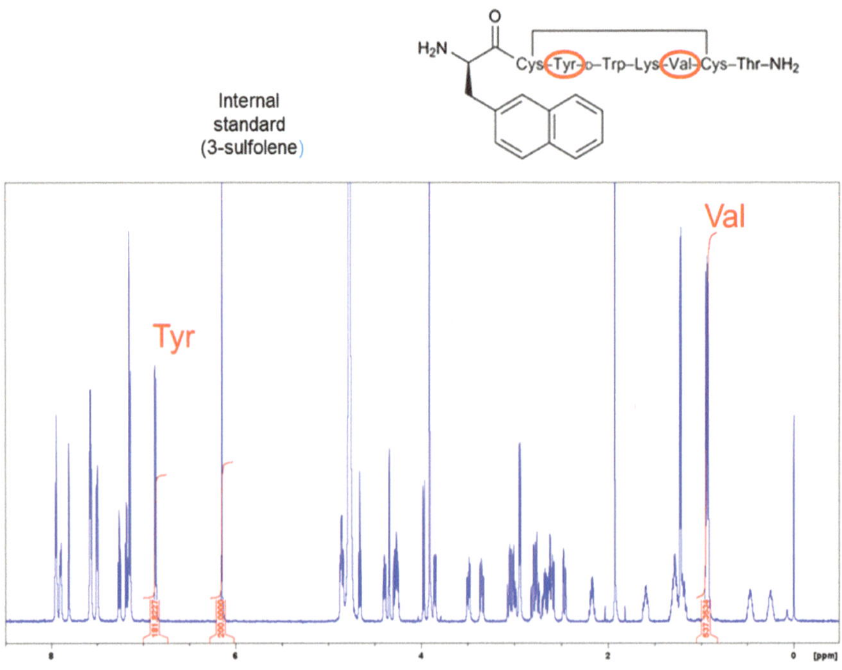

Figure 11.15 NMR spectrum used to determine the concentration of lanreotide (structure shown at top right) vials applying 3-sulfolene as an internal standard. The Tyr and Val signals are suitable for integration.

monomeric insulin show well-resolved NMR spectra that permit quantification. The α-protons of the two His signals of the B chain are adequately resolved at a chemical shift of around 8 ppm. A problem is that the exchangeable NH protons are also situated in this spectral region, resulting in spectral overlap. Protons situated on nitrogen and oxygen are normally exchanged by deuterium in a perdeuterated solvent such as D_2O. For smaller molecules, the H-to-D exchange process is usually immediate or very fast. Unfortunately, this is not the case for insulin, particularly at lower pH. This is shown in Figure 11.16, which illustrates that some of the amide exchangeable protons were not completely exchanged after 7 days.

The H–D exchange rate is much faster at neutral pH and it may be a good idea to keep the solution at a higher pH for some time before pH adjustment to a low pH.

When adequate H–D exchange was obtained, quantification was easily performed using 3-sulfolene as an internal standard, as shown in Figure 11.17.

In principle, either of the two His signals could be integrated, but for this sample His-B10 at 8.9 ppm was chosen owing to the smaller linewidth. Residual ethanol could be determined simultaneously.

^{13}C NMR spectroscopy can also be applied to measure proline peptide bond *cis–trans* isomers, which is very difficult to do otherwise because of the rapid isomerization.[31]

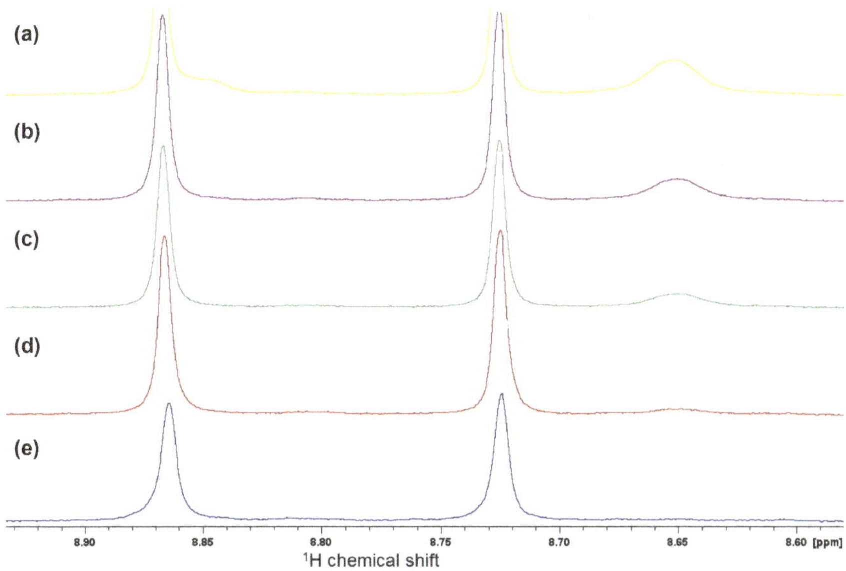

Figure 11.16 H–D exchange for porcine insulin in D_2O–CD_3CN (2 : 1) at pH 3, 25 °C and 600 MHz. The protons signals of the B-chain His-10 (8.87 ppm) and His-5 (8.72 ppm) are suitable for quantification. Very slow H–D exchange was observed for the amide signals at 8.85 and 8.65 ppm in spectra acquired immediately after sample preparation (a) and after 2 days (b), 3 days (c), 7 days (d) and 43 days (e).

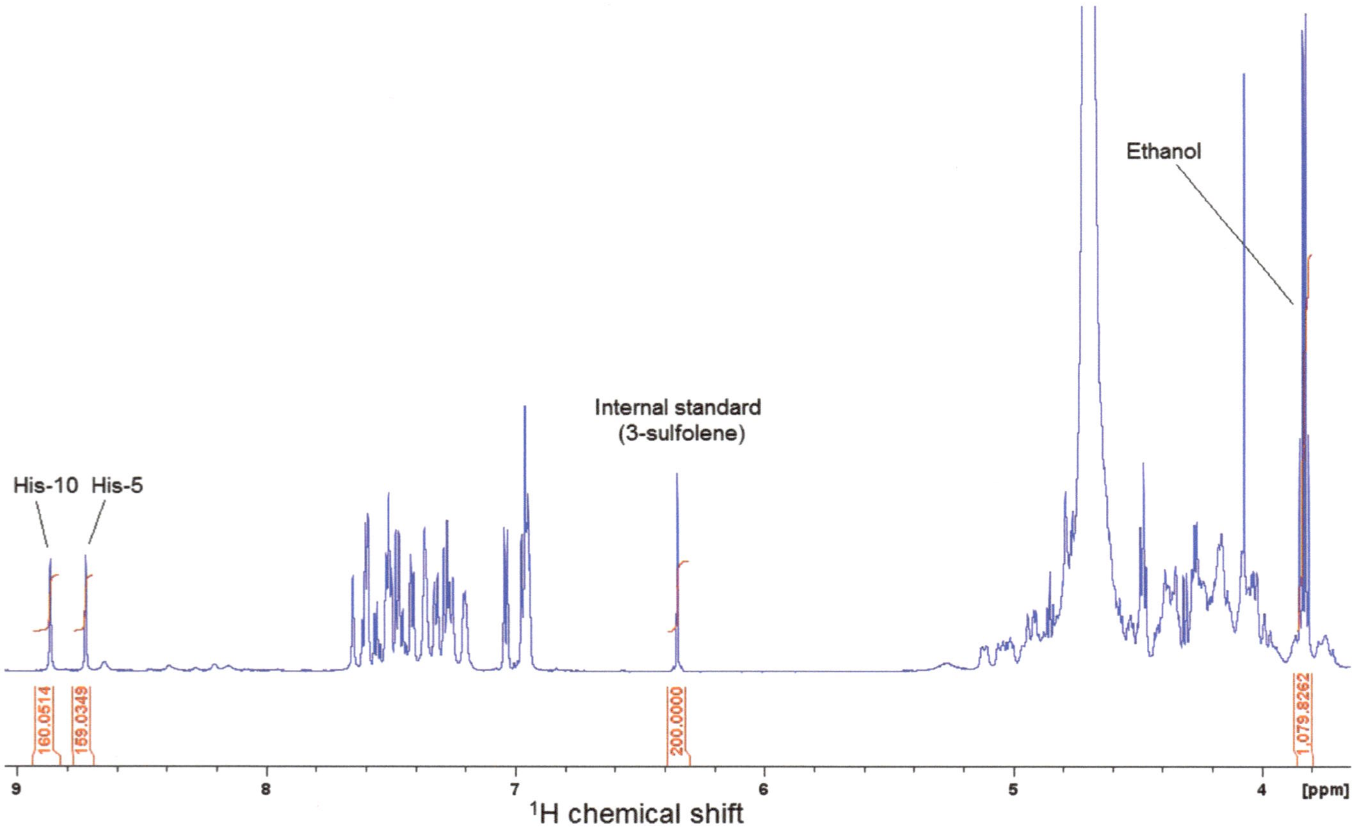

Figure 11.17 Quantification of porcine insulin by ^{1}H NMR spectroscopy in D_2O–acetonitrile-d_3 (2 : 1). The purity was calculated as 91 wt%, using a well-defined amount of 3-sulfolene as the internal standard. The sample contained about 2.5 wt% of residual ethanol.

The major benefit of qNMR spectroscopy lies in its robustness and its long-term precision,[32] and also its selectivity and general applicability.

Quantification of peptide mixtures such as protamine sulfate can be performed by qNMR spectroscopy. The material consists of four slightly different polypeptides A–D, with an undefined molar ratio between them. As a consequence for the calulations, the molecular weight may vary between 4064 and 4319 $g\,mol^{-1}$, with an average of 4217 $g\,mol^{-1}$. The number of amino acid residues varies between materials for Arg (21–22 residues) and Ala (0–1 residues), while all four different peptides contain two Gly residues each. This also affects the number of protons in the respective NMR peak to be used for integration. The measurement error due to the unknown distribution of peptides A–D is only −4 to +2% when the Gly signal is used, but slightly larger when other amino acid peaks are used. In practice, integration of different peaks in the proton spectrum resulted in an underestimated result when the Ala peak was used, compared with when Arg or Gly peaks were used (Table 11.7). The assay is reported as wt% of protamine for three independent determinations, using 3-sulfolene as the internal standard. The content of sulfate and water has to be added to the reported values.

The low value obtained when the Ala peak was used is probably a consequence of less peptides A and/or D in the protamine material, as indicated from analyses by high-performance liquid chromatography-mass spectrometry (HPLC-MS) and HPLC-UV detection. The values for Arg and Gly were similar and reasonably accurate when compared with the results obtained with LC-UV analysis.

Quantification may also be performed by 2D HSQC qNMR. In general, 2D spectra provide much better resolution than 1D proton spectra. A problem with 2D experiments is that they are not truly quantitative, since various compounds show differences in relaxation properties and coupling constants. The method proposed by Hu *et al.*[33] compensates for the differences. The true peak volume of each 2D peak can be obtained by extrapolation from three HSQC experiments with incremented repetition times. Absolute concentrations for each compound can then easily be calculated by the use of an internal standard. Poly(ethylene glycol) (PEG-4000) was selected as the internal standard in order to allow a reasonably short relaxation delay between repetitions (4 s) and thereby provide reasonably short measurement times (4.5 h per experiment). The protamine sulfate concentration was 5–6 $mg\,mL^{-1}$. Practically, three different HSQC experiments were acquired on each sample: $HSQC_1$, $HSQC_2$ and $HSQC_3$, where 1, 2 and 3 define the number of repetitions

Table 11.7 qNMR spectroscopy of protamine sulfate.

	Alanine	Arginine	Glycine
Molecular weight/$g\,mol^{-1}$	4217	4217	4217
NMR peak/ppm	1.40	3.20	3.98
Assay, mean/%	58.2	70.9	68.2
RSD/%	2.4	3.1	3.2

of the HSQC pulse sequence within one cycle. This procedure results in three 2D spectra with decreasing peak amplitudes (Figure 11.18).

For each compound, the respective peak amplitudes 1, 2 and 3 were determined. $HSQC_3$ displayed weak peaks and the protamine content was calculated using the peak volume and amplitude of $HSQC_1$ and the peak amplitude extrapolated from all three experiments according to Hu *et al.*[33] The resulting protamine content as calculated from the glycine signal and the arginine signal is given in Table 11.8 and compared with the respective 1D qNMR results. Similar results were obtained for 2D and 1D qNMR spectra and also irrespective of the NMR signal that was used (Gly *versus* Arg).

2D qNMR spectroscopy may provide an alternative to chemometric methods for calculation of content when the poor resolution in 1D NMR spectra becomes an issue. The fact that a universal internal standard can be employed means that the methodology can be applied in principle to any material. Recently a survey of peptide quantification methods including NMR and comparison of their reproducibility using oxytocin was presented.[34] It concluded that since qNMR has the advantages of simpler

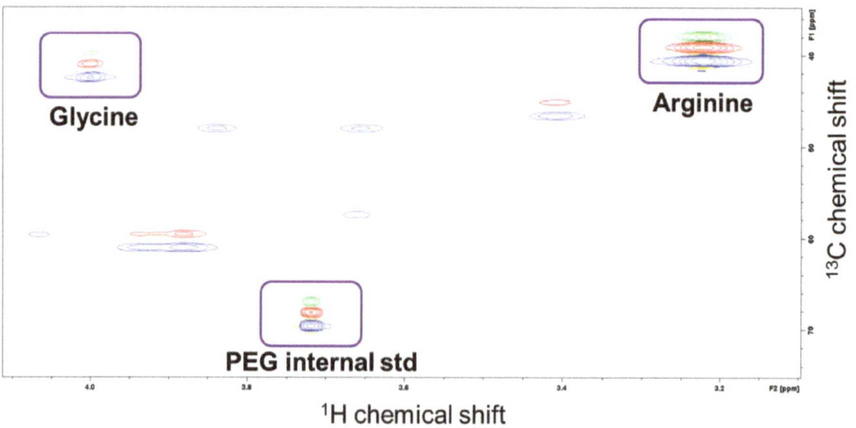

Figure 11.18 HSQC cross-peaks for two different amino acids of protamine sulfate and the internal standard PEG-4000. The peaks are used for calculation of protamine content. Shown are $HSQC_1$ (blue), $HSQC_2$ (red, shifted upwards) and $HSQC_3$ (green, shifted upwards). The peak amplitudes clearly decrease when the HSQC pulse sequence is repeated in $HSQC_2$ (two times) and $HSQC_3$ (three times). The use of 2D spectroscopy considerably increases the distinction between peaks compared with the proton spectrum.

Table 11.8 Protamine content determined by 2D qNMR spectroscopy and comparison with 1D qNMR spectroscopy. Three replicates were analyzed.

	2D HSQC qNMR		1D ^{1}H-qNMR	
	Gly	Arg	Gly	Arg
Content/%	69.3	67.7	69.2	71.9
RSD/%	4.6	2.5	1.3	0.9

operation and shorter analytical time, it deserves further exploration as a primary method for peptide assay and that NMR would need to be introduced as more readily in pharmaceutical quality control (QC) laboratories.

11.4 Peptide Purity

NMR spectroscopy is *not* the method of choice for routine impurity analysis of peptides because of the large number of similar, related impurities at similar, low levels potentially present. However, NMR spectroscopy *is* a powerful, universal detection method to identify unrelated and maybe unforeseen impurities such as process-related impurities, leachables and extractables.[35–37] These impurities may not be easily detected in peptides by HPLC analyses because they lack UV absorbance and/or retention in the chromatographic methods used. Small protonated impurities such as coupling agents and scavengers used in deprotection can be sensitively detected because they tend to have many equivalent protons resulting in sharp signals. Moreover, NMR spectroscopy yields a mole/mole response resulting in a high mass/mass response of these small molecules relative to the larger peptide molecules. Moreover, process-related impurities tend to be significantly smaller than peptides and will therefore have longer relaxation times and faster diffusion, which can be exploited for their detection.[17,38] These qualities can be employed for their detection and characterization. Figure 11.19 shows the ^{1}H NMR spectrum of a monoclonal antibody obtained with and without employing a

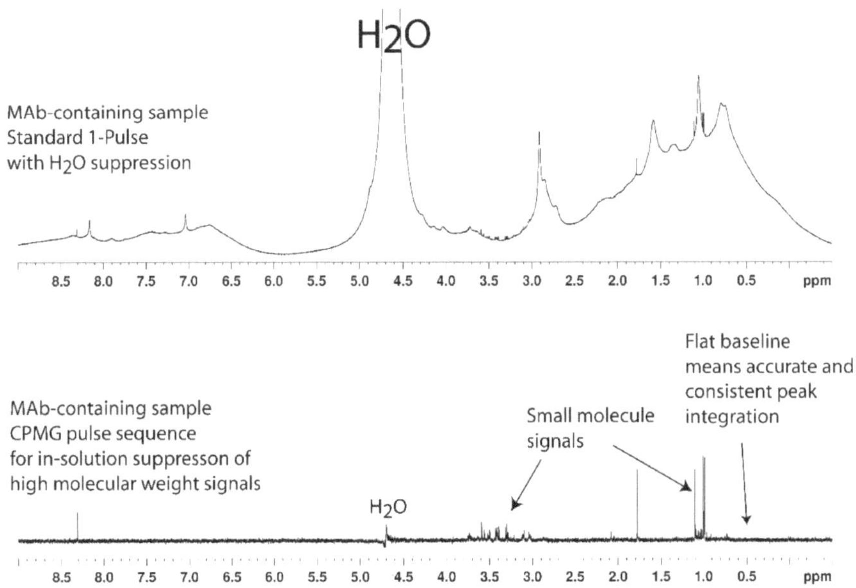

Figure 11.19 Sensitive detection of process-related small-molecule impurities in a solution of a monoclonal antibody (mAb) employing a CPMG sequence to suppress the signals of high molecular weight resonances. Reproduced from ref. 35 with permission from John Wiley and Sons, Copyright © 2012 American Institute of Chemical Engineers (AIChE).

Carr–Purcell–Meiboom–Gill sequence (CPMG) to suppress the signals of high molecular weight resonances. In this way, 1 ppm (mass/volume) levels of propylene glycol could be detected.[34]

In contrast, pulsed field gradient experiments can be used to suppress the signals of low molecular weight, rapidly diffusing molecules as illustrated in Figure 11.20.

Another example in which a gradient stimulated echo pulse sequence effectively suppresses low molecular weight compounds is illustrated in Figure 11.21. The somatotropin spectrum is highlighted, while the strong signals from the excipient, mannitol, and signals from other low molecular weight components are suppressed.

The dipeptide *N*(2)-Ala-Gln is used for glutamine supplementation. A USP monograph is available,[39] with limits for impurities cyclo-(Ala-Gln), alanine, glutamine, Ala-Ala-Gln and Ala-Glu.

The proton spectrum of an Ala-Gln material that has been kept at room temperature for some years showed a number of small peaks from unknown compounds (Figure 11.22). These impurities could be specified or unspecified related substances, degradation products, process related compounds or contaminants.

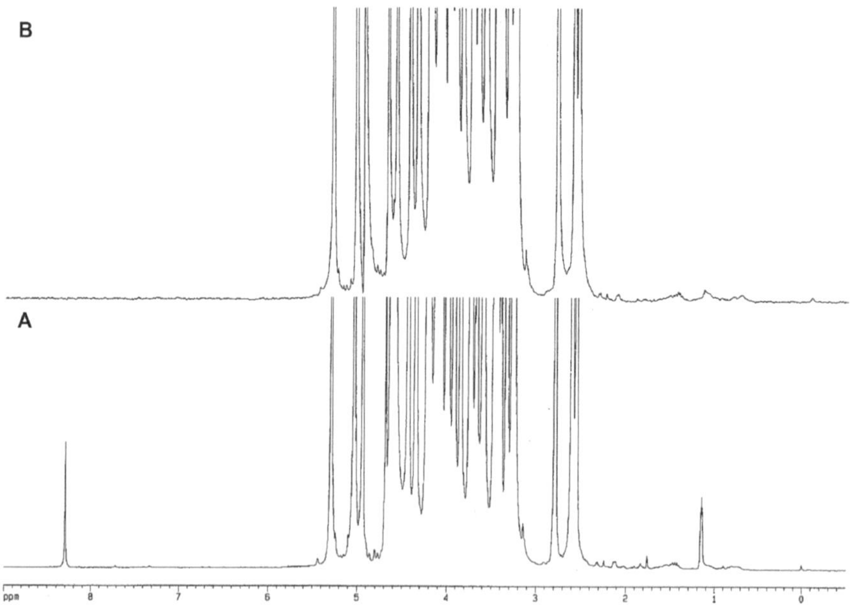

Figure 11.20 Suppression of the NMR signals of low molecular weight, rapidly diffusing molecules by the application of pulsed field gradients. In spectrum B, no gradient was applied and signals of low molecular weight impurities are visible at ~8.3 and 1.1 ppm. The application of two opposite pulsed field gradients separated by a 50 ms delay suppresses the signals of rapidly diffusing, low molecular weight species. Reproduced from ref. 38 with permission from American Chemical Society, Copyright 1999.

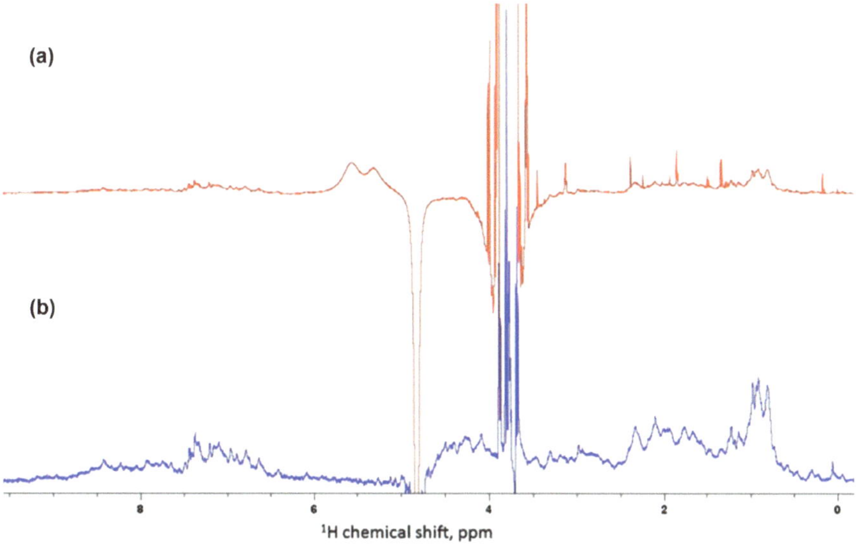

Figure 11.21 Proton NMR spectra of an aqueous solution of somatotropin medicinal product. The ordinary water-suppression proton spectrum is shown in (a) (red), showing mainly signals from mannitol at 3.5–4 ppm and several sharp peaks from other low molecular weight compounds. The gradient-edited spectrum is shown in (b) (blue). Notably, the somatotropin signals are strongly enhanced in the lower spectrum, whereas the peaks of mannitol and other compounds are greatly or completely suppressed.

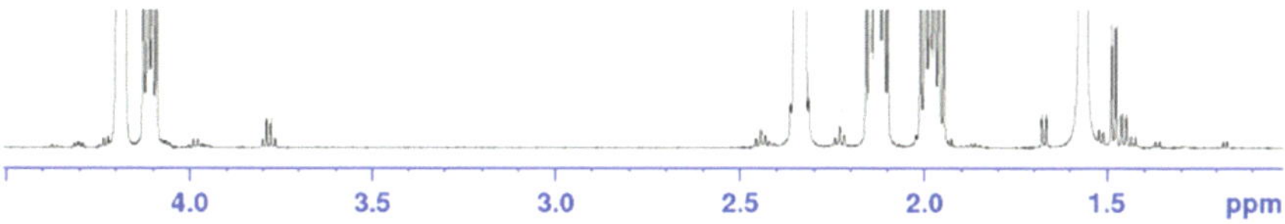

Figure 11.22 Proton spectrum at 600 MHz of Ala-Gln in D_2O at 25 °C. In addition to the ^{13}C satellites on both sides of each main peak, a number of impurity peaks can also be observed.

The methyl region around 1.5 ppm is suitable for a detailed impurity evaluation, since methyl peaks generally are intense and not too split (*J*-coupled). Low-power ^{13}C decoupling can generally be applied to simplify spectra by eliminating ^{13}C satellites, as illustrated in Figure 11.23 for alanine methyl signals.

Comparison with reference spectra of selected impurities showed that the largest impurity doublet peak at 1.48 ppm could be either alanine or cyclo-(Ala-Glu). The methyl signal of alanine residues is *J*-coupled to the respective α-hydrogen and a $^{1}H,^{1}H$-COSY spectrum provides valuable information about the chemical environment of the particular alanine residue. This is illustrated in the expansion of the COSY spectrum of Ala-Gln shown in Figure 11.24. COSY is a fairly sensitive experiment and the second dimension permits the detection of impurity peaks that are hidden under larger

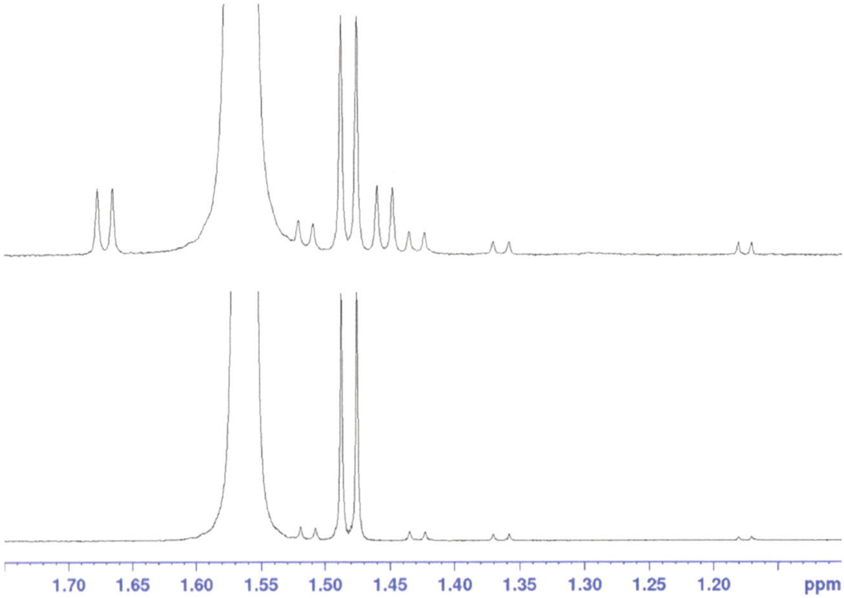

Figure 11.23 The upper spectrum shows the methyl region of Ala-Gln in D_2O. In the lower spectrum, the ^{13}C satellites have been quenched by low-power ^{13}C decoupling during the acquisition. This clearly simplifies the evaluation with respect to the five different impurity peaks.

peaks in the one-dimensional proton experiment. For Ala-Gln, the most abundant impurity showed a cross-peak at 1.48/3.78 ppm, in complete accord with reference NMR shifts for alanine. Smaller impurity cross-peaks were observed at 1.43/4.37 and 1.36/4.15 ppm. Further evaluation of these impurity peaks is possible by recording selective one-dimensional NOESY or ROESY experiments, which may provide sequence information about through-space H–H contacts from Ala to neighboring amino acids.

The cross-peak at 4.02/1.18 ppm is due to small amounts of 2-propanol. Quantification of residual solvents is easily performed by comparison of the integral of the solvent peak with that of the ^{13}C satellite peak of the main compound. The natural abundance of ^{13}C is 1.1%, which means that each ^{13}C satellite signal corresponds to 0.55% of the main peak of the compound. This results in the following equation for the quantification of residual solvent (or any specified impurity) in the compound:

$$\text{Solvent(ppm)} = 10^6 \times \frac{\dfrac{A(\text{solvent}) \times \text{MW}(\text{solvent})}{n\text{H}(\text{solvent})}}{\dfrac{A(\text{compound}) \times \text{MW}(\text{compound})}{0.0055 \times n\text{H}(\text{compound})} + \dfrac{A(\text{solvent}) \times \text{MW}(\text{solvent})}{n\text{H}(\text{solvent})}}$$

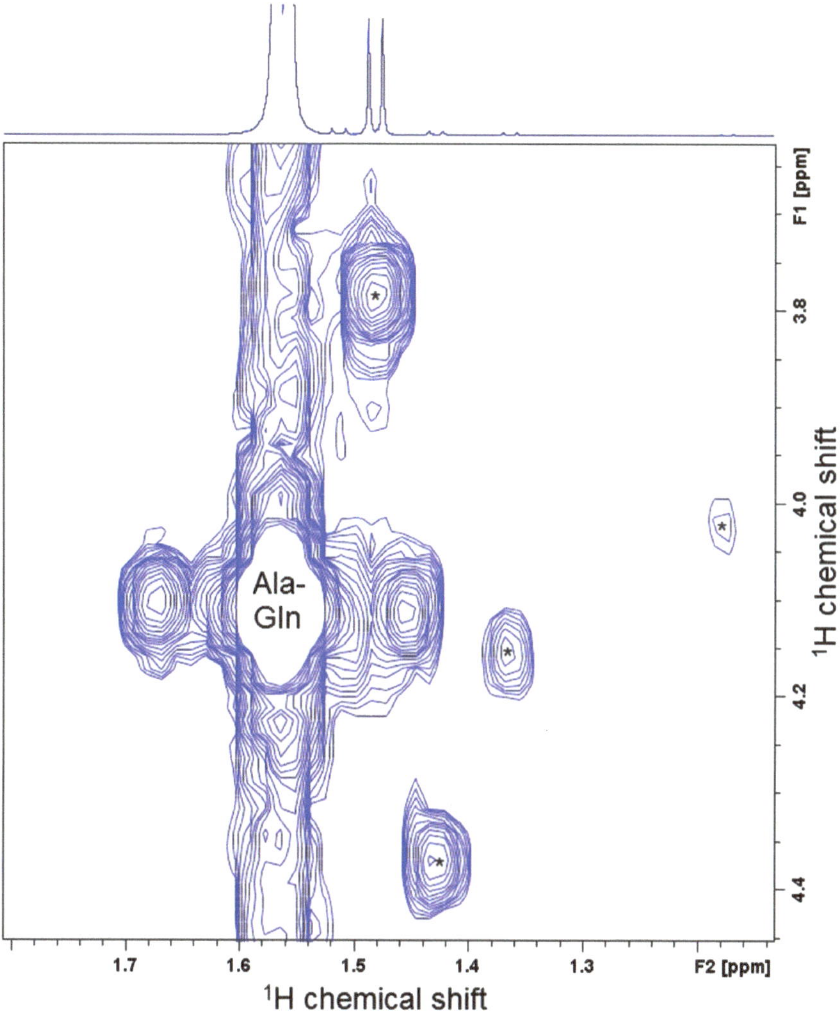

Figure 11.24 The alanine methyl region from a 600 MHz COSY spectrum of Ala-Gln in D_2O solution at 25 °C. Four different cross-peaks related to impurities are observed (see text), as indicated by asterisks.

where A is the peak integral area, MW is the molecular weight (g mol^{-1}) and nH is the number of hydrogens in the peak. For levels lower than ~100 ppm the solvent term in the denominator can be omitted. For the present case, the areas of 2-propanol and Ala-Gln were 13.5 and 100, respectively, which resulted in a 2-propanol level of ~100 ppm (Figure 11.25).

NMR spectroscopy can also be used to elucidate the structure of isolated peptide impurities in conjunction with MS. Furthermore, peptide modifications difficult to elucidate by MS (*e.g.* diastereomeric impurities) may be elucidated by NMR spectroscopy. In stability studies on syringes of a

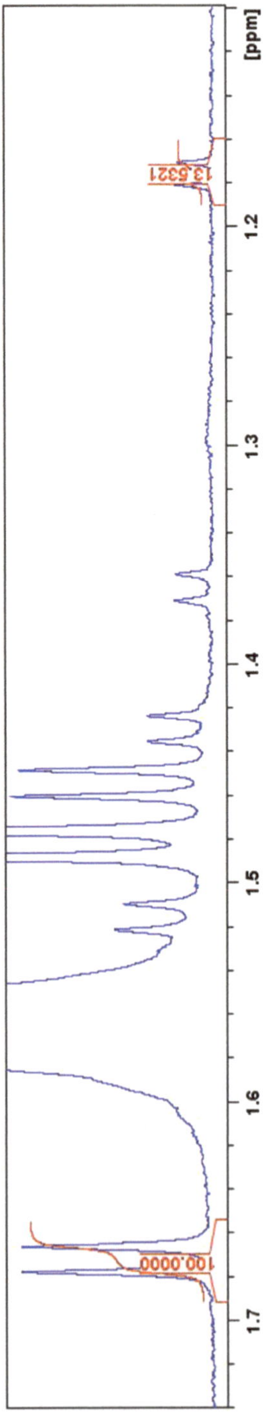

Figure 11.25 Residual 2-propanol at 1.18 ppm in Ala-Gln. The solvent level can be determined from the integral ratio between the solvent peak (six hydrogens) and the ^{13}C satellite peak of the compound (three hydrogens).

pyridylalanine-containing peptide, an HPLC peak of a degradant impurity was detected that increased with time. MS analysis of this peak show that it had the same mass as the peptide itself. Based on this information, all diastereomeric impurities of the peptide were synthesized, but none of them coeluted with the degradant impurity. NMR analysis of a larger amount of the isolated impurity immediately revealed, from the chemical shift of the aromatic pyridyl signals, that the pyridyl nitrogen had reacted with residual acrylic acid monomers from the glue in the syringe to form an *N*-pyridyl derivative by a Michael condensation. Under the conditions of the mass spectrometer, this impurity reverted to the original peptide by a β-elimination. This clearly demonstrates the strength of the nondestructive nature of NMR spectroscopy. NMR spectroscopy is therefore an excellent technique for root-cause analysis and troubleshooting rather than routine analysis.[40]

11.5 Higher-order Structure

Most small linear hydrophilic peptides fail to adopt a defined three-dimensional structure in water.[4] (Deuterated) trifluoroethanol,[41,42] cryoprotective solvents[43–45] and micelles[46,47] have been widely used to induce/stabilize higher-order structures of peptides that otherwise would exist as a random coil. Basically, four indicators of the presence of secondary and tertiary structure in peptides can be detected by NMR spectroscopy:[4,5]

1. Non-random coil chemical shifts (chemical shift dispersion): ^{1}H NMR chemical shifts are extremely sensitive to peptide conformation due to anisotropy. An example is given in Figure 11.26.

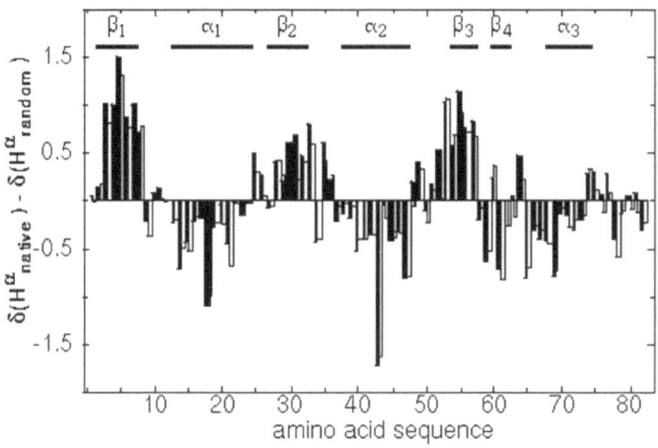

Figure 11.26 Plot of the differences between the observed α-proton chemical shifts and the corresponding random coil values *versus* the amino acid sequence of glutaredoxin 3.
Reproduced from ref. 48 with permission from American Society for Biochemistry and Molecular Biology, Copyright © 1996 by The American Society for Biochemistry and Molecular Biology, Inc.

2. Non-average three-bond *J* couplings, (chemical shift dispersion): exclusively averaged *J* couplings are expected to occur in random coil peptides. Non-random coil *J* couplings indicate secondary structure.
3. (2D) NOE/ROE contacts: these arise if protons are $< \sim 5$ Å apart. If these protons are more than one residue apart in the peptide sequence, this suggests the presence of secondary and tertiary structure elements.
4. (Amide) NH and OH protons will readily exchange with deuterium in a deuterated solvent containing exchangeable protons such as D_2O. Residual signals due to unexchanged protons can be due to NH and/or OH protons protected from exchange because they are involved in strong hydrogen bonds or protected from exchange because they are inaccessible to exchange since they are buried in the interior of the structure.

The determination of the three-dimensional structure of peptides and proteins by heteronuclear, multidimensional NMR spectroscopy has reached maturity and has been the subject of several books[3,11] and reviews[14,49] and is outside the scope of this chapter. The reader is referred to the literature cited.

Aggregation of peptides has been evaluated by concentration dependence, relaxation times and decreased diffusion rates using pulsed field gradients.[50–52]

The 600 MHz ^{1}H ^{1}D spectrum of a 17 mg ml^{-1} solution of oxytocin shown in Figure 11.27 displayed small, additional signals at *ca.* 7 ppm. These signals increased with concentration, which rules out proline *cis*/*trans* isomers and/or impurities.

The 2D ^{1}H DOSY shown in Figure 11.28 obtained on the sample showed that the additional signals corresponded to species with a longer diffusion time. Also different NOE buildup curves were observed for the assumed aggregate (Figure 11.29).

The concentration dependence, the increased diffusion time and NOE buildups of the additional signals are all consistent with the presence of aggregated oxytocin material in this samples.

11.6 Conclusion

This chapter has described the application of NMR spectroscopy as a comprehensive technique to determine the identity, content, purity and structure of synthetic therapeutic peptides, including a number of practical examples. However, NMR spectroscopy still suffers from a number of popular prejudices with "traditional" analytical chemists, regretfully limiting its use. NMR spectroscopy is thought to be expensive, insensitive, slow, difficult, specialist and not quantitative. However, the following points should be taken into account:[40,53]

- *Cost*:
  Thanks to highly developed automation, 24/7 equipment operation. Electronics can be upgraded.

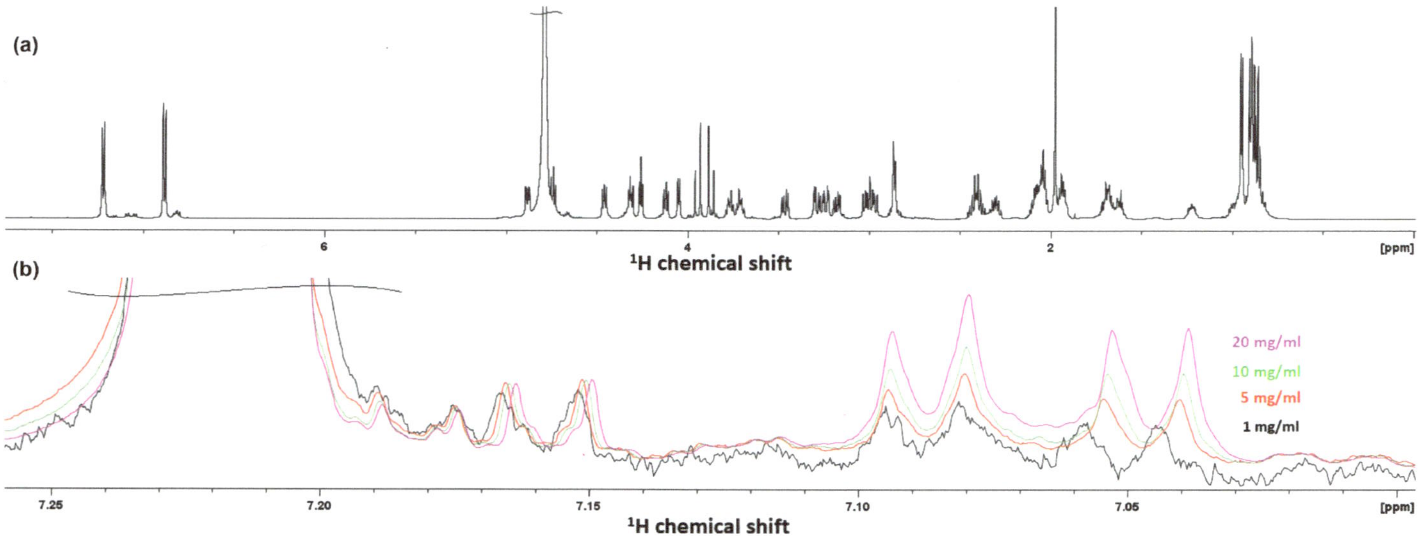

Figure 11.27 (a) The 600 MHz ^{1}H 1D spectrum of a 17 mg ml^{-1} solution of oxytocin. Additional, minor signals were noted at ~7 ppm. In (b), overlay spectra acquired at different concentrations show that the small signals at 7.03–7.10 ppm clearly increase with increasing sample concentration, while other minor peaks remain unaffected. The Tyr main peak at 7.22 ppm was used for normalization.

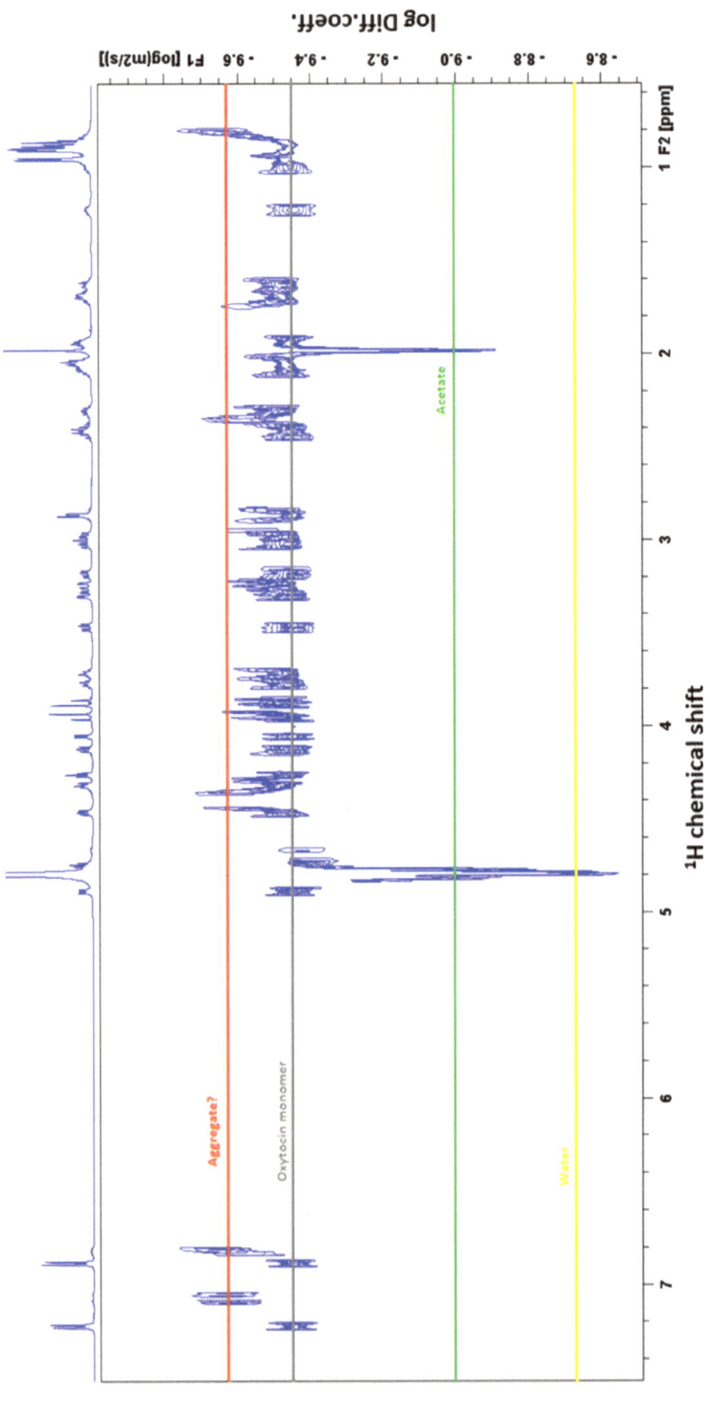

Figure 11.28 2D ^{1}H DOSY spectrum obtained on the sample from Figure 11.27.

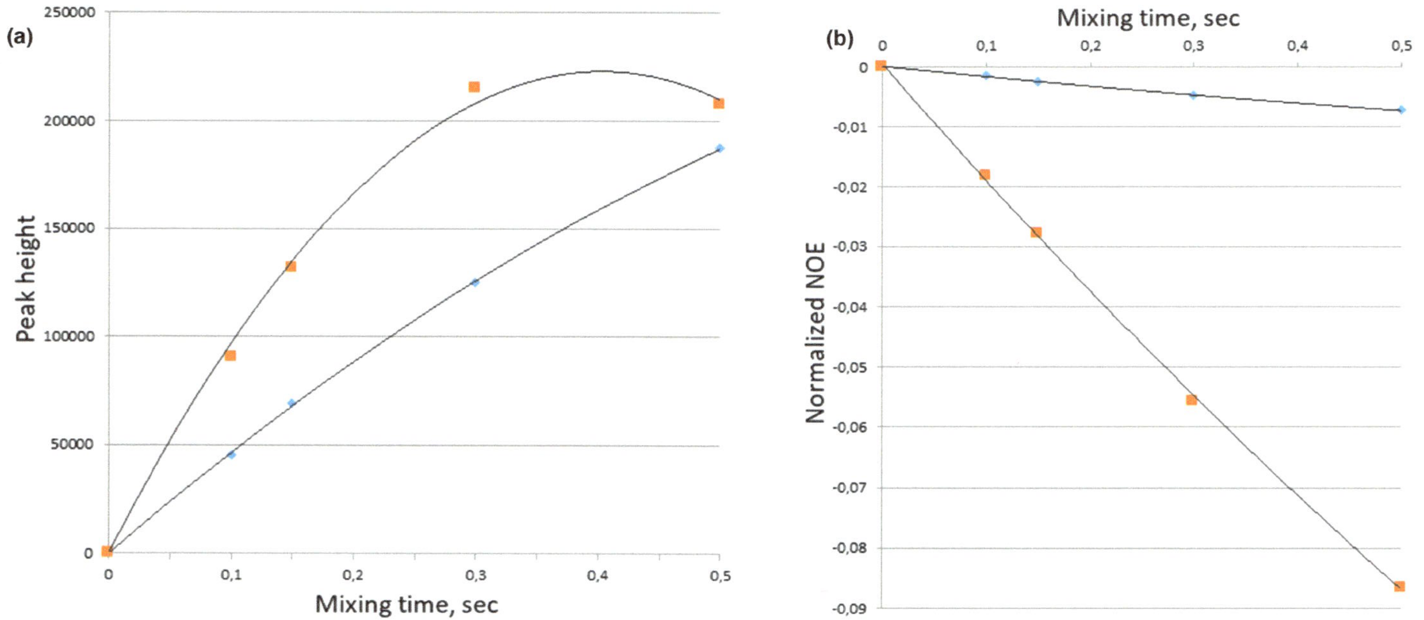

Figure 11.29 NOE build-up curves for oxytocin in D_2O at 600 MHz and 25 °C: Tyr-Ar (~7 ppm) to Tyr-CH_2 (~3 ppm). Peak heights for monomer (orange) and aggregate (blue) cross-peaks at mixing times of 100, 150, 300, 500 ms. (a) shows absolute intensities and (b) shows relative intensities that have been normalized against the auto-peak. Note that the normalized NOE factors are negative by definition; close to zero for oxytocin monomer but distinctly negative for the aggregate.

Magnet lasts for decades.
No or limited sample pretreatment.

- *Complexity*:
  Analysis of intact samples *without* extensive method development.
  In-depth knowledge not required for application of NMR spectroscopy.
- *Quantitation*:
  Accuracy and precision as good as in chromatographic techniques.
  Long-term robustness excellent.
  No problems due to variations introduced by the use of different columns.
  No response factors required.
- *Sensitivity*:
  There are many examples of NMR detection of trace molecules where NMR spectroscopy not only confers sensitivity but also selectivity. NMR spectra can routinely be obtained on milligrams of sample but also, with some more effort, on sample quantities several orders of magnitude lower.
- *Speed*:
  Many 1D and 2D NMR techniques can yield spectra within minutes due to technological developments (*e.g.* cryoprobes).

Hence NMR spectroscopy is an excellent and general tool for the determination of the identity, content and purity of therapeutic peptides. It is expected that in the future other, more advanced NMR methods will be implemented (*e.g.*, multidimensional and/or heteronuclear) enabling further application of the use of NMR in the analysis of therapeutic peptides and their formulations.

References

1. K. Wüthrich, *NMR in Biological Research: Peptides and Proteins*, North-Holland, Amsterdam, 1st edn, 1976.
2. R. A. Dwek, *Nuclear Magnetic Resonance in Biochemistry*, ed. W. Harrington and A. R. Peacock, Clarendon Press, Oxford, 1st edn, 1973, pp. 44–54.
3. J. Cavanagh, W. J. Fairbrother, A. G. Palmer III, M. Rance and N. J. Skelton, *Protein NMR Spectroscopy: Principles and Practice*, Academic Press, New York, 2nd edn, 2010.
4. J. G. Beck, A. O. Frank and H. Kessler, *NMR of Biomolecules: Towards Mechanistic Systems Biology*, ed. I. Bertini, K. S. McGreevy and G. Parigi, Wiley and Blackwell, Chichester, 1st edn, 2012, pp. 328–344.
5. B. Valentine, A. Steinschneider, D. Dhawan, M. I. Burgar, T. S. T. Amour and D. Fiat, *Int. J. Pept. Protein Res.*, 1985, **25**, 56.
6. R. Hoffmann, I. Reichert, W. O. Wachs, M. Zeppezauer and H. R. Kalbitzer, *Int. J. Pept. Protein Res.*, 1994, **44**, 193.
7. J. T. Gerig, *Prog. Nucl. Magn. Reson. Spectrosc.*, 1994, **26**, 293.

8. P. Mykhailiuk, S. Afonin, G. V. Palamarchuk, O. V. Shishkin, A. S. Ulrich and I. V. Komarov, *Angew. Chem., Int. Ed. Engl.*, 2008, **47**, 5765–5767.
9. S. Afonin, R. W. Glaser, M. Berditchevskaia, P. Wadhwani, K.-H. Gührs, U. Möllmann, A. Perner and A. S. Ulrich, *ChemBioChem*, 2003, **4**, 1151–1163.
10. D. H. Spackman, W. H. Stein and S. Moore, *Anal. Chem.*, 1958, **30**, 1190.
11. European Pharmacopoeia 9.8 General Chapter 2.2.64 Peptide Identification by Nuclear Magnetic Resonance Spectroscopy.
12. United States Pharmacopeia 41 Oxytocin and Goserelin Acetate monographs.
13. E. Kellenbach, *et al.*, *NMR Spectroscopy in Pharmaceutical Analysis*, ed. U. Holzgrabe, I. Wawer and B. Diehl, Elsevier, Amsterdam, 1st edn, 2008, ch. 5, pp. 429–436; E. Kellenbach, *et al.*, *Pharmeuropa*, 2008, **2008**, 1–8.
14. K. Wüthrich, *NMR of Proteins and Nucleic Acids*, John Wiley and Sons, New York, 1st edn, 1986.
15. A. Bundi and K. Wüthrich, *Biopolymers*, 1979, **18**, 285.
16. D. S. Wishart and B. D. Sykes, *Methods Enzymol.*, 1994, **239**, 363.
17. G. Maniara, K. Rajamoorthi, S. Rajan and G. W. Stockon, *Anal. Chem.*, 1998, **70**, 4921.
18. F. Malz and H. Jancke, *J. Pharm. Biomed. Anal.*, 2005, **38**, 813.
19. Y. B. Monakhova and B. W. K. Diehl, *J. Pharm. Biomed. Anal.*, 2018, **154**, 332.
20. A. Preiß, J. Kruppa, J. Buschmann and C. Mügge, *J. Pharm. Biomed. Anal.*, 1998, **16**, 1381.
21. D. I. Freedberg, Y. Aubin and C. Jones, *BioPharm Int*, 2010, **6**(Supplement August), 28.
22. C. W. Funke, J.-R. Mellema, P. Salemink and G. N. Wagenaars, *J. Pharm. Pharmacol.*, 1988, **40**, 78.
23. J. A. Hoffmann, R. E. Chance and M. G. Johnson, *Protein Expression Purif.*, 1990, **1**, 127.
24. A. C. Gucinski, M. T. Boyne II and D. A. Keire, *Anal. Bioanal. Chem.*, 2015, **407**, 749.
25. U. Holzgrabe, R. Deubner, C. Schollmayer and B. Waibel, *J. Pharm. Biomed. Anal.*, 2005, **38**, 806; M. Malet-Martino and U. Holzgrabe, *J. Pharm. Biomed. Anal.*, 2011, **55**, 1.
26. C. K. Larive, D. Jayawickrama and L. Orfi, *Appl. Spectrosc.*, 1997, **51**, 1531.
27. European Pharmacopoeia 9.8 General Chapter 2.2.33. Nuclear magnetic resonance spectrometry.
28. United States Pharmacopoeia 41 General Chapter: <761> Nuclear Magnetic Resonance Spectroscopy.
29. D. J. Craik and K. A. Higgins, *Pept. Res.*, 1991, **4**, 177.
30. A. D. Kline and R. M. Justice Jr., *Biochemistry*, 1990, **29**, 2906.
31. C. Grathwohl and K. Wüthrich, *Biopolymers*, 1976, **15**, 2025.
32. W. de Wildt, H. Kooijman, C. Funke, B. Üstün, A. Leika, M. Lunenburg, F. Kaspersen and E. Kellenbach, *Molecules*, 2017, **22**, 1362.

33. K. Hu, W. M. Westler and J. L. Markley, *J. Am. Chem. Soc.*, 2011, **133**, 1662–1665.
34. C. Li, *et al.*, *J. Pharm. Biomed. Anal.*, 2019, **166**, 105.
35. K. Skidmore, D. Hewitt and Y.-H. Kao, *Biotechnol. Prog.*, 2012, **28**, 1526.
36. N. Magarian, K. Lee, K. Nagpal, K. Skidmore and E. Mahajan, *Biotechnol. Prog.*, 2016, **32**, 718.
37. Y.-H. Kao, J. Bender, A. Hagewiesche, P. Wong, Y. Huang and M. Vanderlaan, *PDA J. Pharm. Sci. Technol.*, 2001, **55**, 268.
38. E. Kellenbach, M. Burgering and F. Kaspersen, *Org. Process Res. Dev.*, 1999, **3**, 141.
39. United States Pharmacopoeia 41. L-Alanyl-L-glutamine monograph.
40. E. Kellenbach and P. Dani, *NMR in Pharmaceutical Sciences*, ed. J. R. Everett, R. K. Harris, J. C. Lindon and I. D. Wilson, John Wiley and Sons, Chichester, 2015, ch. 31, pp. 441–451.
41. M. Buck, *Q. Rev. Biophys.*, 1998, **31**, 297.
42. A. Jasanoff and A. R. Fersht, *Biochemistry*, 1994, **33**, 2129.
43. A. Motta, D. Picone, T. Tancredi and P. A. Temussi, *Tetrahedron*, 1988, **44**, 975.
44. R. Spadaccini, O. Crescenzi, D. Picone, T. Tancredi and P. A. Temussi, *J. Pept. Sci.*, 1999, **5**, 306.
45. D. Picone, A. D'Ursi, A. Motta, T. Tancredi and P. A. Temussi, *Eur. J. Biochem.*, 1990, **192**, 433.
46. D. A. Kallick, M. R. Tessmer, C. R. Watts and C. Y. Li, *J. Magn. Reson., Ser. B*, 1995, **109**, 60–65.
47. M. Lindberg, J. Jarvet, Ü. Langel and A. Gräslund, *Biochemistry*, 2001, **40**, 3141.
48. F. Åslund, K. Nordstrand, K. D. Berndt, M. Nikkola, T. Bergman, H. Ponstingl, H. Jörnvall, G. Otting and A. Holmgren, *J. Biol. Chem.*, 1996, **271**, 6736.
49. D. Wishart, *Curr. Pharm. Biotechnol.*, 2005, **6**, 105.
50. S. L. Mansfield, D. A. Jayawickrama, J. S. Timmons and C. K. Larive, *Biochim. Biophys. Acta, Protein Struct. Mol. Enzymol.*, 1998, **1382**, 257.
51. S. L. Mansfield, A. J. Gotch, G. S. Harms, C. K. Johnson and C. K. Larive, *J. Phys. Chem. B*, 1999, **103**, 2262.
52. M. Iqbal and P. Balaram, *Biopolymers*, 1982, **21**, 1427.
53. A. Phillips, *Genotoxic Impurities: Strategies for Identification and Control*, ed. A. Teasdale, John Wiley and Sons, Hoboken, New Jersey, 1st edn, 2010, ch. 13, 351–384.

CHAPTER 12

Transport, Stability and Delivery Considerations for the Design of Peptide Drugs

TOMI SAWYER,[a] JEROME HOCHMAN,[b] SUZANNE M. D'ADDIO,[a] BRAD SHERBORNE[a] AND MARIAN E. GINDY*[b]

[a] MRL, Merck & Co., Inc., 33 Avenue Louis Pasteur, Boston, MA 02115, USA; [b] MRL, Merck & Co., Inc., 770 Sumneytown Pike, West Point, PA 19486, USA

*Email: marian_gindy@merck.com

12.1 Introduction to Peptide Drugs

12.1.1 Therapeutic Opportunities

Several hundred peptide molecules have been entered into clinical trials for various therapeutic indications[1–12] (*e.g.* endocrine, metabolic, cardiovascular, cancer, immune and central nervous system diseases) and many have been approved by the US Food and Drug Administration (FDA) (Table 12.1). Furthermore, several peptide drugs have achieved significant market success, including Lantus® (insulin glargine), Novolog® (insulin aspart), Humalog® (insulin lispro), Levemir® (insulin detemir), Humulin® (insulin), Victoza® (liraglutide), Lupron® (leuprolide), Zoladex® (goserelin acetate), Sandostatin® (octreotide), Byetta® (exenatide) and Forteo ® (teriparatide). Beyond the typical challenges to understanding the structure–activity relationships of such

Drug Discovery Series No. 72
Peptide Therapeutics: Strategy and Tactics for Chemistry, Manufacturing and Controls
Edited by Ved Srivastava

Published by the Royal Society of Chemistry, www.rsc.org

Table 12.1 Some marketed peptide drugs.

	Therapeutic target				
Trade name (INN)	Extracellular/receptor	Intracellular	Antibiotic peptides that disrupt bacterial membranes	Primary use	Approval
DDAVP (desmopressin)	Vasopressin receptor			Diabetes insipidus	1978 (FDA)
Sandimmune (ciclosporin)		Cyclophilin/ calcineurin		Immunotherapy	1983 (FDA)
Lupron (leuprorelin)	GnRH receptor			Oncology	1985 (FDA)
Zoladex (goserelin)	GnRH receptor			Oncology	1989 (FDA)
Invirase (saquinavir)		HIV-1 protease		Infectious disease	1996 (FDA)
Copaxone (glatiramer)	T-cell function			Allergy, immunology	1996 (FDA)
Crixivan (indinavir)		HIV-1 protease		Infectious disease	1996 (FDA)
Viracept (nelfinavir)		HIV-1 protease		Infectious disease	1997 (FDA)
GlucaGen (recombinant glucagon)	Glucagon receptor			Metabolic	1998 (FDA)
Integrilin (eptifibatide)	Integrin receptor			Cardiovascular	1998 (FDA)
Sandostatin (octreotide)	Somatostatin receptor			Acromegly	1998 (FDA)
Angiomax (bivalirudin)	Thrombin			Hematology	2000 (FDA)
Agenerase (amprenavir)		HIV-1 protease		Infectious disease	1999 (FDA)
Cetrotide (cetrorelix)	GnRH receptor			Endocrinology	2000 (FDA)
Trelstar (triptorelin)	GnRH receptor			Oncology	2000 (FDA)
Natrecor (nesiritide)	Natriuretic peptide receptor			Cardiovascular	2001 (FDA)
Byetta (exenatide)	Glucagon-like peptide-1 receptor			Metabolic	2002 (FDA)
Forteo (teriparatide)	Parathyroid hormone receptor			Metabolic	2002 (FDA)
Neulasta (pegfilgrastim)	G-CSF receptor			Oncology	2002 (FDA)
Reyataz (atazanavir)		HIV-1 protease		Infectious disease	2003 (FDA)
Cubicin (daptomycin)			Bacterial cell membrane	Antibacterial	2003 (FDA)

Fuzeon (enfuvirtide)	gp41 of HIV fusion complex		Infectious disease	2003 (FDA)
Plenaxis (abarelix)	GnRH receptor		Oncology	2003 (FDA)
Velcade (bortezomib)		26S proteasome	Oncology	2003 (FDA)
Prialt (ziconotide)	N-type calcium channel		Central nervous system	2004 (FDA)
Symlin (pramlintide)	Amylin receptor		Metabolic	2005 (FDA)
Vantas (histrelin)	GnRH receptor		Oncology	2005 (FDA)
Prezista (darunavir)		HIV-1 protease	Infectious disease	2006 (FDA)
Somatuline (lanreotide)	Somatostatin receptor		Endocrinology	2007 (FDA)
Firmagon (degarelix)	GnRH receptor		Oncology	2009 (FDA)
Victoza (liraglutide)	Glucagon-like peptide-1 receptor		Metabolic	2010 (FDA)
Linzess (linaclotide)	Guanylate cyclase receptor		Gastrointestinal	2012 (FDA)
Signifor (pasireotide)	Somatostatin receptor		Cushing's disease	2012 (FDA)
Gattex teduglutide	Glucagon-like peptide-2 receptor		Gastrointestinal	2012 (FDA)
Kyprolis carfilzomib)		Proteasome	Oncology	2012 (FDA)
Scenesse (afamelanotide)	Melanocortin-1 receptor		Skin pigmentation	2014 (EMA)
Afrezza (insulin inhaled)	Insulin receptor		Diabetes	2014 (FDA)
Saxendra (liraglutide)	Glucagon-like peptide-1 receptor		Metabolic	2014 (FDA)
Trulicity (dulglutide)	Glucagon-like peptide-1 receptor		Metabolic	2014 (FDA)
Ninlaro (ixazomib)		Proteasome	Oncology	2015 (FDA)
PABAL (carbetocin)	Oxytocin receptor		Obstetrics	2015 (EMA)
Natpara (parathyroid hormone)	Parathyroid hormone receptor		Hypocalcemia	2015 (FDA)
Toujeo (insulin glargine)	Insulin receptor		Diabetes	2015 (FDA)
Tresiba (insulin degludec)	Insulin receptor		Diabetes	2015 (FDA)
Adlyxin (lixisenatide)	Glucagon-like peptide-1 receptor		Metabolic	2016 (FDA)
Zepatier (grazoprevir)		HCV protease	Infectious disease	2016 (FDA)

molecules at their specific targets and the *in vitro* to *in vivo* translation in preclinical disease models, the discovery of peptide drugs involves overcoming metabolic instability (*e.g.* degradation by proteolytic enzymes), undesirable pharmacokinetic properties, extremely poor intestinal permeability relative to oral delivery and/or cellular permeability in terms of prosecuting intracellular targets. Despite these challenges, there is an extraordinary future prospect for peptide drugs in terms of expanding druggable target space using design strategies that leverage the chemical diversity available to this class of compounds while incorporating emerging predictive "rules" for structural modifications conferring metabolic stability and/or enhanced cellular permeability properties.

12.1.1.1 Receptor Target Space for Peptide Agonists and Antagonists

Unquestionably, the major impact of peptide drug discovery to date has been realized for receptor target space. For example, G protein-coupled receptors (GPCRs, *e.g.* class A and B GPCRs) have provided the first and continuing surge of GPCR-targeted peptide therapeutics (Figure 12.1).[1–12] In retrospect, ground-breaking studies on class A GPCR peptides include oxytocin, vasopressin, α-melanocyte-stimulating hormone (α-MSH), gonadotropin-releasing hormone (GnRH), somatotropin-release inhibiting factor (SRIF, somatostatin) and the opioid peptides (*e.g.* enkephalin, β-endorphin, dynorphin) during the 1970s–1990s. More recently, another pulse of GPCR-targeted peptide drug discovery has successfully extended to class B GPCRs as exemplified by glucagon, glucagon-like peptide-1 (GLP-1), islet amyloid polypeptide (IAPP, amylin), glucagon-like peptide-2 (GLP-2) and parathyroid hormone (PTH). In the representative case of the GLP-1 receptor, several peptide agonists have reached the market and representative examples include exenatide liraglutide, lixisenatide and semaglutide.

SCENESSE® (*afamelanotide*)
Ac-Ser-Tyr-Ser-Nle-Glu-His-*D*-Phe-Arg-Trp-Gly-Lys-Pro-Val-NH_2

LUPRON® (*leuprorelin*)
<Glu-His-Trp-Ser-Tyr-*D*-Leu-Leu-Arg-Pro-NHEt
(<Glu = pyroglutamyl)

SANDOSTATIN® (*octreotide*)
Phe-Cys-Phe-*D*-Trp-Lys-Thr-Cys-Ser-ol

BYETTA® (*exenatide*)
His-Gly-Glu-Gly-Thr-Phe-Thr-Ser-Asp-Leu-Ser-Lys-Gln-Met-Glu-Glu-Glu-Ala-Val-Arg-Leu-Phe-Ile-Glu-Trp-Leu-Lys-Asn-Gly-Gly-Pro-Ser-Ser-Gly-Ala-Pro-Pro-Pro-Ser-NH_2

FORTEO® (*teriparatide*)
Ser-Val-Ser-Glu-Ile-Gln-Leu-Met-His-Asn-Leu-Gly-Lys-His-Leu-Asn-Ser-Met-Glu-Arg-Val-Glu-Trp-Leu-Arg-Lys-Lys-Leu-Gln-Asp-Val-His-Asn-Phe-OH

GATTEX® (*teduglutide*)
His-Gly-Asp-Gly-Ser-Phe-Ser-Asp-Glu-Met-Asn-Thr-Ile-Ile-Leu-Asp-Asn-Leu-Ala-Ala-Arg-Asp-Phe-Ile-Asn-Trp-Ile-Gln-Thr-Lys-Ile-Thr-Asp-OH

Figure 12.1 Some examples of GPCR class A and class B peptide drugs highlighting secondary structure motifs (*e.g.* β-turn or α-helix) that are key for their cognate receptor binding.

Relative to drug design, these early class A and B GPCR peptide agonist and/or antagonist structure–activity studies provided insight into understanding the intrinsic peptide conformational properties and also predictive 3D pharmacophore models.[7–12] Unfortunately, such early studies were not empowered by high-resolution X-ray crystallographic structures of class A and class B GPCRs until more recently.[13,14] Instead, a systematic analysis of peptide structure–activity relationships (*e.g.* analog modifications by D-amino acids, *N*-alkylamino acids, Cα-alkylamino acids and/or macrocyclization) and also biophysical characterization (*e.g.* by NMR and circular dichroism spectroscopy) revealed the secondary structure motifs essential for GPCR molecular recognition with such molecules. As illustrated by afamelanotide, leuprolide and octreotide, incorporation of D-amino acids regiospecifically within their peptide sequences has been demonstrated to stabilize β-turn (reverse-turn) conformations of their respective class A GPCR agonist peptides (Figure 12.1).[15–17] On the other hand, exenatide, teriparatide and teduglutide share a relatively high propensity for α-helix conformations that have been correlated with binding and activation of their respective class B GPCRs (Figure 12.1).[18–20] Of significance to class B GPCRs, the first X-ray crystallographic structures of GLP-1 receptor extracellular domain complexes with GLP-1 peptide analogs have provided the 3D molecular insight into their receptor binding mechanism.[21]

Beyond GPCRs, there have been both significant achievements and increasing momentum in the development of peptide modulators of growth factor receptors (*e.g.* the insulin receptor and bioengineered insulin agonists thereof) and also cytokine receptors, integrins and ion channels to expand the scope of receptor target space for peptide drug discovery. Noteworthy for such targets are their more structurally complex peptide agonists and antagonists, including those having multiple disulfide bridges such as for insulin analogs, linaclotide, ziconotide and ProTx-II, which exemplify high specificity for the insulin receptor, guanylate cyclase-C, N-type Ca^{2+} channel and voltage-gated Na^{+} channel targets, respectively.[22–25] Unquestionably, insulin analogs as a class dominate the existing peptide drug marketplace,[26] including long-acting insulin agonists (*e.g.* glargine, determir and degludec), intermediate-acting insulin agonists (*e.g.* NPH insulin, Humulin) and rapid-acting insulin agonists (*e.g.* insulin lispro, insulin aspart and insulin glulisine).

12.1.1.2 Intracellular Target Space for Cell Permeable Peptide Modulators

In the 1980s, the discovery of the macrocyclic peptide and natural product cyclosporine A (CsA) and its modulation of cyclophilin–calcineurin complex formation[27] demonstrated the potential of leveraging the peptide modality for intracellular targets (Figure 12.2). The concurrent design of peptidomimetic inhibitors of proteases[28–30] (*e.g.* HIV protease) and the identification of cell-penetrating peptides (CPPs) provided further impetus in this second wave of peptide drug discovery to advance the prosecution of compelling intracellular target space.[31–33]

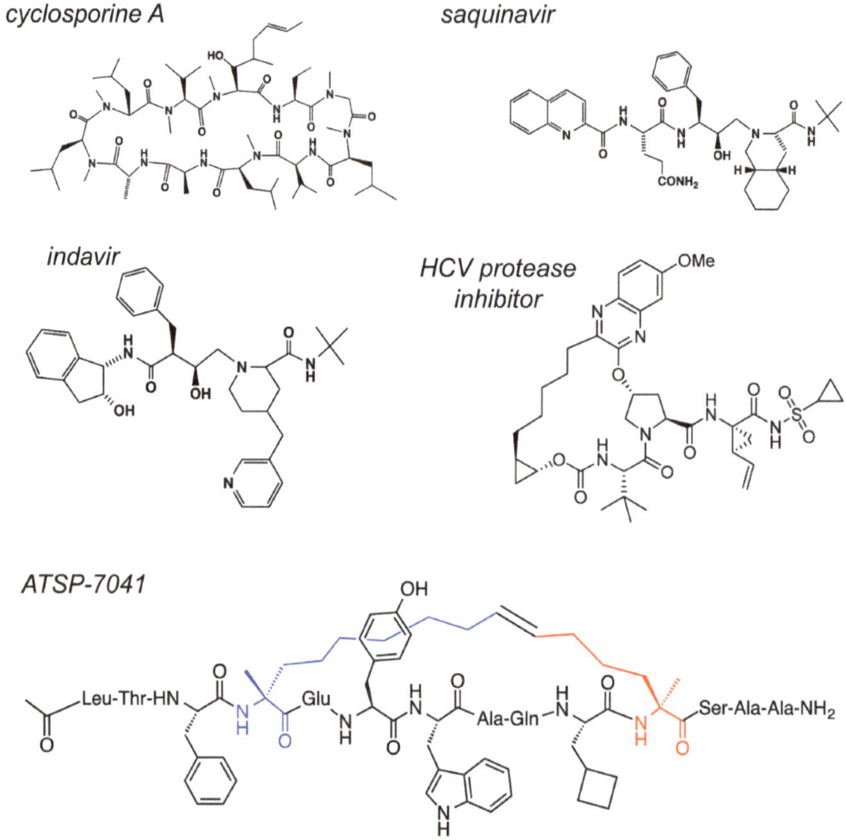

Figure 12.2 Some examples of peptides and peptidomimetics for intracellular targets, including protein–protein interactions and proteases.

Importantly, and especially for the intracellular target space, the convergence of *de novo* identification of peptide ligands using phage, mRNA and synthetic library technologies with sophisticated structure-based design is enabling the third wave of peptide drug discovery.[1,34–44] These biology- and chemistry-driven methods to advance and potentially define an extraordinarily rich chemical space relative to varying ring size, amino acid composition and design for conformational bias (*e.g.* α-helical, β-strand and reverse-turn secondary structures) have created a focus on macrocyclic peptides (see later). This is balanced with the natural propensity for α-helix and β-strand secondary structures that are widely found mediating the interfaces of protein–protein interactions (PPIs), as has been described[45–50] in comprehensive analyses of the Protein Data Bank (www.rcsb.org). Some notable α-helical protein–protein interactions of the more than 1600 non-redundant, unique high-resolution 3D structures derived from the Protein Data Bank[46–50] include BIM BH3:Mcl-1,[51] BAD:Bcl-Xl,[52] p53:MDM2,[53] MAML:Notch,[54] HIF-1α:p300,[55] Myc-Max,[56] eIF4G:eIF4E,[57] TIF2:GCCR,[58] Scr-1:RORγ[59] and IKKβ:NEMO.[60] Specifically, ALRN-6924 has successfully advanced into clinical development as a novel macrocyclic α-helical peptide antagonist of MDM2 and MDMX for p53 wild-type cancers, marking a milestone for such efforts focused on intracellular PPI targets and overcoming the typical barriers such as cell permeability, stability and pharmacokinetics facing peptide drug discovery.

Early peptidomimetic design and chemistry strategies focused on amide surrogates; for example, replacement of the scissile replacement of the scissile C(=O)NH by $CH(OH)CH_2$, CH_2NH and $CH(OH)CH_2NH$ moieties as incorporated in the first-generation inhibitors of intracellular proteases (*e.g.* HIV protease).[28–30] Further modifications of such protease substrate-based inhibitors led to the identification of the first marketed protease inhibitors (*e.g.* amprenavir, indinavir, sasquinavir, darunavir, atazanavir and nelfinavir), leveraging removal or replacement of additional amides by non-hydrolyzable moieties to reduce molecular size and H-bonding and/or by hydrophilic/ionic functional groups (except for those promoting solubility).[28–30] In recent years, peptidomimetic inhibitors of other intracellular proteases such as proteasome [*e.g.* Velcade® (bortezomib), Kyprolis® (carfilzomib) and Ninlaro® (ixazomib)][61] and HCV protease [*e.g.* Zepatier® (grazoprevir)][62] have been advanced into the clinic and achieved marketing approval.

CPPs leveraging the identified intracellular delivery properties of the trans-activator of transcription protein of human immunodeficiency virus type-1 (Tat) and Antennapedia[63–65] were prototypes of a plethora of recently discovered, naturally occurring or designed CPPs (Table 12.2) that deliver cargos including peptides, small molecules, macromolecules (*e.g.* nucleic acids, proteins), nanoparticles and liposomes into cells by way of chemical conjugation or noncovalent complexation.[66–76] In most cases, the CPP sequence has been 8–30 amino acids in length and either cationic or amphiphilic. Furthermore,

Table 12.2 Some representative CPPs (linear and macrocyclic).[68,98,376]

Name	Sequence	Origin	Class
HIV-1 TAT	RKKRRQRRR	HIV-1 TAT protein	Cationic
Penetratin	RQIKIWFQNRRMKWKK	Antennapedia *Drosophila melanogaster*	Cationic
Poly-Arg	Rn	Synthetic	Cationic
Pep-1	KETWWETWWTEWSQPKKKRKV	Trp-rich cluster/ SV40 T antigen NLS	Amphipathic
pVEC	LLIILRRRIRKQAHAHSK	Vascular endothelial cadherin	Amphipathic
MAP	KLALKLALKALKAALKLA	Synthetic	Amphipathic
Transportan	GWTLNSAGYLLGKINLKALAALAKKIL	Galanin–mastoparan chimer	Amphipathic
C105Y	CSIPPEVKFNKPFVYLI	α_1-Antitrypsin	Hydrophobic
PFVYLI	PFVYLI	Synthetic (C105Y derived)	Hydrophobic
Pep-7	SDLWEMMMVSLACQY	CHL8 peptide phage clone	Hydrophobic
Pericyclic CPP	Cyclo(Ff-βNal-RrRf-Q)[a]	Synthetic	Cationic

[a]Abbreviations: f, D-Phe; r, D-Arg; βNal, 2-naphthylalanine.

several CPP–peptide conjugates have been advanced to clinical trials,[68] including D-JNKI1 (brimapitide), KAI-9803, AZX100, RT001 and AVI-4628. D-JNKI1 is an all-D-amino acid-modified peptide that includes a cargo consisting of the retro-inverso analog of a peptide sequence from the JNK binding domain of JNK-interacting protein, which was conjugated with a retro-inverso analog of TAT-derived CPP.[77] KAI-9803 is a peptide that includes a cargo derived from the dV1-1 sequence of δ-protein kinase C (δ-PKC), which was conjugated by a disulfide-bridged, reversible conjugation with a TAT-derived CPP.[78] AZX100 is comprised of a phosphorylated peptide from heat-shock-related protein-20 as the cargo that is conjugated with PTD-4 as the CPP.[79]

12.1.1.3 Assumptions for Druggability with a Peptide Modality

The diverse physiological and pathophysiological roles of peptide–target interactions provide potential opportunities for therapeutic intervention in a wide range of diseases. In retrospect, peptide drug development faced substantial challenges in achieving its full potential owing to inherent limitations in overcoming susceptibility to proteolytic degradation, cell permeability and achieving suitable *in vivo* pharmacokinetic profiles. Unsurprisingly, in part owing to such challenges, the first wave of peptide drug discovery was essentially restricted to extracellular and receptor target space (*e.g.* G protein-coupled receptor agonists or antagonists and insulin receptor agonists). The second wave of peptide drug discovery further leveraged structure-based design, peptidomimetic chemistry and prototype macrocyclic peptides and CPPs to overcome some of the inherent limitation of peptides and expand the druggability of intracellular target space. Such pioneering work significantly advanced the peptide modality, in terms of both basic science and marketed peptide therapeutics.

The assumptions for further increasing the druggable target space with a peptide modality include the following: (1) a peptide modality has unique properties in terms of high affinity/selectivity, *in vivo* efficacy and safety that are compelling *versus* other modality options; (2) a peptide modality may be optimized through chemical modifications to achieve stability, permeability and related pharmacokinetic properties to achieve a viable and marketable drug for the disease and patient population for which it is intended; and (3) the target being prosecuted may be validated in preclinical *in vitro* and *in vivo* models in proof-of-concept studies that further show the promise and possible superiority of a peptide modality *versus* other therapeutic modalities.

The third wave of peptide drug discovery (see below) will further test such assumptions with yet a greater scale of chemical diversity, structural biology and computational modeling and also drug delivery enabling technologies.

12.1.2 Peptide Ligand Discovery

Perhaps most significantly, there has been an emerging third wave of peptide drug discovery over the past decade that is leveraging structure-based design and/or superdiversity methods to advance macrocyclic peptides (*e.g.* stapled α-helical peptides[50,80–85]), CsA-inspired and related monocyclic peptides,[86–92] bicyclic peptides[93–96] and cyclic CPP-hybrid peptides[97,98] using synthetic and/or biological methods (see later).

12.1.2.1 Naturally Occurring Peptides

Naturally occurring peptides have provided many of the first benchmarks to our understanding of target function in both physiological and disease states. The first and second waves of peptide drug discovery (see above) have exemplified the iterative modification of such peptide sequences to achieve preclinical proof-of-concept studies and further advancement into clinical trials. The sources of such naturally occurring peptides include peptide hormones, neurotransmitters, growth factors, cytokines and related native proteins (extracellular or intracellular) from which a peptide fragment may possess both structural and functional properties that mimic the protein ligand at its cognate target. Beyond endogenous (human) sources, peptide benchmarks from other animal species, microorganisms and plant life have been collectively essential to provide what one may call "natural diversity" with respect to the peptide modality. Examples of nonhuman peptides that provide key starting points for peptide drug discovery include teprotide (angiotensin-converting enzyme inhibitor),[99] pepstatin (aspartyl protease inhibitor),[100] CsA (calcineurin inhibitor),[101] penicillin (antibiotic),[102] vancomycin (antibiotic),[103] daptomycin (antibiotic),[104] ziconotide (N-type Ca^{2+} channel blocker)[105] and ProTx-II (Na^{+} channel Nav1.7 blocker).[106] In many instances, importantly, Nature has provided examples that overcome (or avoid) potential development hurdles, including transport or localization sequences and structural features that achieve broad stability against proteolytic metabolism.

12.1.2.2 mRNA and Phage Display Peptide Libraries

Leveraging biology to create superdiverse libraries with self-identifying tagging and/or self-replicating methods such as DNA, mRNA and phage display has emerged to become an extremely powerful strategy to identify and optimize linear and macrocyclic peptides for an increasing scope of compelling target space.[86,88–91,93–96] In the case of both mRNA and phage display library screening, the *de novo* identification of novel macrocyclic peptides binding to targets in either an orthosteric or unprecedented allosteric binding mode may be achieved.

12.1.2.3 Synthetic and Chemically Encoded Peptide Libraries

Heralding the third wave of peptide drug discovery arising from a convergence of both superdiverse peptide library screening and sophisticated structure-based design technologies, it is noteworthy to point out that the synthetic peptide libraries accessible using methods such as one-bead, one-compound[97,107–110] and chemical encoding (*e.g.* DNA)[87,92] are of immense value. In tandem, mRNA display library methods have been developed to incorporate noncanonical amino acids (*e.g.* D- and *N*-methylamino acids), with the unique advantage of these synthetic peptide library approaches, including the expansive building block chemistry, including not only D-and *N*-methylamino acid, but also Cα-methyl and Cα,α-disubstituted amino acids, *N*-substituted Gly (if an iterative sequence, then commonly referred to as peptoids) and various conformationally constrained amino acids (pro- and numerous mono- and bicyclic analogs thereof), and the opportunity to macrocyclize using many types of chemistries.

12.1.3 Peptide Design and Chemistry

Peptide drug design and chemistry have evolved beyond hierarchical strategies[9–12,42,50,80–89,91,111–124] integrating synthetic chemical modifications of the native peptide (or protein), including amino acid substitutions, amide bond replacements, peptide backbone modification, macrocyclization, secondary structure mimicry and nonpeptidic templates (*e.g. de novo* designed pharmacophore mimics of peptide ligands) (Figure 12.3). Such peptide drug design and synthetic chemistry have expanded our understanding of the secondary structural elements β-turns, γ-turns, β-strands, β-sheets, α-helices and 3_{10}-helices in terms of their inherent accessibility, 3D structural properties and molecular recognition at cognate targets.

It is now understood that the high-fidelity molecular recognition between peptides and their targets is achieved through means of a complementary

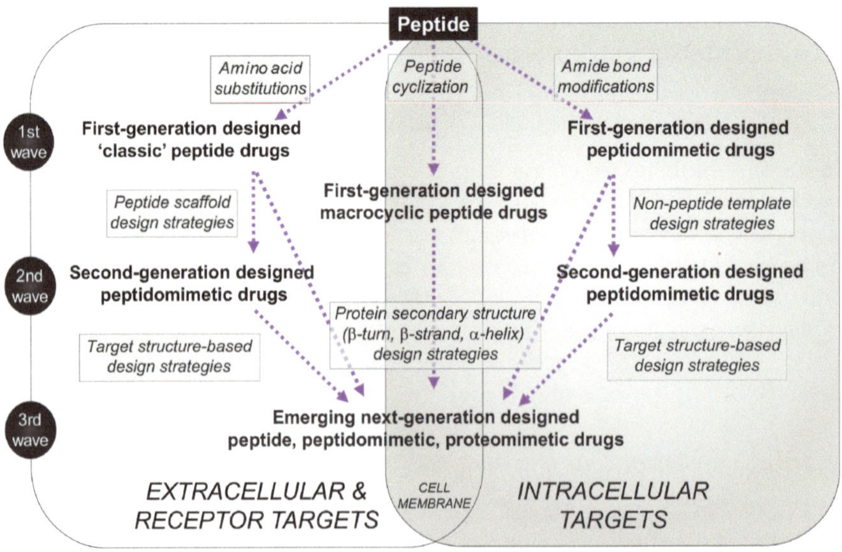

Figure 12.3 Hierarchical strategy of peptide, peptidomimetic and proteomimetic drug discovery highlighting first-, second- and third-generation design strategies (*e.g.* first, second and third waves).

presentation of specific hydrophobic, electrostatic and hydrogen-bonding features to mediate the intermolecular interactions. Although secondary structure is crucial to provide a framework, finding, stabilizing and mimicking critical target–peptide interactions effectively is a key challenge in peptide drug discovery, owing to the molecular size, conformational flexibility and functional group diversity of the peptide and target. Our early knowledge of the molecular recognition of peptide hormones with their cognate GPCR targets evolved from studies focused on 3D pharmacophore models of peptide agonists or antagonists generated from structure–activity relationships and conformational analyses, particularly facilitated through NMR spectroscopy.[9–12,118,119,124] Our current knowledge is rapidly expanding by way of high-resolution 3D structures of peptide–target complexes, leveraging X-ray crystallography and/or NMR spectroscopy, in addition to increasingly sophisticated computational modeling studies.[13,14,19,45–49,115,117,123–140]

12.1.3.1 *Linear Peptides*

Natural and synthetic linear peptides (from dipeptides to, more recently, nearly 100 amino acids) have provided the basis for understanding structure–function and related properties during optimization to become preclinical leads and clinical candidates. In general, linear peptides ranging in length up to <10 amino acids may be relatively simple to modify into peptidomimetics that may incorporate, for example, D-amino acids and/or *N*-methylamino acids for stability. However, simple replacement by amino acids is only the first step of many strategies that may be applied to achieve high-affinity/selectivity binding to a target, in addition to increased metabolic stability and improved pharmacokinetics/bioavailability. Numerous types of amide bond replacements have been described within the scope of peptide and peptidomimetic drug discovery[116,124,141–144] to modify the peptide backbone, including aminomethylene (CH_2NH), ketomethylene [$C(=O)CH_2$], hydroxymethylene [CH(OH)], ethylene (CH_2CH_2), olefin (CH=CH), ether (CH_2O), thioether (CH_2S), tetrazole, thiazole, retroamide (NHCO), thioamide and ester replacements. Both these and the regiospecific incorporation of D-amino acids, *N*-methylamino acids, Cα-methylamino acids, dehydroamino acids, cyclic amino acids and synthetic nonpeptide templates within peptide lead compounds have additionally provided general insights relative to their propensities to induce or stabilize γ-turn, β-turn, β-strand, α-helix and 3_{10}-helix secondary structures.[113–117,120–124,145–178]

12.1.3.2 *Macrocyclic Peptides*

Peptide macrocyclization has resurged as a key driving force of the third wave of peptide drug discovery, leveraging both structure-based design and super-diverse library screening of synthetic peptides to accelerate their generation, optimization and development.[50,80–91,96,97,122,125,127,131–133,140,179–183] The scope of chemical space is remarkable and provides significant opportunity for peptide drug discovery focused on challenging intracellular targets as exemplified by known macrocyclic peptide natural products (*e.g.* CsA[101]), designed macrocyclic peptidomimetic inhibitors of proteases (*e.g.* HCV protease inhibitor[62]), designed stapled α-helical peptides (*e.g.* dual MDM2/X antagonist[184]), CsA-inspired mRNA display library-generated macrocyclic peptides (*e.g.* SIRT2 inhibitor[185]), CsA-inspired DNA encoded library-generated macrocyclic peptides (*e.g.* XIAP antagonist[186]), CPP-hybridized synthetic library-generated macrocyclc peptides (*e.g.* KRAS antagonist[187]) and phage display-generated disulfide-bridged macrocyclic peptides (*e.g.* KRAS antagonist[188]). Collectively, such macrocyclic peptides (Figures 12.2 and 12.4) provide novel chemotypes with unique biophysical and pharmacological properties to tackle so-called "undruggable" targets.

In particular, there has been significant progress in the study of stapled α-helical peptides,[50,80–85,122,179–184,189] including ALRN-6924, the first molecule of this class to advance into clinical trials. The highly cited preclinical progenitor of ALRN-6924, the *in vivo* effective dual MDM2 and MDMX antagonist ATSP-7041, has become a benchmark stapled peptide with respect to its intrinsic biophysical properties (*e.g.* amphipathicity, solubility, cell permeability and metabolic stability[84,184]) that are implicated as key features in the design concepts for this class of macrocyclic α-helical peptides.[81–83,180] Collectively and relative to the first publication[190] of stapled α-helical peptides incorporating α-methyl amino acids as part of the macrocyclization framework, there are now many reported studies illuminating the design, chemistry and biological properties of stapled peptides for a wide range of intracellular protein–protein interaction targets (*e.g.* Bcl-2 family, Notch, estrogen receptor, vitamin D receptor, HIV capsid, HIV integrase, HIF-1, KRAS, Rab, p53, β-catenin, replication protein A, eIF4E, WASF3 and Aurora-A kinase[80,182,190–226]). Nevertheless, there is more to understand with respect to cell permeability and drug delivery of peptides to exploit fully and translate this macrocyclic peptide modality into the clinic.

Beyond the macrocyclic α-helical peptide drug discovery efforts, extraordinary progress has been achieved in recent years with respect to exploiting mRNA display,[86,88–91] DNA encoding,[87,92] phage display[93–96] and synthetic macrocyclic (*e.g.* CPP) peptide libraries.[97,98] In principle, such superdiverse libraries may explore virtually all possible canonical and non-canonical peptide conformational space generally within an 8–20 amino acid size range, albeit there is opportunity to introduce further constraints to molecular flexibility with the inclusion of multiple cyclization designs. In the case of mRNA display, DNA-encoded peptide libraries and other synthetic macrocyclic peptide libraries, unnatural amino acids (*e.g.* *N*-methylamino acids, Cα-methylamino acids, D-amino acids, β-amino acids and novel side-chain functionality) and/or other structurally related building blocks may be incorporated.[86,88–91,93,96,97] In the case of phage display libraries, the use of thioether bridging of Cys residues[93–96] provides the opportunity to explore diversity (*e.g.* peptide sequence, ring size and various linking moieties), and

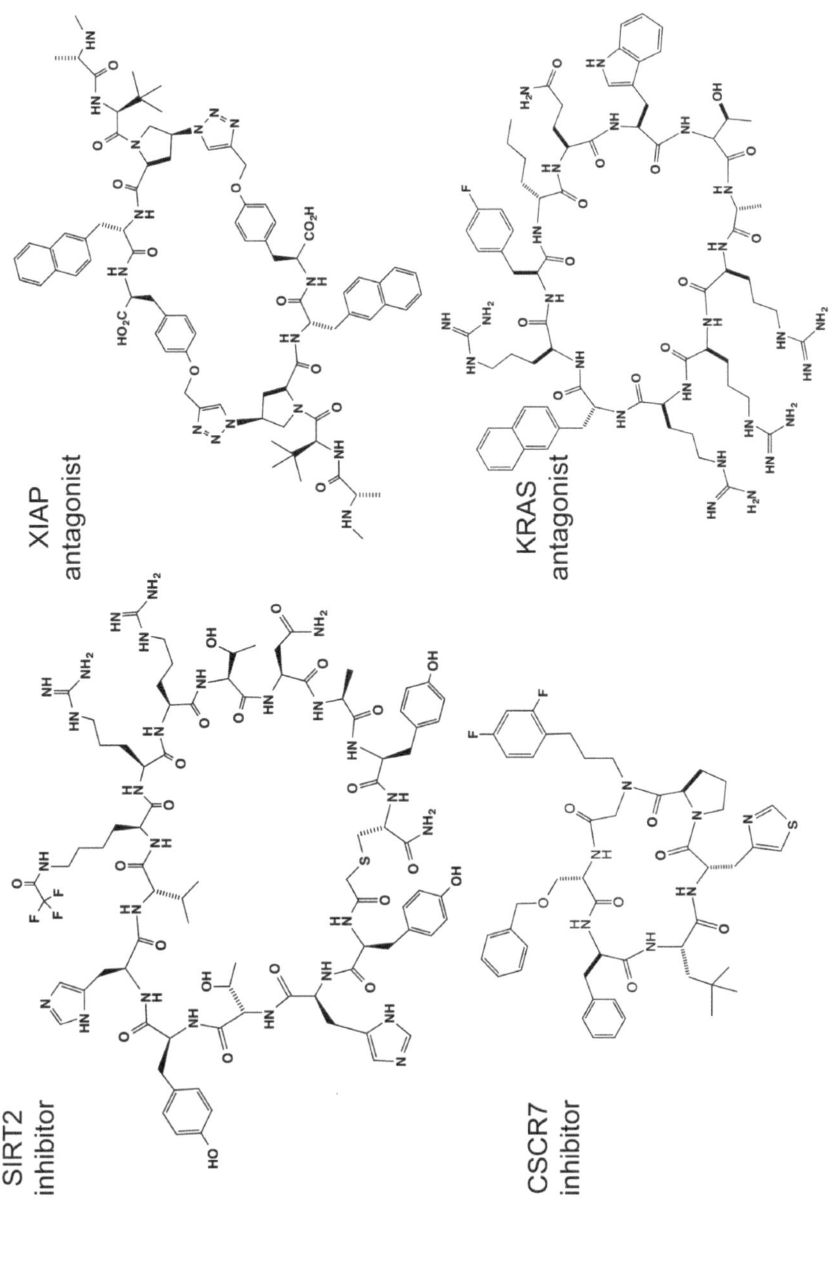

Figure 12.4 Some examples of peptides generated from mRNA and DNAdisplay libraries and structure-based design, including α-helical, CsA-inspired and cyclic-CPP peptide modalities.

more recently the synthesis of phage-generated glycopeptide libraries[96] illustrates further innovative chemistry methods.

Lastly, but not exactly fitting into the realm of superdiverse libraries, are more recent studies on smaller sized macrocyclic peptides (*i.e.* five- and six-membered rings) to explore CsA-like passive permeability through systematic modification of both backbone and side-chain functionalities and leveraging such structure–property relationships to identify peptide scaffolds showing cell permeability, which may then be further modified to achieve molecular recognition for therapeutic targets.[227–232] It is important to emphasize that cyclosporine A provided the first benchmark for designing macrocyclic peptides having oral bioavailability.

12.1.3.3 Conjugated and Cell-penetrating Peptides

For non-oral, receptor-targeted peptide therapeutics, long-lasting exposure levels *in vivo* have been achieved by various approaches,[233–239] including both chemical conjugation [*e.g.* fatty acid, poly(ethylene glycol), antibodies and related recombinant proteins and serum albumin] and sustained-release formulations applicable to parenteral routes of administration (*e.g.* subcutaneous). Specific examples of conjugates include the palmitoyl-modified glucagon-like peptide-1 (GLP-1) agonist liraglutide,[240] a PEGylated GLP-1 agonist/glucagon antagonist hybrid peptide,[241] the PEGylated granulocyte colony-stimulating factor drug pegfilgrastim[242] and a human Fc domain–thrombopoietin peptide agonist conjugate romiplostim.[243] Sustained-release formulations are discussed in Section 12.3.1.1 and Chapter 13. Although the absolute exposure levels are case specific in terms of both time and concentration, the relatively short half-lives (typically a few minutes) of most endogenous (native) peptides provide an opportunity to advance viable drugs with substantially improved metabolic and pharmacokinetic properties.

As part of the strategy focused on prosecuting intracellular targets, Tat and Tat-related polycationic CPPs have each provided benchmarks and tools to understand linear peptides exhibiting cell uptake by endocytosis and/or other translocation mechanisms.[63,64,73,74,76,244–259] Beyond linear CPPs, noteworthy progress is being achieved with stapled α-helical peptides and, more recently, CPP-hybrid macrocyclic peptides in exploring their structure–permeability relationships to optimize cellular potency and modulate intracellular targets and also to understand further their cell uptake mechanisms.[96,98,187] In the specific case of stapled α-helical peptides, some key properties that have been proposed[81–83,180,182,184] to correlate with cellular potency include amphipathicity (hydrophilic and hydrophobic surfaces), lipophilicity (hydrophobicity-driven partitioning into membranes and other hydrophobic compartments), α-helicity (intramolecular H-bonding), proteolytic stability and maintaining high binding affinity to their cognate target (K_d <1–10 nM range) relative to the respective cognate ligand. Accordingly, iterative exploration and analysis of peptide structure–permeability relationships to address intrinsic physicochemical/biophysical properties and also experimental permeability properties using cellular and related noncellular model systems are required to ascertain which of these properties can be exploited. Comparative analysis of linear CPPs, CPP-hybrid macrocyclic peptides and/or stapled α-helical peptides will likely provide insight to benchmark further and understand more deeply peptide cell uptake, including the relationship to endocytosis and/or other translocation mechanisms. Recent studies have addressed this for the comparative analysis of linear CPPs with CPP-hybrid macrocyclic peptides[97,98] and the comparative analysis of linear CPPs with stapled α-helical peptides.[260] Furthermore, it was demonstrated that CPP-hybrid macrocyclic peptides were capable of high cytosolic delivery efficiency and that cell uptake included direct interaction with plasma membrane phosopholipids, endocytosis and subsequent release from endosomes.[96] This study also revealed that such molecules induced membrane curvature on giant unilamellar vesicles and budding of small vesicles that subsequently collapsed into amorphous aggregates of peptide and lipid. In contrast, an evaluation of structurally varied stapled α-helical peptides implicated a correlation of cell uptake *via* a clathrin- and caveolin-independent endocytosis pathway that was suggested to involve, in part, anionic cell surface proteoglycans.[260]

It is essential to have screening tools to understand more deeply peptide cell permeability and drug delivery, including both computational and experimental methods, which can facilitate predictive design rules and iterative testing. Unsurprisingly, there exists a framework of existing screening tools for passive permeability (*e.g.* transcellular permeability using cellular monolayers and proteolytic stability in various biological matrices) in addition to emerging computational and biophysical methods[227–232,261–268] that address exposed polar surface properties, NMR and mass spectrometric (MS) analysis of intramolecular hydrogen bonding, radius of gyration and aqueous–lipid phase partitioning (Table 12.3). Cell-based screening methods[269–273] utilizing MS and/or fluorescence microscopy for unlabeled or fluorescently labeled peptides, respectively, may be implemented to determine cell uptake and to provide correlations with their cellular functional properties (Table 12.3). For nonpassive permeability, it is likely that some (perhaps most) of the same physicochemical and biophysical properties will be important, since facilitated peptide uptake will have similar requirements for hydrophobicity/lipophilicity and desolvation of exposed hydrogen-bonding moieties of the peptide molecule. Certainly, there is ample opportunity for the development of innovative permeability screening tools to enable peptide drug discovery and it is anticipated that such advances will continue to emerge in the future.

12.2 Peptide Discovery: Design for Target Engagement

Peptide drugs derived from either endogenous hormones or natural products have been in therapeutic use for over 50 years. Although the scope of

Table 12.3 Screening tools to study peptide permeability (passive and nonpassive).

Screening tool	Comments
Exposed polar surface area (EPSA)	Experimental EPSA values are determined using supercritical fluid chromatography. Low EPSA values have been shown to correlate with high passive permeability and predicted oral bioavailability
ΔG(insertion)	ΔG(insertion) is a calculated value that refers to the free energy for transferring peptide in a low-dielectric conformation (LDC) from water to a low-dielectric environment (membrane-like) to be predictive of passive permeability
PAMPA	Parallel artificial membrane permeability (PAMPA) uses mixtures of phospholipids infused into lipophilic microfilters with net negative charge (surrogate model system to correlate with experimental oral bioavailablity)
Lipid–water phase partitioning	Both octanol–water phase partitioning and recent modification to incorporate fatty acids and pH gradient as shown for Arg-rich peptides to be predictive of energy-independent translocation (nonpassive permeability)
Cell monolayer transcytosis	Caco-2 cells (or other cell types) used to measure permeability from donor to acceptor compartments and further correlated with experimental oral bioavailablity
Radius of gyration	Radius of gyration is calculated as the root-mean-square distance between the molecule's atoms and its center of gravity. It is an alternative property for MW for beyond "rule-of-five" molecules
NMR analysis of intramolecular H-bonding *versus* solvent H-bonding	Solution NMR studies using hydrogen–deuterium (H–D) exchange experiments to determine amide temperature coefficients and intramolecular H-bonding
Label-free mass spectrometric and fluorescently-tagged cell uptake and permeability analysis	Direct measurement methods for cell uptake of peptides using mass spectrometric methods and/or imaging studies (*e.g.* fluorescence correlation microscopy) to correlate with cell-based functional assays and also non-cell-based surrogate models

these early peptide drugs was limited to empirical observations in phenotypic screens, more recent high-content synthesis and screening approaches[274,275] combined with advances in downstream characterization have allowed the target-directed design of novel therapeutic molecules. Using these approaches, it is not unusual to generate peptides with subnanomolar potency and high specificity to targets. While selection strategies are used to generate potent ligands, the resulting scaffolds generated often lack the structural features needed for intracellular and systemic exposure that are essential for pharmacological activity. In particular, peptides fall outside the "rule of five" criteria[276] established to guide the design of orally bioavailable small-molecule drugs and are susceptible to degradation by peptidase activities ubiquitous to cells and tissues.[277,278] Consequently, overcoming challenging permeability and metabolic barriers is required to achieve the target exposure needed to deliver therapeutic activity.

Regarding barriers that limit target exposure, peptide drugs can be classified based on whether the target is intracellular, requiring permeability and subcellular exposure sufficient to engage the target, or extracellular, if binding to the cell surface or circulating proteins. The primary physiological barriers and kinetic considerations for both target classes can be distinct, each having unique opportunities to optimize pharmacokinetics and pharmacodynamics (PK/PD). In the case of extracellular targets, optimization of peptide PK is focused on enabling bioavailability following subcutaneous (SC) or oral administration and reducing clearance to enhance systemic exposure. In the case of intracellular targets, optimizing intracellular exposure is the first benchmark as it is essential to validate target engagement and cellular activity prior to initiating *in vivo* analysis (*i.e.* without demonstrating cellular activity *in vitro*, it is impossible to interpret pharmacological activity or lack thereof *in vivo*). Most approved peptide drugs are directed to extracellular targets and employ parenteral administration. However, novel formulations[279,280] and greater understanding of peptide drug design principles have made the design of orally absorbed peptides increasingly achievable. Moreover, design strategies to promote membrane permeability and metabolic stability enable targeting of intracellular proteins, expanding therapeutic opportunities to previously undruggable targets.

12.2.1 Oral Absorption *Versus* Intracellular Exposure

Poor permeability is often considered to be the major barrier for peptide therapeutics and is often is ascribed to be the primary reason for the absence of cellular activity for biochemically active peptides. This statement is inaccurate as it groups the ability of a peptide to partition across a biological membrane with other biochemical barriers, limiting peptide association with its target. More precisely, poor exposure to the target is the primary barrier to peptide therapeutic efficacy which is the product of permeability, metabolism and elimination and distribution. For this discussion, we will focus on exposure to the target, where permeability reflecting the partitioning into and across biological membranes and metabolism are discussed separately. Since cytosolic exposure is the first goal needed for achieving biological activity for intracellular targets, permeability and metabolism dictating intracellular exposure are discussed first, followed by factors that control systemic exposure (oral absorption, clearance mechanisms, distribution).

12.2.2 Barriers to Intracellular Exposure and Target Engagement

Pharmacological activity of peptides to intracellular targets is dependent on maintaining cytosolic exposure at levels sufficient to support target occupancy. Factors that determine intracellular exposure are illustrated in Figure 12.5, in which intracellular concentration is determined by the balance between the rate of permeability into the cytosol relative to the rate of

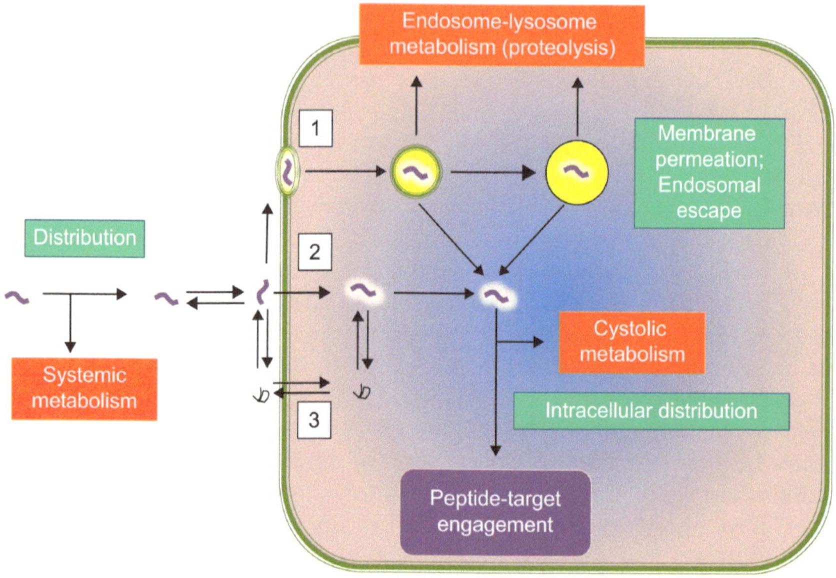

Figure 12.5 Impact of metabolism and permeability on intracellular peptide exposure.

clearance from the cell (primarily by metabolism). In contrast to small molecules, achieving the combination of high permeability and metabolic stability required for target engagement is very challenging, since peptides tend to fall outside the space conventionally associated with high membrane permeability, possessing high hydrogen bond donor (HBD) and acceptor (HBA) counts, high molecular weight and high polar surface area.[281] Simultaneously, peptidase activities are ubiquitous in subcellular compartments.[277,278] Although several strategies exist for overcoming these barriers individually, overcoming both barriers within the bounds of maintaining the pharmacophore and other drug like properties is a formidable task.

12.2.3 Mechanisms for Permeation of Peptides Across Biological Membranes

Passage of peptides across biological membranes can occur by direct permeation across the plasma membrane or by endosomal escape into the cytosol following internalization. In both cases the peptide needs to permeate across the lipid bilayer, which is energetically less favourable than for conventional small-molecule drugs due to peptide size and high HBD and HBA counts. The archetype for membrane-permeable peptides came from evaluation of the natural product CsA, which is both orally absorbed and targets an intracellular protein, cyclophyllin.[282,283] CsA is a large (1.2 kDa), hydrophobic, 11-residue cyclic peptide in which seven of the amide nitrogens are methylated. NMR and crystal structures of CsA revealed that the molecule resides in different conformational states in polar and nonpolar solvents.[284] In nonpolar solvents reflective of the hydrophobic membrane core, the four unmethylated amide bonds are shielded by the formation of intramolecular hydrogen bonds. In polar solvents, conformational changes lead to exposure of the previously intramolecular hydrogen bonds, permitting improved solubility of the CSA through hydrogen bonding with the solvent. This chameleon-like conformational adaptation is postulated to mediate both oral absorption and membrane permeation, enabling cytosolic exposure. Studies by Lokey and co-workers demonstrated that similar chameleon-like properties could be designed into synthetic peptides, including scaffolds supporting binding requirements for novel targets.[285,286] Initial studies on pharmacologically inactive model peptides correlated membrane permeability and oral absorption conformational transitions for a series of synthetic scaffolds.[266,287] Consistent with the involvement of intramolecular hydrogen bonds, the positions of amide methylation, rather than the extent of methylation, had a dramatic influence on the permeability characteristics. Moreover, the rate of membrane permeability was correlated with exposed polar surface area in nonpolar solvents[288] and the calculated Gibbs free energy difference between the peptide in polar and nonpolar environments.[266] Although the original model peptides used to establish these relationships had no pharmacological activity, there is potential to apply these principles within a drug discovery paradigm.[289]

Since structural features required for target engagement constrain the physicochemical properties of peptide drug candidates, the requirements intrinsic to chameleon-like peptide, such as low experimental polar surface area (EPSA), low ΔG for partitioning between solvent and lipid and masking and elimination of hydrogen bond donors, may not be compatible with pharmacological activity for many targets. Since the discovery 20 years ago that TAT-derived peptides could deliver polar macromolecule cargo into cells, considerable attention has been directed to understanding the mechanism of cell uptake and endosomal escape for CPPs, applying the strategy to broaden the accessible target space for peptide drugs.[290–292] Understanding the molecular and biological features associated with membrane translocation opens the potential to incorporate CPP-like features in peptide design to enhance the permeability of polar peptides. Currently, over 800 CPPs have been identified and classified into three categories based on amino acid composition: cationic, amphiphilic and hydrophobic. Although the majority of CPPs contain cationic amino acids, some negatively charged amphiphilic and hydrophobic CPPs have also been identified.[293] Given the diversity of CPPs, it is not surprising that multiple mechanisms have been proposed for membrane penetration.[294,295] For some of these mechanisms, a tight association of peptides with the membrane surface occurs, resulting in mass action-driven restructuring of the lipid packing. In the case of cationic and positively charged amphiphilic peptides, it has been proposed that the phosphate in the lipid headgroup,[296] or terminal carboxylic acids in oxidized lipids and fatty acids,[297,298] form a salt to neutralize the positively charged residues and drag the peptides into the hydrophobic core of the lipid bilayer. Depending on the structure of the peptide, this can lead to translocation across the membrane, formation of a pore or disruption of the lipid membrane. Using combinatorial peptide synthesis combined with high-throughput liposome screening models, Wimley and co-workers[296] identified unique 13-residue peptides capable of spontaneously translocating model membranes without inducing membrane lysis or pore formation in liposomes. In contrast to other CPPs, these peptides spontaneously translocate across membranes in a manner that is not mass action driven, does not require internalization and is not membrane potential driven, suggesting that the peptide–lipid interactions are sufficient to support translocation.[299] The potential to incorporate CPP features into macrocyclic peptides has led to the generation of potent cyclic CPPs. Cyclic CPPs have been designed that efficiently deliver to the cytosol and are metabolically stable.[300] These motifs have been incorporated into mono- and bicyclic KRas inhibitors, generating compounds with reported submicromolar cell potency.[290] Although CPPs have potential for broadening the accessible chemical space for peptide therapeutics, it remains to be seen if requirements for robustness and safety can be met to translate from *in vitro* activity to clinically active drugs.

Intracellular activity has also been observed with stapled peptides such as ATSP-7041. In contrast to macrocycles, stapled peptides tend to display more defined secondary structure and are less likely to undergo solvent-sensitive conformational changes required for passive permeability. In the case of ATSP-7041, this results in EPSA and ΔG values well exceeding values associated with membrane permeation of macrocycles, suggesting a distinct mechanism governing intracellular exposure and target engagement. Although stapling restricts conformational changes associated with chameleon-like passive permeation, introduction of this secondary structure tends to improve metabolic stability and intracellular exposure of stapled peptides. The transport mechanism for stapled peptides is difficult to generalize across sequences since amino acid side-chain composition can impact the route of cell uptake. In the case of ATSP-7041, neutral amino acids predominate and cell uptake appears to be driven by direct membrane partitioning (manuscript in preparation). However, other stapled peptides that incorporate cationic amino acids and CPP-like sequences appear to undergo internalization and endosomal escape.[260,301] Employing different stapling crosslinkers in a cationic amphipathic peptide, Tian *et al.* demonstrated that hydrophobicity, introduced by different crosslinkers corresponded with higher cell uptake, but not α-helix stabilization.[302] Using an azobenzene photoresponsive staple, Kim *et al.* induced *cis–trans* isomerization of the staple, resulting in stabilization of a model LK peptide in an α-helical conformation with UV illumination and destabilization under visible light.[303] The UV-induced conformation of the peptide forms an amphipathic α-helix with lysines sequestered to one side and hydrophobic side chains on the other side of the helix and shows substantially higher cell uptake. Interestingly, the amphipathic helical form of the peptide tended to aggregate, forming 300 nm particles following UV irradiation, but existed in a monomeric state in the visible light-illuminated, helix-disrupted state. The correspondence of cell uptake and peptide aggregation raises the possibility that the formation of multimeric complexes also plays a role in cell uptake and endosomal escape of stapled amphiphilic CPPs.

Given the mechanistic differences in the various routes of membrane permeation, it would be expected that the structural and physical properties associated with permeation will be distinct for different transport mechanisms. As such, optimization of the structure–activity relationship (SAR) for membrane permeation will be different for CPP-like cationic partitioning *versus* passive partitioning (*i.e.* rules for promoting cell transport for one class of peptides may not translate to another class). Ultimately, the physical and structural properties of a peptide drug will be limited by the structural requirements compatible with target binding. Consequently, optimization of membrane permeation and the resulting SAR should be biased towards a mechanism that will fit within the limits of the chemical space defined by the pharmacophore.

12.2.4 Intracellular Metabolism

Peptide metabolism is the other primary barrier regulating the intracellular exposure of peptides. Assuming prolonged extracellular exposure (as would be the case in an *in vitro* cell-based assay or in a chronic drug application setting), steady-state cytosolic peptide concentrations will depend on the balance between the rate of permeation into the cytosol (input rate) and the rate of peptide removal through metabolism (output rate) (Figure 12.5). Although nonpeptidase metabolism can be a factor for some peptides (*e.g.* cyclosporine A), peptidase activity is the predominant route of metabolism for most peptides. In humans, approximately 1100 peptidase genes have been identified (about 5% of the human protein coding genome), although many have very specific sequence requirements and biological functions in protein processing and activation. Peptidases that are more likely to impact peptide drug intracellular exposure are expected to have more generalized functions in protein/peptide turnover and recycling of amino acids. Some proteases that are ubiquitously expressed in cells are listed in Table 12.4.

The mechanism of cellular uptake of peptide drugs can impact the extent and type of metabolic barrier to intracellular exposure. Peptides in which cellular uptake is mediated by endocytosis are preferentially exposed to high concentrations of endosomal and lysosomal proteases prior to cytosolic exposure. During transitions through the endocytic network,[304–306] peptides are exposed to increasing acidity associated with endosome maturation and are subject to proteolytic degradation from cathepsins. Although cathepsin activity is often associated most prominently with lysosome, several cathepsins are also active in early and late endosomes, particularly cathepsins B, D and H, and have critical functions in peptide hormone degradation,

Table 12.4 Proteases mediating peptide clearance.

Distribution	Class	Name	Specificity	Properties	EC No.
Cytosol	Serine protease	Prolyl endopeptidase	Cleaves on the C-terminal side of prolyl residues	Involved in processing and degradation of peptide hormones	EC 3.4.21.26
	Cystine protease	Calpain I and II	Cleavage is not determined by specific amino acid sequence. Cleavage sites may reflect tertiary structure	Papain-like protease; Ca^{2+} dependent. Calpain I and II have similar reactivity but calpain I has higher affinity for Ca^{2+}	EC 3.4.22.52 EC 3.4.22.53
	Metalloprotease	Insulin-degrading enzyme	Catalyzed cleavage of multiple peptide hormones including, but not limited to, insulin, glucagon and bradykinin, in addition to cleavage of diverse short peptide substrates	Zinc-binding protease. Conformational change associated with ATP binding	EC 3.4.24.56
		CN2 (CNDP2)	Dipeptidase, broad specificity, carnosine and other dipeptides	Dipeptidase bestatin sensitive	EC 3.4.13.20
	M3 metallopeptidase	Thimet oligopeptidase (TOP)	Cleavage of 6–17 amino acid peptides releasing terminal oligopeptides	*N*-Exopeptidase; involved in trimming and degradation of peptides released from the proteasome and also other cytosolic peptides	EC 3.4.24.15

Endosome/lysosome	Aspartate protease	Cathepsin D	Recognition of hydrophobic amino acid regions	Catalyzes protein turnover and processing of peptide hormones. Mediates activation of cathepsin B and L. Overexpressed in cancer cells	EC 3.4.23.5
		Cathepsin B	Carboxydipeptidyl peptidase activity in acidic environments. Broad endopeptidase activity at neutral pH	Most abundant cathepsin in lysosomes endo- and exoprotease activity. Overexpressed in cancer cells	EC 3.4.2.1
		Cathepsin L	Endopeptidase activity requiring a hydrophobic residue at position P2 and an aliphatic chain at P1	Most active lysosomal protease; endo- and exoprotease	EC 3.4.22.15
	Cysteine protease	Cathepsins M, N, S, T	Specificity is distinct for each isoform and from B, H and L		
Plasma membrane	Serine exopeptidase	DPP IV	N-terminal cleavage of dipeptides with proline or alanine at the penultimate position		EC 3.4.14.5
	Metallopeptidase	Nuetral endopeptidase 24.11	Preference for neutral amino acids at P1, P2 and P3′	Widely distributed in the body. Also known as CD10, neprilysin	EC 3.4.24.11

antigen processing and presentation.[307] In the case of synthetic peptide drugs in which cell uptake is mediated by endocytosis, stability to endosomal and lysosomal proteases is then a prerequisite for effective endosomal escape and intracellular bioavailability. For peptides that are absorbed directly through permeation of the plasma membrane, cytosolic proteases are expected to mediate clearance of intracellular peptide rather than preventing cytosolic delivery. As such, clearance by cytosolic proteases will counterbalance the rate of permeation into the cell and lower the steady-state peptide concentrations.

Given the numerous proteases capable of degrading peptide drug candidates, tools for analyzing peptide stability generally utilize biological matrices or subcellular fractions. Although these systems may not accurately represent the assortment of peptidases and their respective specificity acting on peptides, they may nevertheless correctly rank peptidase susceptibility and provide some information on structural properties that confer greater proteolytic stability.[278,308] Susceptibility to proteolysis is conferred both by specific amino acid residues and by the conformational properties of peptides. Primary amino acid sequences may define cleavage site specificity for proteases but are not sufficient to confer proteolytic sensitivity. Flexibility to conform to the shape of the protease binding site, accessibility of the cleavage sites and residues stabilizing interactions in protease binding sites all contribute significantly to protease susceptibility.[309] Computational analysis of 266 protease inhibitors binding to 48 different proteases (representing aspartic, serine, cysteine and metallo proteases) shows consistent adoption of an extended β-strand in protease-active sites, suggesting conserved structural motifs for peptide binding to protease cleavage sites.[310] Evaluation of peptide backbone modifications proximal to chymotrypsin cleavage sites suggests that modulation of peptide conformation can significantly reduce susceptibility to protease cleavage.[311] The impact of conformation on proteolytic susceptibility could account for the increased stability conferred by stapling and other modifications that introduce conformational rigidity to peptides. Given the diversity of proteases that cleave peptides, in cells and *in vivo*, stabilization of peptide through modulation of peptide conformation and physical properties may be more effective in early discovery than direct sequence optimization. This could potentially explain why selection of stable peptides in an mRNA display library using pronase (a mixture of bacterial proteases from *Streptomyces griseus*) selected for peptides with improved stability to mammalian proteases.[278]

12.2.5 Systemic Exposure

The discussion up to this point has focused on barriers to activity against intracellular targets in a closed system. Once criteria are met to support biochemical and cellular activity, conventional ADME (absorption, distribution, metabolism and excretion) properties must be optimized to ensure sufficient systemic exposure to support target engagement and drive pharmacological activity. In general, the physiological processes that define peptide disposition are similar to those regulating small-molecule exposure. However, the unique chemical space can introduce subtle differences in factors underlying absorption, distribution and clearance mechanisms. Reduced passive membrane permeability conferred from the larger size and greater polar content can bias peptide exposure primarily to extracellular compartments, thus reducing the volume of distribution and potentially altering predominant absorption pathways. Also, the metabolic pathways responsible for systemic clearance are increased for peptides. Consequently, although absorption and clearance principles guiding systemic exposure of peptides are similar to more conventional drug modalities, underlying mechanisms for these processes may be different.

12.2.6 Absorption

Peptide drugs are typically administered through oral, subcutaneous or intravenous administration. As suggested by the plethora of publications on the subject, oral administration is the preferred but elusive route for administration.[312,313] It should be emphasized that although properties associated with cell membrane permeability were discussed earlier, gut wall and cell membrane permeability are related but distinct processes. Many peptides have biological activity against intracellular targets but are not orally absorbed (*e.g.* ATSP-7041), while other peptides with oral exposure may be limited to extracellular target space (*e.g.* formulation-enhanced liraglutide).

Differences in kinetic constraints and intestinal physiology distinguish permeability requirements for intracellular exposure from those required for oral absorption. During oral absorption, drug in the intestinal lumen permeates across the across the intestinal epithelium, a single cell layer separating the intestinal lumen from the systemic vasculature. Enterocytes, the primary cells comprising the epithelial layer, are bound by tight junctions that connect neighboring cells, restricting passage of molecules between the cells. Thus, transit across this layer into the circulation occurs primarily by transcellular permeation involving partitioning across the apical plasma membrane, diffusion through the cell and permeation of the basolateral membrane into the systemic compartment. Alternatively, some absorption can occur *via* paracellular transport between cells, although size restrictions posed by tight junctions restrict this pathway to relatively small compounds. Most drug absorption occurs during a 1.5–2 h window in which the drug transits through the small intestine.[314,315] The time window limits this first-order transcellular intestinal absorption to peptides with relatively high membrane permeability. Perhaps as a result, many peptides absorbed through this route so far seem to be restricted to cyclosporine-like, passively permeable nonpolar peptides. For more polar peptides, paracellular transport is feasible, although requiring formulation-enabled disruption of tight junctions to increase paracellular passage of peptides in the 1–2 kDa molecular weight range. Currently there are no marketed drugs that directly employ absorption enhancers to facilitate drug absorption, but late-stage clinical trials have been reported for oral semaglutide[279] and oral octreotide,[316,317] showing improved oral efficacy/exposure and safety. Although 5–10-fold enhancement of absorption was observed in these studies, the overall extent of absorption (%*F*) was low compared with more conventional oral drugs. However, for potent peptides with prolonged exposure and a large therapeutic margin, low bioavailability may be commercially acceptable. Further developments may resolve whether absorption enhancer-facilitated oral absorption of peptides can be further improved and generalized to other classes of peptide drugs, thus vastly expanding the range of oral peptide therapeutics.

In addition to the physical barrier posed by the intestine, proteolytic digestive enzymes in the lumen can degrade peptides during absorption (Table 12.4). The pancreatic proteases trypsin, chymotrypsin, papain and elastase catalyze the bulk of peptide metabolism in the intestine. Enterocyte membrane proteases [(PPIV (dipeptidyl peptidase IV), amino- and carboxypeptidase] and intracellular proteases further degrade peptides prior to release into the systemic circulation. As in the case of intracellular proteases, susceptibility to protease cleavage is determined both by residue specificity and conformational flexibility. Thus, targeted sequence-directed modifications and conformational restrictions introduced through cyclization, crosslinking and staples can introduce resistance to peptidase activity and improve intestinal absorption.

Given the challenges to designing orally bioavailable peptides, subcutaneous injection is currently the predominant method for the administration of peptides drugs. Similarly to oral absorption, incomplete bioavailability is a factor for subcutaneous injection.[318,319] Following injection into the subcutaneous space (hypodermis), small peptides (<20 kDa) can reach the systemic circulation through direct passage across vascular beds. Compounds >20 kDa are not able to transit across the vascular endothelium but instead transit through the lymphatic system and are released into the bloodstream through the thoracic duct. Prior to release into the lymphatic or systemic circulation, peptides reside in the interstitial space of the hypodermis where they can be subject to protease degradation and potential aggregation. Transport within the interstitial space and, in turn, the residence time at the injection site are driven by diffusion and convection.[318,320] Since significant proteolysis can occur during residence in the interstitial space, factors impacting convection (*e.g.* movement, massage) may also contribute to variability in bioavailability for metabolically labile peptides.

12.2.7 Clearance

Following absorption, peptide exposure is limited by systemic metabolism and hepatobiliary and renal metabolism and elimination. Since proteases are expressed ubiquitously on cell surfaces and in body fluids, significant clearance can occur during systemic exposure, independent of hepatic and renal clearance. *In vitro* plasma, serum, cell extracts and isolated proteases have been used to make surrogate measures of systemic metabolism and to identify sequence modifications that may confer enhanced systemic exposure.[321,322] The impact of this approach is best illustrated by the introduction of proteolytic stability to somatostatin in the development of octreotide.[323] Using sequential modifications to somatostatin, the cyclic peptide was reduced from 14 to eight amino acids and targeted modifications were introduced to increase the stability to proteases in kidney cell extracts. The stabilized peptide had a systemic half-life of approximately 2 h compared with 1–2 min observed for somatostatin. As a result of the enhanced exposure, octreotide shows a 70-fold improved *in vivo* pharmacological potency in spite of only modest enhancement of *in vitro* potency.

Renal and hepatic clearance can also limit peptide drug exposure. As a guide, unbound molecules with molecular weight <60 kDa are subject to glomerular filtration resulting in rapid renal elimination. For such peptides, extension of systemic exposure has been achieved through increased size *via* conjugation with polymers[324,325] or modifications to promote plasma protein binding.[326] Peptides filtered through the glomerulus can be excreted intact or as proteolytic fragments generated by renal brush border hydrolases. In contrast to optimizing proteolytic stability of peptides to systemic degradation, optimization of renal peptide stability post-glomerular filtration should have little, if any, impact on systemic exposure since it occurs after removal of the peptide from the circulation.

The biochemistry and physiology of the liver are uniquely adapted to function in maintaining blood homeostasis and removal of endogenous and exogenous toxins from the circulation. Hepatic blood flow represents 25% of cardiac output and virtually all the blood draining from the intestine passes through the liver prior to entering the systemic circulation. Additionally, hepatocytes express transporters and drug metabolism enzymes that can catalyze metabolic clearance and hepatobiliary elimination of peptide drugs. Consequently, peptides can be subject to conventional xenobiotic clearance mechanisms in addition to the proteolytic clearance mechanisms ubiquitously expressed in the body. The methylamide-modified peptides cyclosporine A, cyclosporine D (PSC833)[327] and enniatin B[328] are subject to CYP 3A metabolism with additional metabolism of enniatin B catalyzed by CYP 1A2 and CYP 2C19. Similarly, hepatic transporters can facilitate hepatic uptake and hepatobiliary elimination of several peptides.[329] Octreotide, a cyclic cationic peptide, and BQ123, a cyclic anionic peptide, are both excreted intact through hepatobiliary elimination. Hepatic uptake of the cyclic cationic peptide octreotide is mediated by OATP-1B1 and 1B3[330,331] and BQ123[332] uptake is mediated by OATP 1B3 and NTCP, with MRP-2 mediating efflux of both compounds into the bile. BQ485 a linear homolog of BQ123, is also a substrate of sodium-dependent and -independent liver uptake, suggesting that uptake by these hepatic transporters is not restricted to cyclic peptides. In summary, although much of the focus on peptide exposure is directed towards protease stability and membrane permeability, achieving and maintaining systemic exposure may also entail introducing resistance to ADME processes associated with conventional small-molecule drugs.

12.3 Peptide Development: Routes of Administration and Dosage Form Properties

The diversity of molecular properties and therapeutic sites of action for approved peptide products has resulted in peptides being formulated into a wide variety of drug product dosage forms. These include standard parenteral forms such as solutions and lyophilized powders, complex parenteral formulations such as device-based combination products or sustained-release formulations and formulations for noninjectable routes of administration including oral capsules and nasal sprays.[9] This section discusses unique formulation considerations of peptide products with recently approved products provided as illustrative examples.

12.3.1 Routes of Administration

The target product profile (TPP) for a peptide in development sets requirements on the route of administration, constraining the formulation options that can be explored. Safety and efficacy are the most critical formulation attributes to consider in developing a dosage form, although weight is also given to patient preference on image and economic factors.[333] The technical skills and training of the individual administering the dose places practical limitations on the complexity of the drug product preparation. For example, liquid or lyophilized peptide formulations intended for intravenous administration are generally limited to indications where a physician or trained healthcare provider is available to prepare and administer the dose. For peptide drug products that are to be administered by the patient or caregiver, oral dosages or ready-to-use liquid formulations that can be administered subcutaneously are preferred.

12.3.1.1 Subcutaneous and Intramuscular Injection

Subcutaneous injection remains the most common route of administration for approved peptide products.[9] The high potency of most peptides makes self-administration in small injection volumes feasible at the frequency necessary to maintain therapeutic levels. Nine of the 15 peptide products approved from 2010 to 2017 are administered subcutaneously, compared with just three by intravenous administration (Table 12.5). The number of therapeutic peptide products intended for intramuscular injection is significantly smaller, with none approved in the current decade and comprising only 11% of all peptide products approved in the past 65 years.[9]

Table 12.5 Drug products approved from 2010 to 2017[9] and pharmaceutical characteristics retrieved from the package label[a].

Drug	Therapeutic area	Peptide size[b]	Route[c]	Frequency[d]	Formulation/device
Tesamorelin	Metabolic	44 AA, C_6 conjugate	s.c.	q.d.	Lyophilized powder, vial
Pegin esatide	Hematology	21 AA dimer, PEG conjugate	i.v., s.c.	q.m.	Single- and multiple-use solutions; vial or syringe
Lucinactant	Respiratory	21 AA	i.t.	q.6.h.	Suspension
Linaclotide	Gastroenterology	14 AA	p.o.	q.d.	Multiparticulate filled capsules
Pasireotide	Endocrinology	6 AA, cyclic	s.c.	b.i.d.	Solution
Teduglutide	Gastroenterology	33 AA	s.c.	q.d.	Lyophilized powder
Carfilzomib	Oncology	4 AA	i.v.	2× per week on sequential days	Lyophilized powder
Lixisenatide	Metabolic	44 AA	s.c.	q.d.	Solution in pen injector
Afamelanotide	Dermatology	13 AA	s.c.	Every other month, up to 4× per year	PLGA implant
Albiglutide	Metabolic	30 AA dimer, albumin conjugate	s.c.	q.w.	Lyophilized powder in pen injector
Dulaglutide	Metabolic	30 AA dimer, Fc conjugate	s.c.	q.w.	Solution in prefilled syringe or pen injector
Etelcalcetide	Endocrinology	8 AA	i.v.	3 times per week	Solution
Plecanatide	Gastroenterology	16 AA	p.o.	q.w.	Tablet
Abaloparatide	Bones and connective tissues	34 AA	s.c.	q.d.	Solution in pen injector

[a]Information from the Product Label approved by the FDA[356] was referenced for all compounds except afamelanotide, which is under review by the US regulatory agency at the time of writing this chapter. Formulation and dosing information was obtained from the Committee for Medicinal Products for Human Use (CHMP) assessment report for SCENESSE (afamelanotide).[353]
[b]AA, amino acid residues.
[c]s.c., subcutaneous; i.v., intravenous; i.t., intratracheal; p.o., oral.
[d]q.d., daily; q.m., monthly; q.6.h., every six hours; b.i.d., twice daily; q.w., weekly.

Subcutaneous and intramuscular injections into tissues are easier to administer than intravenous injections but also present additional biological hurdles to delivering the peptide to the target site of action. Absorption into the systemic circulation is slowed *via* these routes and the bioavailability can vary widely owing to stability, solubility and distribution considerations. In the subcutaneous space, peptide physicochemical properties including stability, solubility and molecular weight impact the rate and extent of absorption after injection. As noted previously, exposure to proteases may degrade the peptide prior to it entering the circulation, and compounds that have poor solubility may precipitate, form local depots at the injection site and slow the rate of absorption, particularly if solubilizing excipients used in the formulation diffuse from the injection site more rapidly than the peptide drug. There is also evidence that the route of drug absorption, directly into the circulation or first through the lymphatic system, depends on the molecular weight, solubility, lipophilicity and charge of the molecule: most peptides with molecular weights <16 kDa are of a size that can be absorbed through capillaries and directly into systemic circulation whereas higher molecular weight proteins and particulate systems, such as drug-loaded nanoparticles, are absorbed into the lymph after subcutaneous injection, prior to systemic distribution.[334,335]

Beyond the inherent peptide molecular properties, the pharmaceutical properties of the formulated peptide can also influence absorption, bioavailability and efficacy. Clodfelter *et al.* characterized the noncovalent self-association of octanoyl acylation and observed loss of α-helical conformation, increases in aggregate size and greater strength of self-association as a result of increases in temperature and formulation ionic strength.[336] These changes were shown to reduce significantly the bioavailability of a subcutaneous dose of the drug in beagles. The presence of aggregates or other particulates in subcutaneous formulations may also increase risks of immunogenicity, the impact of which can range from simply binding or altering bioavailability to adversely impacting the safety profile of the drug.[337,338] Generally, factors that affect immunogenicity may be patient or product specific. From a regulatory perspective, FDA guidance related to immunogenicity assessment pertains to nonsynthetic compounds greater than 40 amino acids in length.[339,340] In the context of antibody production, immunogenicity is correlated with molecular complexity and size and peptides are generally considered to be too small to elicit inherently a strong immune response. Holfeld *et al.* reported a lack of immunogenicity for the proline-rich antimicrobial peptides Onc72 and Api88, whereas conjugates of these peptides to carrier proteins resulted in an immunogenic reaction.[341] Nevertheless, the development of anti-drug antibodies (ADAs) is assessed in patients during clinical trials, with the frequency of occurrence and impact on efficacy reported on the product labels. Sequence complexity and increased size due to aggregation may increase immunogenicity, as has been reported for subunit peptide vaccines deliberately designed to self-assemble.[342–344,341]

To improve the patient compliance for peptide drugs that are self-administered by frequent injection or administered for extended periods of time, combination products of peptide formulations in pen injector devices provide a convenient and discrete presentation that typically can be used multiple times over the course of a few weeks. Such products have become a standard presentation for peptide drugs targeting the GLP-1 receptor to treat type 2 diabetes mellitus. In 2005, Byetta® (exenatide) was the first peptide GLP-1 receptor agonist to be approved and is provided as a ready-to-use solution in a multiuse pen injector device for twice daily (b.i.d.) dosing.[345] Multiuse formulations must have demonstrated antimicrobial effectiveness,[346] presenting a significant challenge in excipient selection, as common antimicrobial agents have been shown to be physically incompatible with peptides and proteins, inducing aggregation.[347–349] Since the approval of Byetta® (exenatide), self-administration from a pen injector device has become a preferred presentation for GLP-1 therapies, regardless of whether the formulation is multiuse, with more recently approved, long-acting GLP-1R agonists being dosed from disposable pen injectors, as single-use formulations administered once weekly, including albiglutide, deluglutide and semaglutide.

Formulating peptides with short half-lives into injectable controlled-release formulations is an alternative approach to improving patient compliance for extended duration or chronic therapies, by providing a consistent input of drug with reduced injection frequency.[350] For example, exenatide was reformulated into poly(lactide-*co*-glycolide) (PLGA) microspheres, which are packaged as a dry powder in a pen injector device for suspension in an aqueous vehicle prior to subcutaneous administration.[351] Upon subcutaneous injection, the drug-loaded microparticles form a depot from which exenatide elutes in a controlled manner over the course of 1 week as the polymer matrix degrades, reducing the number of self-injections from 14 per week to one. The increasing complexity of such a formulation requires additional development efforts to optimize peptide stability, polymer degradation rates and dose preparation by reconstitution.[351] More recently, afamelanotide was approved in Europe and is under review in the USA as an extruded, solid PLGA drug to reduce the sensitivity of skin to UV exposure.[352] The implant is administered in the clinic *via* a catheter and provides drug release over the course of 2 months. During reviews, the European regulatory authority made the recommendation that an applicator device be developed to aid clinicians in implanting the device in the subcutaneous space.[353]

12.3.1.2 Intravenous Injection

Maximum exposure is achieved by direct intravenous injection of the drug product into the systemic circulation, circumventing solubility, stability and absorption limitations for the subcutaneous and intramuscular space noted above. Intravenous administration is practically limited to acute or periodic dosing by a healthcare professional and often requires a pharmacist for the preparation of the drug product for administration. The drug product can be presented in several forms, including solution, lyophilizate or powder-in-vial for reconstitution as a bolus injection or dilution into a bag for infusion. Etelcalcetide, for example, is indicated for the treatment of secondary hyperparathyroidism and is administered as a bolus intravenous injection, leveraging the venous line of a dialysis circuit to administer to patients at the end of hemodialysis treatment.[354,355] The drug product is provided as a ready-to-use, single-use solution with critical quality attributes (CQAs) identified in the product risk assessment, including assay, impurities/degradants, appearance, sterility, endotoxin, viscosity, solubility, particulates, osmolality, pH and leachables/extractables for the primary container.[356] Owing to direct input into the circulatory system, there are increased safety risks associated with products intended for intravenous infusion or injection, such as stringent sterility requirements and particulate control. Lack of quality assurance in these areas led to recalls of eight intravenous drug products through October 2018, as reported by the FDA.

12.3.1.3 Oral Delivery

The transport of peptide drugs through the gastrointestinal tract and into the systemic circulation has traditionally been poor owing to degradation caused by low gastric pH and digestive enzymes and poor permeability across the epithelium due to high molecular weight, hydrophilicity and charge. In addition to molecular design[228,264,357,358] approaches, drug delivery[359–361] options have also been investigated to mitigate these physiological barriers and permit more efficient oral bioabsorption of peptide drugs.[312,359]

For example, linaclotide is an orally delivered guanylate cyclase C (GC-C) agonist that acts locally in the lower gastrointestinal tract (Table 12.5). Linaclotide and an active metabolite were shown to be stable under hydrolytic and proteolytic conditions until reduction of internal disulfide bonds in the small intestine.[362] Although linaclotide has sufficient stability in the stomach, Ironwood Pharmaceuticals is developing a controlled-release formulation in order to modulate the pharmacological properties of the formulation, targeting release to the ileoceal junction of the colon and thus treating abdominal pain without effecting fluid secretion.[363–365]

There are limited clinical examples of systemic delivery of peptides by the oral route, which has typically required a combination of enteric coating, permeation enhancers, enzyme inhibitors and very high doses to achieve equivalent PK to a parenteral dose. To develop an oral octreotide formulation, D-amino acid substitutions improved resistance to degradation by proteolytic enzymes and a lipid suspension containing sodium caprylate as a permeation enhancer was advanced into clinical trials. Early studies showed that the systemic absorption was variable;[366] subsequent reformulation into enteric coated capsules reduced the variability[367] but the formulation also carried a significant food effect. Currently, a second Phase III study is under way.[368] Novo Nordisk has shown that limited oral bioavailability of peptides can be compensated for with highly potent and long-circulating peptides, such as was demonstrated with semaglutide in a Phase III clinical trial.[369]

Nanocarrier formulations of peptides, which typically leverage encapsulation of a peptide in a matrix or shell structure, have been proposed to present optimal physicochemical properties for transit through the gastrointestinal tract, preserve the stability of peptides for delivery by the oral route and improve absorption of peptides.[370] It is hypothesized that mechanisms by which nanocarriers can achieve enhanced bioavailability are tunable based on the physicochemical properties of the nanocarriers, including size, charge, polarity, bioadhesion and specific surface functionalization. However, to date, such formulations have not progressed beyond testing in preclinical species.

12.3.1.4 Alternative Delivery Routes: Pulmonary, Nasal and Transdermal

None of the peptide products approved in the past decade are delivered by alternative routes of administration. Nevertheless, with a sustained interest in pain-free delivery and the tunability of peptide properties through molecular design, consideration of the physicochemical properties that are favorable for bioavailability of drugs administered by alternative routes is of value to differentiated drug discovery and development efforts.

For pulmonary delivery, inhaled therapies require engineering of the aerodynamic size of the particles or droplets of solid or liquid formulations in conjunction with a co-developed device for delivery of the formulation, such as metered dose inhalers or dry powder inhalers.[371] The stability advantages for dry powders for inhalation are similar to those for lyophilized powders; however, the manufacturing processes to obtain such powders can expose peptides to various stresses, including high temperature and interfaces during spray drying. Solubility in spray drying solvents can pose practical limitations to drug loading, throughput or the selection of green solvents. Nevertheless, these challenges have been overcome and the safety and efficacy of the inhalation route for insulin have been demonstrated in several clinical trials.[372] A number of contributing factors have been cited as possible challenges to the success of inhaled insulin therapies,[373] including market acceptance of a new dosing approach away from the well-accepted subcutaneous injections.[373]

Intranasal delivery of peptides has been demonstrated in preclinical species to achieve exposure in the central nervous system, bypassing the blood–brain barrier, although the mechanism of transport is still not well understood.[374] Limited examples of direct PK measurements and PD readouts in preclinical species have been reported for peptides <1 kDa and as large as 6.5 kDa.[374] Clinical reports have found indirect PD evidence that supports delivery of therapeutics to the brain by the intranasal route, compared with the intravenous route.[374] Nanocarrier formulations have also been proposed as an approach to tailor physicochemical properties, including

Table 12.6 Future perspective: opportunities and associated challenges with the progression of therapeutic peptide modalities.

Opportunities	Challenges
Continue to leverage chemical and biological diversity (*e.g.* learn from Nature in peptide design, including localization sequences, post-translational modifications, special peptides, venoms)	No general set of "rules" that can be applied to molecular design across multiple classes of peptides (*e.g.* similar to "rule of five" for small molecules)
Increase druggability of starting peptides	Develop bio-relevant screening assays and integrate appropriate selection methods early in discovery
Exploit improved structure-based design and computational modeling	A lack of sufficient data from published studies to build out target molecular properties from algorithm- or modeling-based approaches (*e.g.* artificial intelligence)
Apply novel formulations and delivery devices	Peptide molecular properties or therapeutic target may preclude the use of certain formulations or delivery routes (*e.g.* limited drug stability in gastrointestinal tract, intracellular permeability, half-life)

surface charge, size and additional permeation enhancer to facilitate delivery of peptides to the brain.[375]

12.4 Future Perspective

Significant research advances over the last two decades have established peptides as an attractive modality to expand druggable target space. Continued progress in peptide screening *via* mRNA and phage display and synthetic and chemically encoded libraries will enable increasingly better peptide drug discovery. Similarly, more sophisticated characterization approaches to elucidate peptide absorption, distribution and metabolic barriers, systemically and relative to intracellular exposure, will be required to support more robust molecular design SAR. Finally, as peptide design and chemistry advance to enable better pharmaceutical properties, new peptide drug delivery and formulation approaches will provide more expanded options to support increasingly more patient-friendly and cost-effective dosage forms (Table 12.6).

References

1. T. K. Sawyer, in *Peptide-based Drug Discovery: Challenges and New Therapeutics*, The Royal Society of Chemistry, 2017, pp. 1–34.
2. K. Fosgerau and T. Hoffmann, *Drug Discovery Today*, 2015, **20**, 122–128.
3. P. Vlieghe, V. Lisowski, J. Martinez and M. Khrestchatisky, *Drug Discovery Today*, 2010, **15**, 40–56.

4. P. M. Saladin, B. D. Zhang and J. M. Reichert, *IDrugs*, 2009, **12**, 779–784.
5. A. Wise, S. C. Jupe and S. Rees, *Annu. Rev. Pharmacol. Toxicol.*, 2004, **44**, 43–66.
6. J. P. Overington, B. Al-Lazikani and A. L. Hopkins, *Nat. Rev. Drug Discovery*, 2006, **5**, 993–996.
7. J. J. Nestor, Jr, *Curr. Med. Chem.*, 2009, **16**, 4399–4418.
8. V. J. Hruby, *Nat. Rev. Drug Discovery*, 2002, **1**, 847–858.
9. J. L. Lau and M. K. Dunn, *Bioorg. Med. Chem.*, 2018, **26**, 2700–2707.
10. R. M. Jones, P. D. Boatman, G. Semple, Y. J. Shin and S. Y. Tamura, *Curr. Opin. Pharmacol.*, 2003, **3**, 530–543.
11. K. Sriram and P. A. Insel, *Mol. Pharmacol.*, 2018, **93**, 251–258.
12. M. Erak, K. Bellmann-Sickert, S. Els-Heindl and A. G. Beck-Sickinger, *Bioorg. Med. Chem.*, 2018, **26**, 2759–2765.
13. K. Hollenstein, C. de Graaf, A. Bortolato, M. W. Wang, F. H. Marshall and R. C. Stevens, *Trends Pharmacol. Sci.*, 2014, **35**, 12–22.
14. S. M. Lee, J. M. Booe and A. A. Pioszak, *Eur. J. Pharmacol.*, 2015, **763**, 196–205.
15. D. K. Laimou, M. Katsara, M. T. Matsoukas, V. Apostolopoulos, A. N. Troganis and T. V. Tselios, *Amino Acids*, 2010, **39**, 1147–1160.
16. T. K. Sawyer, P. J. Sanfilippo, V. J. Hruby, M. H. Engel, C. B. Heward, J. B. Burnett and M. E. Hadley, *Proc. Natl. Acad. Sci. U. S. A.*, 1980, **77**, 5754–5758.
17. H. Widmer, A. Widmer and W. Braun, *J. Biomol. NMR*, 1993, **3**, 307–324.
18. U. C. Marx, K. Adermann, P. Bayer, W. G. Forssmann and P. Rosch, *Biochem. Biophys. Res. Commun.*, 2000, **267**, 213–220.
19. S. Runge, S. Schimmer, J. Oschmann, C. B. Schiodt, S. M. Knudsen, C. B. Jeppesen, K. Madsen, J. Lau, H. Thogersen and R. Rudolph, *Biochemistry*, 2007, **46**, 5830–5840.
20. K. C. Venneti and C. M. Hewage, *FEBS Lett.*, 2011, **585**, 346–352.
21. S. Runge, H. Thogersen, K. Madsen, J. Lau and R. Rudolph, *J. Biol. Chem.*, 2008, **283**, 11340–11347.
22. J. P. Mayer, F. Zhang and R. D. DiMarchi, *Biopolymers*, 2007, **88**, 687–713.
23. G. Hannig, B. Tchernychev, C. B. Kurtz, A. P. Bryant, M. G. Currie and I. Silos-Santiago, *Front. Mol. Neurosci.*, 2014, **7**, 31.
24. C. I. Schroeder and D. J. Craik, *Future Med. Chem.*, 2012, **4**, 1243–1255.
25. S. T. Henriques, E. Deplazes, N. Lawrence, O. Cheneval, S. Chaousis, M. Inserra, P. Thongyoo, G. F. King, A. E. Mark, I. Vetter, D. J. Craik and C. I. Schroeder, *J. Biol. Chem.*, 2016, **291**, 17049–17065.
26. M. A. Hossain and R. A. D. Bathgate, *Bioorg. Med. Chem.*, 2018, **26**, 2827–2841.
27. S. L. Schreiber and G. R. Crabtree, *Immunol. Today*, 1992, **13**, 136–142.
28. H. M. Abdel-Rahman, G. S. Al-karamany, N. A. El-Koussi, A. F. Youssef and Y. Kiso, *Curr. Med. Chem.*, 2002, **9**, 1905–1922.

29. A. A. Agbowuro, W. M. Huston, A. B. Gamble and J. D. A. Tyndall, *Med. Res. Rev.*, 2018, **38**, 1295–1331.
30. A. K. Ghosh, *J. Med. Chem.*, 2009, **52**, 2163–2176.
31. R. Brock, *Bioconjugate Chem.*, 2014, **25**, 863–868.
32. K. Kurrikoff, M. Gestin and U. Langel, *Expert Opin. Drug Delivery*, 2016, **13**, 373–387.
33. L. Peraro and J. A. Kritzer, *Angew. Chem., Int. Ed.*, 2018, **57**, 11868–11881.
34. R. D. Taylor, M. Rey-Carrizo, T. Passioura and H. Suga, *Drug Discovery Today Technol.*, 2017, **26**, 17–23.
35. R. Liu, X. Li and K. S. Lam, *Curr. Opin. Chem. Biol.*, 2017, **38**, 117–126.
36. Z. Qian, P. G. Dougherty and D. Pei, *Curr. Opin. Chem. Biol.*, 2017, **38**, 80–86.
37. P. G. Dougherty, Z. Qian and D. Pei, *Biochem. J.*, 2017, **474**, 1109–1125.
38. C. Heinis and G. Winter, *Curr. Opin. Chem. Biol.*, 2015, **26**, 89–98.
39. G. Li, W. L. Zheng, Y. Liu and X. Y. Li, *Curr. Opin. Chem. Biol.*, 2015, **26**, 25–33.
40. A. Mollica, R. Costante, A. Stefanucci and E. Novellino, *J. Enzyme Inhib. Med. Chem.*, 2015, **30**, 575–580.
41. T. Morioka, N. D. Loik, C. J. Hipolito, Y. Goto and H. Suga, *Curr. Opin. Chem. Biol.*, 2015, **26**, 34–41.
42. T. Passioura and H. Suga, *Chem Commun*, 2017, **53**, 1931–1940.
43. L. Rahbarnia, S. Farajnia, H. Babaei, J. Majidi, K. Veisi, V. Ahmadzadeh and B. Akbari, *J. Drug Targeting*, 2017, **25**, 216–224.
44. A. Tavassoli, *Curr. Opin. Chem. Biol.*, 2017, **38**, 30–35.
45. B. N. Bullock, A. L. Jochim and P. S. Arora, *J. Am. Chem. Soc.*, 2011, **133**, 14220–14223.
46. A. L. Jochim and P. S. Arora, *Mol. BioSyst.*, 2009, **5**, 924–926.
47. A. L. Jochim and P. S. Arora, *ACS Chem. Biol.*, 2010, **5**, 919–923.
48. O. Keskin, A. Gursoy, B. Ma and R. Nussinov, *Chem. Rev.*, 2008, **108**, 1225–1244.
49. H. Remaut and G. Waksman, *Trends Biochem. Sci.*, 2006, **31**, 436–444.
50. D. A. Guarracino, B. N. Bullock and P. S. Arora, *Biopolymers*, 2011, **95**, 1–7.
51. E. Fire, S. V. Gulla, R. A. Grant and A. E. Keating, *Protein Sci.*, 2010, **19**, 507–519.
52. B. Ku, J. S. Woo, C. Liang, K. H. Lee, H. S. Hong, E. Xiaofei, K. S. Kim, J. U. Jung and B. H. Oh, *PLoS Pathog.*, 2008, **4**, e25.
53. P. H. Kussie, S. Gorina, V. Marechal, B. Elenbaas, J. Moreau, A. J. Levine and N. P. Pavletich, *Science*, 1996, **274**, 948–953.
54. Y. Nam, P. Sliz, L. Y. Song, J. C. Aster and S. C. Blacklow, *Cell*, 2006, **124**, 973–983.
55. S. A. Dames, M. Martinez-Yamout, R. N. De Guzman, H. J. Dyson and P. E. Wright, *Proc. Natl. Acad. Sci. U. S. A.*, 2002, **99**, 5271–5276.
56. S. K. Nair and S. K. Burley, *Cell*, 2003, **112**, 193–205.
57. J. Marcotrigiano, A. C. Gingras, N. Sonenberg and S. K. Burley, *Mol. Cell*, 1999, **3**, 707–716.

58. R. K. Bledsoe, V. G. Montana, T. B. Stanley, C. J. Delves, C. J. Apolito, D. D. McKee, T. G. Consler, D. J. Parks, E. L. Stewart, T. M. Willson, M. H. Lambert, J. T. Moore, K. H. Pearce and H. E. Xu, *Cell*, 2002, **110**, 93–105.
59. L. H. Jin, D. Martynowski, S. Y. Zheng, T. Wada, W. Xie and Y. Li, *Mol. Endocrinol.*, 2010, **24**, 923–929.
60. M. Rushe, L. Silvian, S. Bixler, L. L. Chen, A. Cheung, S. Bowes, H. Cuervo, S. Berkowitz, T. Zheng, K. Guckian, M. Pellegrini and A. Lugovskoy, *Structure*, 2008, **16**, 798–808.
61. P. M. Cromm and C. M. Crews, *ACS Cent. Sci.*, 2017, **3**, 830–838.
62. J. A. McCauley and M. T. Rudd, *Curr. Opin. Pharmacol.*, 2016, **30**, 84–92.
63. A. D. Frankel and C. O. Pabo, *Cell*, 1988, **55**, 1189–1193.
64. M. Green and P. M. Loewenstein, *Cell*, 1988, **55**, 1179–1188.
65. A. Joliot, C. Pernelle, H. Deagostinibazin and A. Prochiantz, *Proc. Natl. Acad. Sci. U. S. A.*, 1991, **88**, 1864–1868.
66. S. El-Andaloussi, T. Holm and U. Langel, *Curr. Pharm. Des.*, 2005, **11**, 3597–3611.
67. C. Foerg and H. P. Merkle, *J. Pharm. Sci.*, 2008, **97**, 144–162.
68. G. Guidotti, L. Brambilla and D. Rossi, *Trends Pharmacol. Sci.*, 2017, **38**, 406–424.
69. H. J. Johansson, S. E. Andaloussi and U. Langel, *Methods Mol. Biol.*, 2011, **683**, 233–247.
70. R. M. Johnson, S. D. Harrison and D. Maclean, *Methods Mol. Biol.*, 2011, **683**, 535–551.
71. R. L. Juliano, R. Alam, V. Dixit and H. M. Kang, *Wiley Interdiscip. Rev.: Nanomed. Nanobiotechnol.*, 2009, **1**, 324–335.
72. E. Vives, J. Schmidt and A. Pelegrin, *Biochim. Biophys. Acta*, 2008, **1786**, 126–138.
73. R. Fischer, M. Fotin-Mleczek, H. Hufnagel and R. Brock, *ChemBioChem*, 2005, **6**, 2126–2142.
74. M. Fotin-Mleczek, R. Fischer and R. Brock, *Curr. Pharm. Des.*, 2005, **11**, 3613–3628.
75. M. Grdisa, *Curr. Med. Chem.*, 2011, **18**, 1373–1379.
76. J. M. Gump and S. F. Dowdy, *Trends Mol. Med.*, 2007, **13**, 443–448.
77. L. Hirt, J. Badaut, J. Thevenet, C. Granziera, L. Regli, F. Maurer, C. Bonny and J. Bogousslavsky, *Stroke*, 2004, **35**, 1738–1743.
78. K. Inagaki, L. Chen, F. Ikeno, F. H. Lee, K. Imahashi, D. M. Bouley, M. Rezaee, P. G. Yock, E. Murphy and D. Mochly-Rosen, *Circulation*, 2003, **108**, 2304–2307.
79. L. B. Lopes, E. J. Furnish, P. Komalavilas, C. R. Flynn, P. Ashby, A. Hansen, D. P. Ly, G. P. Yang, M. T. Longaker, A. Panitch and C. M. Brophy, *J. Invest. Dermatol.*, 2009, **129**, 590–598.
80. G. H. Bird, F. Bernal, K. Pitter and L. D. Walensky, *Methods Enzymol.*, 2008, **446**, 369–386.
81. P. M. Cromm, J. Spiegel and T. N. Grossmann, *ACS Chem. Biol.*, 2015, **10**, 1362–1375.

82. V. Guerlavais and T. K. Sawyer, in *Annual Reports in Medicinal Chemistry*, ed. M. C. Desai, Academic Press, 2014, vol. 49, pp. 331–345.
83. T. K. Sawyer, V. Guerlavais, K. Darlak and E. Feyfant, in *Macrocycles in Drug Discovery*, The Royal Society of Chemistry, 2014, pp. 339–366.
84. T. K. Sawyer, A. W. Partridge, H. Y. K. Kaan, Y. C. Juang, S. Lim, C. Johannes, T. Y. Yuen, C. Verma, S. Kannan, P. Aronica, Y. S. Tan, B. Sherborne, S. Ha, J. Hochman, S. Y. Chen, L. Surdi, A. Peier, B. Sauvagnat, P. J. Dandliker, C. J. Brown, S. Ng, F. Ferrer and D. P. Lane, *Bioorg. Med. Chem.*, 2018, **26**, 2807–2815.
85. G. L. Verdine and L. D. Walensky, *Clin. Cancer Res.*, 2007, **13**, 7264–7270.
86. N. K. Bashiruddin and H. Suga, *Curr. Opin. Chem. Biol.*, 2015, **24**, 131–138.
87. W. H. Connors, S. P. Hale and N. K. Terrett, *Curr. Opin. Chem. Biol.*, 2015, **26**, 42–47.
88. K. Josephson, A. Ricardo and J. W. Szostak, *Drug Discovery Today*, 2014, **19**, 388–399.
89. S. W. Millward, S. Fiacco, R. J. Austin and R. W. Roberts, *ACS Chem. Biol.*, 2007, **2**, 625–634.
90. T. Passioura, T. Katoh, Y. Goto and H. Suga, *Annu. Rev. Biochem.*, 2014, **83**, 727–752.
91. Y. V. Schlippe, M. C. Hartman, K. Josephson and J. W. Szostak, *J. Am. Chem. Soc.*, 2012, **134**, 10469–10477.
92. Z. R. Zhu, A. Shaginian, L. C. Grady, T. O'Keeffe, X. G. E. Shi, C. P. Davie, G. L. Simpson, J. A. Messer, G. Evindar, R. N. Bream, P. P. Thansandote, N. R. Prentice, A. M. Mason and S. Pal, *ACS Chem. Biol.*, 2018, **13**, 53–59.
93. V. Baeriswyl and C. Heinis, *ChemMedChem*, 2013, **8**, 377–384.
94. K. Deyle, X. D. Kong and C. Heinis, *Acc. Chem. Res.*, 2017, **50**, 1866–1874.
95. S. S. Kale, C. Villequey, X. D. Kong, A. Zorzi, K. Deyle and C. Heinis, *Nat. Chem.*, 2018, **10**, 715–723.
96. S. Ng and R. Derda, *Org. Biomol. Chem.*, 2016, **14**, 5539–5545.
97. Z. Qian, P. Upadhyaya and D. Pei, *Methods Mol. Biol.*, 2015, **1248**, 39–53.
98. Z. Q. Qian, A. Martyna, R. L. Hard, J. Wang, G. Appiah-Kubi, C. Coss, M. A. Phelps, J. S. Rossman and D. H. Pei, *Biochemistry*, 2016, **55**, 2601–2612.
99. A. A. Patchett and E. H. Cordes, *Adv. Enzymol. Relat. Areas Mol. Biol.*, 1985, **57**, 1–84.
100. T. Aoyagi, S. Kunimoto, H. Morishima, T. Takeuchi and H. Umezawa, *J. Antibiot.*, 1971, **24**, 687–694.
101. R. E. Handschumacher, M. W. Harding, J. Rice, R. J. Drugge and D. W. Speicher, *Science*, 1984, **226**, 544–547.
102. D. J. Tipper and J. L. Strominger, *Proc. Natl. Acad. Sci. U. S. A.*, 1965, **54**, 1133–1141.
103. D. H. Williams and J. P. Waltho, *Biochem. Pharmacol.*, 1988, **37**, 133–141.

104. R. H. Baltz, V. Miao and S. K. Wrigley, *Nat. Prod. Rep.*, 2005, **22**, 717–741.
105. G. P. Miljanich, *Curr. Med. Chem.*, 2004, **11**, 3029–3040.
106. R. E. Middleton, V. A. Warren, R. L. Kraus, J. C. Hwang, C. J. Liu, G. Dai, R. M. Brochu, M. G. Kohler, Y. D. Gao, V. M. Garsky, M. J. Bogusky, J. T. Mehl, C. J. Cohen and M. M. Smith, *Biochemistry*, 2002, **41**, 14734–14747.
107. M. Lebl, V. Krchnak, N. F. Sepetov, B. Seligmann, P. Strop, S. Felder and K. S. Lam, *Biopolymers*, 1995, **37**, 177–198.
108. O. H. Aina, R. Liu, J. L. Sutcliffe, J. Marik, C. X. Pan and K. S. Lam, *Mol. Pharm.*, 2007, **4**, 631–651.
109. V. V. Komnatnyy, T. E. Nielsen and K. Qvortrup, *Chem. Commun.*, 2018, **54**, 6759–6771.
110. R. Liu, X. Li, W. Xiao and K. S. Lam, *Adv. Drug Delivery Rev.*, 2017, **110–111**, 13–37.
111. G. Abbenante and D. P. Fairlie, *Med Chem*, 2005, **1**, 71–104.
112. J. Chatterjee, C. Gilon, A. Hoffman and H. Kessler, *Acc. Chem. Res.*, 2008, **41**, 1331–1342.
113. Y. Che and G. R. Marshall, *Expert Opin. Ther. Targets*, 2008, **12**, 101–114.
114. M. Eguchi and M. Kahn, *Mini-Rev. Med. Chem.*, 2002, **2**, 447–462.
115. S. Fletcher and A. D. Hamilton, *Curr. Opin. Chem. Biol.*, 2005, **9**, 632–638.
116. L. Gentilucci, R. De Marco and L. Cerisoli, *Curr. Pharm. Des.*, 2010, **16**, 3185–3203.
117. S. J. Hershberger, S. G. Lee and J. Chmielewski, *Curr. Top. Med. Chem.*, 2007, **7**, 928–942.
118. V. J. Hruby, F. Alobeidi and W. Kazmierski, *Biochem. J.*, 1990, **268**, 249–262.
119. G. R. Marshall, *Tetrahedron*, 1993, **49**, 3547–3558.
120. G. R. Marshall, D. J. Kuster and Y. Che, *Methods Mol. Biol.*, 2009, **575**, 123–158.
121. R. P. McGeary and D. P. Fairlie, *Curr. Opin. Drug Discovery Dev.*, 1998, **1**, 208–217.
122. M. Pelay-Gimeno, A. Glas, O. Koch and T. N. Grossmann, *Angew. Chem., Int. Ed. Engl.*, 2015, **54**, 8896–8927.
123. A. S. Ripka and D. H. Rich, *Curr. Opin. Chem. Biol.*, 1998, **2**, 441–452.
124. T. K. Sawyer, in *Structure-Based Drug Design: Diseases, Targets, Techniques and Developments*, ed. P. Veerapandian, Mercel Dekker, New York, 1999, pp. 559–634.
125. M. J. de Vega, M. Martin-Martinez and R. Gonzalez-Muniz, *Curr. Top. Med. Chem.*, 2007, **7**, 33–62.
126. W. L. DeLano, *Curr. Opin. Struct. Biol.*, 2002, **12**, 14–20.
127. D. J. Diller, J. Swanson, A. S. Bayden, M. Jarosinski and J. Audie, *Future Med. Chem.*, 2015, **7**, 2173–2193.
128. D. C. Fry and L. T. Vassilev, *J. Mol. Med.*, 2005, **83**, 955–963.

129. B. Ma, T. Elkayam, H. Wolfson and R. Nussinov, *Proc. Natl. Acad. Sci. U. S. A.*, 2003, **100**, 5772–5777.
130. B. Ma and R. Nussinov, *Curr. Top. Med. Chem.*, 2007, **7**, 999–1005.
131. P. Matsson, B. C. Doak, B. Over and J. Kihlberg, *Adv. Drug Delivery Rev.*, 2016, **101**, 42–61.
132. S. M. McHugh, J. R. Rogers, S. A. Solomon, H. Yu and Y. S. Lin, *Curr. Opin. Chem. Biol.*, 2016, **34**, 95–102.
133. J. L. Medina-Franco, O. Mendez-Lucio and K. Martinez-Mayorga, *Adv. Protein Chem. Struct. Biol.*, 2014, **96**, 1–37.
134. I. M. Nooren and J. M. Thornton, *EMBO J.*, 2003, **22**, 3486–3492.
135. J. C. Obenauer and M. B. Yaffe, *Methods Mol. Biol.*, 2004, **261**, 445–468.
136. Y. Pommier and J. Cherfils, *Trends Pharmacol. Sci.*, 2005, **26**, 138–145.
137. D. P. Ryan and J. M. Matthews, *Curr. Opin. Struct. Biol.*, 2005, **15**, 441–446.
138. A. Schreyer and T. Blundell, *Chem. Biol. Drug Des.*, 2009, **73**, 157–167.
139. A. Stein and P. Aloy, *PLoS Comput. Biol.*, 2010, **6**, e1000789.
140. A. M. Watkins and P. S. Arora, *Eur. J. Med. Chem.*, 2015, **94**, 480–488.
141. J. M. Ahn, N. A. Boyle, M. T. MacDonald and K. D. Janda, *Mini-Rev. Med. Chem.*, 2002, **2**, 463–473.
142. P. M. Fischer, *Curr. Protein Pept. Sci.*, 2003, **4**, 339–356.
143. M. Hagihara, N. J. Anthony, T. J. Stout, J. Clardy and S. L. Schreiber, *J. Am. Chem. Soc.*, 1992, **114**, 6568–6570.
144. S. Matej, B. Jiri, C. Stanislav, P. Javed, S. Renata, K. Viliam, A. Lukas and J. Tomas, *Ann. Oncol.*, 2008, **19**, 64.
145. M. J. Adler, A. G. Jamieson and A. D. Hamilton, *Curr. Top. Microbiol. Immunol.*, 2011, **348**, 1–23.
146. A. D. Bautista, C. J. Craig, E. A. Harker and A. Schepartz, *Curr. Opin. Chem. Biol.*, 2007, **11**, 685–692.
147. Y. Che and G. R. Marshall, *J. Med. Chem.*, 2006, **49**, 111–124.
148. R. P. Cheng, S. H. Gellman and W. F. DeGrado, *Chem. Rev.*, 2001, **101**, 3219–3232.
149. C. G. Cummings and A. D. Hamilton, *Curr. Opin. Chem. Biol.*, 2010, **14**, 341–346.
150. J. M. Davis, L. K. Tsou and A. D. Hamilton, *Chem. Soc. Rev.*, 2007, **36**, 326–334.
151. R. F. Hirschmann, K. C. Nicolaou, A. R. Angeles, J. S. Chen and A. B. Smith, 3rd, *Acc. Chem. Res.*, 2009, **42**, 1511–1520.
152. W. S. Horne and S. H. Gellman, *Acc. Chem. Res.*, 2008, **41**, 1399–1408.
153. W. A. Loughlin, J. D. Tyndall, M. P. Glenn and D. P. Fairlie, *Chem. Rev.*, 2004, **104**, 6085–6117.
154. A. Moretto, M. Crisma, F. Formaggio and C. Toniolo, *Biopolymers*, 2010, **94**, 721–732.
155. A. B. Smith, 3rd, A. K. Charnley and R. Hirschmann, *Acc. Chem. Res.*, 2011, **44**, 180–193.
156. N. G. Angelo and P. S. Arora, *J. Am. Chem. Soc.*, 2005, **127**, 17134–17135.

157. N. P. Chongsiriwatana, J. A. Patch, A. M. Czyzewski, M. T. Dohm, A. Ivankin, D. Gidalevitz, R. N. Zuckermann and A. E. Barron, *Proc. Natl. Acad. Sci. U. S. A.*, 2008, **105**, 2794–2799.
158. J. T. Ernst, J. Becerril, H. S. Park, H. Yin and A. D. Hamilton, *Angew. Chem., Int. Ed.*, 2003, **42**, 535–539.
159. R. M. Freidinger, D. F. Veber, D. S. Perlow, J. R. Brooks and R. Saperstein, *Science*, 1980, **210**, 656–658.
160. J. Garner and M. M. Harding, *Org. Biomol. Chem.*, 2007, **5**, 3577–3585.
161. R. Hirschmann, K. C. Nicolaou, S. Pietranico, E. M. Leahy, J. Salvino, B. Arison, M. A. Cichy, P. G. Spoors, W. C. Shakespeare, P. A. Sprengeler, P. Hamley, A. B. Smith, T. Reisine, K. Raynor, L. Maechler, C. Donaldson, W. Vale, R. M. Freidinger, M. R. Cascieri and C. D. Strader, *J. Am. Chem. Soc.*, 1993, **115**, 12550–12568.
162. K. S. Kee and S. D. Jois, *Curr. Pharm. Des.*, 2003, **9**, 1209–1224.
163. J. A. Kritzer, O. M. Stephens, D. A. Guarracino, S. K. Reznik and A. Schepartz, *Bioorg. Med. Chem.*, 2005, **13**, 11–16.
164. K. Leonard, J. J. Marugan, P. Raboisson, R. Calvo, J. M. Gushue, H. K. Koblish, J. Lattanze, S. Y. Zhao, M. D. Cummings, M. R. Player, A. C. Maroney and T. B. Lu, *Bioorg. Med. Chem. Lett.*, 2006, **16**, 3463–3468.
165. U. Nagai and K. Sato, *Tetrahedron Lett.*, 1985, **26**, 647–650.
166. S. T. Phillips, M. Rezac, U. Abel, M. Kossenjans and P. A. Bartlett, *J. Am. Chem. Soc.*, 2002, **124**, 58–66.
167. J. A. Robinson, *Acc. Chem. Res.*, 2008, **41**, 1278–1288.
168. N. T. Ross, W. P. Katt and A. D. Hamilton, *Philos. Trans. Royal Soc. A Math. Phys. Eng. Sci.*, 2010, **368**, 989–1008.
169. T. J. Sanborn, C. W. Wu, R. N. Zuckerman and A. E. Barron, *Biopolymers*, 2002, **63**, 12–20.
170. A. B. Smith, M. C. Guzman, P. A. Sprengeler, T. P. Keenan, R. C. Holcomb, J. L. Wood, P. J. Carroll and R. Hirschmann, *J. Am. Chem. Soc.*, 1994, **116**, 9947–9962.
171. A. B. Smith, S. D. Knight, P. A. Sprengeler and R. Hirschmann, *Bioorg. Med. Chem.*, 1996, **4**, 1021–1034.
172. Y. C. Tang and C. M. Deber, *Biopolymers*, 2002, **65**, 254–262.
173. P. Tosovska and P. S. Arora, *Org. Lett.*, 2010, **12**, 1588–1591.
174. J. D. A. Tyndall, B. Pfeiffer, G. Abbenante and D. P. Fairlie, *Chem. Rev.*, 2005, **105**, 793–826.
175. B. Vaz and L. Brunsveld, *Org. Biomol. Chem.*, 2008, **6**, 2988–2994.
176. P. N. Wyrembak and A. D. Hamilton, *J. Am. Chem. Soc.*, 2009, **131**, 4566–4567.
177. D. Y. Xin, L. M. Perez, T. R. Ioerger and K. Burgess, *Angew. Chem., Int. Ed.*, 2014, **53**, 3594–3598.
178. H. Yin, G. I. Lee, K. A. Sedey, O. Kutzki, H. S. Park, B. P. Orner, J. T. Ernst, H. G. Wang, S. M. Sebti and A. D. Hamilton, *J. Am. Chem. Soc.*, 2005, **127**, 10191–10196.

179. F. Bernal and S. G. Katz, *Methods Mol. Biol.*, 2014, **1176**, 107–114.
180. G. H. Bird, E. Mazzola, K. Opoku-Nsiah, M. A. Lammert, M. Godes, D. S. Neuberg and L. D. Walensky, *Nat. Chem. Biol.*, 2016, **12**, 845–852.
181. Y. W. Kim, T. N. Grossmann and G. L. Verdine, *Nat. Protoc.*, 2011, **6**, 761–771.
182. L. D. Walensky and G. H. Bird, *J. Med. Chem.*, 2014, **57**, 6275–6288.
183. A. M. White and D. J. Craik, *Expert Opin. Drug Discovery*, 2016, **11**, 1151–1163.
184. Y. S. Chang, B. Graves, V. Guerlavais, C. Tovar, K. Packman, K. H. To, K. A. Olson, K. Kesavan, P. Gangurde, A. Mukherjee, T. Baker, K. Darlak, C. Elkin, Z. Filipovic, F. Z. Qureshi, H. Cai, P. Berry, E. Feyfant, X. E. Shi, J. Horstick, D. A. Annis, A. M. Manning, N. Fotouhi, H. Nash, L. T. Vassilev and T. K. Sawyer, *Proc. Natl. Acad. Sci. U. S. A.*, 2013, **110**, E3445–E3454.
185. J. Morimoto, Y. Hayashi and H. Suga, *Angew. Chem., Int. Ed.*, 2012, **51**, 3423–3427.
186. B. A. Seigal, W. H. Connors, A. Fraley, R. M. Borzilleri, P. H. Carter, S. L. Emanuel, J. Fargnoli, K. Kim, M. Lei, J. G. Naglich, M. E. Pokross, S. L. Posy, H. Shen, N. Surti, R. Talbott, Y. Zhang and N. K. Terrett, *J. Med. Chem.*, 2015, **58**, 2855–2861.
187. P. Upadhyaya, Z. Q. Qian, N. G. Selner, S. R. Clippinger, Z. R. Wu, R. Briesewitz and D. H. Pei, *Angew. Chem., Int. Ed.*, 2015, **54**, 7602–7606.
188. K. Sakamoto, Y. Kamada, T. Sameshima, M. Yaguchi, A. Niida, S. Sasaki, M. Miwa, S. Ohkubo, J. Sakamoto, M. Kamaura, N. Cho and A. Tani, *Biochem. Biophys. Res. Commun.*, 2017, **484**, 605–611.
189. Y. S. Tan, D. P. Lane and C. S. Verma, *Drug Discovery Today*, 2016, **21**, 1642–1653.
190. C. E. Schafmeister, J. Po and G. L. Verdine, *J. Am. Chem. Soc.*, 2000, **122**, 5891–5892.
191. F. Bernal, A. F. Tyler, S. J. Korsmeyer, L. D. Walensky and G. L. Verdine, *J. Am. Chem. Soc.*, 2007, **129**, 2456–2457.
192. N. N. Danial, L. D. Walensky, C. Y. Zhang, C. S. Choi, J. K. Fisher, A. J. A. Molina, S. R. Datta, K. L. Pitter, G. H. Bird, J. D. Wikstrom, J. T. Deeney, K. Robertson, J. Morash, A. Kulkarni, S. Neschen, S. Kim, M. E. Greenberg, B. E. Corkey, O. S. Shirihai, G. I. Shulman, B. B. Lowell and S. J. Korsmeyer, *Nat. Med.*, 2008, **14**, 144–153.
193. L. K. Henchey, J. R. Porter, I. Ghosh and P. S. Arora, *ChemBioChem*, 2010, **11**, 2104–2107.
194. Y. H. Lau, P. de Andrade, S. T. Quah, M. Rossmann, L. Laraia, N. Skold, T. J. Sum, P. J. E. Rowling, T. L. Joseph, C. Verma, M. Hyvonen, L. S. Itzhaki, A. R. Venkitaraman, C. J. Brown, D. P. Lane and D. R. Spring, *Chem. Sci.*, 2014, **5**, 1804–1809.
195. A. Muppidi, Z. Y. Wang, X. L. Li, J. D. Chen and Q. Lin, *Chem. Commun.*, 2011, **47**, 9396–9398.
196. F. Wachter, A. M. Morgan, M. Godes, R. Mourtada, G. H. Bird and L. D. Walensky, *Oncogene*, 2017, **36**, 2184–2190.

197. H. Zhang, Q. Zhao, S. Bhattacharya, A. A. Waheed, X. Tong, A. Hong, S. Heck, F. Curreli, M. Goger, D. Cowburn, E. O. Freed and A. K. Ebnath, *J. Mol. Biol.*, 2008, **378**, 565–580.
198. H. T. Zhang, F. Curreli, X. H. Zhang, S. Bhattacharya, A. A. Waheed, A. Cooper, D. Cowburn, E. O. Freed and A. K. Debnath, *Retrovirology*, 2011, **8**, 28.
199. L. K. Henchey, S. Kushal, R. Dubey, R. N. Chapman, B. Z. Olenyuk and P. S. Arora, *J. Am. Chem. Soc.*, 2010, **132**, 941–943.
200. M. M. Madden, A. Muppidi, Z. Y. Li, X. L. Li, J. D. Chen and Q. Lin, *Bioorg. Med. Chem. Lett.*, 2011, **21**, 1472–1475.
201. R. E. Moellering, M. Cornejo, T. N. Davis, C. Del Bianco, J. C. Aster, S. C. Blacklow, A. L. Kung, D. G. Gilliland, G. L. Verdine and J. E. Bradner, *Nature*, 2009, **462**, 182–U157.
202. A. Muppidi, X. L. Li, J. D. Chen and Q. Lin, *Bioorg. Med. Chem. Lett.*, 2011, **21**, 7412–7415.
203. A. Patgiri, K. K. Yadav, P. S. Arora and D. Bar-Sagi, *Nat. Chem. Biol.*, 2011, **7**, 585–587.
204. C. Phillips, L. R. Roberts, M. Schade, R. Bazin, A. Bent, N. L. Davies, R. Moore, A. D. Pannifer, A. R. Pickford, S. H. Prior, C. M. Read, A. Scott, D. G. Brown, B. Xu and S. L. Irving, *J. Am. Chem. Soc.*, 2011, **133**, 9696–9699.
205. K. Pitter, F. Bernal, J. LaBelle and L. D. Walensky, *Programmed Cell Death, the Biology and Therapeutic Implications of Cell Death, Part B*, 2008, **446**, 387–408.
206. J. R. Porter, M. R. Helmers, P. Wang, J. L. Furman, S. T. Joy, P. S. Arora and I. Ghosh, *Chem. Commun.*, 2010, **46**, 8020–8022.
207. D. O. Sviridov, I. Z. Ikpot, J. Stonik, S. K. Drake, M. Amar, D. O. Osei-Hwedieh, G. Piszczek, S. Turner and A. T. Remaley, *Biochem. Biophys. Res. Commun.*, 2011, **410**, 446–451.
208. A. D. Bautista, J. S. Appelbaum, C. J. Craig, J. Michel and A. Schepartz, *J. Am. Chem. Soc.*, 2010, **132**, 2904–2906.
209. F. Bernal, M. Wade, M. Godes, T. N. Davis, D. G. Whitehead, A. L. Kung, G. M. Wahl and L. D. Walensky, *Cancer Cell*, 2010, **18**, 411–422.
210. C. R. Braun, J. Mintseris, E. Gavathiotis, G. H. Bird, S. P. Gygi and L. D. Walensky, *Chem. Biol.*, 2010, **17**, 1325–1333.
211. S. M. Q. Chee, J. Wongsantichon, Q. S. Tng, R. Robinson, T. L. Joseph, C. Verma, D. P. Lane, C. J. Brown and F. J. Ghadessy, *PLoS One*, 2014, **9**, e104914.
212. H. K. Cui, B. Zhao, Y. H. Li, Y. Guo, H. Hu, L. Liu and Y. G. Chen, *Cell Res.*, 2013, **23**, 581–584.
213. Y. Demizu, S. Nagoya, M. Shirakawa, M. Kawamura, N. Yamagata, Y. Sato, M. Doi and M. Kurihara, *Bioorg. Med. Chem. Lett.*, 2013, **23**, 4292–4296.
214. A. O. Frank, B. Vangamudi, M. D. Feldkamp, E. M. Souza-Fagundes, J. W. Luzwick, D. Cortez, E. T. Olejniczak, A. G. Waterson, O. W. Rossanese, W. J. Chazin and S. W. Fesik, *J. Med. Chem.*, 2014, **57**, 2455–2461.

215. E. Gavathiotis, M. Suzuki, M. L. Davis, K. Pitter, G. H. Bird, S. G. Katz, H. C. Tu, H. Kim, E. H. Y. Cheng, N. Tjandra and L. D. Walensky, *Nature*, 2008, **455**, 1076–U1076.
216. T. N. Grossmann, J. T. H. Yeh, B. R. Bowman, Q. Chu, R. E. Moellering and G. L. Verdine, *Proc. Natl. Acad. Sci. U. S. A.*, 2012, **109**, 17942–17947.
217. E. S. Leshchiner, A. Parkhitko, G. H. Bird, J. Luccarelli, J. A. Bellairs, S. Escudero, K. Opoku-Nsiah, M. Godes, N. Perrimon and L. D. Walensky, *Proc. Natl. Acad. Sci. U. S. A.*, 2015, **112**, 1761–1766.
218. Y. Q. Long, S. X. Huang, Z. Zawahir, Z. L. Xu, H. Y. Li, T. W. Sanchez, Y. Zhi, S. De Houwer, F. Christ, Z. Debyser and N. Neamati, *J. Med. Chem.*, 2013, **56**, 5601–5612.
219. J. Spiegel, P. M. Cromm, A. Itzen, R. S. Goody, T. N. Grossmann and H. Waldmann, *Angew. Chem., Int. Ed.*, 2014, **53**, 2498–2503.
220. M. L. Stewart, E. Fire, A. E. Keating and L. D. Walensky, *Nat. Chem. Biol.*, 2010, **6**, 595–601.
221. L. D. Walensky, K. Pitter, J. Morash, K. J. Oh, S. Barbuto, J. Fisher, E. Smith, G. L. Verdine and S. J. Korsmeyer, *Mol. Cell*, 2006, **24**, 199–210.
222. S. J. Wei, T. Joseph, S. Chee, L. Li, L. Yurlova, K. Zolghadr, C. Brown, D. Lane, C. Verma and F. Ghadessy, *PLoS One*, 2013, **8**, e81068.
223. D. Lama, S. T. Quah, C. S. Verma, R. Lakshminarayanan, R. W. Beuerman, D. P. Lane and C. J. Brown, *Sci. Rep.*, 2013, **3**, 3451.
224. Y. Teng, H. Y. Qin, A. Bahassan, N. G. Bendzunas, E. J. Kennedy and J. K. Cowell, *Cancer Res.*, 2016, **76**, 5133–5142.
225. Y. Teng, A. Bahassan, D. Y. Dong, L. E. Hanold, X. O. Ren, E. J. Kennedy and J. K. Cowell, *Cancer Res.*, 2016, **76**, 965–973.
226. Y. K. Rennie, P. J. McIntyre, T. Akindele, R. Bayliss and A. G. Jamieson, *ACS Chem. Biol.*, 2016, **11**, 3383–3390.
227. C. L. Ahlbach, K. W. Lexa, A. T. Bockus, V. Chen, P. Crews, M. P. Jacobson and R. S. Lokey, *Future Med. Chem.*, 2015, **7**, 2121–2130.
228. A. T. Bockus, K. W. Lexa, C. R. Pye, A. S. Kalgutkar, J. W. Gardner, K. C. R. Hund, W. M. Hewitt, J. A. Schwochert, E. Glassey, D. A. Price, A. M. Mathiowetz, S. Liras, M. P. Jacobson and R. S. Lokey, *J. Med. Chem.*, 2015, **58**, 4581–4589.
229. A. Furukawa, C. E. Townsend, J. Schwochert, C. R. Pye, M. A. Bednarek and R. S. Lokey, *J. Med. Chem.*, 2016, **59**, 9503–9512.
230. W. M. Hewitt, S. S. F. Leung, C. R. Pye, A. R. Ponkey, M. Bednarek, M. P. Jacobson and R. S. Lokey, *J. Am. Chem. Soc.*, 2015, **137**, 715–721.
231. J. Schwochert, Y. T. Lao, C. R. Pye, M. R. Naylor, P. V. Desai, I. C. G. Valcarcel, J. A. Barrett, G. Sawada, M. J. Blanco and R. S. Lokey, *ACS Med. Chem. Lett.*, 2016, **7**, 757–761.
232. J. Schwochert, R. Turner, M. Thang, R. F. Berkeley, A. R. Ponkey, K. M. Rodriguez, S. S. F. Leung, B. Khunte, G. Goetz, C. Limberakis, A. S. Kalgutkar, H. Eng, M. J. Shapiro, A. M. Mathiowetz, D. A. Price, S. Liras, M. P. Jacobson and R. S. Lokey, *Org. Lett.*, 2015, **17**, 2928–2931.
233. Z. Antosova, M. Mackova, V. Kral and T. Macek, *Trends Biotechnol.*, 2009, **27**, 628–635.

234. P. Jarver, I. Mager and U. Langel, *Trends Pharmacol. Sci.*, 2010, **31**, 528–535.
235. J. H. Lin, *Curr. Drug Metab.*, 2009, **10**, 661–691.
236. D. K. Malik, S. Baboota, A. Ahuja, S. Hasan and J. Ali, *Curr. Drug Delivery*, 2007, **4**, 141–151.
237. C. L. Stevenson, *Curr. Pharm. Biotechnol.*, 2009, **10**, 122–137.
238. F. M. Veronese, *Biomaterials*, 2001, **22**, 405–417.
239. M. Werle and A. Bernkop-Schnurch, *Amino Acids*, 2006, **30**, 351–367.
240. D. Verge and X. Lopez, *Curr. Diabetes Rev.*, 2010, **6**, 191–200.
241. C. Q. Pan, J. M. Buxton, S. L. Yung, I. Tom, L. Yang, H. Chen, M. MacDougall, A. Bell, T. H. Claus, K. B. Clairmont and J. P. Whelan, *J. Biol. Chem.*, 2006, **281**, 12506–12515.
242. B. B. Yang and A. Kido, *Clin. Pharmacokinet.*, 2011, **50**, 295–306.
243. D. J. Kuter, J. B. Bussel, R. M. Lyons, V. Pullarkat, T. B. Gernsheimer, F. M. Senecal, L. M. Aledort, J. N. George, C. M. Kessler, M. A. Sanz, H. A. Liebman, F. T. Slovick, J. T. de Wolf, E. Bourgeois, T. H. Guthrie, Jr, A. Newland, J. S. Wasser, S. I. Hamburg, C. Grande, F. Lefrere, A. E. Lichtin, M. D. Tarantino, H. R. Terebelo, J. F. Viallard, F. J. Cuevas, R. S. Go, D. H. Henry, R. L. Redner, L. Rice, M. R. Schipperus, D. M. Guo and J. L. Nichol, *Lancet*, 2008, **371**, 395–403.
244. S. Console, C. Marty, C. Garcia-Echeverria, R. Schwendener and K. Ballmer-Hofer, *J. Biol. Chem.*, 2003, **278**, 35109–35114.
245. S. Deshayes, M. C. Morris, G. Divita and F. Heitz, *Curr. Pharm. Des.*, 2005, **11**, 3629–3638.
246. R. Fischer, K. Kohler, M. Fotin-Mleczek and R. Brock, *J. Biol. Chem.*, 2004, **279**, 12625–12635.
247. S. Sandgren, F. Cheng and M. Belting, *J. Biol. Chem.*, 2002, **277**, 38877–38883.
248. J. P. Richard, K. Melikov, H. Brooks, P. Prevot, B. Lebleu and L. V. Chernomordik, *J. Biol. Chem.*, 2005, **280**, 15300–15306.
249. F. Duchardt, M. Fotin-Mleczek, H. Schwarz, R. Fischer and R. Brock, *Traffic*, 2007, **8**, 848–866.
250. M. M. Fretz, N. A. Penning, S. Al-Taei, S. Futaki, T. Takeuchi, I. Nakase, G. Storm and A. T. Jones, *Biochem. J.*, 2007, **403**, 335–342.
251. I. M. Kaplan, J. S. Wadia and S. F. Dowdy, *J. Controlled Release*, 2005, **102**, 247–253.
252. M. Kosuge, T. Takeuchi, I. Nakase, A. T. Jones and S. Futaki, *Bioconjugate Chem.*, 2008, **19**, 656–664.
253. G. Ter-Avetisyan, G. Tunnemann, D. Nowak, M. Nitschke, A. Herrmann, M. Drab and M. C. Cardoso, *J. Biol. Chem.*, 2009, **284**, 3370–3378.
254. G. Tunnemann, R. M. Martin, S. Haupt, C. Patsch, F. Edenhofer and M. C. Cardoso, *FASEB J.*, 2006, **20**, 1775–1784.
255. W. P. Verdurmen, M. Thanos, I. R. Ruttekolk, E. Gulbins and R. Brock, *J. Controlled Release*, 2010, **147**, 171–179.
256. J. He, W. B. Kauffman, T. Fuselier, S. K. Naveen, T. G. Voss, K. Hristova and W. C. Wimley, *J. Biol. Chem.*, 2013, **288**, 29974–29986.

257. W. B. Kauffman, T. Fuselier, J. He and W. C. Wimley, *Trends Biochem. Sci.*, 2015, **40**, 749–764.
258. W. P. Verdurmen and R. Brock, *Trends Pharmacol. Sci.*, 2011, **32**, 116–124.
259. W. C. Wimley, *Methods Mol. Biol.*, 2015, **1324**, 89–106.
260. Q. Chu, R. E. Moellering, G. J. Hilinski, Y. W. Kim, T. N. Grossmann, J. T. H. Yeh and G. L. Verdine, *MedChemComm*, 2015, **6**, 111–119.
261. G. H. Goetz, L. Philippe and M. J. Shapiro, *ACS Med. Chem. Lett.*, 2014, **5**, 1167–1172.
262. C. R. W. Guimaraes, A. M. Mathiowetz, M. Shalaeva, G. Goetz and S. Liras, *J. Chem. Inf. Model.*, 2012, **52**, 882–890.
263. H. D. Herce, A. E. Garcia and M. C. Cardoso, *J. Am. Chem. Soc.*, 2014, **136**, 17459–17467.
264. D. S. Nielsen, R. J. Lohman, H. N. Hoang, T. A. Hill, A. Jones, A. J. Lucke and D. P. Fairlie, *ChemBioChem*, 2015, **16**, 2289–2293.
265. S. B. Rafi, B. R. Hearn, P. Vedantham, M. P. Jacobson and A. R. Renslo, *J. Med. Chem.*, 2012, **55**, 3163–3169.
266. T. Rezai, J. E. Bock, M. V. Zhou, C. Kalyanaraman, R. S. Lokey and M. P. Jacobson, *J. Am. Chem. Soc.*, 2006, **128**, 14073–14080.
267. T. Rezai, B. Yu, G. L. Millhauser, M. P. Jacobson and R. S. Lokey, *J. Am. Chem. Soc.*, 2006, **128**, 2510–2511.
268. C. K. Wang, S. E. Northfield, J. E. Swedberg, B. Colless, S. Chaousis, D. A. Price, S. Liras and D. J. Craik, *Eur. J. Med. Chem.*, 2015, **97**, 202–213.
269. S. Aubry, B. Aussedat, D. Delaroche, C. Y. Jiao, G. Bolbach, S. Lavielle, G. Chassaing, S. Sagan and F. Burlina, *Biochim. Biophys. Acta*, 2010, **1798**, 2182–2189.
270. L. J. Gordon, M. Allen, P. Artursson, M. M. Hann, B. J. Leavens, A. Mateus, S. Readshaw, K. Valko, G. J. Wayne and A. West, *J. Biomol. Screen*, 2016, **21**, 156–164.
271. T. Holm, S. E. Andaloussi and U. Langel, *Methods Mol. Biol.*, 2011, **683**, 207–217.
272. J. R. LaRochelle, G. B. Cobb, A. Steinauer, E. Rhoades and A. Schepartz, *J. Am. Chem. Soc.*, 2015, **137**, 2536–2541.
273. R. Rezgui, K. Blumer, G. Yeoh-Tan, A. J. Trexler and M. Magzoub, *Biochim. Biophys. Acta Biomembr.*, 2016, **1858**, 1499–1506.
274. K. Josephson, A. Ricardo and J. W. Szostak, *Drug Discovery Today*, 2014, **19**, 388–399.
275. R. Liu, X. Li, W. Xiao and K. S. Lam, *Adv. Drug Delivery Rev.*, 2017, **110–111**, 13–37.
276. C. A. Lipinski, F. Lombardo, B. W. Dominy and P. J. Feeney, *Adv. Drug Delivery Rev.*, 2001, **46**, 3–26.
277. J. S. Bond and P. E. Butler, *Annu. Rev. Biochem.*, 1987, **56**, 333–364.
278. S. M. Howell, S. V. Fiacco, T. T. Takahashi, F. Jalali-Yazdi, S. W. Millward, B. Hu, P. Wang and R. W. Roberts, *Sci. Rep.*, 2014, **4**, 6008.

279. S. T. Buckley, T. A. Baekdal, A. Vegge, S. J. Maarbjerg, C. Pyke, J. Ahnfelt-Ronne, K. G. Madsen, S. G. Scheele, T. Alanentalo, R. K. Kirk, B. L. Pedersen, R. B. Skyggebjerg, A. J. Benie, H. M. Strauss, P. O. Wahlund, S. Bjerregaard, E. Farkas, C. Fekete, F. L. Sondergaard, J. Borregaard, M. L. Hartoft-Nielsen and L. B. Knudsen, *Sci. Transl. Med.*, 2018, **10**, eaar7047.
280. A. F. B. Rader, M. Weinmuller, F. Reichart, A. Schumacher-Klinger, S. Merzbach, C. Gilon, A. Hoffman and H. Kessler, *Angew Chem Int., Ed.*, 2018, **57**, 14414–14438.
281. A. T. Bockus, C. M. McEwen and R. S. Lokey, *Curr. Top. Med. Chem.*, 2018, **13**, 821–836.
282. J. F. Borel, G. Baumann, I. Chapman, P. Donatsch, A. Fahr, E. A. Mueller and J. M. Vigouret, *Adv. Pharmacol.*, 1996, **35**, 115–246.
283. C. K. Wang and D. J. Craik, *Biopolymers*, 2016, **106**, 901–909.
284. J. Lautz, H. Kessler, R. Kaptein and W. F. van Gunsteren, *J. Comput.-Aided Mol. Des.*, 1987, **1**, 219–241.
285. C. K. Wang, S. E. Northfield, B. Colless, S. Chaousis, I. Hamernig, R. J. Lohman, D. S. Nielsen, C. I. Schroeder, S. Liras, D. A. Price, D. P. Fairlie and D. J. Craik, *Proc. Natl. Acad. Sci. U. S. A.*, 2014, **111**, 17504–17509.
286. M. Boehm, K. Beaumont, R. Jones, A. S. Kalgutkar, L. Zhang, K. Atkinson, G. Bai, J. A. Brown, H. Eng, G. H. Goetz, B. R. Holder, B. Khunte, S. Lazzaro, C. Limberakis, S. Ryu, M. J. Shapiro, L. Tylaska, J. Yan, R. Turner, S. S. F. Leung, M. Ramaseshan, D. A. Price, S. Liras, M. P. Jacobson, D. J. Earp, R. S. Lokey, A. M. Mathiowetz and E. Menhaji-Klotz, *J. Med. Chem.*, 2017, **60**, 9653–9663.
287. T. Rezai, B. Yu, G. L. Millhauser, M. P. Jacobson and R. S. Lokey, *J. Am. Chem. Soc.*, 2006, **128**, 2510–2511.
288. G. H. Goetz, L. Philippe and M. J. Shapiro, *ACS Med. Chem. Lett.*, 2014, **5**, 1167–1172.
289. M. Boehm, K. Beaumont, R. Jones, A. S. Kalgutkar, L. Zhang, K. Atkinson, G. Bai, J. A. Brown, H. Eng, G. H. Goetz, B. R. Holder, B. Khunte, S. Lazzaro, C. Limberakis, S. Ryu, M. J. Shapiro, L. Tylaska, J. Yan, R. Turner, S. S. F. Leung, M. Ramaseshan, D. A. Price, S. Liras, M. P. Jacobson, D. J. Earp, R. S. Lokey, A. M. Mathiowetz and E. Menhaji-Klotz, *J. Med. Chem.*, 2017, **60**, 9653–9663.
290. P. Upadhyaya, Z. Qian, N. G. Selner, S. R. Clippinger, Z. Wu, R. Briesewitz and D. Pei, *Angew. Chem., Int. Ed.*, 2015, **54**, 7602–7606.
291. T. B. Trinh, P. Upadhyaya, Z. Qian and D. Pei, *ACS Comb. Sci.*, 2016, **18**, 75–85.
292. G. Guidotti, L. Brambilla and D. Rossi, *Trends Pharmacol. Sci.*, 2017, **38**, 406–424.
293. I. Martin, M. Teixido and E. Giralt, *ChemBioChem*, 2011, **12**, 896–903.
294. W. C. Wimley and K. Hristova, *J. Membr. Biol.*, 2012, **239**, 27–34.
295. W. B. Kauffman, T. Fuselier, J. He and W. C. Wimley, *Trends Biochem. Sci.*, 2015, **40**, 749–764.

296. J. R. Marks, J. Placone, K. Hristova and W. C. Wimley, *J. Am. Chem. Soc.*, 2011, **133**, 8995–9004.
297. T. Y. Wang, Y. Sun, N. Muthukrishnan, A. Erazo-Oliveras, K. Najjar and J. P. Pellois, *J. Biol. Chem.*, 2016, **291**, 7902–7914.
298. H. D. Herce, A. E. Garcia and M. C. Cardoso, *J. Am. Chem. Soc.*, 2014, **136**, 17459–17467.
299. M. Shirangi, M. Najafi, D. T. S. Rijkers, R. J. Kok, W. E. Hennink and C. F. van Nostrum, *Bioconjugate Chem.*, 2016, **27**, 576–585.
300. Z. Qian, A. Martyna, R. L. Hard, J. Wang, G. Appiah-Kubi, C. Coss, M. A. Phelps, J. S. Rossman and D. Pei, *Biochemistry*, 2016, **55**, 2601–2612.
301. L. Dietrich, B. Rathmer, K. Ewan, T. Bange, S. Heinrichs, T. C. Dale, D. Schade and T. N. Grossmann, *Cell. Chem. Biol.*, 2017, **24**, 958–968.
302. Y. Tian, Y. Jiang, J. Li, D. Wang, H. Zhao and Z. Li, *ChemBioChem*, 2017, **18**, 2087–2093.
303. G. C. Kim, J. H. Ahn, J. H. Oh, S. Nam, S. Hyun, J. Yu and Y. Lee, *Biomacromolecules*, 2018, **19**, 2863–2869.
304. C. Collinet, M. Stoter, C. R. Bradshaw, N. Samusik, J. C. Rink, D. Kenski, B. Habermann, F. Buchholz, R. Henschel, M. S. Mueller, W. E. Nagel, E. Fava, Y. Kalaidzidis and M. Zerial, *Nature*, 2014, **464**, 243–249.
305. F. Authier, M. Metioui, S. Fabrega, M. Kouach and G. Briand, *J. Biol. Chem.*, 2002, **277**, 9437–9446.
306. F. Authier, P. H. Cameron, C. Merlen, M. Kouach and G. Briand, *Endocrinology*, 2003, **144**, 5353–5364.
307. F. Authier, B. I. Posner and J. J. Bergeron, *FEBS Lett.*, 1996, **389**, 55–60.
308. L. Di, *AAPS J.*, 2015, **17**, 134–143.
309. S. J. Hubbard, R. J. Beynon and J. M. Thornton, *Protein Eng.*, 1998, **11**, 349–359.
310. D. P. Fairlie, J. D. Tyndall, R. C. Reid, A. K. Wong, G. Abbenante, M. J. Scanlon, D. R. March, D. A. Bergman, C. L. Chai and B. A. Burkett, *J. Med. Chem.*, 2000, **43**, 1271–1281.
311. H. M. Werner, C. C. Cabalteja and W. S. Horne, *ChemBioChem*, 2016, **17**, 712–718.
312. E. Moroz, S. Matoori and J. C. Leroux, *Adv. Drug Delivery Rev.*, 2016, **101**, 108–121.
313. A. F. B. Rader, M. Weinmuller, F. Reichart, A. Schumacher-Klinger, S. Merzbach, C. Gilon, A. Hoffman and H. Kessler, *Angew. Chem., Int. Ed. Engl.*, 2018, **57**, 14414–14438.
314. H. Lennernas, *Xenobiotica*, 2007, **37**, 1015–1051.
315. S. K. Kim, *Am. J. Roentgenol., Radium Ther. Nucl. Med.*, 1968, **104**, 522–524.
316. S. Tuvia, D. Pelled, K. Marom, P. Salama, M. Levin-Arama, I. Karmeli, G. H. Idelson, I. Landau and R. Mamluk, *Pharm. Res.*, 2014, **31**, 2010–2021.
317. S. Tuvia, J. Atsmon, S. L. Teichman, S. Katz, P. Salama, D. Pelled, I. Landau, I. Karmeli, M. Bidlingmaier, C. J. Strasburger, D. L. Kleinberg, S. Melmed and R. Mamluk, *J. Clin. Endocrinol. Metab.*, 2012, **97**, 2362–2369.

318. W. F. Richter, S. G. Bhansali and M. E. Morris, *AAPS J.*, 2012, **14**, 559–570.
319. W. F. Richter and B. Jacobsen, *Drug Metab. Dispos.*, 2014, **42**, 1881–1889.
320. H. Kim, H. Park and S. J. Lee, *Sci. Rep.*, 2017, **7**, 9613.
321. R. Bottger, R. Hoffmann and D. Knappe, *PLoS One*, 2017, **12**, e0178943.
322. M. Werle and A. Bernkop-Schnurch, *Amino Acids*, 2006, **30**, 351–367.
323. W. Bauer, U. Briner, W. Doepfner, R. Haller, R. Huguenin, P. Marbach, T. J. Petcher and Pless, *Life Sci.*, 1982, **31**, 1133–1140.
324. W. C. Zamboni, *Pharmacotherapy*, 2003, **23**, 9S–14S.
325. B. B. Yang, P. K. Lum, M. M. Hayashi and L. K. Roskos, *J. Pharm. Sci.*, 2004, **93**, 1367–1373.
326. K. Bellmann-Sickert, C. E. Elling, A. N. Madsen, P. B. Little, K. Lundgren, L. O. Gerlach, R. Bergmann, B. Holst, T. W. Schwartz and A. G. Beck-Sickinger, *J. Med. Chem.*, 2011, **54**, 2658–2667.
327. V. Fischer, A. Rodriguez-Gascon, F. Heitz, R. Tynes, C. Hauck, D. Cohen and A. E. Vickers, *Drug Metab. Dispos.*, 1998, **26**, 802–811.
328. L. Ivanova, C. K. Faeste and S. Uhlig, *Anal. Bioanal. Chem.*, 2011, **400**, 2889–2901.
329. D. Kovacsics, I. Patik and C. Ozvegy-Laczka, *Expert Opin. Drug. Metab. Toxicol.*, 2017, **13**, 409–424.
330. T. Yamada, K. Niinuma, M. Lemaire, T. Terasaki and Y. Sugiyama, *J. Pharmacol. Exp. Ther.*, 1996, **279**, 1357–1364.
331. M. Visentin, B. Stieger, M. Merz and G. A. Kullak-Ublick, *J. Pharmacol. Exp. Ther.*, 2015, **355**, 145–151.
332. S. Akhteruzzaman, Y. Kato, H. Kouzuki, H. Suzuki, A. Hisaka, B. Stieger, P. J. Meier and Y. Sugiyama, *J. Pharmacol. Exp. Ther.*, 1999, **290**, 1107–1115.
333. J. F. Jin, L. L. Zhu, M. Chen, H. M. Xu, H. F. Wang, X. Q. Feng, X. P. Zhu and Q. Zhou, *Patient Prefer. Adherence*, 2015, **9**, 923–942.
334. L. Kagan, P. Gershkovich, A. Mendelman, S. Amsili, N. Ezov and A. Hoffman, *Eur. J. Pharm. Biopharm.*, 2007, **67**, 759–765.
335. C. J. H. Porter and S. A. Charman, *J. Pharm. Sci.*, 2000, **89**, 297–310.
336. D. K. Clodfelter, A. H. Pekar, D. M. Rebhun, K. A. Destrampe, H. A. Havel, S. R. Myers and M. L. Brader, *Pharm. Res.*, 1998, **15**, 254–262.
337. Q. Yang and S. K. Lai, *Wires Nanomed. Nanobi.*, 2015, **7**, 655–677.
338. J. A. Pedras-Vasconcelos, The immunogenicity of therapeutic proteins-what you don't know can hurt YOU and the patient, 2014, (last accessed 2 Feb 2019). https://www.fda.gov/downloads/Drugs/DevelopmentApprovalProcess/SmallBusinessAssistance/UCM408709.pdf.
339. *Draft Guidance for Industry: Biosimilars: Questions and Answers Regarding Implementatioin of the Biologics Price Competition and Innovation Act of 2009*, U.S. Department of Health and Human Services, Food and Drug Administration, CDER, CBER, 2012.
340. *Guidance for the Industry: Immunogenicity Assessment for Therapeutic Protein Products*, U.S. Department of Health and Human Services, Food and Drug Administration, CDER, CBER, 2014.

341. L. Holfeld, N. Herth, D. Singer, R. Hoffmann and D. Knappe, *Future Med. Chem.*, 2015, **7**, 1581–1596.
342. Z. H. Huang, L. Shi, J. W. Ma, Z. Y. Sun, H. Cai, Y. X. Chen, Y. F. Zhao and Y. M. Li, *J. Am. Chem. Soc.*, 2012, **134**, 8730–8733.
343. J. S. Rudra, S. Mishra, A. S. Chong, R. A. Mitchell, E. H. Nardin, V. Nussenzweig and J. H. Collier, *Biomaterials*, 2012, **33**, 6476–6484.
344. J. Zepeda-Cervantes and L. Vaca, *Expert Rev. Vaccines*, 2018, **17**, 723–738.
345. B. L. Furman, *Toxicon*, 2012, **59**, 464–471.
346. B. K. Meyer, A. Ni, B. H. Hu and L. Shi, *J. Pharm. Sci.*, 2007, **96**, 3155–3167.
347. P. Heljo, A. Ross, I. E. Zarraga, A. Pappenberger and H. C. Mahler, *Pharm. Res.*, 2015, **32**, 3201–3212.
348. R. L. Hutchings, S. M. Singh, J. Cabello-Villegas and K. M. G. Mallela, *J. Pharm. Sci.*, 2013, **102**, 365–376.
349. Y. F. Maa and C. C. Hsu, *Int. J. Pharmaceut*, 1996, **140**, 155–168.
350. I. T. Degim and N. Celebi, *Curr. Pharm. Design*, 2007, **13**, 99–117.
351. M. B. DeYoung, L. MacConell, V. Sarin, M. Trautmann and P. Herbert, *Diabetes Technol. Ther.*, 2011, **13**, 1145–1154.
352. E. I. Minder, J. Barman-Aksoezen and X. Y. Schneider-Yin, *Clin. Pharmacokinet.*, 2017, **56**, 815–823.
353. *Scenesse: EPAR – Public Assessment Report* EMA/CHMP/709396/2014 Rev.1, Committee for Medicinal Products for Human Use (CHMP), 2014.
354. H. A. Blair, *Drugs*, 2016, **76**, 1787–1792.
355. *PARSABIV (etelcalcetide) US Drug Label*, 2017.
356. Chemistry Review(s), 2017, US FDA, Drugs@FDA. (last accessed January 2019). https://www.accessdata.fda.gov/scripts/cder/daf/.
357. C. K. Wang and D. J. Craik, *Biopolymers*, 2016, **106**, 901–909.
358. C. K. Wang, S. E. Northfield, B. Colless, S. Chaousis, I. Hamernig, R. J. Lohman, D. S. Nielsen, C. I. Schroeder, S. Liras, D. A. Price, D. P. Fairlie and D. J. Craik, *Proc. Natl. Acad. Sci. U. S. A.*, 2014, **111**, 17504–17509.
359. T. A. S. Aguirre, D. Teijeiro-Osorio, M. Rosa, I. S. Coulter, M. J. Alonso and D. J. Brayden, *Adv. Drug Delivery Rev.*, 2016, **106**, 223–241.
360. P. Li, H. M. Nielsen and A. Mullertz, *Expert Opin. Drug Delivery*, 2012, **9**, 1289–1304.
361. J. O. Morales, K. R. Fathe, A. Brunaugh, S. Ferrati, S. Li, M. Montenegro-Nicolini, Z. Mousavikhamene, J. T. McConville, M. R. Prausnitz and H. D. C. Smyth, *AAPS J.*, 2017, **19**, 652–668.
362. R. W. Busby, M. M. Kessler, W. P. Bartolini, A. P. Bryant, G. Hannig, C. S. Higgins, R. M. Solinga, J. V. Tobin, J. D. Wakefield, C. B. Kurtz and M. G. Currie, *J. Pharmacol. Exp. Ther.*, 2013, **344**, 196–206.
363. P. Taylor, Ironwood's colonic delivery platform a 'pipeline within a pipeline,' say analysts, 2017, FierceBiotech.com. (last accessed 07 Nov 2018). https://www.fiercebiotech.com/biotech/ironwood-s-colonic-delivery-platform-a-pipeline-within-a-pipeline-say-analysts.

364. Ironwood Begins New Phase III Study to Expand Linzess Label, 2018, Zacks Equity Research, zacks.com, Nasdaq.com. (last accessed 07 Nov 2018). https://www.nasdaq.com/article/ironwood-begins-new-phase-iii-study-to-expand-linzess-label-cm992323.
365. Ironwood Pharmaceuticals Presents Data Further Elucidating Linaclotide's Effect on Pain at Digestive Disease Week® 2017, 2017, Ironwood Pharmaceuticals, Inc., (last accessed 17 Jul 2018). http://news.ironwoodpharma.com/news-releases/news-release-details/ironwood-pharmaceuticals-presents-data-further-elucidating.
366. S. Tuvia, J. Atsmon, S. L. Teichman, S. Katz, P. Salama, D. Pelled, I. Landau, I. Karmeli, M. Bidlingmaier, C. J. Strasburger, D. L. Kleinberg, S. Melmed and R. Mamluk, *J. Clin. Endocrinol. Metab.*, 2012, **97**, 2362–2369.
367. S. Tuvia, D. Pelled, K. Marom, P. Salama, M. Levin-Arama, I. Karmeli, G. H. Idelson, I. Landau and R. Mamluk, *Pharm. Res.*, 2014, **31**, 2010–2021.
368. Chiasma Achieves 50% of Target Patients Randomized in CHIASMA OPTIMAL Phase 3 Clinical Trial of Octreotide Capsules in Patients with Acromegaly, 2018, (last accessed July 18, 2018). http://ir.chiasmapharma.com/phoenix.zhtml?c=254057&p=irol-newsArticle&ID=2354661.
369. A. J. Ahmann, M. Capehorn, G. Charpentier, F. Dotta, E. Henkel, I. Lingvay, A. G. Holst, M. P. Annett and V. R. Aroda, *Diabetes Care*, 2018, **41**, 258–266.
370. B. T. Griffin, J. F. Guo, E. Presas, M. D. Donovan, M. J. Alonso and C. M. O'Driscoll, *Adv. Drug Delivery Rev.*, 2016, **106**, 367–380.
371. A. H. L. Chow, H. H. Y. Tong, P. Chattopadhyay and B. Y. Shekunov, *Pharm. Res.*, 2007, **24**, 411–437.
372. J. S. Patton, J. Bukar and S. Nagarajan, *Adv. Drug Delivery Rev.*, 1999, **35**, 235–247.
373. J. Oleck, S. Kassam and J. D. Goldman, *Diabetes Spectrum*, 2016, **29**, 180–184.
374. J. J. Lochhead and R. G. Thorne, *Adv. Drug Delivery Rev.*, 2012, **64**, 614–628.
375. E. Samaridou and M. J. Alonso, *Bioorg. Med. Chem.*, 2018, **26**, 2888–2905.
376. D. Pei and M. Buyanova, *Bioconjugate Chem.*, 2019, **30**, 273–283.

CHAPTER 13

Assessing the Impact of Functional Excipients on Peptide Drug Product Attributes During Pharmaceutical Development

SUZANNE M. D'ADDIO,*[a] JAMESON R. BOTHE,[b] ERIKA WALSH[a] AND MARIAN E. GINDY[c]

[a] MRL, Merck & Co., Inc., 33 Avenue Louis Pasteur, Boston, MA 02115, USA; [b] MRL, Merck & Co, Inc., 126 East Lincoln Avenue, Rahway, NJ 07065, USA; [c] MRL, Merck & Co, Inc., 770 Sumneytown Pike, West Point, PA 19486, USA
*Email: suzanne.daddio2@merck.com

13.1 Introduction

During pharmaceutical drug product development, scientists must design, develop and characterize formulations and manufacturing processes that reliably produce a drug product that meets predetermined performance criteria as established in the Quality Target Product Profile (QTPP), with regulatory expectations delineated in the International Council on Harmonisation of Technical Requirements for Registration of Pharmaceuticals for Human Use (ICH) Q8 (R2) guidance.[1] In a Quality-by-Design (QbD) framework for drug product development, the criteria in the QTPP include considerations for the intended therapeutic use, including clinical route of administration, desirable

Drug Discovery Series No. 72
Peptide Therapeutics: Strategy and Tactics for Chemistry, Manufacturing and Controls
Edited by Ved Srivastava

Published by the Royal Society of Chemistry, www.rsc.org

dosage strength, container closure system, drug-specific liabilities and other quality criteria associated with the intended drug product dosage form (Figure 13.1).[2] Potential Critical Quality Attributes (pCQAs) are identified at the onset of product development, based on the QTPP and prior knowledge of drug molecule-specific risks. Knowledge is derived from experience on related compounds, external reports, patent literature and CMC risks assessed during early-stage de-risking of the peptide (Figure 13.2). These pCQAs are those thought to have a significant impact on the efficacy, safety and quality of the final drug product. Drug product dosage forms and formulations are thus evaluated against criteria defined in the QTPP; for example, options for injectable, oral and inhaled routes of administration include vials, ready-to-use injectors, tablets, injectable depots, implantable devices and nasal sprays. Outcomes of these assessments inform on the fundamental understanding of a drug product formulation, relating variations in raw materials, processing and environmental conditions to the final product quality. Based on the level of risk, CQAs can be finalized, a formulation design space established and a control strategy implemented in order to ensure quality. Within the scope of formulation definition, a description and justification of all excipients included in the final drug product are required in section P.2 of the Common Technical Document (CTD) to support the identity, concentration and compatibility of excipients in the formulation.

Regardless of when development of the final drug product formulation intended for market approval begins, activities assessing product-specific risks should begin at the Discovery stage to inform the QTPP (Figure 13.2). Pharmaceutical Discovery teams assess the formulation developability of a molecule, first exploring stability liabilities that have the potential to adversely affect the quality and/or safety of the peptide drug product. Through this process, identification of significant development risks may trigger molecular redesign of the peptide active substance to arrive at a more suitable entity. Often, pharmaceutical development liabilities that are not considered "show stoppers" are deferred for evaluation during early preformulation assessments. The outcomes of these studies are leveraged to narrow or exclude the selection of a particular formulation type. For example, preformulation evaluations may inform on excessive chemical instability of a peptide in solution, suggesting instead a focus on lyophilized formulation development. Remaining formulation risks must be prioritized for mitigation during development, ranked by the magnitude of the impact on the quality of the final drug product. Once these determinations have been made, the process of evaluating and selecting excipients that mitigate specific risks begins.

This chapter provides a framework for peptide formulation development across phases of Discovery and Development to filing (Figure 13.2), with a focus on considerations to support excipient selection and evaluation. As noted, this process begins with physicochemical characterization and evaluation of peptide stability risks during lead identification and optimization in Discovery, followed by the mechanistic assessment of stability liabilities and evaluation of various excipients during preformulation studies. This

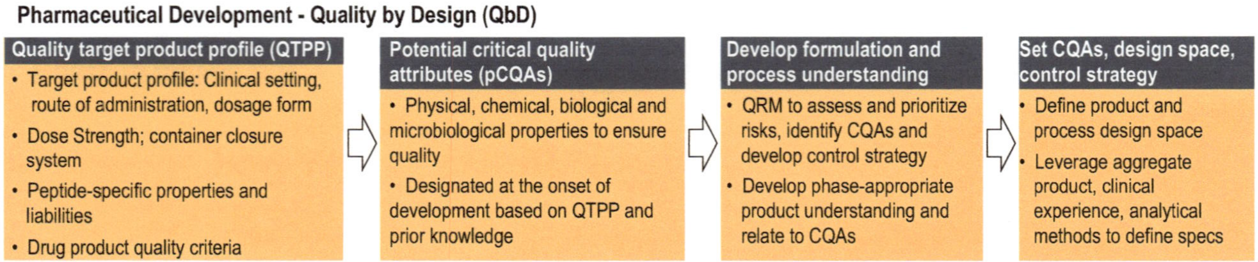

Figure 13.1 Process for pharmaceutical development with implementing QbD principles, adapted from ICH Q8 (R2).

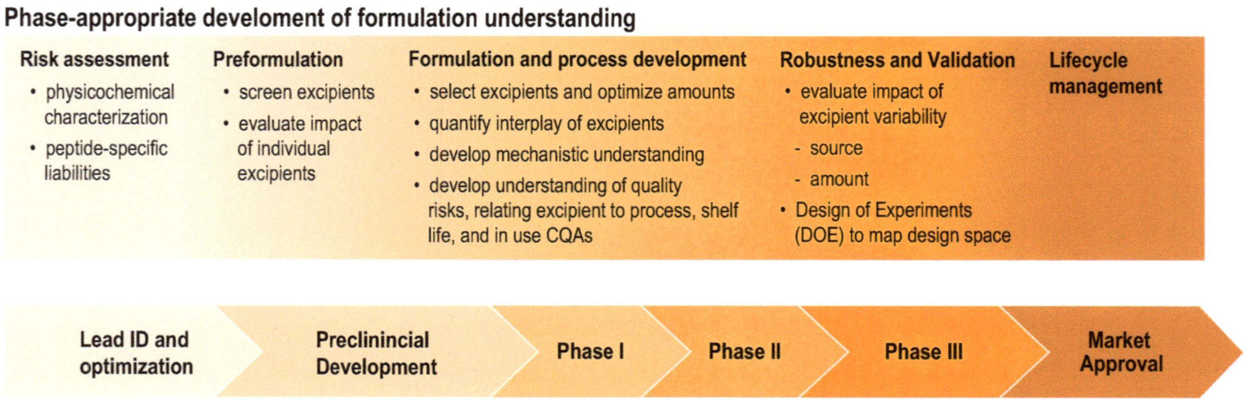

Figure 13.2 Development of compound assessment and understanding across the phases of Discovery and Development, which inform and guide formulation development and excipient selection.

knowledge is leveraged during formulation development and optimization to arrive at robust formulations that meet predefined quality criteria. The critical attributes of a given peptide drug product can be both peptide and formulation specific and a comprehensive summary of pCQAs for drug product types will not be presented. Other chapters in this book, and also the references in Table 13.1, provide additional, detailed examples and considerations for specific applications, including immunogenicity concerns and approaches for the preparation of micro- and nanoparticulate formulations.

We review available sources of information relevant to the selection of excipients for peptide formulations and discuss considerations for their evaluation, including a brief focus on novel excipients. Illustrative examples of peptide formulation development and characterization are discussed, although a comprehensive review is beyond the scope of this chapter. Case studies from the literature and patents are used to emphasize a focus on establishing a fundamental understanding of a formulation and its robustness in relation to excipients. Given that most approved peptide drug products are ready-to-use solutions or freeze-dried solids for parenteral administration, many examples are focused on quality attributes, instability mechanisms and excipient selections relevant to those formulations. Nevertheless, the underlying QbD process for assessing risks, identifying instabilities, developing formulations and defining a design space are applicable regardless of the dosage form and route of administration.

13.2 Identification of Excipients in Peptide Drug Products

There are several excipient-specific resources available to the pharmaceutical scientist that can be leveraged during the design and development of peptide formulations. In this section, a description of several references is provided to inform on the use of excipients in a formulation, guidance for their evaluation in pharmaceutical products, commonly used amounts and underlying mechanism(s) by which an excipient functions in the formulation.

In the USA, the Food and Drug Administration (FDA) maintains the Inactive Ingredient Database, which tabulates routes of administration and the maximum concentration of an excipient in drug products approved by the FDA.[3] This information can be used to identify that excipients are approved for an intended dosing route and the level that serves as a practical upper limit for initial evaluations. As most peptide drug products are dosed parenterally,[4] existing precedence for use is an important benchmark to de-risk safety and tolerability in a formulation. Health Canada maintains the Natural Health Products Ingredients Database, which contains chemical information, functions, monographs and other information on nonmedicinal ingredients.[5] The Japanese Pharmaceutical Excipients Council similarly compiles the Japanese Pharmaceutical Excipients Dictionary (JPED), which lists excipients used in previously approved products in the Japanese market and monographs.[6] The JPED also includes the allowable maximum daily doses for some excipients by route of administration. Consultation of such information early in excipient evaluation is critical in developing formulations for the global market; for instance, there are some instances of tighter restrictions on excipient use in Japan compared with the USA. The use of calcium disodium edetate is one such example. The FDA Inactive Ingredient Database currently lists a maximum concentration of 0.34% for subcutaneously administered products, whereas the JPED places a limit of not more than 0.09 mg of calcium disodium edetate per person per day by the subcutaneous route, although as much as 30 mg may be given intravenously. Hence, for a hypothetical formulation that contains 0.3% calcium disodium edetate in a subcutaneous formulation, the maximum dose volume of that formulation that could be given per day in Japan is 30 μL or less. Depending on the dose, solubility, stability and clinical setting, dosing of such low volumes to limit daily calcium disodium edetate exposure may not be feasible.

Although there is no longer a database maintained by the European Medicines Agency (EMA) on allowable excipients, a recent review by Gervasi *et al.* included a statistical analysis of the frequency of use of the most common excipients and a supplemental list of the excipients used in all parenteral protein formulations approved in the European Union from 1995 to 2018.[7]

As most therapeutic peptides are administered parenterally, primarily *via* intravenous and subcutaneous routes,[4] reviews of the excipients used in such formulations for nonpeptide active substances, including small molecules and proteins, can serve as useful references. Excipients in solution and suspension and lyophilized formulations of small-molecule drugs were reviewed by Pramanick *et al.*, with references to specific marketed drug products and typical levels for different excipients, although maximum allowable concentrations and the role of route of administration limiting the excipient concentration were not considered.[8] Similarly, Akers reviewed common excipients used in formulations of small-molecule and protein parenteral drugs, with a focus on excipient-mediated stabilizing and destabilizing effects.[9]

Mechanisms of peptide instability and the stabilizing functions of excipients in formulations have been reviewed in a number of publications.[8–13] Many more reviews have focused on excipient use in the formulation of proteins and antibodies, which are subject to many of the same quality risks as peptides. Jorgensen *et al.* provided a list of excipient functions in typical parenterally administered biopharmaceutical products, including anti-adsorption agents, antioxidants, buffers, stabilizers, tonicity modifiers and bulking agents, with extensive references to examples and brief commentary on regulatory considerations related to such excipients.[10] Chi presented a comprehensive review of chemical degradation and physical instability mechanisms and associated stabilizing excipients for use in solution and lyophilized formulations of protein products for parenteral administration.[13] In a review of excipients emphasizing physical stabilization of protein drugs, Kamerzell *et al.* stressed the value of developing a fundamental understanding of physical instability to inform on stabilizing strategies and described a variety of analytical techniques suitable for characterizing the solution behavior of proteins and their interactions with excipients, including dynamic light scattering, differential scanning calorimetry, isothermal titration calorimetry, intrinsic and extrinsic fluorescence, circular dichroism (CD) and UV–Vis spectroscopy.[12] Many of the described characterization techniques also have applicability for peptides, while other commonly used protein characterization techniques lose meaning or sensitivity owing to the lower molecular weight and lack of higher order structure in peptides. For example, capillary electrophoresis (CE) is used to analyze the distribution of charged species in pharmaceutical formulations of monoclonal antibodies, but Varga demonstrated the potential challenges of resolving peptide monomer or dimer peaks, stemming from the peptide size being close to the lower end of resolution for the systems, and noted that CE is not a universally applicable method for the analysis of peptides.[14]

While the cited reviews focused on overcoming instabilities in common injectable formulations, peptides can also be formulated as solutions and suspensions for nasal and ocular administration, capsules and tablets for

Table 13.1 Peptide formulation types, potential excipient functions within each formulation and relevant references to reviews and research articles.

Formulation	Potential excipient functions	Ref.
Solutions and lyophilized solids (i.m., s.c., i.v., nasal and nebulized pulmonary administration)	Solvent, buffer, co-solvent, tonicity modifier, solubilizer, stabilizer, bulking agent, lyoprotectant, antimicrobial preservative, propellant (metered-dose inhaler)	Reviews: 8–10, 12, 41, 48, 114–116 Peptide formulations: 29–31, 35, 42, 46, 58, 64, 66, 67, 79, 95, 98, 104, 105, 117, 118
Depot and implants	Controlled-release matrix, buffer, tonicity modifier, chemical stabilizer, processing aid, solvent, complexing agent	Reviews: 17, 70, 75, 101 Peptide formulations: 33, 34, 69, 70, 73, 106, 110, 119–121
Solid oral, oral suspensions or liquid-filled capsules	Chemical stabilizer, bulking agent, lubricant, permeation enhancer, enteric coating, phospholipid, oil	Reviews: 102, 122, 123 Peptide formulations: 16, 59, 124–126
Ocular	Solvent, antimicrobial preservative, permeation enhancer, controlled-release matrix or carrier	Reviews: 127–130 Peptide formulations: 131, 132
Pulmonary	Carrier particles (dry powder), stabilizer	Reviews: 133, 134 Peptide formulations: 135–138

oral administration and as inhaled particulates for pulmonary delivery, and several such products have received marketing approval.[4] Formulations in these settings are engineered with a broad range of materials, including polymers, surfactants and lipids, to optimize the peptide physiochemical properties or improve absorption and bioavailability upon administration.[15] A notable example of success in designing a suspension-based formulation to enhance peptide bioavailability by the oral route is a microemulsion formulation of cyclosporine A (Sandimmune®).[16] Similarly, polymer-based implants and particulate drug depots[17] for systemic peptide delivery have also advanced to the market, including Scenesse® (afamelanotide) and Bydureon® (exenatide). A summary of peptide drug product dosage forms and some examples of excipient functions in such formulations are given in Table 13.1.

13.3 Quality Risk Management and Risk Assessment

ICH Q9 defines Quality Risk Management (QRM) as "a systematic process for the assessment, control, communication and review of risk to the quality of the drug."[18] The QRM framework is readily applied to the process of excipient evaluation and selection to ensure a robust understanding of product quality as a result of the selection and amount of and variations in excipients. It is prudent to initiate the QRM process as early as possible by identifying and evaluating the peptide- and formulation-specific risks and choosing to accept, mitigate or defer risks. Risks, probabilities of occurrence and consequences are weighted in a manner that is appropriate for the phase of development.[2] Specifically, at the Discovery stage, it is often sufficient to highlight the existing risks and defer mitigation until chemical matter is defined and enters development. The risk assessment facilitates prioritization of development activities and is revisited as formulation design and development progress, augmented with learnings and experimentation in order to develop a foundational understanding of probabilities and impacts of risks.

Attention to pCQAs corresponding to peptide molecular structure should be evaluated in Discovery risk assessments and used to inform on chemical redesigns that prevent or mitigate risks, where possible. Developability risks and pCQAs are typically assessed through stress studies that are relevant for the peptide, dosage form and intended route of administration.[19] As demonstrated by Evers and co-workers, the development and adoption of a pharmaceutical risk assessment during molecular design allowed the identification and mitigation of key risks pertaining to peptide activity and stability in the presence of antimicrobial excipients.[20,21] The following discussion is focused on mechanisms of instability that require consideration of unique risks during risk assessments of peptide formulations. Many peptide products conform to the chemical stability expectations for small molecules and physical stability requirements for large molecules over the shelf life, although ICH guidance documents state that testing and quality requirements do not apply to peptide products.[22] A draft guidance issued by the FDA outlines expectations for assessing the safety of degradants and impurities applicable to synthetic peptides in generic formulations of five peptides of rDNA origin: glucagon, liraglutide, nesiritide, teriparatide and teduclutide.[23] Currently, regulatory agency expectations are evolving with increasing numbers of peptide products being submitted for registration, making interactions between sponsors and regulatory agencies during development important to ensure that phase-appropriate characterization of drug product attributes is provided.

13.3.1 Chemical Stability

The mechanisms of chemical degradation in peptides and proteins in solution and solid formulations have been studied extensively, in the context of both pharmaceutically relevant peptides and short peptide fragments to model degradation-prone sequences in larger proteins.[10,24–38] Commonly encountered chemical degradation pathways for biologics include oxidation, deamidation, hydrolysis, isomerization, covalent dimer formation and disulfide scrambling, with numerous studies of the mechanisms and formulation parameters that adversely impact stability having been reviewed.[24,36] However, although peptides and proteins are made of the same constituent amino acid building blocks, the globular structure of larger proteins provides additional chemical stability by shielding some amino acids in the hydrophobic core and providing steric hindrance. In contrast, most peptides in solution have only secondary structure or are random coils, resulting in maximal solvent exposure of nearly all peptide residues. This, in turn, makes the achievement of optimal chemical stability through formulation a significant challenge. Expectations for control of degradants and degradant levels over the shelf-life of a peptide product are typically aligned with expectations for small molecules.

Some peptide chemical degradation pathways are listed in Table 13.2, with common molecular design strategies to mitigate them also indicated. When a particular chemical degradation liability is identified that cannot be resolved solely through peptide molecular design, formulation, processing and storage-directed approaches such as those given in Table 13.2 can be explored. In the context of formulation, initial buffer and pH screens provide direction to optimizing these parameters, while the inclusion of excipients that modify solution properties such as dielectric constant and viscosity may reduce degradation reaction rates, *e.g.* deamidation.[24,29,32,39] The addition of stabilizing excipients, such as sacrificial antioxidants and chelators to complex metal ion contaminants, can significantly inhibit the degradation of susceptible residues in the formulation.[10,12,40] If such measures do not stabilize the peptide in a liquid formulation sufficiently over the shelf-life, lyophilization is another strategy to improve chemical stability, immobilizing the peptide in a dry, solid matrix, requiring consideration of excipients to ensure peptide stability, and also providing additional performance criteria specific to a lyophilized product, such as reconstitution time and

Table 13.2 Potential chemical degradation mechanisms in peptide therapeutics and mitigation strategies.

Chemical degradation mechanism	Mitigation strategy Molecular design[26]	Formulation, process and storage
Deamidation, *e.g.* Asn/Gln	• Replace Asn • Avoid N–G, N–A, N–S, N–D	• Formulation 4 < pH < 6 • Optimize formulation buffer • Excipients • Store product at 2–8 °C • Lyophilization
Isomerization, *e.g.* Asp	• Avoid D–G, D–P, D–A, D–S, D–D • Increase steric constraints	• Formulation 4 < pH < 6 • Screen buffers • Excipients • Store product at 2–8 °C • Lyophilization
Oxidation, *e.g.* Met/Cys	• Replace Met	• Free Met • Metal chelators, *e.g.* EDTA • Nonreducing sugars • Avoid excipients with oxidative impurities • Purge container with N_2 • Eliminate headspace • Avoid UV light • Lyophilize

elegance.[39,41,42] Process contact parts, ambient conditions and exposure to light must all be evaluated during risk evaluation for process development, and mitigation strategies such as controlled hold times, purging with nitrogen, refrigerated storage and protection from light should be implemented, where necessary.

13.3.2 Peptide Solubility

Peptide solubility in aqueous media can impact the injection volume needed to deliver an efficacious dose, where injection volumes of 1.5 mL or less are desirable for subcutaneous injection, although higher volumes may be feasible.[43] Table 13.3 presents some molecular design and formulation strategies to mitigate poor solubility and molecular self-association behavior for peptide therapeutics. Peptide solubility is dependent on the primary peptide sequence and the propensity to self-associate in solution. Generally, polar and charged residues improve solubility through enhanced interaction with water and electrostatic repulsion. The primary sequence of short peptides with ionizable residues results in pH-dependent solubility profiles and the isoelectric point (p*I*), or the pH at which the net charge on a peptide is neutral, is the pH at which the solubility is poorest.

Induced amphiphilicity from the peptide sequence[55] or resulting from synthetic modifications[44] can also contribute to solubility challenges or complex self-association behavior, which may vary across the pH spectrum.[45,46]

Table 13.3 Potential solubility and physical instability mitigation strategies.

Formulation risk	Mitigation strategy	
	Molecular design	Formulation
Poor solubility	• Shift p*I* away from desired formulation pH • Increase charged residues • Attach hydrophilic moieties	• Optimize buffer selection • Co-solvents • Solubilizing excipients (surfactants, complexing agents) • Shift formulation pH away from p*I*
Conformational instability	• Incorporation of α-helical-promoting residues • Increase rigidity with intramolecular staples	• Excipients • Store liquid formulation at low temperatures • Lyophilize
Aggregation	• Improve conformational stability • Synthesize prodrugs	• Lyophilize • Decrease peptide concentration • Excipients • Increase viscosity

For example, Wang *et al.* characterized the oligomerization of liraglutide, bearing a C_{16} palmitoyl chain to extend the *in vivo* half-life, in buffered solutions.[44] The oligomer size was found to increase from an 8-mer to a 12-mer when the pH was decreased below 6.9, towards the p*I*, driven by a process of monomer dissociation and reassembly with equilibration occurring slowly over the course of days at 22 °C. In order to achieve a solution formulation of a sufficient concentration, formulation excipients such as co-solvents, surfactants and complexing agents may be evaluated to determine the effectiveness to improve solubility, and a number of solubilizing excipients for parenteral formulations have been reviewed.[9,47] When solubilization is needed to increase achievable peptide concentrations in formulations, preformulation studies are carried out to quantify the impact of excipients on solubility, and also to assess any unintended secondary effects on peptide chemical and physical stability.

13.3.3 Physical Stability

Physical stability of peptides has become an increasingly challenging hurdle to the development of stable peptide formulations as the complexity of peptides has increased with primary sequence length, secondary structure and pharmacokinetics (PK)-enhancing modifications. Conformational changes in peptide secondary structure, such as the loss of α-helical structure, may result in structures that are more prone to specific or nonspecific aggregation. Stress-induced changes in secondary structure have been reported to result in nonspecific self-association or aggregation or the formation of peptide fibrils.[48–50] Ghodke *et al.* leveraged extrinsic and intrinsic fluorescence measurements, far-UV CD spectroscopy and electron microscopy to provide specific mechanistic evidence for stability loss by the fibrillation pathway.[51]

Similarly, Zhang *et al.* revealed significant pH and temperature effects on the kinetics of fibril formation of a 29-amino acid peptide using CD spectroscopy, extrinsic fluorescence spectroscopy and measurements of turbidity and soluble peptide concentrations to track changes in secondary structure and the subsequent loss of physical stability.[50]

In the absence of conformational transformations, colloidal stability of peptide monomers or oligomers may be driven by intramolecular repulsive electrostatic or steric forces, which depend on solution and environmental conditions, such as pH, temperature, buffer strength, ionic strength and excipient additives.[52,53] Variations in these properties can reduce the strength of stabilizing repulsive forces and lead to net attractive forces or repulsive forces that are easily overcome by modest stresses to the system, leading to aggregation, which may or may not be reversible. As an example, high ionic strength conditions that screen the repulsive electrostatic forces between charged peptides can lead to flocculation or bulk precipitation; in some cases, these events may be reversed upon dilution of the formulation or warming to physiological temperature. Formulation conditions play a critical role in determining the dynamic equilibrium of peptide species that are present in solution.[54] Probing peptide structural and aggregation tendencies under relevant stresses and in the presence of different stabilizing excipients helps to build a fundamental understanding of what the relevant instability mechanisms are and how best to impart stability.

13.3.4 High-throughput, Material-sparing Methods for Risk Assessment

The complexity of peptide synthesis and purification significantly limits the amount of material available for characterization in the Discovery stage. High-throughput, material-sparing assays in a multi-well plate format that screen the effects of pH, ionic strength and temperature on the chemical stability, physical stability and solubility of peptides are readily incorporated into work flows evaluating the physicochemical properties of new chemical entities and their formulations for early *in vivo* studies and to identify potential development risks. In stressed studies, the characterization techniques employed must be appropriate for elucidating the changes undergone by the peptide. Recent reviews have described analytical approaches to characterize the chemical and physical stability of peptides in parenteral drug products.[12,55–57] Orthogonal characterization techniques are typically required to build a comprehensive understanding of the peptide stability in a formulation. Briefly, hyphenated liquid chromatography (LC) and mass spectrometric (MS) techniques are used to identify degradants and track the growth of chemical degradants in stressed studies. Nuclear magnetic resonance (NMR) techniques and near- and far-UV spectroscopy are useful tools for characterizing peptide conformational changes, and static, dynamic and multi-angle light scattering, size-exclusion chromatography, fluorescence spectroscopy and electron microscopy are commonly used to develop an understanding of self-association and aggregate formation and properties.[56,57]

An illustrative example of a high-throughput, material-sparing approach to the estimation of the solubility limit of CsA in aqueous buffer uses only 30 μg of material, as shown in Figure 13.3a. To obtain these data, a stock solution of CsA in dimethyl sulfoxide (DMSO) was titrated into buffer at several concentrations and the absorbance at 375 nm was read in a 384-well plate after overnight equilibration. The turbidity in each well decreases with decreasing concentration of peptide, corresponding to an increasing fraction of the peptide being solubilized, until ~22 μg mL^{-1}, at which point the turbidity corresponds to a peptide-free well. The apparent solubility limit obtained in this assay approximates the published 23 μg mL^{-1} solubility

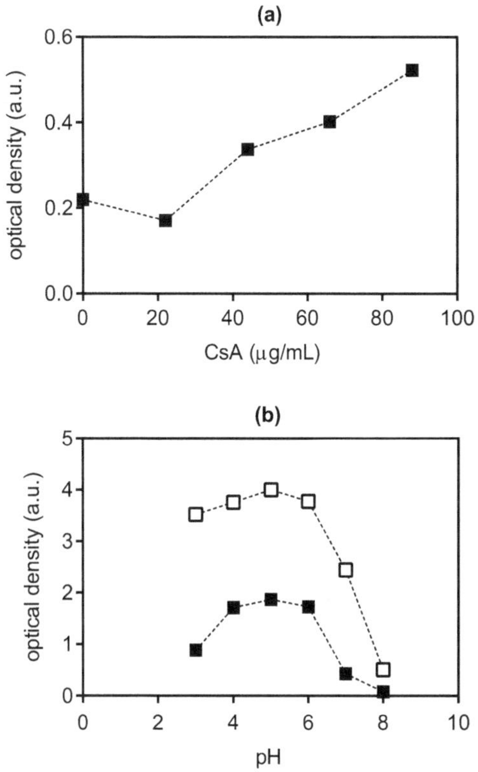

Figure 13.3 (a) Optical density measurements of 45 μL solutions or suspensions of cyclosporine A (CsA) in aqueous buffer, revealing the equilibrium solubility limit with low material requirements and without the need for separation steps. (b) Optical density measurements for 6 mg mL^{-1} peptide mixtures in various buffers as 100 μL solutions in a 96-well plate (■) or 45 μL solutions in a 384-well plate (□) are shown to be a material-sparing approach to screening for suitable formulation conditions. Increased optical density is indicative of precipitation in the well and, hence, the presence of aggregates or precipitates.

limit for CsA obtained by equilibrating excess CsA (5 mg) in 5 mL of aqueous medium for 4 days at 25 °C.[58,59] Multiple iterations of such a screen can be used to refine the apparent solubility value. In addition to being material sparing, an advantage of the turbidity output is that the measurement does not rely on separation by filtration of insoluble peptide, as reversed-phase ultra-high-performance liquid chromatographic (RP-UPLC) or absorbance measurements do. Filtration may skew the measured solubility values results due to peptide membrane binding or the presence of nanometer-sized species that pass through typical filters. Complementing turbidity measurements with dynamic light scattering data and nephelometry can provide additional insights into the self-association propensity of peptides.

A similar approach can be used to screen and identify favorable buffer conditions. In Figure 13.3b, increased optical density corresponds to the presence of precipitates or aggregates of a novel 31-amino acid peptide. The plate was prepared by titrating a DMSO stock solution into buffers with pH increasing from 3 to 8 in 1 pH unit increments. The experiment was run in a 96-well plate with 100 μL per well and a 384-well plate with 45 μL per well and a total peptide concentration of 6 mg mL^{-1}. A wider dynamic range is obtained in the 384 well-plate owing to an increased fill depth in each well relative to the 96-well plate. The smaller well size also requires roughly half the mass of peptide needed in the 96-well plate, allowing such screens to be run with a few milligrams of material, rapidly identifying buffers and other formulation components that may improve solubility. The presence of significant DMSO fractions can lead to overestimation of solubility, so the results should be interpreted appropriately. Nevertheless, the trends observed in the full sample sets provide valuable information that facilitates additional study or understanding of molecule performance in *in vitro* and *in vivo* studies.

High-throughput evaluation of chemical and physical stability in buffers as a function of other stresses such as heat, light, shaking and freeze–thaw cycles may be carried out in a similar way, using methods amenable to high-throughput analysis to characterize the stressed samples.[12] Empirical phase diagrams (EPDs) have been leveraged to evaluate the biophysical stability of proteins and peptides by summarizing large datasets in a visual a stability map.[12] This approach was adopted Nonoyoma *et al.* to visualize structural stability findings for the 37-amino acid pramlintide as a function of pH and temperature.[60] Owing to the lack of distinct structural transitions observed for bulk measurements in the solution phase, few conformational transformations were observed beyond an aggregation tendency above pH 6 and at higher concentrations.

13.4 Preformulation

Once a peptide candidate has been selected, preformulation studies are performed to provide a detailed evaluation of the candidate's physicochemical properties, instability pathways, the impact of stresses on key formulation attributes and risks identified in the risk assessment. A broad discussion on the role of peptide and protein preformulation studies within the context of discovery and development was presented by Bandopadhyay *et al.*[61] The impacts of buffer species, ionic strength, temperature and pH on the formulation properties, such as viscosity, solubility, chemical degradation, peptide conformation and aggregation, are evaluated under unstressed and stressed conditions that may include high temperature, shear or mixing, freeze–thaw cycles, acidic and basic conditions and oxidizing conditions.[61] Required or acceptable forced degradation conditions to support characterization are not described by regulatory agencies. A recent review by Tamizi and Jouyban is a comprehensive reference with reported forced degradation conditions for proteins and peptides from 1992 to 2005, focusing on chemical and physical instabilities and the impacts on protein and peptide attributes both in formulation and under biological conditions.[62]

Preformulation studies may include titration of individual excipients to evaluate the impact of each potential formulation component on the peptide pCQAs, such as a screen for the impact of excipients on the solubility of a poorly soluble therapeutic peptide as illustrated in Figure 13.4. Each excipient was titrated into buffered solutions of the peptide and the equilibrium peptide solubility was quantified by reversed-phase high-performance liquid chromatography (RP-HPLC); it was found that the solubility impact and magnitude vary with each excipient. Profiling the influence of each excipient on solubility and other quality attributes provides an understanding of overall formulation quality during development.

In early prototype formulations, chemical stability for peptide formulations can be projected and compared by conducting accelerated stability studies at elevated temperatures to model the kinetics of degradation. Prior to the evaluation of the ability of excipients to inhibit deamidation in a

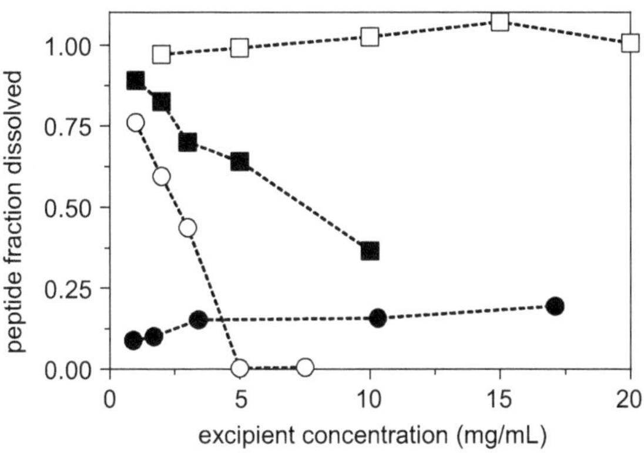

Figure 13.4 The fraction of a therapeutic peptide dissolved in aqueous media in the presence of four different excipients was quantified as a function of the amount of each excipient added. Each curve corresponds to a different excipient. The initial concentration of the model peptide was held constant.

6-mer peptide, Stratton *et al.* modeled the kinetics of deamidation in Tris buffer at pH 9 at temperatures from 5 to 65 °C and peptide concentrations from 0.1 to 100 mg mL^{-1}, obtaining the activation energy for the reaction and reporting a decreasing deamidation rate as the peptide concentration increased.[30] Such studies should be paired with the identification of major degradants to inform the rational design of a stabilizing formulation, which may include inhibitory molecules, such as sacrificial antioxidants, nonpolar molecules to reduce the solvent dielectric constant, viscosity modifiers and high-purity excipients to limit the introduction of reactive species (Table 13.2).

Preformulation studies evaluating the propensity, mechanism and inhibition of fibril formation involve a combination of screening studies and residue-level evaluation of molecular interactions. To support formulation development, Zhang *et al.* evaluated the change in lag time to the onset of fibrillation as a function of peptide concentration, pH, buffer species, ionic strength and temperature by kinetic thioflavin T fluorescence, coupled with measurements of soluble peptide loss, turbidity and conformational stability, and subsequently modeled the lag time under different stress conditions to make projections about the lag time under refrigerated conditions.[50] Using similar studies to establish the kinetics of glucagon fibrillation under stress conditions, Moorthy *et al.* demonstrated the use of hydrogen–deuterium exchange coupled with mass spectrometric analysis (HDX-MS) to detect aggregation *via* reduced backbone solvent accessibility and to identify residue-level "hot spots" where deuterium exchange was significantly inhibited compared with the rest of the structure.[49] Highlighting the unusual instability behavior that can be associated with peptide primary structure, Zapadka *et al.* found that the relationship between peptide concentration and onset of fibrillation inverted with decrease in pH, observing an increasing onset time as the peptide concentration decreased at pH 7.7 and 8.2, but a decreasing onset time as the peptide concentration decreased at pH 7.5.[63] In addition to the unusual nature of these observations, the results further highlight the sensitivity of peptide behavior in an extremely narrow pH range.

13.5 Risk Evaluation and Mitigation in Formulation Development

With the identification of risks, instability pathways and potential excipients that impart the desired performance functions to the formulation, the formulation development team must define a strategy to select the necessary excipients and optimal levels to be included in the drug product formulation, considering quality impacts. A fundamental understanding of formulation design with increasing levels of complexity is rooted in continuously evaluating the impact of the combination of excipients on all pCQAs. We shall now review examples of peptide formulation risk assessment and mitigation approaches with a focus on excipient selection and evaluation.

13.5.1 Chemical Degradation

Preformulation chemical stability studies reported by Knepp *et al.* included a screen of several antioxidants to overcome liabilities that arise from dissolved oxygen, metal contaminants or peroxides.[64] The objective of the study was to rank order stabilizing antioxidants suitable for parenteral delivery in combination with two recombinant human proteins, ciliary neurotrophic factor and nerve growth factor. Whereas some antioxidants, such as glutathione and cysteine, effectively inhibited oxidation of the proteins in the presence of peroxides, others, including ascorbic acid and cysteine, increased the rates of oxidation. A complementary study by Li *et al.*, characterizing the impact of ascorbic acid on short, model peptides, elucidated the potential for this excipient to serve as either an antioxidant or a prooxidant, depending on the presence of catalytic metal ions in solution.[25] Taken together, these studies illustrate the role of excipient screening in rational formulation design to overcome known degradation risks, but also the importance of thorough characterization to understand the impact that an individual excipient may have on a unique peptide formulation.

Mitigation of deamidation through control of pH, dielectric constant and viscosity has been characterized.[24,29,32,65] Excipient-based strategies towards inhibiting deamidation leverage excipients that have a physical interaction with peptides. After determination of the kinetics of deamidation in a 6-mer peptide containing the sequence Pro-Asn-Gly described in Section 13.4, Stratton *et al.* evaluated the effect of sucrose and poloxamer 407 on inhibiting the rate of deamidation.[30] A quantitative assessment of stabilization was performed by using an Arrhenius model to obtain the increased activation energy for deamidation in the presence of sucrose. Decreased deamidation in the presence of sucrose was attributed to a particular peptide conformation measured by CD spectroscopy that inhibited the formation of the cyclic intermediate, highlighting the important interrelationship of chemical and structural rigidity in peptides. A patent granted to Anderson in 2016 described the use of divalent organic anions such as *N*-acetyltryptophan with a pK_a of 0.5–3.5 to form complexes with proteins that inhibit acid- and base-catalyzed deamidation and permit the production of stable liquid formulations near physiological pH.[37]

Excipients included in order to mitigate a particular instability pathway may have an adverse effect on other quality attributes. Hydroxypropyl-β-cyclodextrin (HPβCD) has been shown to stabilize the secondary structure of deslorelin, inhibiting digestion by enzymes,[66] and thymopentin, allowing the preparation of a room-temperature stable, ready-to-use solution.[67] In contrast, we have found that HPβCD can negatively impact the chemical stability of a 29-amino acid peptide. In Figure 13.5, the formation of two Asp degradation products formed by deamidation is tracked in the presence of 0, 10 and 15% HPβCD at relevant formulation pHs. The formation of both degradation products is enhanced at 10 and 15% HPβCD, with Isomer 1 showing no pH dependence, whereas Isomer 2 shows pH dependence in all three formulations.

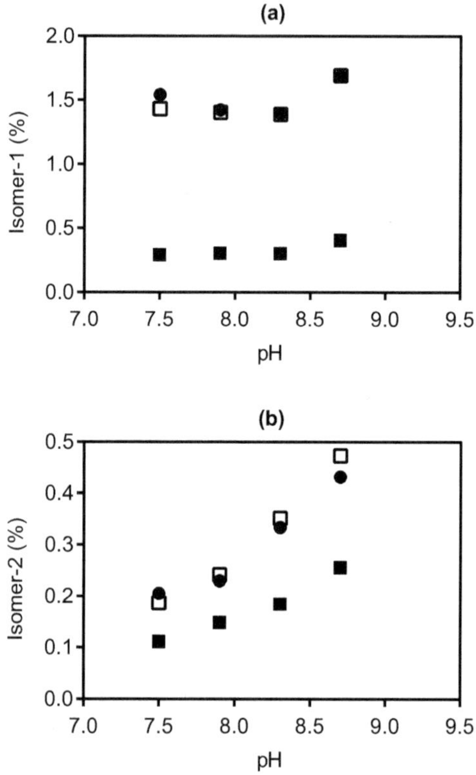

Figure 13.5 Impact of hydroxypropyl-β-cyclodextrin at 0% (■), 10% (□) and 15% (●) on the formation of two different Asp degradants, (a) Isomer 1 and (b) Isomer 2, in a 29-amino acid peptide.

Polymer-based delivery of peptides from degradable polyesters such as poly(lactide-*co*-glycolide) (PLGA), polylactide (PLA) and polycaprolactone (PCL) is an attractive approach to control the release of peptides, extending the therapeutic exposure and reducing dosing frequency.[68] However, encapsulation in degradable polymer excipients can expose the peptide to low local pH resulting from polymer hydrolysis, moisture and elevated temperatures for long periods of time. Peptide backbone hydrolysis,[69] covalent dimerization,[70] deamidation, oxidation and acylation[34,71,72] have been observed for polymer-encapsulated peptides during release. The selection of alternative polymer chemistries can inhibit chemical reaction of the peptide in a controlled-release matrix. Octreotide is an octapeptide containing two primary amines that may undergo acylation when formulated in PLGA microspheres; Murty *et al.* showed that selection of PLGA microsphere matrices with a higher lactic acid content results in decreased formation of octreotide acetate impurities, potentially due to greater steric inhibition by α-methyl groups of the lactide groups.[73] To protect free amines in octreotide, Shirangi *et al.* used a pH-dependent *O*-4-nitrophenyl-*O*′-4-acetoxybenzyl carbonate protecting group that blocked adduct formation under acidic conditions but was eliminated upon release from the matrix into neutral buffer.[74] Formulation in the polymer with buffering or basic excipients, such as $Mg(OH)_2$, also may to limit acid-induced degradation and promote complete release of the payload,[70] although the processing methods can also impact the generation of a local acidic microenvironment during release and the need for such excipients.[75] Houchin *et al.* performed an analysis of a range of pH-modifying additives in PLGA films of a model 6-mer peptide.[33] The study revealed complex effects on peptide and polymer stability, with peptide acylation increased in the presence of all pH-modifying excipients, but generally reduced deamidation and peptide cleavage relative to a control film without pH modifier. Magnesium hydroxide and ammonium acetate both increased the initial pH in the hydrated polymer film, but also increased the extent of deamidation and peptide cleavage, respectively.

13.5.2 Peptide Self-association and Aggregation

Despite a lack of well-defined structure in solution, it is believed that the conformation of a peptide is vital to its activity when bound to the intended target.[76] For example, adoption of a particular peptide conformation upon interaction with membranes has been shown to be important for anti-bacterial peptide activity, often resulting in induced amphiphilicity, with membrane specificity related to phospholipid composition.[77] The lipopeptide antibiotic daptomycin oligomerizes in a calcium-dependent manner upon intercalation in Gram-positive cell membranes, disrupting the structural integrity. Consistent with this biotransformation, daptomycin, which is provided as a lyophilized powder containing a minimal amount of sodium hydroxide for pH adjustment prior to freeze-drying, was characterized in the presence of increasing amounts of calcium to develop an understanding of peptide aggregation propensity across a biologically relevant pH range.[78] Qiu and Kirsh[78] attributed daptomycin aggregation to induced amphiphilicity in the molecule as the charge state changed, allowing for a framework to harmonize various characterization studies performed on daptomycin previously. Since the first approval of daptomycin in 2003, a new lyophilized formulation was approved in 2016, containing sucrose as a bulking agent to provide for room-temperature storage and additional sodium hydroxide to provide a neutral pre-lyophilization pH and ensure more rapid reconstitution.[79]

The risk of the aggregation and the effectiveness of stabilizing excipients should both be evaluated during preformulation studies, but re-evaluation is needed as formulations proceed through the Development stage in primary packaging materials. Hogstedt *et al.* evaluated the aggregation propensity of carbetocin using Fourier transform infrared (FTIR), thioflavin T fluorescence, HPLC and surface tension measurements to inform on excipient inhibition strategies.[80] They hypothesized that aggregation was driven by hydrophobic self-association due to a lack of molecular charge and evaluated excipients that might bind to hydrophobic surfaces and increase surface charge. In the presence of different stabilizers, they found various aggregation mechanisms under forced stress conditions, and also a dependence on fill volume and contact interfaces in the vial.

Although glucagon is prone to form fibrils and gels in reconstituted commercial formulations at room temperature, the stability is sufficient for rapid reconstitution to treat hypoglycemia. There have been a number of efforts to extend the in-use period and/or shelf-life of glucagon in solution for additional therapeutic applications, such as the artificial pancreas,[81] by mitigating fibrillation risk in solution through both formulation efforts[81–85] and molecular design.[86–88] Exploration of novel excipients that can provide robust stability of a ready-to-use glucagon solution is an area of active research. Steiner *et al.* developed a micellar formulation of glucagon in order specifically to inhibit intramolecular glucagon interactions to prevent fibrillation, including a reducing sugar, lysolecithin and an alcohol.[82] They demonstrated improved physical stability over the course of 5 days at 30 °C relative to a glucagon rescue kit, prepared and stored under the sample conditions, by tracking the optical density of the formulations. Ferulic acid, a degradant of curcumin, was identified by Bakhtiani *et al.* to be an effective excipient for inhibiting glucagon fibrillation at alkaline pH, using transmission electron microscopy to monitor for the characteristic morphology of fibrils, and *in vitro* and *in vivo* assays were used to confirm potency and efficacy, respectively.[84] At the end of a 7-day shelf-life for continuous administration of a ferulic acid-stabilized formulation *via* an artificial pancreas, glucagon was found to be 94% intact.[83] Although the authors cited its use in food to indicate a lower risk of toxicity, the use of ferulic acid as a novel excipient in a parenteral formulation would require safety assessment in addition to the development of an excipient grade that is suitable for parenteral use.

Insulin formulations prepared *via* complexation with zinc produce conformationally stable hexamers that dissociate upon direct administration in the blood and which replicate the rapid onset of action for insulin released by B cells in healthy individuals.[89,90] In formulations, the formation of zinc hexamer complexes and inclusion of phenol or *m*-cresol reduce insulin flexibility and improve conformational stability,[91] which is critical to inhibiting the formation of insulin fibrils. However, administration in the subcutaneous space results in a delayed onset of action due to dissociation and absorption, which presents an inconvenience to those with diabetes, who need to offset the use of meal-time insulin from the start of a meal.[77] Faster acting insulin replacement therapies have been achieved through molecular design, by the introduction of amino acid substitutions that destabilize the insulin hexamers and result in more rapid dissociation in the subcutaneous space, as in the case of insulin lispro.[90] The identification of excipients that stabilize insulin monomers and dimers also results in faster acting insulins, as in the case of insulin aspart prepared with arginine and nicotinamide.[90]

13.5.3 Multi-use Formulations

A limited number of antimicrobial preservatives are used in multi-use formulations in order to meet criteria set by the US, Japanese and European Pharmacopeias for preservative effectiveness testing.[92] At relevant concentrations, conformational changes and aggregation have been observed for some peptides and proteins,[46,93,94] and there are also reports of no measurable impact of preservatives on other peptides,[95] requiring a fundamental understanding of the impacts of preservatives on the peptide and *vice versa* in order to determine an optimization strategy to ensure a quality drug product. The following examples are discussed to highlight the multiple interactions that may occur with preservative excipients in formulations.

A detailed exploration of formulations of a model acylated 41-amino acid peptide with three preservatives was carried out by Heljo *et al.*,[46] where antimicrobial preservative-induced self-association measured by light scattering was correlated with a loss of preservative effectiveness, particularly against *Escherichia coli*. These findings give indirect evidence of interactions between preservatives and peptides, which can reduce the free concentration of the preservative available for antimicrobial effectiveness. NMR and hydrogen–deuterium exchange measurements were performed by Alford *et al.* to evaluate the mechanism by which the preservative benzyl alcohol accelerated aggregation in recombinant human interleukin-1.[94] The studies revealed that solvent-exposed residues are involved in binding benzyl alcohol, implying that the preservative does not diffuse into the hydrophobic core of the protein. Furthermore, the exposed residues were graded as hydrophilic, indicating that hydrophobic interactions were not driving the instability, prompting further investigation. Costantino *et al.* evaluated salmon calcitonin formulation similarity when 0.25 $mg\,mL^{-1}$ of chlorobutanol was used to replace the 0.1% benzalkonium chloride in the marketed Miacalcin® nasal spray formulation.[95] A comparison of peptide structure, dynamics, stability under thermal and pH stress and binding activity in the presence of either preservative with a reference standard without a preservative led to the conclusion that neither preservative interacted with the peptide, promoted association or degradation or impacted stability or activity.

In addition to providing the required antimicrobial effectiveness, characterization of the impacts of preservatives in insulin formulations has shown beneficial impacts to both chemical and physical stability. A process for the crystallization of insulin in the presence of zinc was described in 1934,[96] with later studies identifying that three insulin dimers complexed with zinc to form hexamers. With the preparation and analysis of multi-use formulations of insulin, it was found that the addition of phenolic preservatives enhances the stability of the formulations by promoting the formation of a hexameric insulin,[91] with reports of increased helicity of zinc complexes[97] and inhibited deamidation and intermolecular crosslinking.[98]

Beyond interactions of antimicrobial preservatives in multi-use formulations with the therapeutic peptide, other excipients added to a formulation matrix can have effects on the overall antimicrobial effectiveness of a formulation. Surfactants, such as polysorbate 80 (PS-80), are used to improve physical stability and inhibit surface adsorption of active peptides and proteins, but have been reported in the *Handbook of Pharmaceutical Excipients* to cause the precipitation of *m*-cresol and phenol and reduce the effectiveness of benzyl alcohol.[99] Nevertheless, several marketed formulations contain combinations of nonionic surfactants and antimicrobial preservatives (Table 13.4),[7] with Omnitrope®, Ovaleap® and Refib® each containing both PS-80 and benzyl alcohol in multi-use formulations. Other excipients in formulations have been shown to mitigate[9,100] or impact the effectiveness of antimicrobial preservatives.[92]

13.5.4 Bioperformance

Excipients are considered inactive ingredients because, on their own, there is typically no pharmacological effect. However, the role of excipients in formulations can extend beyond quality attributes related to the stability of the peptide in the formulation, where the selection of excipients can actively influence peptide absorption, *in vivo* stability or membrane permeability. Excipients also have the potential to impact adversely the profile of the drug product after administration, particularly if related to adverse reactions.[101] The reader is encouraged to explore in-depth reviews focused on the development of formulation types to achieve drug delivery by various administration routes, such as oral,[102] intranasal[103] and examples cited in Table 13.1. This section briefly illustrates the interplay between formulation excipients and bioperformance with formulation approaches that have been explored to extend PK profiles and reduce the administration frequency for peptides with very short circulation half-lives.

Noncovalent physical association with excipients *via* complexation or adsorption has been shown to provide control over the physicochemical properties of peptides and modulate release and circulation *in vivo*. Using a graft copolymer composed of poly(ethylene glycol) (PEG) and stearic acid grafted to a poly-L-lysine backbone [protected graft copolymer (PGC)] to bind

Table 13.4 Examples of marketed peptide and protein products containing benzyl alcohol and surface-active excipients.

Product	Benzyl alcohol (BA)	Surface-active excipients
Icrelex® (mecasermin) multidose vial	0.9% BA	Polysorbate 80
Omnitrope® [somatropin (rDNA origin)]	1.5% BA	Poloxamer 188
Ovaleap® (follitropin alfa)	1.0% BA and 0.002% benzalkonium chloride	Polysorbate 20
Pegasys® (peginterferon alfa-2α)	1.0% BA	Polysorbate 80
Puregon® (follitropin beta)	1.0% BA	Polysorbate 20
Follistim® AQ cartridge (follitropin beta injection)	1.0% BA	Polysorbate 80

vasoactive intestinal peptide (VIP), Reichstetter *et al.* demonstrated that peptide–polymer complexes formed with a hydrodynamic diameter of 17–22 nm resulted in an increased mean residence time of 11.2 h and maximum tolerated dose of 1250 μg VIP kg^{-1} compared with 0.52 h and 175 μg kg^{-1}, respectively, when dosed without the polymer in BALB/c mice.[104] Similar PK extension resulted when formulations of PGC and native GLP-1 were dosed subcutaneously in Zucker diabetic fatty rats and a once-weekly dose reduced HbA1c values as effectively as exendin-4 dosed twice per day.[105] Multiple strategies to provide controlled release of a hydrophilic 11-amino acid GLP-1 peptide were demonstrated by Qian *et al.*, first forming reversible, low-solubility complexes with zinc with an associated half-life of 8.5 h after subcutaneous administration in dogs and subsequently formulating the complex into erodible polymer microspheres with nearly zero-order release kinetics over an 18-day period.[106] In addition to control of release through noncovalent interactions and polymer erosion, the formation of degradable covalent linkages between a peptide and a desired excipient can allow for control of drug release through the degradation rate of the linker, as has been developed by ProLyx with a tetra-PEG hydrogel drug delivery system with half-life extension demonstrated for both octreotide[107] and exenatide.[108]

As controlled-release formulations increase in complexity, the design space should be mapped out to connect variations in formulation attributes, such as size, to impacts on CQAs, such as release rates. The development and optimization of a once-weekly exenatide formulation using the Medisorb® PLGA microsphere technology resulted in an improved tolerability profile relative to b.i.d. solution dosing of the immediate-release formulation, which was attributed to the shorter time required to achieve a steady state.[109] *In vitro* drug release testing was used to determine the impact of the drug particle size and the relative weight ratios of the drug and the polymer, with smaller particles providing slower initial release rates and better sustained release, and drug loads below 10% yielding more robust release profiles with minor variations in drug loading.[110] Importantly, the developer looked at the impact of the polymer molecular weight on multiple quality attributes, including the initial release rates and the mechanism of release, finding that small particles had slower initial release rates but better sustained release over longer times. The magnitude of the impact of polymer physicochemical properties was explored using the same *in vitro* release assay, with the chemical composition of the polymer emerging as an important parameter, whereas polymer molecular weight significantly impacted the release rate only for lower molecular weight polymers.

13.6 Novel Excipients

The use of excipients is generally limited by the precedent for use in previously approved products and other regulatory limits on levels and route of administration. Approval of novel excipients for peptide formulations requires review and approval of a therapeutic peptide formulation in which a novel excipient has been used and must be accompanied by a comprehensive dossier characterizing safety and quality.[111] Despite these hurdles, the complex challenges encountered during formulation development of peptide products has prompted the exploration of excipient novelty in terms of use at higher levels, administered by new routes or entirely new chemical entities.

There are tiers of risk associated with the novel use of excipients, which should be carefully considered in the design of a drug product. Beginning with the patient, one should consider the particulars of the product being developed, including the therapeutic need, potential quality impacts, acute *versus* chronic administration and any pediatric considerations. Next, any existing information or use that pertains to the use of the excipient can de-risk adverse safety impacts, based on the level of experience with a given novel excipient. Introducing an excipient at a higher level than has been dosed previously, or by a different route, presents a lower risk of safety concerns than the introduction of an entirely new excipient. Potential impacts of extended duration or increased frequency of dosing compared with prior use should also be considered. Furthermore, some novel uses of excipients have progressed into the clinic but have not yet been approved for use in marketed products owing to termination of the clinical program for the molecule. Manufacturers of novel excipients may provide such information, and also on any existing preclinical toxicology studies that have been performed.

Practical concerns of supply cannot be overlooked, as previously discussed by Jorgensen *et al.*[10] As peptides are typically administered parenterally, multiple sources of GMP-grade material with high purity and low in endotoxins are needed to de-risk product supply chain interruptions, ensuring that the excipient remains available throughout the life of the product. In the context of cost, the need for a licensing agreement can present additional challenges, requiring time for negotiations on potential exclusivity, cost and supply.

Despite the risks associated with the use of novel excipients to formulate peptide therapeutics, the synthesis and evaluation of such excipients have shown promise in overcoming the unique challenges in the formulation of peptide formulations, such as chemical stability, aggregation and permeability. The practical evaluation of novel excipients can present additional complexities, especially when explored in the context of drug product development. An example of the impact of novel excipient quality is presented in Figure 13.6. A promising stabilizing, reagent-grade, novel excipient was identified to inhibit fibrillation in a ready-to-use solution formulation of a novel peptide. However, when the formulation, including the novel excipient, was stored at 25 °C for 1 week, significant levels of new degradants were observed in the chromatogram, eluting from 35 to 40 min (Figure 13.6). Based on these data, the levels observed would preclude the use of the excipient in the drug product formulation. Indeed, when the same formulation was stored at 5 °C over 1 week, the same degradants were formed. Evaluation of the stability of the placebo formulation showed that the degradants were not related to the excipient itself and no new peaks were observed in the chromatogram. Subsequently, a GMP-grade source of the novel excipient was identified and the 25 °C stability study was repeated. In this instance, the new degradants that had previously been observed at retention times of 35–40 min were not produced. This example highlights the importance of the quality of excipients that are evaluated during drug product research and development. In the absence of quality standards, reagent-grade and custom-synthesized peptides should be characterized and purified to determine accurately the potential of such excipients to enable shelf-stable formulations of peptides to be produced.

Novel excipients provide an opportunity to overcome existing quality challenges and transform the delivery of peptide therapeutics, as many drug delivery systems are based on materials and excipients that would qualify as new excipients. The promises of drug delivery, and nanotechnology in particular, may provide enhanced delivery by targeting particular tissues, improving bioavailability and providing better stability, relying heavily upon materials science and engineering.[112–114] As these approaches continue to develop, they may provide a strong justification to bring forth novel materials to permit the delivery of peptide therapeutics.

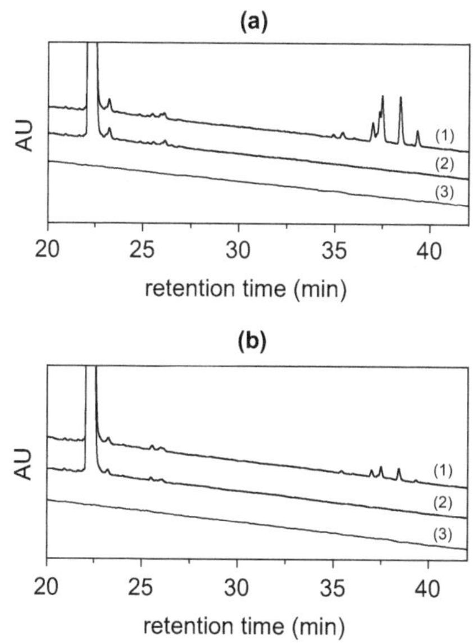

Figure 13.6 Overlaid RP-HPLC traces for a novel 29-amino acid peptide formulated with (1) the reagent-grade stabilizing novel excipient or (2) the GMP-quality stabilizing novel excipient. The peptide was formulated with the excipient at a level that had been determined to inhibit fibrillation in separate experiments and then stored (a) at 25 °C for 1 week and (b) at 5 °C for 1 week. Chromatogram (3) corresponds to the placebo formulations with the reagent-grade excipient.

13.7 Conclusion

There is great potential for novel peptide therapeutics to overcome therapeutic challenges that cannot be met by other modalities. However, the increasing complexity of therapeutic peptides presents unique challenges owing to their physicochemical properties. Evaluating development risks, understanding instability pathways and developing optimized, robust formulations that deliver on a QTPP must be executed with phase-appropriate investment. The development of knowledge about the molecule-specific risks, combined with consideration of external guidance, regulatory requirements and best practices, is key to the identification of excipients that mitigate risk and ensure safety, efficacy and quality of the final drug product.

Acknowledgements

The perspective provided in this chapter could only be achieved through the significant contributions of many talented scientists at Merck Sharp & Dohme Corporation, a subsidiary of Merck & Co., Inc., Kenilworth, NJ, USA. Additionally, the authors thank Sundeep Dhareshwar for the evaluation of the impact of excipients on peptide solubility and Alexandra Andrews for chemical stability analysis. Thanks are due to Maria Cueto for assistance with references and Lee Klein, Yogita Krishnamachari, Anthony Leone, Sachin Mittal, Claudia Neri, Yongchao Su, Paul Walsh, Daniel Yin and Jingtao Zhang for scientific discussions.

References

1. *ICH Q8(R2) Pharmaceutical Development*, International Conference on Harmonization, 2009.
2. A. S. Rathore, *Trends Biotechnol.*, 2009, **27**, 546–553.
3. Inactive Ingredient Search for Approved Products, US Food & Drug Administration, Center for Drug Evaluation and Research, (last accessed 05 Aug 2018). https://www.accessdata.fda.gov/scripts/cder/iig/index.cfm.
4. J. L. Lau and M. K. Dunn, *Bioorg. Med. Chem.*, 2018, **26**, 2700–2707.
5. J. P. Mayer, F. Zhang and R. D. DiMarchi, *Biopolymers*, 2007, **88**, 687–713.
6. *Japan Pharmaceutical Excipients Dictionary*, Japanese Pharmaceutical Excipients Council, Ministry of Health, Labor and Welfare, 1996.
7. V. Gervasi, R. Dall Agnol, S. Cullen, T. McCoy, S. Vucen and A. Crean, *Eur. J. Pharm. Biopharm.*, 2018, **131**, 8–24.
8. S. Pramanick, D. Singodia and V. Chandel, *Pharmatimes*, 2013, **45**, 65–77.
9. M. J. Akers, *J. Pharm. Sci.*, 2002, **91**, 2283–2300.
10. L. Jorgensen, S. Hostrup, E. H. Moeller and H. Grohganz, *Expert Opin. Drug Delivery*, 2009, **6**, 1219–1230.
11. M. C. Manning, D. K. Chou, B. M. Murphy, R. W. Payne and D. S. Katayama, *Pharm. Res.*, 2010, **27**, 544–575.

12. T. J. Kamerzell, R. Esfandiary, S. B. Joshi, C. R. Middaugh and D. B. Volkin, *Adv. Drug Delivery Rev.*, 2011, **63**, 1118–1159.
13. E. Y. Chi, in *Pharmaceutical Excipients: Properties, Functionality and Applications in Research and Industry*, 1st edn., ed. O. M. Y. Koo, John Wiley & Sons, Inc., 2017, pp. 145–198.
14. R. Varga, 2017 USP Workshop on Synthetic Therapeutic Peptides, Rockville, MD, 2017.
15. A. Banerjee and H. Onyuksel, *Wiley Interdiscip. Rev. Nanomed. Nanobiotechnol.*, 2012, **4**, 562–574.
16. E. A. Mueller, J. M. Kovarik, J. B. Vanbree, W. Tetzloff, J. Grevel and K. Kutz, *Pharm. Res.*, 1994, **11**, 301–304.
17. I. T. Degim and N. Celebi, *Curr. Pharm. Design*, 2007, **13**, 99–117.
18. *ICH Q9 Quality Risk Management*, International Conference on Harmonization, 2006.
19. A. Bak, D. Leung, S. E. Barrett, S. Forster, E. C. Minnihan, A. W. Leithead, J. Cunningham, N. Toussaint and L. S. Crocker, *AAPS J.*, 2015, **17**, 144–155.
20. S. P.-M. Andreas Evers, M. Bossart, C. Heubel, U. Stock, G. Tiwari, B. Gebauer, B. Elshorst, A. Pfenninger, U. Lukasczyk, G. Hessler, W. Kamm and M. Wagner, *J. Pharm. Sci.*, 2018, **108**, 1404–1414.
21. A. Evers, M. Bossart, S. Pfeiffer-Marek, R. Elvert, H. Schreuder, M. Kurz, S. Stengelin, M. Lorenz, A. Herling, A. Konkar, U. Lukasczyk, A. Pfenninger, K. Lorenz, T. Haack, D. Kadereit and M. Wagner, *J. Med. Chem.*, 2018, **61**, 5580–5593.
22. L. C. Wu, F. Chen, S. L. Lee, A. Raw and L. X. Yu, *Int. J. Pharm.*, 2017, **518**, 320–334.
23. *ANDAs for Certain Highly Purified Synthetic Peptide Drug Products That Refer to Listed Drugs of rDNA Origin Guidance for Industry; Draft Guidance.*, U.S. Department of Health and Human Services, Food and Drug Administration, CDER, 2017.
24. T. Geiger and S. Clarke, *J. Biol. Chem.*, 1987, **262**, 785–794.
25. S. H. Li, C. Schoneich, G. S. Wilson and R. T. Borchardt, *Pharm. Res.*, 1993, **10**, 1572–1579.
26. M. F. Powell, *Formulation and Delivery of Proteins and Peptides*, 1994, vol. 567, pp. 100–117.
27. M. C. Lai, M. J. Hageman, R. L. Schowen, R. T. Borchardt and E. M. Topp, *J. Pharm. Sci.*, 1999, **88**, 1073–1080.
28. M. C. Lai, M. J. Hageman, R. L. Schowen, R. T. Borchardt, B. B. Laird and E. M. Topp, *J. Pharm. Sci.*, 1999, **88**, 1081–1089.
29. R. Li, A. J. D'Souza, B. B. Laird, R. L. Schowen, R. T. Borchardt and E. M. Topp, *J. Pept. Res.*, 2000, **56**, 326–334.
30. L. P. Stratton, R. M. Kelly, J. Rowe, J. E. Shively, D. D. Smith, J. F. Carpenter and M. C. Manning, *J. Pharm. Sci.*, 2001, **90**, 2141–2148.
31. B. Li, M. H. O'Meara, J. W. Lubach, R. L. Schowen, E. M. Topp, E. J. Munson and R. T. Borchardt, *J. Pharm. Sci.*, 2005, **94**, 1723–1735.

32. A. A. Wakankar and R. T. Borchardt, *J. Pharm. Sci.*, 2006, **95**, 2321–2336.
33. M. L. Houchin, S. A. Neuenswander and E. M. Topp, *J. Controlled Release*, 2007, **117**, 413–420.
34. M. L. Houchin, K. Heppert and E. M. Topp, *J. Controlled Release*, 2006, **112**, 111–119.
35. K. Wisniewski, J. Finnman, M. Flipo, R. Galyean and C. D. Schteingart, *Biopolymers*, 2013, **100**, 408–421.
36. R. Torosantucci, C. Schoneich and W. Jiskoot, *Pharm. Res.*, 2014, **31**, 541–553.
37. D. Anderson, U. S. Pat., 9,394,339 B2, 2016.
38. A. A. Kosky, U. O. Razzaq, M. J. Treuheit and D. N. Brems, *Protein Sci.*, 1999, **8**, 2519–2523.
39. L. N. Bell, *Biotechnol. Prog.*, 1997, **13**, 342–346.
40. M. Z. Dion, D. Leiske, V. K. Sharma, C. L. Z. de Zafra and C. M. Salisbury, *Pharm. Res.*, 2018, **35**, 222.
41. H. R. Costantino, *Lyophilization of Biopharmaceuticals*, 2004, vol. 2, p. 139.
42. W. J. Fang, W. Qi, J. Kinzell, S. Prestrelski and J. F. Carpenter, *Pharm. Res.*, 2012, **29**, 3278–3291.
43. R. Mathaes, A. Koulov, S. Joerg and H. C. Mahler, *J. Pharm. Sci.*, 2016, **105**, 2255–2259.
44. Y. Wang, A. Lomakin, S. Kanai, R. Alex and G. B. Benedek, *Mol. Pharm.*, 2015, **12**, 411–419.
45. T. M. Frederiksen, P. Sonderby, L. A. Ryberg, P. Harris, J. T. Bukrinski, A. M. Scharff-Poulsen, M. N. Elf-Lind and G. H. Peters, *Biophys. J.*, 2015, **109**, 1202–1213.
46. P. Heljo, A. Ross, I. E. Zarraga, A. Pappenberger and H. C. Mahler, *Pharm. Res.*, 2015, **32**, 3201–3212.
47. R. G. Strickley, *Pharm. Res.*, 2004, **21**, 201–230.
48. R. W. Payne and M. C. Manning, *Innovations Pharm. Technol.*, 2009, **28**, 64–68.
49. B. S. Moorthy, H. T. Ghomi, M. A. Lill and E. M. Topp, *Biophys. J.*, 2015, **108**, 937–948.
50. J. T. Zhang, X. P. Mao and W. Xu, *Mol. Pharm.*, 2018, **15**, 5591–5601.
51. S. Ghodke, S. B. Nielsen, G. Christiansen, H. A. Hjuler, J. Flink and D. Otzen, *FEBS J.*, 2012, **279**, 752–765.
52. J. S. Philo and T. Arakawa, *Curr. Pharm. Biotechnol.*, 2009, **10**, 348–351.
53. C. J. Roberts, *Trends Biotechnol.*, 2014, **32**, 372–380.
54. E. Y. Chi, S. Krishnan, T. W. Randolph and J. F. Carpenter, *Pharm. Res.*, 2003, **20**, 1325–1336.
55. S. Ohtake, Y. Kita, R. Payne, M. Manning and T. Arakawa, *Protein Pept. Lett.*, 2013, **20**, 1308–1323.
56. S. M. D'Addio, J. R. Bothe, C. Neri, P. L. Walsh, J. T. Zhang, E. Pierson, Y. Mao, M. Gindy, A. Leone and A. C. Templeton, *J. Pharm. Sci.*, 2016, **105**, 2989–3006.

57. Y. Su, J. Song, K. J. Smith, E. E. Pierson, J. K. Cheung, W. Xu and A. C. Templeton, *AAPS News Magazine*, 2018.
58. K. Miyake, F. Hirayama and K. Uekama, *J. Pharm. Sci.*, 1999, **88**, 39–45.
59. Y. H. Zhang, X. R. Li, Y. X. Zhou, Y. T. Fan, X. N. Wang, Y. Q. Huang and Y. Liu, *Mol. Pharm.*, 2010, **7**, 1169–1182.
60. A. Nonoyama, J. S. Laurence, L. Garriques, H. Qi, T. Le and C. R. Middaugh, *J. Pharm. Sci.*, 2008, **97**, 2552–2567.
61. S. Bandopadhyay, N. Bandyopadhyay, P. K. Deb, C. Singh, R. K. Tekade and R. K. Tekade, in *Dosage Form Design Considerations*, Academic Press, 2018, pp. 401–433.
62. E. Tamizi and A. Jouyban, *Eur. J. Pharm. Biopharm.*, 2016, **98**, 26–46.
63. K. L. Zapadka, F. J. Becher, S. Uddin, P. G. Varley, S. Bishop, A. L. G. dos Santos and S. E. Jackson, *J. Am. Chem. Soc.*, 2016, **138**, 16259–16265.
64. V. M. Knepp, J. L. Whatley, A. Muchnik and T. S. Calderwood, *PDA J. Pharm. Sci. Technol.*, 1996, **50**, 163–171.
65. T. V. Brennan and S. Clarke, *Protein Sci.*, 1993, **2**, 331–338.
66. K. N. Koushik, N. Bandi and U. B. Kompella, *Pharm. Dev. Technol.*, 2001, **6**, 595–606.
67. N. D. Brown, D. L. Butler and P. K. Chiang, *J. Pharm. Pharmacol.*, 1993, **45**, 666–667.
68. S. Mohammadi-Samani and B. Taghipour, *Pharm. Dev. Technol.*, 2015, **20**, 385–393.
69. G. Crotts and T. G. Park, *J. Microencapsul.*, 1998, **15**, 699–713.
70. G. Z. Zhu, S. R. Mallery and S. P. Schwendeman, *Nat. Biotechnol.*, 2000, **18**, 52–57.
71. J. L. Cleland, A. Mac, B. Boyd, J. Yang, E. T. Duenas, D. Yeung, D. Brooks, C. Hsu, H. Chu, V. Mukku and A. J. S. Jones, *Pharm. Res.*, 1997, **14**, 420–425.
72. M. L. Houchin and E. M. Topp, *J. Pharm. Sci.*, 2008, **97**, 2395–2404.
73. S. B. Murty, J. Goodman, B. C. Thanoo and P. P. DeLuca, *AAPS PharmSciTech*, 2003, **4**, E50.
74. M. Shirangi, M. Najafi, D. T. S. Rijkers, R. J. Kok, W. E. Hennink and C. F. van Nostrum, *Bioconjugate Chem.*, 2016, **27**, 576–585.
75. S. P. Schwendeman, *Crit. Rev. Ther. Drug Carrier Syst.*, 2002, **19**, 73–98.
76. H. J. Dyson and P. E. Wright, *Annu. Rev. Biophys. Biophys. Chem.*, 1991, **20**, 519–538.
77. E. T. Kaiser and F. J. Kezdy, *Annu. Rev. Biophys. Biophys. Chem.*, 1987, **16**, 561–581.
78. J. Qiu and L. E. Kirsch, *J. Pharm. Sci.*, 2014, **103**, 853–861.
79. C. Frankenfeld, S. Mittal, Y. Melendez, L. Mendez-Vigo, K. C. Lamp, K. N. Keller and S. R. Bertolami, *Drug Des., Dev. Ther.*, 2018, **12**, 1953–1958.
80. U. B. Hogstedt, J. Ostergaard, T. Weiss, H. Sjogren and M. van de Weert, *J. Pharm. Sci.*, 2018, **107**, 838–847.

81. M. A. Jackson, N. Caputo, J. R. Castle, L. L. David, C. T. Roberts and W. K. Ward, *Curr. Diabetes Rep.*, 2012, **12**, 705–710.
82. S. S. Steiner, M. Li, R. Hauser and R. Pohl, *J. Diabetes Sci. Technol.*, 2010, **4**, 1332–1337.
83. N. Caputo, M. A. Jackson, J. R. Castle, J. El Youssef, P. A. Bakhtiani, C. P. Bergstrom, J. M. Carroll, M. E. Breen, G. L. Leonard, L. L. David, C. T. Roberts and W. K. Ward, *Diabetes Technol. Ther.*, 2014, **16**, 747–758.
84. P. A. Bakhtiani, N. Caputo, J. R. Castle, J. E. Youssef, J. M. Carroll, L. L. David, C. T. Roberts and Q. K. Ward, *J. Diabetes Sci. Technol.*, 2015, **9**, 17–23.
85. B. Newswanger, S. Ammons, N. Phadnis, W. K. Ward, J. Castle, R. W. Campbell and S. J. Prestrelski, *J. Diabetes Sci. Technol.*, 2015, **9**, 24–33.
86. J. Chabenne, M. D. Chabenne, Y. Zhao, J. Levy, D. Smiley, V. Gelfanov and R. DiMarchi, *Mol. Metab.*, 2014, **3**, 293–300.
87. P. A. Mroz, D. Perez-Tilve, F. Liu, J. P. Mayer and R. D. DiMarchi, *ACS Chem. Biol.*, 2016, **11**, 3412–3420.
88. U. Hovelmann, B. V. Bysted, U. Mouritzen, F. Macchi, D. Lamers, B. Kronshage, D. P. E. V. Moller and T. Heise, *Diabetes Care*, 2018, **41**, 531–537.
89. T. Tsalkova, F. C. Mei, S. Li, O. G. Chepurny, C. A. Leech, T. Liu, G. G. Holz, V. L. Woods and X. D. Cheng, *Proc. Natl. Acad. Sci. U. S. A.*, 2012, **109**, 18613–18618.
90. C. Mathieu, P. Gillard and K. Benhalima, *Nat. Rev. Endocrinol.*, 2017, **13**, 385–399.
91. J. Brange and L. Langkjoer, *Pharm. Biotechnol.*, 1993, **5**, 315–350.
92. B. K. Meyer, A. Ni, B. H. Hu and L. Shi, *J. Pharm. Sci.*, 2007, **96**, 3155–3167.
93. Y. F. Maa and C. C. Hsu, *Int. J. Pharm.*, 1996, **140**, 155–168.
94. J. R. Alford, A. C. Fowler, D. S. Wuttke, B. A. Kerwin, R. F. Latypov, J. F. Carpenter and T. W. Randolph, *J. Pharm. Sci.*, 2011, **100**, 4215–4224.
95. H. R. Costantino, H. Culley, L. Chen, D. Morris, M. Houston, S. Roth, M. J. Phoenix, C. Foerder, J. S. Philo, T. Arakawa, L. Eidenschink, N. H. Andersen, G. Brandt and S. C. Quay, *J. Pharm. Sci.*, 2009, **98**, 3691–3706.
96. D. A. Scott, *Biochem. J.*, 1934, **28**, 1592–1602, 1591.
97. A. Wollmer, B. Rannefeld, B. R. Johansen, K. R. Hejnaes, P. Balschmidt and F. B. Hansen, *Biol. Chem. H-S*, 1987, **368**, 903–911.
98. J. Brange and L. Langkjaer, *Acta Pharm. Nord.*, 1992, **4**, 149–158.
99. *Handbook of Pharmaceutical Excipients*, Pharmaceutical Press, London, Chicago.
100. J. Proprzan and M. G. DeNavarre, *J. Soc. Cosmet. Chem.*, 1959, **10**, 81–87.
101. W. Chen, B. C. Yung, Z. Y. Qian and X. Y. Chen, *Adv. Drug Delivery Rev.*, 2018, **127**, 20–34.

102. P. Li, H. M. Nielsen and A. Mullertz, *Expert Opin. Drug Delivery*, 2012, **9**, 1289–1304.
103. J. J. Lochhead and R. G. Thorne, *Adv. Drug Delivery Rev.*, 2012, **64**, 614–628.
104. S. Reichstetter, G. M. Castillo, I. Rubinstein, A. Nishimoto-Ashfield, M. S. Lai, C. C. Jones, A. Banjeree, A. Lyubimov, D. C. Bloedow, A. Bogdanov and E. M. Bolotin, *Pharm. Res.*, 2013, **30**, 670–682.
105. G. M. Castillo, S. Reichstetter and E. M. Bolotin, *Pharm. Res.*, 2012, **29**, 306–318.
106. F. Qian, N. N. Ni, L. S. Burton, Y. F. Wang, S. Desikan, M. Hussain and R. L. Smith, *Int. J. Pharm.*, 2009, **374**, 46–52.
107. E. L. Schneider, J. Henise, R. Reid, G. W. Ashley and D. V. Santi, *Bioconjugate Chem.*, 2016, **27**, 1638–1644.
108. E. L. Schneider, J. Henise, R. Reid, G. W. Ashley and D. V. Santi, *Bioconjugate Chem.*, 2016, **27**, 1210–1215.
109. V. R. Aroda and M. B. DeYoung, *Postgrad. Med.*, 2011, **123**, 228–238.
110. M. B. DeYoung, L. MacConell, V. Sarin, M. Trautmann and P. Herbert, *Diabetes Technol. Ther.*, 2011, **13**, 1145–1154.
111. Guidance For Industry: Nonclinical Studies for the Safety Evaluation of Pharmaceutical Excipients, U.S. Department of Health and Human Services, Food and Drug Administration, CDER, 2005.
112. M. Witting, K. Obst, W. Friess and S. Hedtrich, *Biotechnol. Adv.*, 2015, **33**, 1355–1369.
113. B. T. Griffin, J. F. Guo, E. Presas, M. D. Donovan, M. J. Alonso and C. M. O'Driscoll, *Adv. Drug Delivery Rev.*, 2016, **106**, 367–380.
114. E. Samaridou and M. J. Alonso, *Bioorg. Med. Chem.*, 2018, **26**, 2888–2905.
115. E. Marttin, J. C. Verhoef and F. W. H. M. Merkus, *J. Drug Targeting*, 1998, **6**, 17–36.
116. E. T. Maggio and D. J. Pillion, *Drug Delivery Transl. Res.*, 2013, **3**, 16–25.
117. S. Chandrasekhar and E. M. Topp, *J. Pharm. Sci.*, 2015, **104**, 1291–1302.
118. Z. W. Huang, H. Wu, B. B. Yang, L. K. Chen, Y. Huang, G. L. Quan, C. N. Zhu, X. Li, X. Pan and C. B. Wu, *Drug Delivery*, 2017, **24**, 527–538.
119. G. Y. Zhang, J. L. Li, T. Wang, L. J. Gao and D. Q. Quan, *Curr. Pharm. Biotechnol.*, 2015, **16**, 187–193.
120. M. Parent, I. Clarot, S. Gibot, M. Derive, P. Maincent, P. Leroy and A. Boudier, *Int. J. Pharm.*, 2017, **521**, 357–360.
121. M. P. Even, S. Bobbala, B. Gibson, S. Hook, G. Winter and J. Engert, *Eur. J. Pharm. Biopharm.*, 2017, **114**, 79–87.
122. B. F. Choonara, Y. E. Choonara, P. Kumar, D. Bijukumar, L. C. du Toit and V. Pillay, *Biotechnol. Adv.*, 2014, **32**, 1269–1282.
123. S. Maher, R. J. Mrsny and D. J. Brayden, *Adv. Drug Delivery Rev.*, 2016, **106**, 277–319.
124. S. Maher, T. W. Leonard, J. Jacobsen and D. J. Brayden, *Adv. Drug Delivery Rev.*, 2009, **61**, 1427–1449.

125. J. Heade, S. Maher, S. B. Bleiel and D. J. Brayden, *J. Pharm. Sci.*, 2018, **107**, 1648–1655.
126. E. Pallagi, R. Ismail, T. L. Paal and I. Csoka, *Eur. J. Pharm. Sci.*, 2018, **122**, 160–169.
127. A. Mandal, D. Pal, V. Agrahari, H. M. Trinh, M. Joseph and A. K. Mitra, *Adv. Drug Delivery Rev*, 2018, **126**, 67–95.
128. A. Patel, K. Cholkar, V. Agrahari and A. K. Mitra, *World J. Pharmacol.*, 2013, **2**, 47–64.
129. R. Gaudana, H. K. Ananthula, A. Parenky and A. K. Mitra, *AAPS J.*, 2010, **12**, 348–360.
130. S. P. Vyas, R. Paliwal, S. R. Paliwal and C. Van Der Walle, in *Peptide and Protein Delivery*, Academic Press, Boston, 2011, pp. 87–103.
131. C. Newton, B. M. Gebhardt and H. E. Kaufman, *Invest. Ophthalmol. Visual Sci.*, 1988, **29**, 208–215.
132. L. Lajavardi, A. Bochot, S. Camelo, B. Goldenberg, M. C. Naud, F. Behar-Cohen, E. Fattal and Y. de Kozak, *Invest. Ophthalmol. Visual Sci.*, 2007, **48**, 3230–3238.
133. F. Wan, M. J. Maltesen, S. Bjerregaard, C. Foged, J. Rantanen and M. Yang, *J. Drug Delivery Sci. Technol.*, 2013, **23**, 355–363.
134. V. Agarwal, V. Khattry, G. R. Biswas and S. B. Majee, *Res. J. Pharm., Biol. Chem. Sci.*, 2017, **8**, 470–476.
135. A. Hussain, Q. H. Majumder and F. Ahsan, *Pharm. Res.*, 2006, **23**, 138–147.
136. F. Qian, N. Mathias, P. Moench, C. Chi, S. Desikan, M. Hussain and R. L. Smith, *Int. J. Pharm.*, 2009, **366**, 218–220.
137. M. Yang, H. Yamamoto, H. Kurashima, H. Takeuchi, T. Yokoyama, H. Tsujimoto and Y. Kawashima, *Eur. J. Pharm. Sci.*, 2012, **46**, 374–380.
138. S. H. Surasarang, G. Florova, A. A. Komissarov, S. Shetty, S. Idell and R. O. Williams, *Drug Dev. Ind. Pharm.*, 2018, **44**, 184–198.

CHAPTER 14

Formulations of Microspheres and Nanoparticles for Peptide Delivery

JYOTHI THUNDIMADATHIL

Bachem Americas, Inc, Torrance CA 90505, USA,
Email: jyothi.thundimadathil@bachem.com; jyothitm@yahoo.com

14.1 Introduction

In recent years, incredible progress has been made in designing peptide- and protein-based drugs targeting a variety of diseases.[1,2] However, the delivery of such drugs to specific targets has been a challenging task owing to a number of factors.[3] The parenteral route is the most common method of administration for therapeutic peptides and proteins. However, owing to the short *in vivo* half-life of most peptide drugs, frequent injections are required, thereby resulting in poor patient compliance.[4] Therefore, noninvasive drug delivery routes such as nasal, transdermal, pulmonary and oral have attracted substantial interest in the past decade. Oral delivery of proteins and peptides is highly challenging owing to the presence of proteolytic enzymes in the gastrointestinal tract (GIT) and their intrinsic physicochemical and biological properties such as poor stability at the low pH of gastric fluid (1.5–3.5), large molecular size and poor permeation across the gastro-intestinal membrane.[4] The large size and proteolytic instability are major factors in the poor absorption of therapeutic proteins across nasal and pulmonary mucosal surfaces. Moreover, physiological barriers such as

Drug Discovery Series No. 72
Peptide Therapeutics: Strategy and Tactics for Chemistry, Manufacturing and Controls
Edited by Ved Srivastava

Published by the Royal Society of Chemistry, www.rsc.org

mucociliary clearance may further limit protein and peptide absorption.[5] Similarly, hydrophilicity and large molecular size also limit transdermal peptide/protein delivery. Owing to their large molecular size, peptide therapeutics usually require chemical or physical interventions to increase skin permeability transiently.[6] To overcome the above problems during drug delivery, the use of microparticles and nanoparticles has been emerging as an alternative solution for peptide delivery.[7–17] Encapsulation of peptides or proteins in such carrier systems during formulation could potentially lead to a better drug profile. These systems could be used to mitigate the solubility problems of poorly soluble drugs and also for long-acting injectable depot formulations and specific drug targeting options.

By definition, microencapsulation comprises the coating of solid, liquid or gaseous materials with a film of polymer or fat to generate free-flowing micrometric particles. The products obtained by this process are called "microparticles," "microcapsules" or "microspheres," which are differentiated by morphology and internal structure.[14] When these particles are smaller than 1 μm, they are known as nanoparticles, nanocapsules or nanospheres.[15] Microspheres are small spherical particles with diameters in the micrometer range typically from 1 to 1000 μm (1 mm).

Microencapsulation was first described in 1931, with gelatin microcapsules being generated by a procedure called "coacervation."[18] In the 1940s, ink encapsulation for carbon paper manufacture became its most important application. During the following years, microencapsulation was extended to other fields of application. Microencapsulation was originally applied to encapsulate drugs and control their release by the pharma industry in the late 1950s. Aspirin was among the first microencapsulated drugs, intended to avoid gastric irritation. In spite of the delayed application and recognition of this method, it was subsequently readily and widely introduced in the field of pharmaceuticals. Microencapsulated products currently on the market include aspirin, theophylline derivatives, vitamins, pancrelipase, progesterone, potassium chloride, Zoladex® (goserelin peptide), Trelstar® (triptorelin peptide) and Lupron Depot® (leuprolide peptide). Because microspheres are larger than nanoparticles, they are able to encapsulate larger amounts of drugs. Although they cannot easily be used for intravenous or systemic delivery (because they could agglomerate and cause clotting), they are effective for local delivery, such as by subcutaneous injection, and can be used in sustained-release systems.

The applications of nanoparticles (NPs) have attracted unprecedented attention in broad areas of science and engineering in the last two decades and new technologies are still emerging in this area.[19–26] NPs are defined as solid, colloidal particles with a size range from 10 nm to <1000 nm. The extremely small size of NPs results in a high ratio of surface area to volume, which promotes a high degree of surface adsorption by drugs, proteins and other molecules. It also allows for increased interaction with other particles, which leads to changes in physical properties. NPs can be made aggregation free, making them useful for intravenous or systemic drug delivery of drug molecules that are otherwise difficult to handle. NPs are more likely to bypass the reticuloendothelial system and also biological barriers such as the blood–brain barrier (BBB).[25] They can additionally enter all cells *via* pinocytosis. Currently, NPs can be manufactured and produced in sterile form. Several nanosized drug particle products are marketed commercially, including Rapamune®, Emend®, Tricor®, Megace ES®, Invega Sustenna® and Triglide®. These products are administered intramuscularly or orally.

This chapter focuses on the formulation of microspheres and NPs for peptide delivery. Different types of NP and microsphere formulations are commercially available from various manufacturers.

14.2 Microspheres for Peptide Delivery

In the last 20 years, biodegradable microspheres composed of biocompatible polymers have been studied as systems for the controlled release of proteins and peptides. Generally, the drug is distributed through the polymeric matrix and released by two main mechanisms: diffusion through the matrix and polymer degradation, which erodes the particles.[27] Major advantages of the use of microspheres for drug delivery include controlled release of a drug to specific target sites, protection of unstable drug and the ability to manipulate the *in vivo* action, pharmacokinetic profile, tissue distribution and cellular interaction of the drug. The use of peptide/protein-encapsulated microspheres could potentially reduce the frequency of administration, with better acceptance by the patient, increase therapeutic benefit, eliminating fluctuations in serum concentrations of the drug, decrease the total dosage required for the treatment owing to a higher bioavailability of the dose administered and decrease adverse events owing to the decrease in the amount of drug delivered in the body at the moment of administration. A list of natural and synthetic polymers used in the preparation of microspheres is provided in Table 14.1.

During the manufacture of microspheres, important physicochemical characteristics such as particle size and distribution, polymer molecular

Table 14.1 Natural and synthetic polymers used in the preparation of microspheres.

Type of material	Examples
Natural materials – proteins	Albumins, gelatin, collagen
Natural materials – carbohydrates	Starch, agarose, carrageenan, chitosan
Chemically modified – carbohydrates	Poly(acryl dextran), poly(acryl starch), DEAE-cellulose
Biodegradable – synthetic	Lactides, poly(alkyl cyanoacrylate)s, polyanhydrides
Nonbiodegradable – synthetic	Poly(methyl methacrylate) (PMMA), glycidyl methacrylate, acrolein, epoxy polymers

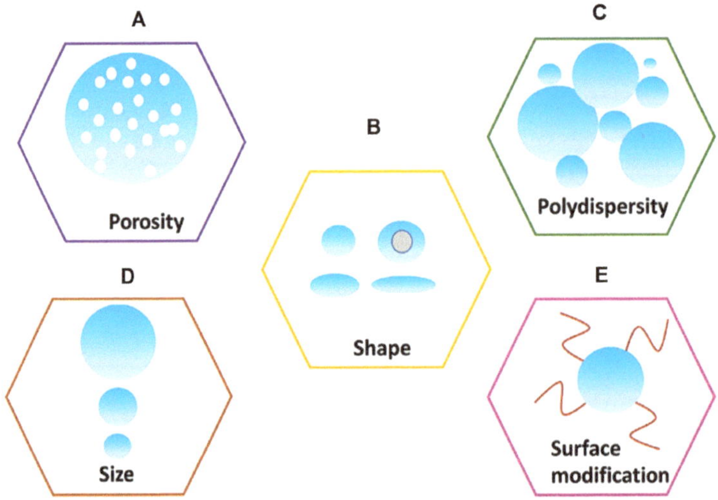

Figure 14.1 Important parameters of polymer microparticle design. The chemical functionalities of the polymer affect essentially all aspects of microparticle performance, including the efficiency of drug encapsulation, the rate of polymer degradation and drug release and toxicity at the injection site. (A) The porosity of the polymer microsphere affects the penetration of molecules, degradation of the polymer, drug release and diffusion of the drug. (B) The shape of the polymer affects the interactions of particles with macrophages; elongated particles exhibit orientation-dependent internalization by macrophages. (C) Polydispersity of microparticles could influence drug release rates. (D) The size of the microsphere particles affects the duration of drug release (in general, larger particles lead to more prolonged release) and the size of the needle required for administration (smaller needles are required for smaller particles). (E) The surface properties of particles affect their interactions with the surroundings at the injection site, especially immune cells. Modification of the surface with polymers such as poly(ethylene glycol) (PEG) is used to modulate the interactions of the microsphere with immune cells.

weight, ratio of drug to polymer and total mass of drug and polymer can be controlled. Key parameters of polymeric microparticles are highlighted in Figure 14.1.

14.2.1 Preparation of Microspheres

Several methods are available for the preparation of microspheres, including single-emulsion techniques, double-emulsion techniques, polymerization techniques such as normal polymerization or interfacial polymerization, coacervation phase separation techniques, spray drying, spray congealing and solvent extraction.[28] Peptide and protein molecules are fragile under experimental conditions, accounting for the main limitations of their microencapsulation conditions. Therefore, among the different techniques available, only a few are frequently used to obtain microspheres loaded with peptide substances. One of them, the simple coacervation method with the addition of a non-solvent compound, increases the efficiency of encapsulation, but solvent residues in the final product and difficulties in scaling up have limited its application to obtain biopharmaceutical products.

The *spray drying method* increases the encapsulation efficiency, is relatively easy to scale up and facilitates conducting the process under aseptic conditions.[29] Although several groups have addressed the properties of the particles obtained by this method, only a few data have been published on the stability of the encapsulated protein. During the process, the active ingredient is exposed to harsh conditions (organic solvents and high temperatures) that could considerably damage the activity of the encapsulated proteins, requiring the verification of their biological activity.

The *interfacial polymerization method* advantageously preserves all or most of the biological activity of the protein, owing to its mild microencapsulation conditions. Nevertheless, trace monomers in the final product have limited its application.

The *solvent evaporation method*, starting from a simple emulsion, has been mainly applied to encapsulate peptides. It was favored by the introduction of a water-in-oil-in-water emulsion for highly hydrophilic drugs (including proteins), being the preferred method at the laboratory scale for research purposes and also at the industrial scale to obtain the majority of the commercialized products. This is due to the simplicity of the method and the equipment required, but the main limitations arise from the process conditions. The exposure to organic solvents, in addition to the vigorous shaking used to generate emulsions, could reduce the biological activity of the drug. However, several proteins have been successfully encapsulated with adequate preservation of their biological activity.

In the *double-emulsion method*, a solution containing a polymer dissolved in an organic solvent is used to emulsify or disperse a dissolved or solid drug. A second aqueous phase is added to the previous emulsion to form a double emulsion, then the mixture is further stirred to extract/evaporate the solvent and facilitate the formation of microspheres, which are collected, washed and dried. A variant of the double-emulsion/solvent evaporation method at liquid nitrogen temperature allowed the encapsulation of the human growth hormone with an almost intact biological activity; this is also known as the *cryogenic method*. Because of the complex equipment required, this method has not yet been generally implemented. The solvent evaporation method remains the most widely studied and applied method for peptides and proteins.[30]

14.2.2 Methods of Characterization of Microspheres

Physiochemical characterization is an important aspect of formulated microspheres,[8,28,29] and confirms the quality and the quantity of the drug (such as drug loading, absence of incompatibility, stability and crystallinity) encapsulated in the microspheres, quality of the microspheres (such as particle size and shape and percentage entrapment) and quality of the microencapsulation process (such as yield, percentage encapsulation and particle size distribution). Therefore, for the determination of the percentage drug loading, entrapment/encapsulation efficiency and particle size distribution of microsphere formulations, optical microscopy, scanning electron microscopy, Fourier transform infrared (FTIR) spectroscopy, differential scanning calorimetry (DSC) and X-ray diffraction (XRD) are routinely employed. The drug present in the microspheres must be completely extracted by a suitable method during the determination of drug content. Controlled-release microspheres formulated for parenteral administration should be free-flowing powders and should contain spherical particles less than 250 μm and ideally less than 125 μm in diameter. The desired size of microspheres to be injected depends mainly on the route of administration. In general, particles less than 5 μm in size are used for the intravenous route and less than 125 μm for the intra-arterial and intra-articular routes. The compact nature, shape and particle size of microspheres can be gauged by optical microscopy and scanning electron microscopy. FTIR spectroscopy is used to confirm the identity of molecular species and functional groups that are part of the microspheres. DSC and XRD are valuable tools for assessing the amorphous/crystalline nature of microsphere particles.

14.2.3 Examples of Peptide Microsphere Formulations

Commercially available peptide drugs that use microsphere-based formulations include leuprolide, triptorelin, octreotide, lanreotide, human growth hormone, buserelin, abarelix and exenatide. A literature survey revealed numerous studies investigating the delivery of peptides using microspheres.[31–35] The commercially available long-acting, sustained-release preparations of triptorelin are microsphere injections, including products for administration once per 4 weeks, once per 12 weeks and once per 24 weeks. It is recognized that triptorelin must be administered at large doses for a long period of time in order to decrease the gonadal hormone secretion capacity of the gonads and the atrophy of sexual organs, thus achieving the purpose of treating diseases such as hormone-dependent prostate cancer, *etc.* Accordingly, unlike most other pharmaceutical microspheres, for which the smaller the initial release the better, an ideal long-acting microsphere preparation of triptorelin, after its injection, needs a relatively large initial release, so as to maintain the pharmaceutical effects until a later stage of drug delivery. Several commercially available luteinizing hormone-releasing hormone (LHRH) analog microspheres are mostly of this drug release mode. For example, the microspheres of leuprorelin have an initial release up to >20% in 1–2 days , whereas the triptorelin palmoate microspheres from Debiopharm have a drug release of up to >40% in 1–2 days, during which time the concentration of testosterone in the blood plasma rises in the early stage of drug delivery, reaches a maximum by about the fourth day and then decreases to a castration level, thus producing the pharmaceutical effects thereof.[36]

Poly(lactic-*co*-glycolic acid) (*PLGA*) polymers are the most common material used for the encapsulation of peptides or proteins, as they are mechanically strong, hydrophobic and biocompatible and degrade into toxicologically acceptable products that are eliminated from the body. PLGA polymers have been widely used in marketed products, for example, for delivering risperidone (Risperdal®; Johnson & Johnson) to treat schizophrenia and naltrexone for treating alcoholism and narcotics addiction. They have also been used to deliver peptides such as LHRH analogs for treating advanced prostate cancer and endometriosis, glucagon-like peptides (GLPs) for treating type 2 diabetes and proteins such as human growth hormone for treating pituitary dwarfism.[37] PLGA acts as an adjuvant for weakly immunogenic antigens and mimics booster responses after a single dose, thereby serving as a single-shot vaccine delivery vehicle. Efficient delivery of tumor antigen-derived peptide to antigen-presenting cells (APCs) has been made possible by the use of a PLGA particle-based vaccine delivery system. Also, the plasma half-life of the PLGA-encapsulated vaccine increases as it is protected from degradation, prior to its further release. PLGAs are reported to be catabolized into individual nontoxic units once inside the host and are further degraded *via* normal metabolic pathways.[38]

Goserelin acetate is a potent synthetic decapeptide analog of LHRH, also known as a gonadotropin-releasing hormone (GnRH) agonist analog. Goserelin-releasing biodegradable polymeric implants (Zoladex®) with 1 and 3 months' sustained release of goserelin were successfully developed in the past for the treatment of hormone-sensitive prostate and breast cancer. The implants are injected subcutaneously by means of an applicator equipped with either a 14- or 16-gauge needle, which is inconvenient for the patient. *Goserelin acetate PLGA microspheres* were designed and prepared successfully through the emulsification/extraction method. The results for formulations revealed that the drug-to-polymer ratio, PVA concentration and stirring speed are imperative to acquire sustained release and entrapment efficiency. The formulations described in this study released goserelin acetate constantly for 4 weeks.[39] The *in vitro* goserelin acetate release study from PLGA microspheres demonstrated that the microspheres had the properties of an ideal sustained-release formulation for anticancer therapy.[39]

Lupron Depot® (AbbVie) is a microsphere delivery system for administering leuprolide (leuprorelin) acetate. Continued leuprorelin administration results in suppression of gonadal steroid synthesis, resulting in pharmacological castration. Since leuprorelin is a peptide, it is orally inactive and generally given subcutaneously or intramuscularly. Sustained-release parenteral depot formulations, in which the hydrophilic leuprorelin is entrapped in biodegradable highly lipophilic synthetic polymer microspheres, have been developed to avoid daily injections. The peptide drug is released from these depot formulations at a functionally constant daily rate for 1, 3 or 4 months, depending on the polymer type [poly(lactic-*co*-glycolic acid) (PLGA) for a 1-month depot and poly(lactic acid) (PLA) for a depot of >2 months], with doses ranging between 3.75 and 30 mg. Mean peak plasma leuprorelin concentrations (C_{max}) of 13.1, 20.8–21.8, 47.4, 54.5 and 53 μg L^{-1} occur within 1–3 h of depot subcutaneous administration of 3.75, 7.5, 11.25, 15 and 30 mg, respectively, compared with 32–35 μg L^{-1} at 36–60 min after a subcutaneous injection of 1 mg of a non-depot formulation. Sustained drug release from the PLGA microspheres maintains plasma concentrations between 0.4 and 1.4 μg L^{-1} over 28 days after single 3.75, 7.5 or 15 mg depot injections. Mean areas under the concentration–time curve (AUCs) are similar for subcutaneous and intravenous injection of short-acting leuprorelin 1 mg; a significant dose-related increase in the AUC from 0 to 35 days is noted after depot injection of leuprorelin 3.75, 7.5 and 15 mg. The mean volume of distribution of leuprorelin is 37 L after a single subcutaneous injection of 1 mg and 36, 33 and 27 L after depot administration of 3.75, 7.5 and 15 mg, respectively. Total body clearance is 9.1 L h^{-1} and the elimination half-life is 3.6 h after a subcutaneous 1 mg injection; corresponding values after intravenous injection are 8.3 L h^{-1} and 2.9 h. A 3-month depot PLA formulation of leuprorelin acetate 11.25 mg ensures a C_{max} of around 20 μg L^{-1} at 3 h after subcutaneous injection and continuous drug concentrations of 0.43–0.19 μg L^{-1} from day 7 until before the next injection. Recently, an implant that delivers leuprorelin for 1 year has been evaluated. Serum leuprorelin concentrations remained at a steady mean of 0.93 μg L^{-1} until week 52, suggesting zero-order drug release from the implant. In general, regular or depot leuprorelin treatment is well tolerated. Local reactions are more common after application of the 3- or 4-month depot in comparison with the 1-month depot.[40]

A microsphere system of *thyrotropin-releasing hormone* (TRH) prepared using PLGA could continuously release the drug for 2–4 weeks. Using these systems effectively reduced the required dose compared with that needed with daily injection owing to more continuous receptor hits on the target organs and could improve patient compliance. Chemoembolization using PLGA microspheres containing an angiogenesis inhibitor, TNP-470, resulted in dramatic regression of VX-2 carcinoma in rabbits.[41] Mehta *et al.* prepared microspheres containing the peptide salmon calcitonin (sCT) by solvent extraction techniques using temperature or dilution. sCT was also incorporated into preformed microspheres by adsorption. sCT has a strong hydrophobic region and hence can bind to the polymer by a combination of hydrophobic and ionic forces, resulting in high incorporation from an aqueous medium. The microspheres with entrapped sCT exhibited *in vivo* release of sCT between days 5 and 9, whereas microspheres with adsorbed sCT showed low but detectable *in vivo* release for 3–4 days.[42]

Preparation of *vancomycin* (VA)–PLGA microspheres by an emulsification/spray-drying technique was reported for the topical ocular delivery of the drug. The drug encapsulation efficiencies were close to the theoretical values (84.2–99.5%); the average particle size was about 11 μm. The microspheres were able to modulate the *in vitro* drug release of VA with a behavior dependent on their composition: the highest drug content corresponded to the highest release rate. *In vivo* studies were carried out by assessing the pharmacokinetic profile of VA in the aqueous humor of rabbits after topical administration of aqueous suspensions of microspheres. High and prolonged VA concentrations and increased AUC values (twofold) with respect to an aqueous solution of the drug were observed.[43]

Exenatide (Bydureon®) is a US Food and Drug Administration (FDA)-approved glucose-lowering peptide drug for the treatment of type 2 diabetes by subcutaneous injection. Bydureon® is the only once-weekly treatment of its kind, with a unique continuous-release delivery system that gradually releases the medicine (exenatide) from microspheres. The Bydureon® microspheres are prepared based on Medisorb® microsphere technology (Alkermes). The water-in-oil (W/O) solvent evaporation method is employed. Exenatide is dissolved in water, then mixed with a dichloromethane solution containing PLGA. Following this, the W/O emulsion is sonicated for several minutes. After a coacervate formation process undertaken to form embryonic microspheres with silicone oil, the embryonic microspheres are transferred into a heptane–ethanol solvent mixture to harden them.[44] The diameter of the microspheres is 0.06 mm and they have a typical pinched-raisin shape and dense surface layer. In comparative trials, compared with twice-daily exenatide, Bydureon® microspheres improved hemoglobin A1c (HbA1c) and reduced fasting plasma glucose, with fewer gastrointestinal side effects, no increased risk of hypoglycemia and with similar reductions in body weight.

Other researchers have also investigated exenatide PLGA microspheres, aiming to develop a better formulation than the Bydureon® microspheres. For instance, Shi *et al.* investigated how the molecular weight of PLGA affects the properties of exenatide microspheres.[45] Three different molecular weights were studied (10, 20 and 30 kDa). The results showed that as the molecular weight increased, the encapsulation efficiency increased. PLGA with a molecular weight of 30 kDa could achieve an encapsulation rate of $78.1 \pm 8.8\%$ (mean $\pm$ standard deviation). The 10 kDa PLGA microspheres showed a severe initial burst, while the release from the other two microspheres lasted for 30 days and the release profiles were similar to that of the Bydureon® microsphere. Kwak *et al.* prepared a biweekly dosage formulation of exenatide microspheres by single-emulsion solvent evaporation using PLGA.[46] Liu *et al.* formulated 1-monthly PLGA microspheres of exenatide by the double-emulsion solvent evaporation method. The size of the prepared microspheres ranged from 5.8 to 13.6 μm.[47] Kim *et al.* fabricated a porous PLGA microsphere loaded with palmityl-acylated exenatide for pulmonary delivery.[48] The palmityl-acylation was to aid exenatide absorption onto the hydrophobic surface of PLGA microspheres. Yang *et al.* prepared polysaccharide microspheres of exenatide by a water-in-oil-in-water ($W_1/O/W_2$) emulsion method with three different degrees of acetylation (0.8, 1.5 and 2.3) of polysaccharide. The mean particle size of the microspheres was 35–110 μm. A higher degree of acetylation (2.3) showed a higher encapsulation efficiency (90.3%). The sustained release lasted for 21 days and the fact that there was no decrease in the molecular weight of exenatide showed that the exenatide was well protected in the polysaccharide microspheres.[49]

To address the issues on the inconvenience for patient use and the difficulty of oral administration of peptide drugs, chemical crosslinking of two pH-responsive biomaterials, alginate and hyaluronate, was carried out to prepare a new material for the encapsulation of exenatide as a form of microspheres.[50] The exenatide-loaded microspheres exhibited spherical structures with excellent loading and release behaviors in the simulated GIT environments. After oral administration of the microspheres in mice, the maximum plasma concentration of exenatide appeared at 4 h and blood glucose was effectively reduced to a normal level within 2 h and maintained for a further 4 h. The bioavailability of the exenatide-loaded microspheres, relative to subcutaneous injection of exenatide, reached 10.2%. Collectively, this study demonstrated the feasibility of orally delivering exenatide with the new crosslinked biomaterial and formulation and showed therapeutic potential for clinical applications.[50]

14.2.4 Factors Affecting Drug Release From Microspheres

Encapsulation, biocompatibility and peptide/protein release from microspheres are closely related to their structural properties.[8,51,52] For example, the release rates of proteins depend on the degradation of the polymer and/or diffusion of the protein from the microspheres, which in turn depends on the molecular mass of the polymer and the protein, the lactic-to-glycolic acid molar ratio, particle size and the loading of proteins within the microspheres, in addition to microsphere size and porosity. The shape of a particle also influences its behavior, in particular its interactions with macrophages. Because the diffusional transport of proteins through PLGA is limited, the rate of polymer degradation has an important role in the mechanism and rate of release of the drug from microspheres. In addition, diffusion of the protein through the biological fluid surrounding the microspheres is dependent on the size of the protein, and this will affect the rate of absorption after release.

14.2.5 Limitations of Microsphere Delivery Systems

Although clear advances have been made in the use of microparticles for delivering biopharmaceutical drugs, challenges still remain. Disadvantages of microspheres include the difficulty of large-scale manufacturing, inactivation of the drug during fabrication and poor control of drug release rates. Specifically, further advances in controlling the burst release can increase the duration of drug release. In addition, methods that provide active control of the rate of drug release, including on-demand termination of release, could open up new opportunities. A reduction in the size of the microspheres (but still within the micron range) could lead to a subsequent reduction in the needle size required for administration while maintaining the rate of sustained release. Additional unmet needs include the requirement for a robust manufacturing process that can be achieved at a reasonable cost, the need for stabilization and sterilization strategies to be broadly applicable to many biopharmaceutical drugs and the need for preclinical tools and/or models for addressing the risk of immunogenicity. In addition, it is important to consider issues such as ease of injection – that is, addressing the reconstitution of the lyophilized product and the clogging of needles with drugs. Many of these issues are not unique to protein- and peptide-based drugs, however, but are in fact common to all microsphere-based drugs.

Nutropin Depot®, comprising Genentech's recombinant human growth hormone (rhGH) encapsulated within poly(D,L-lactide-*co*-glycolide) (PLG) microspheres using Alkermes' proprietary encapsulation technology, was pulled from the market in 2004 because manufacturing and production costs were too high.

Proteins can lose their structure and biological activity upon prolonged incubation with biological fluids under physiological conditions. For example, aggregation and incomplete release from the microsphere have been observed with growth hormones and a covalent dimer formed in a microsphere formulation of darbepoetin alfa (Aranesp®; Amgen). In addition, the local degradation of polymers at the injection site can lower the pH inside the microspheres, which can further add to the potential for protein inactivation. The addition of magnesium hydroxide and other antacids within the microspheres has been shown to negate the adverse effects of low pH.[53–55]

14.3 Nanoparticles for Peptide Delivery

Nanoparticles made from natural and synthetic polymers (biodegradable and nonbiodegradable) have attracted considerable attention as they can be customized for the targeted delivery of drugs, improve bioavailability and provide controlled release of the drug from a single dose.[56–58] Owing to their small size and large surface area, drug NPs show increased solubility and thus enhanced bioavailability, with additional ability to cross the BBB, enter the pulmonary system and be absorbed through the tight junctions of endothelial cells of the skin. Different types of NPs are available, such as liposomes, polymeric micelles, polymeric NPs, nanoemulsions, nanogels, dendrimers, fullerenes, carbon nanotubes, magnetic NPs, metal NPs and quantum dots. Micelles and liposomes offer an important avenue for the delivery of chemotherapeutic agents. Additionally, micelles also provide a useful means to make insoluble drugs soluble owing to their hydrophobic core and hydrophilic shell. If the micelle's surface is further PEGylated, it increases the ability of the nanocarriers to pass through the fenestrated vasculature of tumors and inflamed tissue through passive transport, thus resulting in a higher drug concentration in tumors. Currently, several polymeric micelles containing anticancer drugs, NK012, NK105, NK911, NC-6004 and SP1049C, are in clinical trials and one such system, Genexol® PM (paclitaxel), is approved for breast cancer patients.[59,60] Dendrimers are highly branched macromolecules with many functional groups available for the attachment of drugs.[61] Nanoparticle therapeutics based on dendrimers can improve the therapeutic index of cytotoxic drugs by employing biocompatible components and surface derivatization with PEGylation, acetylation, glycosylation and various amino acids.[62]

Among nanocarriers, polymeric NPs have demonstrated significant advantages over other delivery systems.[63] Polymeric NPs exhibit high stability in biological fluids compared with liposomes and solid lipid NPs. The properties of polymeric NPs are significantly affected by the nature of the polymers, either natural or synthetic, and the method of preparation. A few examples of commonly employed natural polymers include chitosan (CS), gelatin and alginate. Among natural polymers, CS has shown the most interesting potential, which is attributed to its better solubility at intestinal pH, improving mucoadhesivness and permeation enhancement. In the small intestine, CS NPs can adhere to and infiltrate into mucus layers and open the tight junctions between contiguous epithelial cells. Furthermore, pH-sensitive CS NPs can disintegrate and release the encapsulated drugs, which then penetrate through the opened paracellular pathway.

Examples of synthetic polymers investigated for the delivery of proteins and peptides are poly(D,L-lactide-*co*-glycolide) (PLGA), polycaprolactone (PCL), poly(lactic acid) (PLA), poly(alkyl cyanoacrylate), poly(acrylic acid) (PAA) and poly(methyl methacrylate) (PMMA).[64] Among synthetic polymers, polyesters, alone or in combination, are the most relevant and most frequently studied for protein delivery. In contrast to natural polymers, synthetic polymers permit adjustable controlled drug release for a period of several days to weeks. Polymeric NPs can offer an alternative strategy to improve the bioavailability of encapsulated proteins and peptides by providing protection against degradation in the gastrointestinal environment, enhancing cellular contact with the intestinal membrane and promoting absorption in the small intestine. Possible interactions of NPs with the intestinal barrier are depicted in Figure 14.2.[57]

There are two distinct mechanisms for the transport of NPs across the intestinal epithelium: the paracellular pathway (between adjacent cells) and the transcellular route. The paracellular space represents only about 0.01–0.1% of the total absorptive intestinal surface with dimensions in the order of 10 Å. Therefore, transport of macromolecules and also NPs is severely restricted from crossing the mucosal epithelia through the paracellular route. To improve the paracelluar transport, various enhancers such as starch, thiolated polymers and calcium chelators have been successfully employed to open tight junctions reversibly. The transcellular route for proteins and peptides includes passive transcellular diffusion,

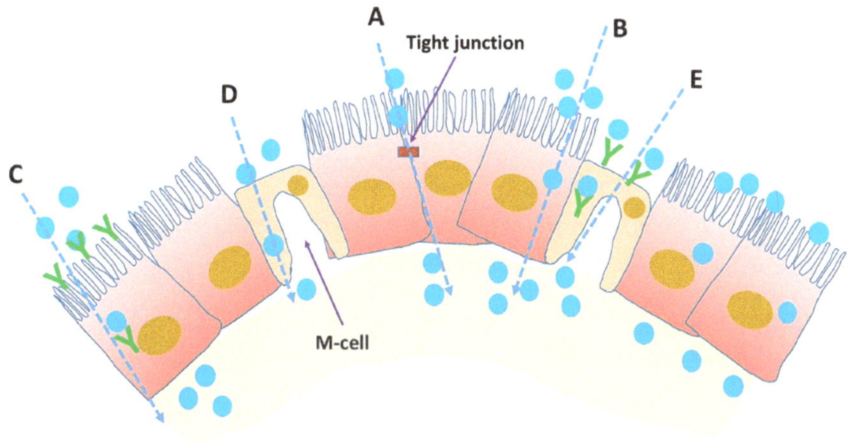

Figure 14.2 Plausible transport mechanisms of nanoparticles (NPs) through the intestinal barrier. (A) Paracellular transport; (B) transcellular transport; (C) receptor-mediated transcellular endocytosis; (D) M cell-mediated phagocytosis; (E) receptor- and M cell-mediated phagocytosis.

carrier-mediated transport, transcytosis by normal enterocytes (with/without being receptor mediated) and phagocytosis by M cells (with/without being receptor mediated). Transcellular transport of NPs across the intestinal epithelium generally involves the last two pathways. Transcytosis of NPs across the intestinal epithelial cells involves two steps: first, an endocytic process occurring at the apical membrane, which is followed by transport processes that allow particles to be delivered across the basolateral membrane. The intestinal epithelium is primarily composed of enterocytes. Translocation of NPs through enterocytes is severely restricted owing to the low endocytic capacity of the enterocytes M cells, that represent only 1% of the total intestinal cells and approximately 5% of the human follicle-associated epithelium (FAE). M cell-mediated phagocytosis is a major route of translocation for orally administered proteins and peptides and also NPs owing to their high transcytotic activity. Several studies have reported that the majority of particles are translocated in the FAE.

Proteins/peptides can be encapsulated, adsorbed or chemically linked to the surface of polymeric NPs. Protein incorporation in polymeric NPs can be achieved by various methods. Natural polymers are generally more sensitive to the processing conditions. Therefore, NPs with natural polymers are generated using mild techniques including ionic gelation, polyelectrolyte complexation and coacervation. NPs composed of synthetic polymers are normally prepared by more extensive techniques such as interfacial polymerization, emulsification–polymerization, emulsification–solvent evaporation, nanoprecipitation, salting out, supercritical fluids and emulsification–solvent diffusion. Preparation methods for NPs have been widely reviewed in the literature.[65–67]

14.3.1 Examples of Peptide Nanoparticle Formulations

NP-based medicines currently on the market include doxorubicin liposomes (Doxil®; Janssen) for the treatment of ovarian cancer and AIDS-associated Kaposi's sarcoma and paclitaxel-containing albumin NPs (Abraxane®; Celgene) for second-line treatment of patients with breast cancer. There are several examples of NP-mediated delivery of small-molecule chemotherapeutic agents at the clinical and preclinical stages. There are fewer examples of the use of NPs for delivering proteins and peptides and these agents are in the early stages of development. An example of NP-mediated protein delivery that shows promise for clinical application is the delivery of cytokines for tumor immunotherapy.[68] Specifically, immunosuppressive factors associated with the tumor microenvironment were targeted using a combination of a small-molecule transforming growth factor-β (TGFβ) inhibitor and the immunostimulant interleukin-2 (IL-2). Core–shell NPs were developed to encapsulate both the water-soluble protein (IL-2) and the hydrophobic small-molecule cargo (the TGFβ inhibitor) into a single NP that was designed to enable the simultaneous release of both drugs.[68] This method led to sustained local delivery, thereby reducing the toxicity of IL-2 and the combination therapy delayed tumor growth and increased survival in a mouse model of melanoma. NPs have been actively used for delivering vaccines and as vaccine adjuvants. The primary advantage of nanoparticulate vaccines is effective antigen presentation to dendritic cells, which yields a stronger immune response than with soluble antigens. A detailed mechanism of nanomaterial action on the regulation of immunological functions and several major applications of NPs for tumor peptide vaccine were reported by Wang *et al.*[69] Some examples of polymeric NP-based delivery of proteins and peptides are given in Table 14.2.

Several factors need to be considered to develop efficient nanocarriers for oral delivery.

Uptake of NPs through the intestinal epithelium is dependent on particle size, surface charge, polymer hydrophobicity, mucoadhesivity and the presence or absence of surface ligands. In addition to the change in surface properties, transport of NPs across oral mucosa can be improved by application of bioadhesive polymers and surface modification with targeting ligands. In a recent study, a novel nanocarrier based on an alginate–dextran sulfate core, complexed with a CS–PEG–albumin shell, improved the oral delivery of insulin.[70] NPs entered owing to high adhesion to enterocytes and especially the Peyer's patch region. *In vivo* oral studies demonstrated a significant reduction in glycemia for 24 h. In another study, carboxylated CS-grafted PMMA NPs were developed for oral insulin delivery. This nanocarrier exhibited sustained release at lower pH (2.0) relative to higher pH values such as 6.8 and 7.4. At 25 IU kg^{-1}, the pharmacological bioavailability was 9.7% with desirable tissue and blood compatibility.[71]

Various polymeric NPs have been examined for their efficacy in improving protein and peptide absorption across nasal mucosa. Chitosan–*N*-acetyl-L-cysteine (CS–NAC) NPs have been investigated for nasal delivery of insulin.[72]

Table 14.2 Examples of polymeric nanoparticle-based delivery of proteins and peptides.

Peptide/protein	Delivery route	Nanoparticle type
Insulin	Nasal	CS–*N*-acetylcysteine (NAC)
Insulin	Nasal	Hybrid polyoligosaccharide NPs comprising CS and cyclodextrins
Insulin	Nasal	Chitosan–alginate blended NPs
Leuprolide	Nasal	Thiolated CS–thioglycolic acid (CS–TGA)
Neurotoxin-I	Nose to brain	PLA NPs (NT-I-NPs)
Thyrotropin-releasing hormone (TRH)	Nose to brain	PLGA
α_1-Antitrypsin	Pulmonary	PLGA
Calcitonin	Pulmonary	Glycol and thiolated glycol CS
Urocortin	Parenteral	Lactoferrin-modified PEG–PLGA
Angiopep-2 peptide	Parenteral	PEG–PCL
Neuroprotective peptide	Parenteral	B6 peptide (CGHKAKGPRK)-functionalized PEG–PLA
Insulin	Transdermal	PEG–PCL
C16Y (an integrin-antagonist peptide)	Ocular	PLA/PLA–poly(ethylene oxide) (PEO)
Restasis A®	Ocular	Cyclosporine A-loaded PLGA

The particle size ranged from 140 to 210 nm with a zeta potential of 119.5–131.7 mV. Particles were spherical in shape and had loading efficiencies of 13–42%. CS–NAC NPs displayed an excellent swelling behavior and generated an initial burst followed by slow release of insulin, *in vitro*. *In vivo* studies revealed that CS–NAC NPs were more promising in reducing blood glucose levels relative to CS NPs and insulin control solution. Intranasal administration of insulin-loaded CS–NAC and CS particles decreased the glucose level to 59 and 74% at 30 and 60 min post-dosing, respectively. The total reductions in plasma glucose levels observed with insulin-loaded CS–NAC and CS NPs were 16.2 and 8.3%, respectively, within 5 h. These results clearly signify the potential of thiolated CS in improving the nasal absorption of proteins such as insulin. Hybrid polyoligosaccharide NPs comprising CS and cyclodextrins have been developed and investigated for nasal insulin delivery.[73] These NPs demonstrated high efficacy in reducing the *trans*-epithelial resistance of nasal membranes reversibly. Moreover, intranasal administration of insulin-loaded hybrid NPs generated more than a 35% reduction in glucose levels at 1 h post-dosing in conscious rabbits. In contrast, insulin solution resulted in a 14% reduction in plasma glucose levels 30 min after dose administration.

Chitosan–alginate blended NPs were developed by Goycoolea *et al.*[74] The particle size of insulin-loaded NPs ranged from 273 to 396 nm. *In vitro* release studies demonstrated nearly 80% insulin release within 20 min. The insulin-loaded NPs generated a 35% reduction in the plasma glucose level at 45 min post-intranasal administration. PEGylated trimethyl CS nanocomplexes have demonstrated high potential in improving nasal insulin absorption.[75] These nanocomplexes generated an approximately 34–7% reduction in plasma glucose levels in rats. Thiolated CS–thioglycolic acid (CS–TGA) NPs were developed by Shahnaz *et al.* as a promising drug carrier for leuprolide.[76] The mean particle size of these NPs was 252 ± 82 nm. Leuprolide was released slowly over a period of 6 h. Thiolated CS NPs generated 5.2-fold higher leuprolide transport across porcine nasal mucosa relative to leuprolide solution. These NPs displayed 6.9- and 3.8-fold higher AUCs, respectively, and maximum plasma concentration relative to leuprolide solution,. Cheng *et al.* demonstrated brain delivery of neurotoxin-I with PLA NPs (NT-I-NPs) following intranasal administration.[77] Intranasal administration of NT-I-NPs demonstrated a shorter time to reach maximum concentration (T_{max}) relative to intravenous (i.v.) administration of NT-I-NPs and NT-I solution. The absolute bioavailability produced by intranasal administration was about 160 and 196% higher relative to i.v. administration of NT-I-NPs and NT-I solution, respectively. These results indicate that the brain delivery of peptide-loaded NPs can be significantly improved by intranasal administration relative to i.v. administration.

Xia *et al.* demonstrated the efficacy of low molecular weight protamine (LMWP)-modified PEG–PLA NPs to deliver therapeutic agents to the central nervous system *via* olfactory and trigeminal nerve pathways.[78] Such nanoparticulate formulations could be employed for improving brain protein or peptide absorption. PLGA NPs loaded with thyrotropin-releasing hormone (TRH) of size 108 ± 12 nm were developed by Veronesi *et al.*[79] Intranasal administration of TRH-loaded NPs significantly raised the stimulations required to reach stage V seizures in rats. Moreover, seizure after duration discharge was observed to be significantly reduced in rats treated with TRH-loaded NPs intranasally.

Pulmonary administration of proteins and peptides could provide numerous advantages, including an enormous absorptive surface area (100 m^2), high vascularization, thin alveolar epithelial membrane (0.1–0.2 μm) and low enzymatic activity. Glycol and thiolated glycol CS NPs significantly enhanced the pharmacological availability of calcitonin following pulmonary administration.[80] Calcitonin-loaded thiolated and nonthiolated CS NPs generated availabilities of 40 and 27%, respectively. A marked hypocalcemic activity was observed for a period of 24 and 12 h with thiolated and unmodified CS NPs, respectively. Particles were of size 230–330 nm and demonstrated high calcitonin entrapment efficiency. Following intratracheal administration, thiolated CS NPs displayed twofold higher mucoadhesion relative to glycol CS NPs. Pulmonary delivery of modified gelatin-based NPs has been proposed as an excellent delivery system to enhance insulin absorption. Wanakule *et al.* suggested an enzyme-responsive NP-entrapped microgel system as a promising formulation for pulmonary delivery of proteins and peptides.[81] This delivery system was proposed to overcome challenges such as alveolar macrophage uptake, nonspecificity, low respirable fractions and poor deposition in deeper lung regions.

The parenteral route of drug delivery can avoid first-pass metabolism and provide the highest bioavailability and active targeting for drug delivery. However, properties of NPs such as size, charge, dissolution rate and surface coating can significantly alter the *in vivo* fate. Nanoparticles with size >200 nm or a slow dissolution rate can be rapidly cleared by the mononuclear phagocytic system. This process might significantly enhance nanoparticulate uptake in the liver and spleen. PEGylation has been proposed as a promising approach to avoid reticuloendothelial system uptake. In addition, the use of poly(*N*-vinyl-2-pyrrolidone) as a coating material to improve the blood circulation time of particulate carriers has also been reported.[82] Interestingly, NPs with size between 100 and 300 nm can greatly accumulate in tumors owing to enhanced permeability and retention effect.

In recent years, several polymeric NPs have been investigated for targeting poorly permeable tissues such as the brain and lung following systemic administration. Angiopep-2 peptide- and EGFP–EGF1 protein-functionalized PEG–PCL NPs were developed by Huile *et al.*[83] These functionalized NPs demonstrated a remarkable potential in penetrating the BBB (angiopep-2) and binding to neuroglial cells (EGFP–EGF1). Significantly higher brain accumulation relative to unmodified NPs was observed. Novel CS-based NPs loaded with caspase inhibitor *N*-benzyloxycarbonyl-Asp(OMe)-Glu(OMe)-Val-Asp(OMe)-fluoromethyl ketone (Z-DEVD-FMK) or basic fibroblast growth factor were observed to transport rapidly across the BBB following systemic circulation.[84] Bao *et al.* developed OX26-functionalized hyperbranched polyglycerol-conjugated PLGA to improve the brain delivery of endomorphins.[85] Xia *et al.* demonstrated that penetratin-modified poly(ethylene glycol)–poly(lactic acid) NPs act as an excellent brain drug delivery system compared with protamine (high arginine content)-modified NPs.[86] B6 peptide (CGHKAKGPRK)-functionalized PEG–PLA NPs have been demonstrated to be an excellent system for the brain delivery of neuroprotective peptide-NAPVSIPQ (NAP). NAP-loaded B6 NPs demonstrated a remarkable improvement in learning impairment recovery, loss of hippocampal neurons and cholinergic disruption at significantly lower doses.[87]

Transdermal delivery offers several advantages, such as an enormous surface area, avoidance of hepatic first-pass metabolism, continuous and noninvasive administration and patient compliance. Rastogi *et al.* studied the potential of transdermal electroporation of insulin-loaded nanocarriers for insulin delivery.[88] Insulin-loaded PCL–PEG–PCL NPs were prepared by the double-emulsion solvent evaporation method. The insulin entrapment efficiency was observed to be $32.9 \pm 7.6\%$ with an average NP diameter of about 85 ± 9.4 nm. Electroporation of NPs resulted in fourfold higher insulin deposition in rat skin relative to solution. *In vivo* studies demonstrated $77 \pm 5\%$ (87.2 ± 6.4 mIU mL^{-1}, $t = 2$ h) and $85 \pm 8\%$ (37.8 ± 10.2 mIU mL^{-1}, $t = 4$ h) reductions in blood glucose levels with therapeutic levels maintained for 24 and 36 h for solution and NPs, respectively.

The ocular route has been utilized for both systemic and localized delivery of proteins and peptides.[89] For systemic delivery, the ocular route provides advantages such as ease of administration, avoidance of first-pass metabolism and a comparatively rapid rate of absorption (over oral administration). Systemic absorption of several proteins and peptides, including insulin, calcitonin and enkephalins, has been studied following ocular administration. Currently, the ocular route is mainly used for the delivery of proteins and peptides for the treatment of local ocular disorders such as age-related macular degeneration, dry eye disease or proliferative diabetic retinopathy.[90] Lucentis® and Eylea® are commercially available proteins intended for the treatment of ocular diseases. Polymeric NPs have been successfully designed to overcome these barriers and improve the ocular bioavailability of proteins and peptides. In a study by Yenice *et al.*, coating of cyclosporine A (CsA)-loaded polycaprolactone (PCL)–benzalkonium chloride (BKC) nanospheres with hyaluronic acid resulted in high concentrations of cyclosporine A in the cornea compared with uncoated NPs.[91] Cationic polymers such as Eudragit® have also been employed to increase the precorneal residence time. Aksungur *et al.* evaluated the efficiency of Eudragit® in improving the effectiveness of CsA NP formulations against inflammation of the eye surface.[92] The NPs were prepared using either PLGA alone or a mixture of Eudragit® RL with PLGA or were coated with Carbopol®. The kinetic parameters of tears were determined following topical application of a CsA-loaded NP suspension and Restasis® (drug-loaded emulsion) in rabbit eyes. Even though topical administration is the major route for the anterior segment conditions, it is difficult to deliver proteins and peptides to posterior ocular tissues *via* the topical route. Currently, intravitreal injection is the most commonly employed technique to treat posterior ocular diseases. Several studies have demonstrated the suitability of intravitreally injected NPs in the treatment of posterior segment ocular diseases. Kim *et al.* developed poly(lactic acid)/poly(lactic acid)–poly(ethylene oxide) (PLA/PLA–PEO) NPs for the delivery of C16Y (an integrin-antagonist peptide) to the subretinal space for the treatment of choroidal neovascularization (CNV).[93]

Drug delivery through the colon has emerged as an attractive strategy for delivering large molecular drugs into the systemic circulation.[94] The colon is the preferred absorption site for proteins and peptides for the following reasons: (i) less proteolytic activity of colon mucosa compared with the small intestine; (ii) less endogenous metabolic enzymes; (iii) longer retention time in the colon of 5 days; and (iv) significant enhancement by absorption enhancers due to the slow transit and stirring. However, the colonic drug delivery system of proteins and peptides is also complicated and limited because of the following: (i) the need for advanced technology, multiple formulation steps and skills of manufacturing; (ii) difficulty in accessing the colon at the distal portion of the alimentary canal; (iii) harsh conditions such as different pH values and enzymes in the GIT; (iv) drug loss due to binding to dietary residues, intestinal secretions or mucus in the colon; (v) possible involvement of microflora in the colon, which could affect colonic performance *via* metabolic degradation of protein and peptide drugs; and (vi) smaller surface area and tight junctions in the colon restrict drug transport across the mucosa into the systemic circulation. To overcome these challenges, polymeric NPs have been utilized for colon-targeted protein and peptide delivery. Coco *et al.* formulated three different polymeric NPs, pH-sensitive Eudragit® NPs, mucoadhesive trimethylchitosan (TMC) NPs and PLGA-based NPs with targeting ligands to deliver ovalbumin (OVA) to the colon, for treating inflammatory bowel disease (IBD).[95] The TMC NPs exhibited the highest apparent permeability for OVA in the untreated Caco-2 monolayer model; however, there was no difference among the three NP formulations in the inflamed model. Moreover, the targeted PLGA NPs presented the highest accumulation of OVA in the inflamed mouse colon. These results demonstrate the potential of active targeting of the colon for IBD treatment. Laroui *et al.* demonstrated the efficacy of tripeptide Lys–Pro–Val (KPV)-loaded NPs of size 400 nm in alginate–chitosan hydrogel for the treatment of IBD *via* a mouse model.[96] The hydrogel collapsed in the colon and NPs quickly released the drug on or within the closed area of colonocytes. KPV-loaded NPs also exhibited significant anti-inflammatory efficacy in mice. Moreover, KPV-loaded NPs generated a similar therapeutic efficacy with a 12 000-fold lower concentration in the formulation than that of KPV solution. These observations clearly indicate the potential of drug-loaded NPs as an effective therapeutic approach for the treatment of IBD.

14.3.2 Factors Affecting Drug Release From Nanoparticle Systems

NPs can enter the human body *via* three main routes: direct injection, inhalation and oral intake. Once they have entered the systemic circulation, particle–protein interaction is the first phenomenon taking place before distribution into various organs.[97] Absorption from the blood capillaries allows the lymphatic system to further distribute and eliminate the particles. This system has three main functions, two of which pertain to drug delivery. The first, fluid recovery, involves the filtering of fluids by the lymphatic system from blood capillaries. The second encompasses immunity and may be the most relevant to this topic. As the system recovers excess fluid, it also picks up foreign cells and chemicals from the tissues. As the fluids are filtered back into the blood, the lymph nodes detect any foreign matter passing through.[98] If something is recognized as foreign, macrophages will engulf it and clear it from the body. This tends to be the struggle with NP-based drug delivery; however, clearance can be influenced by the size and surface characteristics of the particles.[99] Figure 14.3 depicts various hurdles involved in the delivery of therapeutic NPs to targeted tissues.

14.3.3 Modified Nanoparticle Systems

Manipulation of the surface characteristics of NP-based drug formulations has potential to generate improved systems.[100] In order to create an

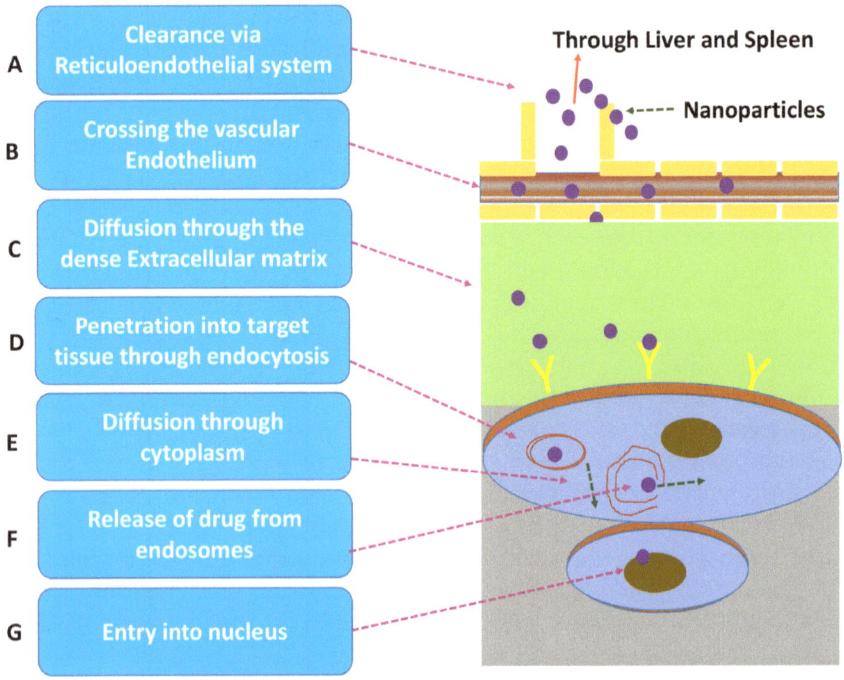

Figure 14.3 Hurdles associated with the delivery of NP–drug systems. When NPs are injected into the bloodstream (intravascular delivery), the barrier manifests in the form of (A) clearance by the reticuloendothelial system, including the liver and spleen, especially by the resident macrophages in these organs, and (B) permeation across the vascular endothelium of the diseased tissue with penetration into the diseased tissue, both of which pose a considerable hurdle. (C) NPs that escape the blood vessel still need to diffuse through the dense extracellular matrix to reach relevant target cells embedded deep within the tissue. (D) Upon arriving at the surface of the target cells, NPs need to enter the cells *via* endocytosis. (E) Diffusion through the cytoplasm. (F) NPs that are internalized by the cells within endosomes need to escape the endosome to release the active drug cargo. (G) Entry into the nucleus, if needed.

optimum NP drug delivery system, the incorporation of appropriate targeting ligands, surface curvature and reactivity is important to address the prevention of aggregation, stability and receptor binding and subsequent pharmacological effects of the drug.[101] First, possible clearance of nanosystems needs to be addressed. Since NPs can be recognized by the lymphatic system, they are subject to the body's natural immune response to foreign matter. The more hydrophobic an NP is, the more likely it is to be cleared due to higher binding of blood components.[102] Therefore, it seems logical to assume that making their surface hydrophilic would increase their time in circulation. In fact, coating the NPs with polymers or surfactants or creating copolymers such as poly(ethylene oxide), poly(ethylene glycol) (prevents hepatic and splenic localization), polyoxamer, poloxamine and polysorbate 80 has been demonstrated to be useful.[103] PEG is a hydrophilic and relatively inert polymer that, when incorporated in the NP surface, hinders the binding of plasma proteins (opsonization), thus preventing substantial loss of the given dose. PEGylated NPs are often referred to as "stealth" NPs, because without opsonization they remain undetected by the reticuloendothelial system.[104]

By creating polymer complexes, clearance issues have been addressed, but aggregation is still a concern with small particles owing to the large surface area. Particularly NPs such as dendrimers, quantum dots and micelles are especially prone to aggregation. Several strategies have been employed to prevent aggregation and require coating of particles with capping agents and altering the surface charges.[105] Overall, these methods and theories can be summarized into one idea: the size of the particle must be large enough to avoid leakage into blood capillaries, but not large enough to become susceptible to macrophage clearance. By manipulating the surface, the extent of aggregation and clearance can be controlled.[106] The size and surface properties of NPs have been explored to optimize bioavailability, decrease clearance and increase stability. By controlling these characteristics, it is possible to get the drug to tissues in the body that may previously have been inaccessible. However, there is no significance to this practice if the drug cannot then be released from the NP matrix. The release of the drug from the NP-based formulation depends on may factors, including pH, temperature, drug solubility, desorption of the surface-bound or adsorbed drug, drug diffusion through the NP matrix, NP matrix swelling and erosion and the combination of erosion and diffusion processes.[107,108]

Depending on the type of NP being used, the release of the drug will differ. The prepared polymeric NPs can be called nanocapsules or nanospheres based on their composition. Nanospheres are a homogeneous system such that the polymer chains arrange in a similar fashion to surfactants in micelle formation (phase separated from the bulk solution), whereas nanocapsules are a heterogeneous system, such that the drug is inside a reservoir composed of the polymer (like vesicles).[109] In relation to nanospheres, which are a matrix system where the drug is physically and uniformly dispersed, drug is released by erosion of the matrix. There is a rapid burst of drug release related to the drug weakly bound to the large surface area of the NPs followed by a sustained release.[110] On the other hand, if nanocapsules are used, the release is controlled by diffusion of the drug through the polymeric layer, hence diffusibility of the drug through that polymer is definitely a determining factor in its deliverability. If there are ionic interactions between the drug and polymer, they will form complexes that inhibit the release of the drug from the capsule. This can be avoided by adding other auxiliary agents such as poly(ethylene oxide)–poly(propylene oxide) (PEO–PPO). This will decrease the interactions between the drug and the capsule matrix, allowing for greater release of the drug to target tissues.[111]

Modifying NPs by absorption or covalent attachment of targeting ligands on the surface is an alternative strategy for enhancing the transport of NPs across intestinal epithelial cells.[112] These targeting ligands can interact with surface receptors on M cells, goblet cells or enterocytes. The targeting moieties studied include lectins, peptides, vitamins and wheat germ agglutinin. Targeted delivery was employed to improve the bioavailability and hypoglycemic activity of insulin *via* oral administration. Insulin-loaded CS NPs were surface modified by conjugation with a synthetic CRTLTVRKC peptide.[113] The peptide significantly enhanced targeting to the epithelium and translocation of CS NPs by interaction with a transmembrane protein, stabilin-2. These interactions facilitate efficient delivery and higher transport of NPs. Choi *et al.* reported the enhancement of oral drug absorption *via* M cell-targeting CS NPs.[114] NPs were linked with a targeting peptide such as CKSTHPLSC (CKS9). The modified CS NPs linked with CKS9 peptide improved uptake at the FAE area of Peyer's patch. In an *in vitro* transcytosis study across rat small intestine, the CKS9 peptide-modified CS NPs showed a more effective transport ability and more specific accumulation into Peyer's patch regions compared with unmodified CS NPs. In another study, goblet cell-targeting NPs of insulin were developed by coupling the cell-targeting peptide CSKSSDYQC (CSK) with the trimethyl CS chloride surface to augment oral absorption of insulin.[115] CSK peptide modification was observed to enhance the transport of NPs in villi and facilitate the transition of drug molecules across the cellular membrane. CSK peptide also induced greater internalization of proteins/peptides through clathrin- and caveolae-dependent endocytosis in HT29-MTX cells. Oral administration of CSK peptide-modified NPs to diabetic rats resulted in a 1.5-fold higher bioavailability and a greater hypoglycemic effect relative to unmodified NPs.[115]

14.3.4 Limitations of Nanoparticle Delivery Systems

It has been discussed how important the NP drug delivery systems are, but these systems would be of no use if the drug is not released or is released ineffectively.[116] As the particle size becomes smaller, the surface area-to-volume ratio becomes larger. This implies that more of the drug is closer to the surface of the particle compared with a larger molecule. Being at or near the surface would lead to faster drug release.[117] It would be beneficial to create NP systems that have a large surface area-to-volume ratio; however, toxicity must always be monitored. As mentioned earlier, the size of the NP determines the biological fate. The vascular and lymph systems are responsible for the filtering and clearance of foreign matter and chemicals. This is yet another factor that must be engineered into the ideal NP system. It has been shown that particles 200 nm or larger tend to activate the lymphatic system and are removed from the circulation faster.[118] Thus, from the literature evaluation and discussion so far, it is clear that the optimum size for an NP is approximately 100 nm. At this size, the particle could pass through the BBB, with a sufficient amount of drug delivery due to the high surface area-to-volume ratio, and avoiding immediate clearance by the lymphatic system. In general, systemically delivered NPs face several physiological barriers before reaching their target. Technological advances for overcoming these hurdles are important to enable future success in this field; these include PEGylation for prolonging the circulation of NPs and enhancing tissue penetration, optimization of particle size and the use of substances to enhance deep tumor penetration of NPs.

14.4 Conclusion

Microparticle formulation has been successfully applied in a number of commercial products containing peptides as active pharmaceutical ingredients. Despite the clear advances made in the use of microparticles for delivering peptides and proteins, challenges still remain. These include the difficulty of large-scale manufacturing, inactivation of the drug during fabrication and poor control of drug release rates. Further advances in increasing the duration of drug release and in methods that provide active control of the rate of drug release, including on-demand termination of release, could open up new opportunities. There are several examples of NP-mediated delivery of small-molecule chemotherapeutic agents at the clinical and preclinical stages. However, the use of NPs for delivering proteins and peptides is still in the early stages of development and is expected to mature in the near future. Recent studies have shown that for the creation of an optimum NP drug delivery system, several factors, such as appropriate targeting ligands, surface curvature and reactivity, are important. This could potentially address the prevention of aggregation, stability and receptor binding and subsequent pharmacological effects of the drug.

It is expected that improved drug delivery systems based on microspheres and NPs will revolutionize the way in which peptide drugs are administered in the future. The availability of a variety of optimal drug delivery systems coupled with the ability to administer peptide drugs *via* different routes will definitely help patients.

References

1. J. Thundimadathil, *J. Amino Acids*, 2012, 1–13. Article ID **967347**.
2. A. M. Thayer, Improving peptides, *Chem. Eng. News*, 2011, **89**, 13–20.
3. L. Otvos, Jr and J. D. Wade, *Front. Chem.*, 2014, **2**, 62.
4. J. Richard, *Ther. Delivery*, 2017, **8**(8), 663.
5. T. R. Kumar, K. Soppimath and S. K. Nachaegari, *Curr. Pharm. Biotechnol.*, 2006, **7**(4), 261.
6. F. Antunes, F. Andrade, D. Ferreira, H. M. Nielsen and B. Sarmento, *Curr. Drug Metab.*, 2013, **14**(1), 4.
7. C. P. Reis, R. J. Neufeld, A. J. Ribeiro and F. Veiga, *Nanomedicine*, 2006, **2**(2), 53.

8. S. Mitragotri, P. A. Burke and R. Langer, *Nat. Rev. Drug. Discovery*, 2014, **13**(9), 655.
9. D. Bhattacharyya and S. Singh, *Int. J. u- and e-Serv. Sci. Technol.*, 2009, **2**(3), 29.
10. V. Saez, J. R. Hernández and C. Peniche, *Biotecnología Aplicada*, 2007, **24**, 108.
11. D. S. Kohane, *Biotechnol. Bioeng.*, 2007, **96**(2), 203.
12. R. Langer and J. Folkman, *Nature*, 1976, **263**, 797.
13. S. Cohen, T. Yoshioka, M. Lucarelli, L. H. Hwang and R. Langer, *Pharm. Res.*, 1991, **8**, 713.
14. S. D. Putney and P. A. Burke, *Nature Biotech.*, 1998, **16**, 153–157.
15. G. Zhu, S. R. Mallery and S. P. Schwendeman, *Nature Biotech.*, 2000, **18**, 52.
16. D. F. Emerich and C. Thanos, *Biomol. Eng.*, 2006, **23**, 171.
17. D. F. Emerich and C. Thanos, *J. Drug Targeting*, 2007, **15**(3), 163.
18. D. J. Burgess and J. E. Carless, *Int. J. Pharm.*, 1985, **27**(1), 61.
19. P. Calvo and C. Remuñan-López, *Pharm. Res.*, 1997, **14**, 1431.
20. R. Singh and J. W. Lillard Jr, *Exp. Mol. Pathol.*, 2009, **86**, 215.
21. X. Zhang and Y. Huang, *Ther. Delivery*, 2014, **5**(1), 53.
22. J. Zhang and M. Saltzman, *Chem. Eng. Prog.*, 2013, **109**(3), 25.
23. G. Tiwari, R. Tiwari, B. Sriwastawa, L. Bhati, S. Pandey, P. Pandey and S. K. Bannerjee, *Int. J. Pharm. Invest.*, 2012, **2**(1), 2.
24. C. E. Mora-Huertas, H. Fessi and A. Elasissari, *Int. J. Pharm.*, 2010, **385**(1–2), 113.
25. E. Blanco, H. Shen and M. Ferrari, *Nat. Biotechnol.*, 2015, **33**(9), 941.
26. J. McMillan, E. Batrakova and H. E. Gendelman, *Prog. Mol. Biol. Transl. Sci.*, 2011, **104**, 563.
27. N. K. Varde and D. W. Pack, *Expert Opin. Biol. Ther.*, 2004, **4**(1), 35.
28. A. Mitra and B. Dey, *Indian J. Pharm. Sci.*, 2011, **73**(4), 355.
29. P. He, S. S. Davis and L. Illum, *Int. J. Pharm.*, 1999, **187**(1), 53.
30. P. B. O'Donnell and J. W. McGinity, *Adv. Drug Delivery Rev.*, 1997, **28**(1), 25.
31. P. Mahlumba, Y. E. Choonara, P. Kumar, L. C. du Toit and V. Pillay, *Molecules*, 2016, **21**(8), 1002.
32. R. Vaishya, V. Khurana, S. Patel and A. K. Mitra, *Expert. Opin. Drug. Deliv.*, 2015, **12**(3), 415.
33. Z. Zhou, Q. Yao, L. Li, X. Zhang, B. Wei, L. Yuan and L. Wang, *Med. Sci. Monit.*, 2018, **24**, 6934.
34. Y. Zhang and S. P. Schwendeman, *J. Controlled Release*, 2012, **162**(1), 119.
35. J. Guo, X. Sun, H. Yin, T. Wang, Y. Li, C. Zhou, H. Zhou, S. He and H. Cong, *Front. Cell. Infect. Microbiol.*, 2018, **8**, 163.
36. S. P. Schwendeman, R. B. Shah, B. A. Bailey and A. S. Schwendeman, *J. Controlled Release*, 2014, **190**, 240.
37. H. K. Makadia and S. J. Siegel, *Polymers*, 2011, **3**(3), 1377–1397.

38. S. Bhatnagar, R. A. Naqvi, R. Ali and D. N. Rao, *Methods Mol. Biol.*, 2014, **1139**, 443.
39. S. M. Kakade and D. M. Hassan, *Asian J. Pharm.*, 2018, **12**(2), S691.
40. P. Periti, T. Mazzei and E. Mini, *Clin. Pharmacokinet.*, 2002, **41**(7), 485.
41. H. Okada, M. Yamamoto, T. Heya, Y. Inoue, S. Kamei, Y. Ogawa and H. Toguchi, *J. Controlled Release*, 1994, **28**(1–3), 121.
42. R. C. Mehta, R. Jeyanthi, S. Calls, B. C. Thanoo, K. W. Burton and P. P. DeLuca, *J. Controlled Release*, 1994, **29**(3), 375.
43. E. Gavini, P. Chetoni, M. Cossu, M. G. Alvarez, M. F. Saettone and P. Giunchedi, *Eur. J. Pharm. Biopharm.*, 2004, **57**(2), 207.
44. Y. Cai, L. Wei, L. Ma, X. Huang, A. Tao, Z. Liu and W. Yuan, *Drug. Des., Dev. Ther.*, 2013, **7**, 963.
45. L. Shi, B. Liu, M. Wang and Y. Chen, *China Modern Medicine*, 2009, **7**, 117.
46. H. H. Kwak, W. S. Shim, S. Hwang, M. K. Son, Y. J. Kim, T. H. Kim, Z. H. Yoon, H. J. Youn, G. I. Lee, S. H. Kang and C. K. Shim, *Pharm. Res.*, 2009, **26**(11), 2504.
47. B. Liu, Q. Dong, M. Wang, L. Shi, Y. Wu, X. Yu, Y. Shi, C. Jiang, X. Zhang, T. Gu, Y. Chen and W. Kong, *Chem. Pharm. Bull.*, 2010, **58**(11), 1474.
48. H. Kim, H. Park, J. Lee, T. H. Kim, E. S. Lee, K. T. Oh, K. C. Lee and Y. S. Youn, *Biomaterials*, 2011, **32**(6), 1685–1693.
49. H. J. Yang, I. S. Park and K. Na, *Colloids Surf., A*, 2009, **340**(1–3), 115.
50. B. Zhang, D. He, Y. Fan, N. Liu and Y. Chen, *PLoS One*, 2014, **9**(1), e86064.
51. F. Alexis, *Polym. Intl.*, 2005, **54**(1), 36.
52. C. Berkland, M. J. Kipper, B. Narasimhan, K. K. Kim and D. W. Pack, *J. Controlled Release*, 2004, **94**(1), 129.
53. B. L. Silverman, S. L. Blethen, E. O. Reiter, K. M. Attie, R. B. Neuwirth and K. M. Ford, *J. Pediatr. Endocrinol. Metab.*, 2002, **2**, 715.
54. P. A. Burke, L. A. Klumb, J. D. Herberger, X. C. Nguyen, R. A. Harrell and M. Zordich, *Pharm. Res.*, 2004, **21**(3), 500.
55. N. K. Varde and D. W. Pack, *J. Controlled Release*, 2007, **124**(3), 172.
56. J. Zhang and M. Saltzman, *Chem. Eng. Prog.*, 2013, **109**(3), 25.
57. A. Patel, M. Patel, X. Yang and A. K. Mitra, *Protein Pept. Lett.*, 2014, **21**(11), 1102.
58. A. Zimmer and J. Kreuter, *Adv. Drug Delivery Rev.*, 1995, **16**, 61.
59. C. Oerlemans, W. Bult, M. Bos, G. Storm, J. F. Nijsen and W. E. Hennink, *Pharm. Res.*, 2010, **27**(12), 2569.
60. M. E. Werner, N. D. Cummings, M. Sethi, E. C. Wang, R. Sukumar, D. T. Moore and A. Z. Wang, *Int. J. Radiat. Oncol., Biol., Phys.*, 2013, **86**(3), 463.
61. S. Somani and C. Dufès, *Nanomedicine*, 2014, **9**(15), 2403.
62. N. Shao, Y. Su, J. Hu, J. Zhang, H. Zhang and Y. Cheng, *Int. J. Nanomedicine*, 2011, **6**, 3361.

63. N. Kamaly, B. Yameen, J. Wu and O. C. Farokhzad, *Chem. Rev.*, 2016, **116**(4), 2602.
64. W. B. Liechty, D. R. Kryscio, B. V. Slaughter and N. A. Peppas, *Annu. Rev. Chem. Biomol. Eng.*, 2010, **1**, 149.
65. J. K. Patra, G. Das, L. F. Fraceto, E. V. R. Campos, M. P. Rodriguez-Torres, L. S. Acosta-Torres, L. A. Diaz-Torres, R. Grillo, M. K. Swamy, S. Sharma, S. Habtemariam and H. Shin, *J. Nanobiotechnol.*, 2018, **16**, 71.
66. T. A. Ahmed and B. M. Aljaeid, *Drug Des,. Dev. Ther.*, 2016, **10**, 483.
67. M. Elsabahy and K. L. Wooley, *Chem. Soc. Rev.*, 2012, **41**(7), 2545.
68. J. Park, S. H. Wrzesinski, E. Stern, M. Look, J. Criscione, R. Ragheb, S. M. Jay, S. L. Demento, A. A. Agawu, P. L. Licona, A. F. Ferrandino, D. Gonzalez, A. Habermann, R. A. Flavell and T. M. Fahmy, *Nat. Mater.*, 2012, **11**(10), 895.
69. J. Wang, X. Hu and D. Xiang, *Drug Delivery*, 2018, **25**(1), 1319.
70. M. Lopes, N. Shrestha, A. Correia, M. A. Shahbazi, B. Sarmento, J. Hirvonen, F. Veiga, R. Seiça, A. Ribeiro and H. A. Santos, *J. Controlled Release*, 2016, **232**, 29.
71. F. Cui, F. Qian, Z. Zhao, L. Yin, C. Tang and C. Yin, *Biomacromolecules*, 2009, **10**(5), 1253.
72. X. Wang, C. Zheng, Z. Wu, D. Teng, X. Zhang, Z. Wang and C. Li, *J. Biomed. Mater. Res., Part B*, 2009, **88B**(1), 150.
73. D. Teijeiro-Osorio, C. Remuñán-López and M. J. Alonso, *Biomacromolecules*, 2009, **10**(2), 243.
74. F. M. Goycoolea, G. Lollo, C. Remuñán-López, F. Quaglia and M. J. Alonso, *Biomacromolecules*, 2009, **10**(7), 1736.
75. S. Mao, O. Germershaus, D. Fischer, T. Linn, R. Schnepf and T. Kissel, *Pharm. Res.*, 2005, **22**(12), 2058.
76. G. Shahnaz, A. Vetter, J. Barthelmes, D. Rahmat, F. Laffleur, J. Iqbal, G. Perera, W. Schlocker, S. Dünnhaput, P. Augustijns and A. Bernkop-Schnürch, *Int. J. Pharm.*, 2012, **428**(1–2), 164.
77. Q. Cheng, J. Feng, J. Chen, X. Zhu and F. Li, *Biopharm. Drug Dispos.*, 2008, **29**(8), 431.
78. H. Xia, X. Gao, G. Gu, Z. Liu, N. Zeng, Q. Hu, Q. Song, L. Yao, Z. Pang, X. Jiang, J. Chen and H. Chen, *Biomaterials*, 2011, **32**(36), 9888.
79. M. C. Veronesi, Y. Aldouby, A. J. Domb and M. J. Kubek, *Brain Res.*, 2009, **1303**, 151.
80. A. Makhlof, M. Werle, Y. Tozuka and H. Takeuchi, *Int. J. Pharm.*, 2010, **397**(1–2), 92.
81. P. Wanakule, G. W. Liu, A. T. Fleury and K. Roy, *J. Controlled Release*, 2012, **162**(2), 429.
82. T. Ishihara, T. Maeda, H. Sakamoto, N. Takasaki, M. Shigyo, T. Ishida, H. Kiwada, Y. Mizushima and T. Mizushima, *Biomacromolecules*, 2010, **11**(10), 2700.
83. G. Huile, P. Shuaiqi, Y. Zhi, C. Shijie, C. Chen, J. Xinguo, S. Shun, P. Zhiqing and H. Yu, *Biomaterials*, 2011, **32**(33), 8669.

84. M. Yemişci, Y. Gürsoy-Özdemir, S. Caban, E. Bodur, Y. Capan and T. Dalkara, *Methods Enzymol.*, 2012, **508**, 253.
85. H. Bao, X. Jin, L. Li, F. Lv and T. Liu, *J. Mater. Sci.: Mater. Med.*, 2012, **23**(8), 1891.
86. H. Xia, X. Gao, G. Gu, Z. Liu, Q. Hu, Y. Tu, Q. Song, L. Yao, Z. Pang, X. Jiang, J. Chen and H. Chen, *Int. J. Pharm.*, 2012, **436**(1–2), 840.
87. Z. Liu, X. Gao, T. Kang, M. Jiang, D. Miao, G. Gu, Q. Hu, Q. Song, L. Yao, Y. Tu, H. Chen, X. Jiang and J. Chen, *Bioconjugate Chem.*, 2013, **24**(6), 997–1007.
88. R. Rastogi, S. Anand and V. Koul, *Drug Dev. Ind. Pharm.*, 2010, **36**(11), 1303.
89. Y. C. Lee, P. Simamora, S. Pinsuwan and S. H. Yalkowsky, *Int. J. Pharm.*, 2002, **233**, 1.
90. J. L. Davis, B. C. Gilger and M. R. Robinson, *Curr. Opin. Mol. Ther.*, 2004, **6**, 195.
91. I. Yenice, M. C. Mocan, E. Palaska, A. Bochot, E. Bilensoy, I. Vural, M. Irkeç and A. A. Hincal, *Exp. Eye Res.*, 2008, **87**(3), 162.
92. P. Aksungur, M. Demirbilek, E. B. Denkbaş, J. Vandervoort, A. Ludwig and N. Unlü, *J. Controlled Release*, 2011, **151**(3), 286.
93. H. Kim and K. G. Csaky, *J. Controlled Release*, 2010, **142**(2), 286.
94. A. K. Philip and B. Philip, *Oman Med. J.*, 2010, **25**(2), 79.
95. R. Coco, L. Plapied, V. Pourcelle, C. Jérôme, D. J. Brayden, Y. J. Schneider and V. Préat, *Int. J. Pharm.*, 2013, **440**(1), 3.
96. H. Laroui, G. Dalmasso, H. T. Nguyen, Y. Yan, S. V. Sitaraman and D. Merlin, *Gastroenterology*, 2010, **138**(3), 843.
97. Q. Mu and G. Jiang, *Chem. Rev.*, 2014, **114**(15), 7740.
98. H. S. Park, S. H. Nam, J. Kim, H. S. Shin, Y. D. Suh and K. S. Hong, *Sci. Rep.*, 2016, **6**, 27407.
99. F. Alexis and E. Pridgen, *Mol. Pharm.*, 2008, **5**(4), 505–515.
100. C. Bantz, O. Koshkina, T. Lang, H. J. Galla, C. J. Kirkpatrick, R. H. Stauber and M. Maskos, *Beilstein J. Nanotechnol.*, 2014, **5**, 1774.
101. R. Khanbabaie and M. Jahanshahi, *Curr. Neuropharmacol.*, 2012, **10**(4), 370.
102. L. Kou and J. Sun, *Asian J. Pharm. Sci.*, 2013, **8**, 1.
103. L. Araujo and R. Lobenberg, *J. Drug Target*, 1999, **6**, 373.
104. S. D. Li and L. Huang, *J. Controlled Release*, 2010, **145**(3), 178.
105. D. Li and R. B. Kaner, *J. Am. Chem. Soc.*, 2006, **128**(3), 968.
106. E. A. Sykes, Q. Dai, C. D. Sarsons, J. Chen, J. V. Rocheleau, D. M. Hwang, G. Zheng, D. T. Cramb, K. D. Rinker and W. C. Chan, *Proc. Natl. Acad. Sci. U. S. A.*, 2016, **113**, E1142.
107. G. H. Son, B. J. Lee and C. W. Cho, *J. Pharmaceut. Invest.*, 2017, **47**(4), 287.
108. S. Mura, J. Nicolas and P. Couvreur, *Nat. Mater.*, 2013, **12**, 991.
109. C. E. Mora-Huertas and H. Fessi, *Int. J. Pharm.*, 2010, **385**(1–2), 113.
110. J. H. Lee and Y. Yeo, *Chem. Eng. Sci.*, 2015, **125**, 75.
111. P. Calvo and C. Remuñan-López, *Pharm. Res.*, 1997, **14**, 1431.

112. L. Guerrini, R. A. Alvarez-Puebla and N. Pazos-Perez, *Materials*, 2018, **11**, 1154.
113. G. Y. Lee, J. H. Kim, G. T. Oh, B. H. Lee, I. C. Kwon and I. S. Kim, *J. Controlled Release*, 2011, **155**(2), 211.
114. J. H. Choi, S. K. Kang, I. K. Park and M. K. Yoo, *Biomaterials*, 2010, **31**(30), 7738.
115. S. Salatin and A. Y. Khosroushahi, *J. Cell. Mol. Med.*, 2017, **21**(9), 1668.
116. M. D. Chavanpatil and A. Khdair, *J. Pharm. Sci.*, 2007, **96**(12), 3379.
117. C. Buzea, I. I. Pacheco and K. Robbie, *Biointerphases*, 2007, **2**(4), MR17.
118. A. Prokop and J. M. Davidson, *J. Pharm. Sci.*, 2008, **97**(9), 3518.

Subject Index

www.ingramcontent.com/pod-product-compliance
Ingram Content Group UK Ltd.
Pitfield, Milton Keynes, MK11 3LW, UK
UKHW062339170926
8624IPUK00029B/417
* 9 7 8 1 7 8 8 0 1 4 3 3 5 *